ENVIRONMENTAL CHEMISTRY OF SOILS

ENVIRONMENTAL CHEMISTRY OF SOILS

Murray B. McBride

New York   Oxford
OXFORD UNIVERSITY PRESS
1994

Oxford University Press

Oxford New York Toronto
Delhi Bombay Calcutta Madras Karachi
Kuala Lumpur Singapore Hong Kong Tokyo
Nairobi Dar es Salaam Cape Town
Melbourne Auckland Madrid

and associated companies in
Berlin Ibadan

Published by Oxford University Press, Inc.
198 Madison Avenue, New York, New York 10016-4314

Library of Congress Cataloging-in-Publication Data
McBride, Murray.
Environmental chemistry of soils / Murray McBride.
p. cm. Includes index.
ISBN 0-19-507011-9
1. Soil chemistry. I. Title.
S592.5.M39 1994 631.4'1—dc20 93-1552

9 8 7 6 5 4 3

Printed in the United States of America
on acid-free paper

TO JAN AND DEVIN

and to the many generations of farmers in my family,

whose inherited knowledge of the soil

is not to be found in a book.

Preface

An ordinary housefly, if placed within a hollow clear glass globe having a single hole for an outlet, generally escapes rather promptly. But a bee, under like circumstances, seldom escapes. This . . . does not indicate superior wisdom on the part of the fly—quite the contrary. For the fly, dashing hither and thither in planless flight, chances upon the exit, whereas the bee, flying steadfastly towards the light, continues to try to force its way through what must seem to it to be queerly resistant atmosphere.

From the preface to *Colloid Chemistry*, Vol. III, J. Alexander, 1931

Science has numerous examples of a deficient or even incorrect theory explaining the observations at hand, only later to be found unsound. Although it has been said that there is nothing so practical as a good theory, faulty theories can be quite damaging to the progress of science. How do we know which models or theories to accept or reject? For nonscientists, there is an expectation that science can provide precise and certain answers to problems of society and questions about the physical world. Those who teach and do research in science know that the answers are usually far less than certain. In fact, some scientists believe that certainty about any scientific theory is never completely achievable. After all, a theory is only a plausible hypothesis that gains scientific acceptance, but further investigation may well replace it with a better theory.

Much of soil science is empirical rather than theoretical in practice. This fact is a result of the extreme complexity and heterogeneity of soils, which are impossible to fully describe or quantify by simple chemical or physical models. It is not unusual for working solutions to be found for soil chemical problems with little, or even fallacious, understanding at a fundamental level. The simplicity of the empirical approach becomes the seed of its undoing, because the primary advantage of the scientific method, predictive capability, is curtailed or lost. On the other hand, the universality of chemical principles and laws (i.e., theories) permit processes to be described in such a way that soil chemical behavior can be understood and predicted in many situations not yet studied.

This book is an attempt to describe soil chemistry within the bounds of established chemical principles. It avoids as much as possible the more empirical descriptions and models, instead stressing concepts that build from our present knowledge of inorganic, organic, and physical chemistry as well as surface science. It is hoped that a consistent, and not too complex, conceptual framework has been constructed that will help to explain the many seemingly disparate facts and observations that constitute the chemical behavior of soils. The emphasis is on environmental as opposed to agricultural topics, compared with most texts in soil chemistry, recognizing that a major challenge of the future is to protect the soil ecosystem from the pollutants of an industrial society.

Regarding the structure of the book, the reader will find numerous sections that are denoted with a smaller type size. These sections elaborate upon certain concepts, but are not considered essential to a broad understanding of the topic. As this book was written with the intention that it be used as a textbook of soil and environmental chemistry, it is suggested that the level of the course being taught would determine whether such sections are appropriate for study. The reader will also notice that rather few references are made to the technical literature. This again reflects the main purpose of the book, to serve as a text rather than a scientific reference. Although many hundreds of scientific papers were consulted in writing this book, it was decided to reference only those from which specific information was used in the text. Reader are encouraged to review the list of suggested additional reading, from which a large body of important references can be obtained.

Some of the new concepts found in this book are original (if any idea can be truly so), but most have been distilled from advanced texts and research papers, or originate from interactions with peers and outstanding graduate students with whom I have worked over the years. I will not try to acknowledge the many individuals who had a strong influence on my thinking. However, I am particularly indebted to Dr. Max Mortland, who instilled in me an interest in understanding clay surface processes at the molecular level.

June 1993 M. B. McB.
Ithaca, N.Y.

Contents

ENVIRONMENTAL CHEMISTRY OF SOILS

1

Review of Chemical Principles

1.1. INTRODUCTION

This chapter reviews those chemical principles that will be particularly useful for a fundamental understanding of soil chemistry. The reader may not wish to read this chapter before going on to the main text dealing with soil chemistry, but may prefer to refer back to specific sections within this chapter when unfamiliar or difficult chemical concepts are encountered in the text.

1.2. THE NECESSARY CONCEPTS

1.2a. Gram Molecular Weight (Mole)

Most equations of theoretical chemistry are expressed in terms of the *number* of atoms or molecules in a system rather than their weight. The same number of atoms, 6.023×10^{23} (Avogadro's number), is contained within one *gram atomic weight* of each element. Similarly, this number of molecules is contained in 1 *gram molecular weight* of each compound. Because gram atomic weight and gram molecular weight are cumbersome terms, they are replaced by the abbreviated term *mole,* which stands for the amount of an element or compound containing Avogadro's number of atoms or molecules. For example, there are Avogadro's number of carbon atoms in 12 grams of pure carbon (graphite or diamond), and there are Avogadro's number of water molecules in 18 grams of pure water. That is, the gram molecular weight of H_2O, determined from the atomic weights of its constituent H and O atoms, is $(2 \times 1) + (1 \times 16) = 18$.

In soil chemistry, mole is also used to denote Avogadro's number of fundamental charge units on surfaces or ions, symbolized by mole(+) or mole(−) depending on the sign of the charge. For example, 1 mole of Ca^{2+} ions carries 2 moles(+) of charge.

1.2b. Types of Chemical Bonding

The important kinds of bonds formed between atoms are summarized below.

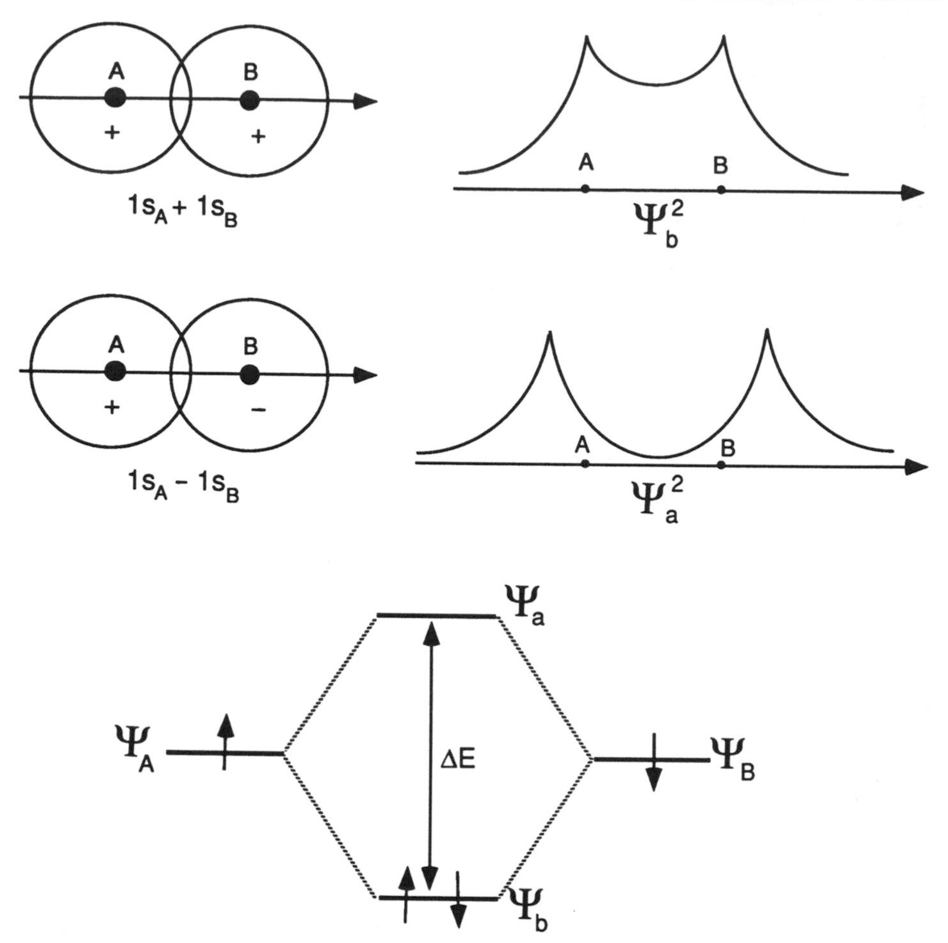

Figure 1.1. Representation of the geometry and energy of the bonding (Ψ_b) and antibonding (Ψ_a) molecular orbitals formed by covalent bonding between the 1*s* atomic orbitals of two H atoms, A and B. The probability function, Ψ^2, indicates electron density along the internuclear A-B axis.

Covalent Bonding. This type of bonding is strongly directional and fairly short-range because electron orbital overlap is required to form it. The bond is typically very strong, on the order of 50 to 100 kilocalories/mole. An example of covalent bonding is given in Figure 1.1, where the atomic orbitals of H atoms overlap to form bonding and antibonding orbitals. It is the bonding orbital that results in the strong (103 kcal/mole) H—H bond. The bond order for a molecular bond is equal to half of the difference between the number of bonding and antibonding electrons. For the hydrogen molecule, H_2, this is $\frac{1}{2}(2 - 0) = 1$. That is, the bond order is unity. As bond order increases, the strength of the covalent bond increases.

Ionic Bonding. This form of bonding results if the outer valence electrons of two atoms, A and B, are of very different energies, so that transfer of one or more electrons occurs:

$$A + B \rightarrow A^{+} + B^{-} \tag{1.1}$$

The attractive energy, E_A, between these two ions is approximated by the equation

$$E_A = \frac{z^{+}z^{-}}{r} \tag{1.2}$$

where z^{+} and z^{-} are the ionic charges and r is the interionic distance. The strength of ionic bonds is about the same as that of covalent bonds, but ionic bonds are non-directional and relatively long-range (energy decreases as $1/r$).

For ionic solids, in which both attractive and repulsive electrostatic forces as well as short-range repulsive forces complicate the description of the overall energy, the Born–Lande equation has been shown to provide an adequate estimate of lattice energy, E_L:

$$E_L = \frac{AN_0 z^{+} z^{-} e^{2}}{r_L}\left(1 - \frac{1}{n}\right) \tag{1.3}$$

Here, e is the electronic charge unit ($e = 4.8 \times 10^{-10}$ esu), N_0 is Avogadro's number, A is the Madelung constant, and n is a parameter arising from repulsive forces that build up as atoms come into contact and atomic orbitals begin to overlap. The distance between ion pairs, r_L, can be determined for an ionic solid from the known radii of the constituent ions. For solid KCl, as an example, the radii of K^{+} and Cl^{-} are 1.33 Å (Angstrom) and 1.81 Å, respectively. The value of r_L is then 1.33 + 1.81 = 3.14 Å, and n is known to be 9 for this particular lattice. Equation 1.3 can then be used to calculate the overall lattice energy of KCl and other ionic solids.

Ion-Dipole Bonding. Many molecules do not possess net positive or negative charge, yet are electrostatically attracted to charged molecules or surfaces because they possess a *dipole,* which is a separation of charge within the molecule. If the charge separation is defined as q and the distance of separation is r, then the *dipole moment,* μ, of the molecule is given by

$$\mu = qr \tag{1.4}$$

Dipoles tend to orient in an electric field, directed along the field gradient. Thus, dipoles are attracted to ions, and ion-dipole forces are in large part (if not solely) responsible for the tendency of cations and anions to hydrate in water, a compound with a large dipole moment. Hydration of cations and anions is depicted in Figure 1.2. The energy of an ion-dipole interaction is given by

$$E_{\mathrm{ID}} = \frac{|z^{+,-}|e\mu}{r^{2}} \tag{1.5}$$

where $z^{+,-}$ is the cation or anion charge and r is the distance between the ion and dipole. From this formula it can be seen that ion-dipole interactions are shorter range and weaker than ionic bonds, since the charge separation q is less than a full electron charge. It also can be seen that small ions with high charge (e.g., Al^{3+}) hydrate much more strongly than large ions with low charge (e.g., K^{+}).

The electric field of ions can induce dipole moments in uncharged, nonpolar molecules by distorting (polarizing) the electron distribution within the molecules.

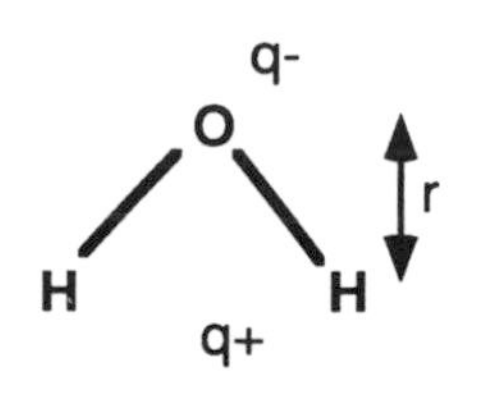

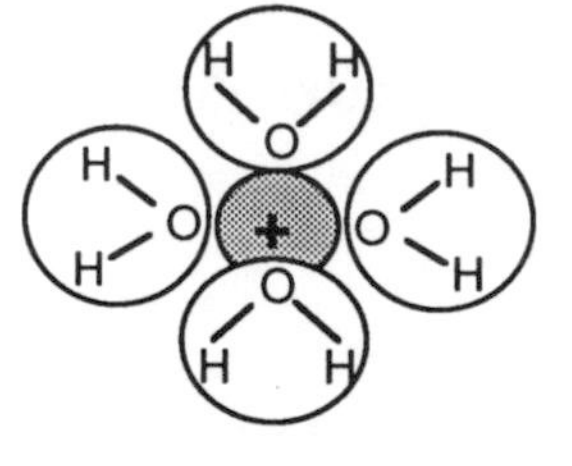

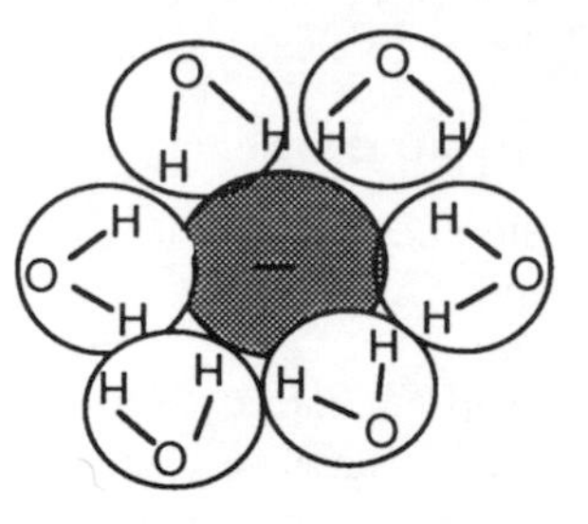

WATER DIPOLE HYDRATED CATION HYDRATED ANION

Figure 1.2. Separation of charge, q, for the H_2O dipole, and the orientation of these water dipoles around cations and anions in solution.

This, however, is a very short-range force, and the energy of such an ion-induced dipole interaction decreases as $1/r^4$.

Dipole-Dipole Bonding. Two dipoles can interact by orienting oppositely charged ends toward one another. The energy of this interaction is given by

$$E_{DD} = \frac{-2\mu_1\mu_2}{r^3} \tag{1.6}$$

where μ_1 and μ_2 are the dipole moments of the two molecules. For example, at room temperature, water molecules align their dipoles to maximize H—O bonding among molecules. The bonding is directional, so that polar liquids such as water are well structured at low temperature. Many of the unique and essential properties of water result from this dipole-dipole bonding.

Hydrogen bonding is a specific form of dipole-dipole interaction and is said to exist when an H atom is bonded to two or more other atoms. In the case of water, the H atom is covalently bonded (short bond) to one oxygen atom and hydrogen bonded (long bond) to another oxygen atom (see Figure 1.3). The hydrogen bond appears to be a dipole-dipole interaction, but the covalent bond is delocalized over both oxygen atoms associated with H, lending unusual strength to the hydrogen bond.

The electric field of a dipole can induce the formation of a dipole in a second molecule, generating net attraction between dipolar and nonpolar molecules. Such forces are, however, weak, and important only at very short intermolecular distances. The energy of this interaction decreases with distance according to $1/r^6$.

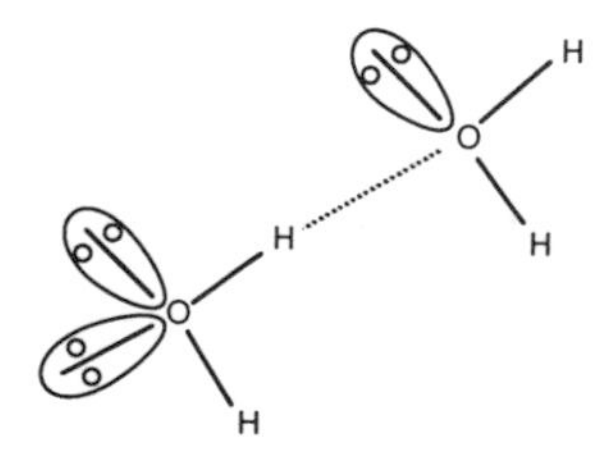

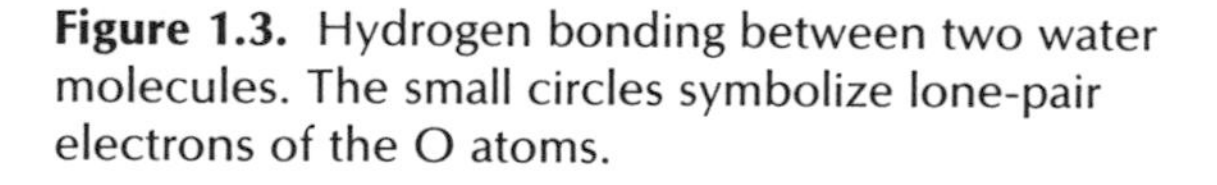

Figure 1.3. Hydrogen bonding between two water molecules. The small circles symbolize lone-pair electrons of the O atoms.

Dispersion (van der Waals) Forces. Even in molecules with no permanent dipole, imbalances in electron distribution produce instantaneous dipoles. If two or more adjacent molecules "synchronize" their electronic motion, electron-electron repulsion can be minimized while electron-nuclear attraction is maximized. The result is an extremely short-range and weak attraction between individual nonpolar molecules, with energy decreasing with distance according to $1/r^6$. These weak attractive forces are referred to as the London dispersion forces or van der Waals attraction. Being additive, they become more significant in large molecules. Furthermore, van der Waals attraction between surfaces is expected to be less short-range than that between individual molecules because of geometric considerations.[1]

Repulsive Forces. As atoms approach one another or surfaces very closely, the electron clouds of the interacting atoms begin to overlap, with the result that repulsion (known as Born repulsion) becomes the dominant force and closer approach becomes impossible. For this reason, the "hard sphere" model can be used to describe ionic solids. The individual atoms have well-defined radii that determine the distances of closest approach in close-packed arrangements. In equation 1.3, this distance is found to be 3.14 Å for KCl.

1.2c. Ions in Solution

Concept of Activity. Electrolytes dissociate in water to form solutions of ions. For the electrolyte AB, dissociation can be written as

$$AB \rightarrow A^+ + B^- \tag{1.7}$$

The *activity product* of this electrolyte is expressed as the product, a_+a_-, where a_+ and a_- are the activity of the cation and anion. Individual ion activities appear only as products or quotients in thermodynamics, never as separate values. The *mean activity* is defined as

$$a_{+/-} = \sqrt{a_+a_-} \tag{1.8}$$

Activities of the ions in water can be considered to be "effective" concentrations, more thermodynamically precise than the actual molar concentrations of ions. The activity of any ion, a_i, is related to the concentration, c_i, by the activity coefficient, γ_i:

$$a_i = \gamma_i c_i \tag{1.9}$$

As a result, the activity product can be reexpressed:

$$a_+a_- = (\gamma_+c_+)(\gamma_-c_-) = \gamma_+\gamma_-[A^+][B^+] \tag{1.10}$$

where the square brackets denote concentrations (in mole per liter units). The *mean activity coefficient* is then given by

$$\gamma_{+/-} = \sqrt{\gamma_+\gamma_-} \tag{1.11}$$

1. Molecular interaction is essentially between two points in space (unless the molecules are very large), while interaction between two flat surfaces results from the many-point attraction across the planes.

If these expressions are applied to a salt such as Na_2CO_3, whose dissociation in water is given by

$$Na_2CO_3 \rightarrow 2Na^+ + CO_3^{2-} \qquad (1.12)$$

then the solubility product is $a_{Na}^2 a_{CO_3}$, and the mean activity and mean activity coefficient are

$$a_{+/-} = [(a_{Na})^2(a_{CO_3})]^{1/3} \qquad (1.13)$$

$$\gamma_{+/-} = [\gamma_{Na}^2 \gamma_{CO_3}]^{1/3} \qquad (1.14)$$

The solubility product is an example of a thermodynamically precise quantity, which can be expressed wholly in terms of the mean activity with no loss of thermodynamic information:

$$\text{Solubility product} = a_{Na}^2 a_{CO_3} = (a_{+/-})^3 \qquad (1.15)$$

This is a consequence of the fact that no thermodynamically exact method is known that allows the determination of individual ion activities. Thermodynamics is the systematic study of energy changes in chemical and physical systems, and as such is independent of models and theories, including the molecular theory of matter. Acknowledging the existence of dissociated ions in solution is necessary for many purposes, but is not necessary to the definition of thermodynamically exact quantities.[1]

The Debye-Huckel Model for Activity Coefficients. In spite of the indifference of classical thermodynamics to the existence of ions, the quantification of soil chemical processes such as ion exchange requires that individual ion activities in solution at least be estimated. This can be done using the Debye-Huckel theory of activities, which recognizes that ions of opposite charge tend to surround and electrostatically stabilize any particular ion in solution. The theory stems from the empirical observation that the mean activity coefficient, $\gamma_{+/-}$, of a strong electrolyte is the same in all dilute solutions of the same ionic strength regardless of ionic composition. It is necessary to define ionic strength here, using the equation

$$I = \tfrac{1}{2} \sum_i c_i z_i^2 \qquad (1.16)$$

where c_i = concentration of the ith ion in solution and z_i = charge of the ith ion in solution. Ionic strength, I, is a measure of the nonideality that the solution imposes on any dissociated electrolyte. Solutions with higher ionic strength are less "ideal" in the sense that individual ions interact more strongly with neighboring ions of opposite charge. This electrostatic attraction generally lowers the effective concentration, or activity, of ions.

The extended Debye-Huckel equation, which applies for ionic strengths of less than 0.1, is given by

$$\log \gamma_{+,-} = -0.5091 z_{+,-}^2 \frac{\sqrt{I}}{1 + \mathring{a}B\sqrt{I}} \qquad (1.17)$$

1. One might protest that pH is an obvious example where a single-ion activity is quantified for the purpose of describing chemical systems. It must be recognized, however, that the operation of the electrodes used to measure pH cannot be described in strictly thermodynamic terms, and the measured pH is not precisely related to a thermodynamically exact quantity.

Table 1.1. Values of the Parameter $\mathring{a}$ Used in the Extended Debye-Huckel Equation

å	Ions
3	OH^-, F^-, Cl^-, Br^-, I^-, NO_2^-, NO_3^-
	K^+, Rb^+, Cs^+, $CH_3NH_3^+$
4	HCO_3^-, $H_2PO_4^-$, CH_3COO^-, SO_4^{2-}, HPO_4^{2-}, PO_4^{3-}
	Na^+, $(CH_3)_4N^+$
5	S^{2-}, CO_3^{2-}, MoO_4^{2-}, $(COO)_2^{2-}$, $(citrate)^{3-}$
	Sr^{2+}, Ba^{2+}, Cd^{2+}, Hg^{2+}, Pb^{2+}
6	$C_6H_5COO^-$
	Li^+, Ca^{2+}, Cu^{2+}, Zn^{2+}, Mn^{2+}, Fe^{2+}, Ni^{2+}, Co^{2+}
8	Mg^{2+}
9	H^+, Al^{3+}, Fe^{3+}, Cr^{3+}

This semiempirical equation utilizes two parameters, $\mathring{a}$ and B, to evaluate $\log \gamma_+$ or $\log \gamma_-$. While B has a value of about 0.33, $\mathring{a}$ is approximated by the size of the hydrated cation (if γ_+ is being evaluated) or anion (if γ_- is being evaluated), expressed in Angstrom units. Values of $\mathring{a}$ for common ions are given in Table 1.1. Since an infinitely dilute solution of electrolyte is considered to be ideal (no electrostatic interactions), the Debye-Huckel equation predicts that $\gamma_+ = \gamma_- = 1$ for $I = 0$. This means that activities and concentrations of ions approach the same numerical value in increasingly dilute solutions. Activities are unitless, however, while concentrations are expressed in moles per liter.

The Debye-Huckel equation (and other empirical expressions that correct measured concentrations to activities) fails to account for specific *ion pairing* and *complexation* in solution, which in some salt solutions may contribute more to the inequality between concentration and activity than the nonspecific electrostatic interactions modeled by the equation. Ion pairing or complexation is likely to become significant in solutions with any of the following characteristics:

1. The cations and/or anions have high charge (≥ 2).
2. Inorganic anions other than NO_3^- or Cl^-, neither of which complexes significantly with common cations, are present.
3. Transition (or heavy) metals and neutral or anionic organic molecules (ligands) are present.
4. The complexing species (ligand) is present in high concentration.
5. The pH is relatively high, and polyvalent cations are present.

Ion pairing and complexation must be accounted for separately from the Debye-Huckel correction for nonideal behavior of solutions. For example, in salt solutions containing Mg^{2+} and SO_4^{2-}, Mg^{2+} tends to pair with SO_4^{2-}:

$$Mg^{2+} + SO_4^{2-} = MgSO_4^0\,(aq) \tag{1.18}$$

The extent of this reaction can be determined from the known association constant, K_{as}:

$$K_{as} = \frac{a_{MgSO_4}}{a_{Mg}a_{SO_4}} = 200 \tag{1.19}$$

In most soil solutions, the ionic strength I is low (<0.01), so that the extended Debye-Huckel equation is applicable for the correction of ionic concentrations to the more thermodynamically meaningful activities. Typically, conductivity measurements are used to estimate the ionic strength, a much less laborious procedure than measurement of each cation and anion present in solution. More problematic, however, is the detection and measurement of complexing anions and molecules (ligands). As will be shown later in this chapter, their presence can result in activities of metal ions in soil solution being much lower than measured concentrations would suggest.

Ion Hydration and Hydrolysis. All ions hydrate to some degree in water. Conceptually, the strength of hydration of an ion could be determined by measuring the heat released when the gaseous form of the ion is immersed in water. For a gaseous metal cation, M^{z+} (g), this reaction would be

$$M^{z+}\ (g) + n\ H_2O \rightarrow M(H_2O)_n^{z+} + E_h \tag{1.20}$$

where E_h is the hydration energy of the cation. In general, ions of small size and high charge hydrate most strongly because of their strong electrical fields, which cause energetic bonding between water dipoles and the ions. In fact, the standard free energy of hydration of cations or anions, ΔG_h^0, can be estimated from the empirical relation

$$\Delta G_h^0 = \frac{-164z^2}{r'} + 1.3\ \text{(kcal/mole)} \tag{1.21}$$

where:

$$r' = \text{cation radius} + 0.72\ \text{Å}$$

or:

$$r' = \text{anion radius} + 0.3\text{–}0.6\ \text{Å}$$

and z is the charge of the ion. This equation is more successful with the lighter non-transition metals such as Na^+ and Ca^{2+} because it does not account for covalent bonding and crystal field effects, which strengthen hydration bonds of heavy and transition metals.

The free energy change associated with the transfer of the ion from the gaseous to the aqueous state arises from the entropy change, ΔS_h^0, as well as from the enthalpy or energy change, ΔH_h^0, according to the equation

$$\Delta G_h^0 = \Delta H_h^0 - T\Delta S_h^0 \tag{1.22}$$

(The next section defines these thermodynamic properties.) When ions are transferred into water, there is a large decrease in the entropy term, especially if the ion is small or highly charged, because of the restrictions to motion imposed on the water molecules by the ion-dipole bonds. This molecular ordering of water around ions is depicted in Figure 1.2. Since entropy is a measure of molecular disorder in a system, ordering produces an entropy decrease.

Cation behavior in aqueous solution can be generalized into three classes, depending on the charge, z, and the radius, r, of the ion, as outlined in Table 1.2.

Table 1.2. Dependence of Cation Behavior in Water on the Ionic Potential, z^2/r, of the Ion

z^2/r	Reaction in Water	Species Formed
Small	$M^{z+} + n\,H_2O \rightarrow M(H_2O)_n^{z+}$ (hydration)	Hydrated cation
Intermediate	$M^{z+} + z\,H_2O \rightarrow M(OH)_z + zH^+$ (hydrolysis)	Insoluble hydroxide
Large	$M^{z+} + y\,H_2O \rightarrow M(OH)_y^{(y-z)-} + y\,H^+$ (hydrolysis)	Soluble oxyanion or hydroxyanion

Table 1.2 reveals that cations having a very large charge to radius ratio tend to polarize water sufficiently strongly to promote *hydrolysis*. This can be thought of as hydration taken to the extreme, where the "ionic potential," z^2/r, of the ion is large enough to rupture O—H bonds. The first step of hydrolysis is visualized as

$$M^{z+}-\underset{\displaystyle H}{\underset{|}{O}}-H \rightarrow M^{z+}-\underset{\displaystyle H}{\underset{|}{O}}-H \rightarrow M(OH)^{(Z-1)+} + H^+ \quad (1.23)$$

In comparing the relative tendency of cations to undergo this first step in hydrolysis, it is necessary to define the hydrolysis constant, K_h^1. Using Al^{3+} as an example, the first hydrolysis reaction is

$$Al^{3+} + H_2O = Al(OH)^{2+} + H^+ \quad (1.24)$$

and the equilibrium constant for the reaction is

$$K_h^1 = \frac{a_{Al(OH)}a_H}{a_{Al}} = 10^{-4.97} \quad (1.25)$$

Consequently, the pK_h^1 (negative logarithm of K_h^1) is 4.97. These pK_h^1 values for a number of metals are plotted in Figure 1.4 as a function of the ionic potential. The straight line in this figure marks the relation between hydrolysis and ionic potential that would be expected if only electrostatic (i.e., ion-dipole) forces were operative. However, crystal field and covalent contributions to bonding increase the tendency for most transition and heavy metals to hydrolyze. The straight line most accurately predicts the hydrolysis behavior of the alkali and alkaline earth metals.

1.2d. Thermodynamic Properties

The internal energy, E, enthalpy, H, entropy, S, and Gibbs free energy, G, are *state functions* of a system. Chemical processes result in changes in these state functions, which are thermodynamic properties of the system. Such properties could, in principle, be calculated for a given temperature T and pressure P from the molecular properties of the system. The molecular disorder of the system is measured by the entropy S, while the energy of the system is measured by E or H. As the system undergoes reaction, these measures of disorder and energy change.

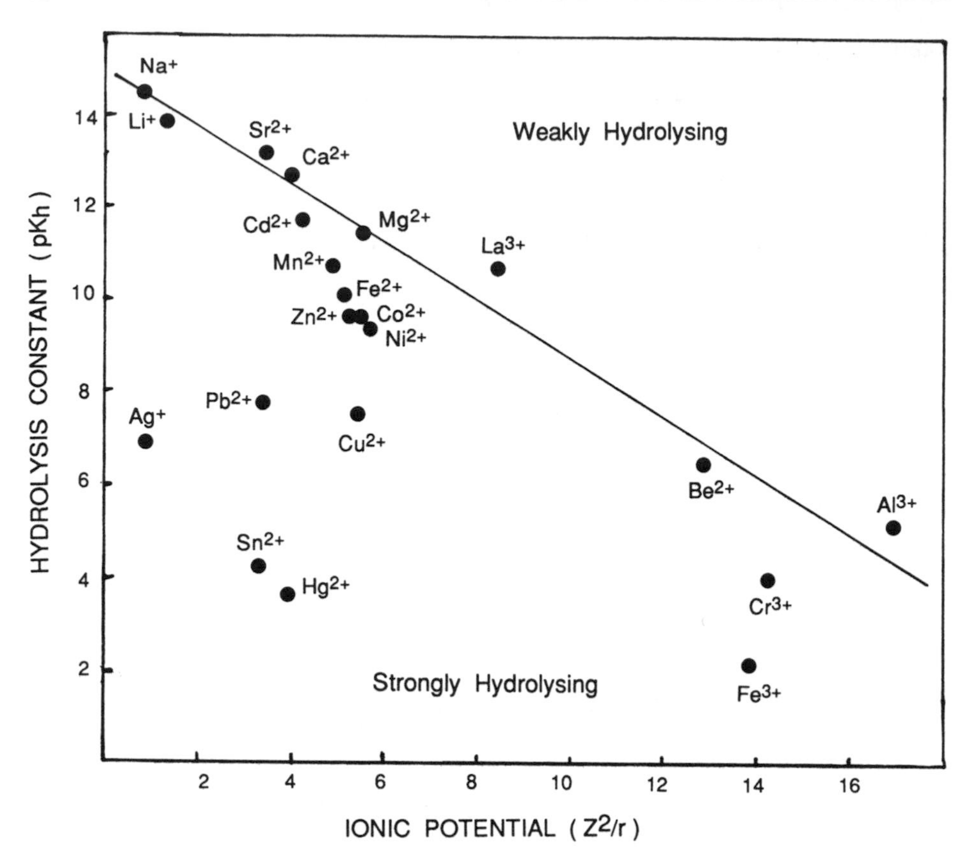

Figure 1.4. Dependence of the tendency of metal cations to hydrolyze on the ionic potentials of the ions. (Data from J. E. Huheey 1972. *Inorganic Chemistry.* New York: Harper & Row.)

The enthalpy is related to energy according to the equation

$$H = E + PV \tag{1.26}$$

where V is the volume of the system. The Gibbs free energy is defined as

$$G = H - TS \tag{1.27}$$

For constant temperature and pressure, $\partial G = 0$ defines the condition of an equilibrium state. The second law of thermodynamics requires for spontaneous processes that

$$\partial E < T\partial S - P\partial V \tag{1.28}$$

This means, for example, that at constant entropy and volume, $\partial E < 0$; that is, the internal energy must decrease for the process to occur spontaneously.

For a chemical reaction at constant T, in which the system changes from one composition to another, say from state 1 to state 2, equation 1.27 can be used to show that the change in free energy, ΔG, will be

$$\Delta G = \Delta H - T\Delta S \tag{1.29}$$

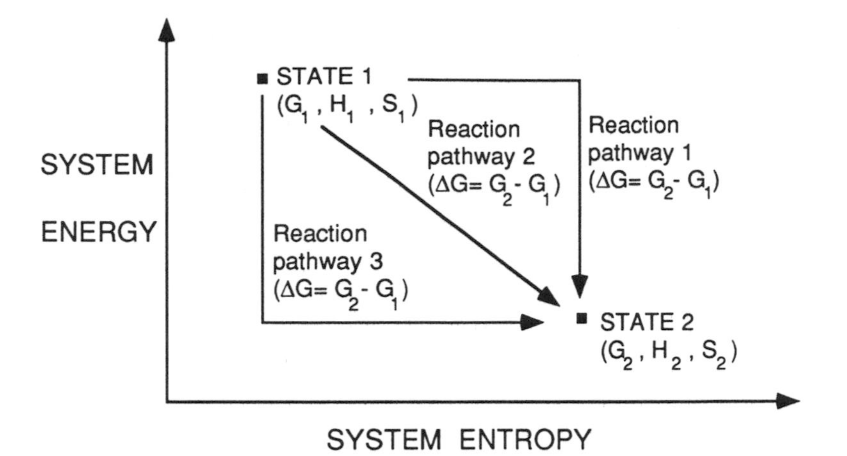

Figure 1.5. Diagram of the change in state functions of a reaction that proceeds from state 1 (initial) to state 2 (final), showing the irrelevance of reaction pathway to ΔG, ΔH, and ΔS of the reaction.

These changes in the system, ΔG, ΔH, and ΔS, depend only on the properties of state 1 and state 2. The reaction pathway followed to get from one state to the other does not alter the values of ΔG, ΔH, and ΔS. This principle of state functions is schematically illustrated in Figure 1.5.

When ΔG is negative, the maximum work that can be done by the spontaneous process within a closed system at constant T and P is given by $-\Delta G$. When ΔG is positive, the process is not spontaneous, and ΔG is then the minimum work that must be provided *to* the system to drive the process. While the reaction pathway followed has no effect on the free energy change of the reaction, the pathway does determine the amount of useful work than can be obtained from a spontaneous reaction process. The free energy change, ΔG, is a measure of the work that could theoretically be obtained from a reaction, but this amount of work could only be realized if the reaction or process were conducted in a reversible manner, that is, at near-equilibrium conditions where there is a very small driving force.

The *chemical potential* of any chemical component,[1] i, of a system is the partial molar free energy:

$$\mu_i = \left[\frac{\partial G}{\partial n_i}\right]_{T,P,n_j} \tag{1.30}$$

where n_i is the number of moles of component i in the system and n_j is the number of moles of all other components. In systems with several phases (a phase is defined as a chemically and physically uniform part of the system), the chemical potential of each component is equal in all phases (1, 2, 3, . . . N) at equilibrium:

$$\mu_i(1) = \mu_i(2) = \mu_i(3) = \cdots = \mu_i(\mathrm{N}) \tag{1.31}$$

1. A chemical component, i, of a system is any chemical species whose quantity can be varied independently of all other components.

The activity of a component i, given by a_i, is defined in terms of chemical potentials:

$$\mu_i = \mu_i^0 + RT \ln a_i \tag{1.32}$$

where μ_i^0 is the standard state chemical potential of i and R is the gas constant. It can then be shown that for any chemical reaction involving m moles of component M, n moles of component N, p moles of component P, and q moles of component Q:

$$m\text{M} + n\text{N} = p\text{P} + q\text{Q} \tag{1.33}$$

that the free energy change, ΔG, for the reaction is

$$\Delta G = \Delta G^0 + RT \ln \left[\frac{(a_P)^p (a_Q)^q}{(a_M)^m (a_N)^n} \right] \tag{1.34}$$

where ΔG^0 is the free energy change for the reactants and products at standard state conditions.[1]

If the chemical reaction proceeds to equilibrium, no further driving force exists, and $\Delta G = 0$. Consequently, equation 1.34 becomes:

$$\Delta G^0 = -RT \ln \left[\frac{(a_P)^p (a_Q)^q}{(a_M)^m (a_N)^n} \right]_{\text{equilibrium}} \tag{1.35}$$

This relationship holds when the component activities in equation 1.35 are those of the system at equilibrium. Since ΔG^0 for any particular reaction at a given temperature has a fixed value, the bracketed activity term in equation 1.35 must have a fixed value at equilibrium. This activity term is denoted by the symbol K, and is called the *equilibrium constant:*

$$K = \frac{(a_P)^p (a_Q)^q}{(a_M)^m (a_N)^n} \tag{1.36}$$

With this definition, equation 1.35 simplifies to

$$\Delta G^0 = -RT \ln K \tag{1.37}$$

an equation that is extremely useful because, beyond indicating the direction in which a reaction will proceed, it allows calculation of the equilibrium state that will finally result.

1.2e. Equilibrium Constants

Dissolved Gases and Volatile Solutes. By analogy with equation 1.32, the chemical potential of a gas in a gas mixture is given by:

$$\mu_i = \mu_i^0(\text{g}) + RT \ln P_i \tag{1.38}$$

where P_i is the partial pressure of the gas and $\mu^0_i(\text{g})$, the chemical potential of the gas in the standard state, is a function of T but independent of P_i. *Raoult's law* states that, for ideal behavior of gases in solutions, the vapor pressure of the dissolved gas

1. Standard state conditions for component i is $a_i = 1$ at standard temperature and pressure.

(or volatile solute), P_i, equals the partial vapor pressure of the pure component, P_i^0, multiplied by the mole fraction of the component dissolved in solution, x_i:

$$P_i = x_i P_i^0 \tag{1.39}$$

Raoult's law is more closely followed by real solutions as x_i approaches 1.

For dilute solutions of ideal gases and volatile solutes in water, the vapor pressure is found to be proportional to the mole fraction of the dissolved component:

$$P_i = K_i x_i \tag{1.40}$$

This is *Henry's law,* and K_i is the Henry's law constant, which must be measured for each gaseous compound of interest that dissolves in water. The law states that the tendency of a volatile compound dissolved in water to escape to another phase (air, surface, etc.) is proportional to its mole fraction in the water. Henry's law behavior is followed most closely as x_i approaches zero, that is, in very dilute solutions of the volatile compound. As x_i approaches zero, the mole fraction becomes linearly proportional to the more conventional measure of concentration, molarity, so that equation 1.40 can be expressed equivalently with molarity replacing mole fraction.

For reactions involving dissolved gases (such as O_2 and CO_2), the activity of the gas could be expressed either in terms of its partial pressure in the gas phase, P_i, or its mole fraction in the solution phase, x_i, assuming that the two phases are at equilibrium. The convention is to express activity as P_i, using partial pressure units. For example, the CO_2 concentration in the atmosphere is 350 parts per million by volume, which is a partial pressure of $350 \times 10^{-6} = 0.000350$ atmospheres. This means that, if all the gaseous molecules except CO_2 could be removed from the air, the remaining pressure would be only 0.000350 atmospheres instead of 1 atmosphere. Equilibrium expressions that call for gas activities generally utilize these partial pressure units.[1]

The standard state of a gas is taken to be 1 atmosphere pressure of the gas at 25°C.

Solubility Products. The activity expression of equation 1.36 can be applied in general to any chemical reaction. In particular, it can be used to describe dissolution reactions for solids such as aluminum hydroxide:

$$Al(OH)_3\,(s) = Al^{3+} + 3OH^- \tag{1.41}$$

Equation 1.36 then defines the solubility product, K_{SO}:

$$K_{SO} = a_{Al3+} \cdot a_{OH^-}^3 = 10^{-34} \tag{1.42}$$

It should be noted that the activity of the solid phase does not appear explicitly in expression 1.42, a result of the fact that the thermodynamic activities of pure solid phases can be taken to be constant and are assigned a value of 1 by convention.[2]

1. The ideal gas law, $P_i V = n_i RT$, establishes the linear relation between pressure of a gas, P_i, and mole quantity, n_i. It follows that the partial pressure is linearly related to the mole fraction in a gas mixture.

2. Solids containing ionic impurities can have solubility products different from those of the pure minerals. For example, substitution of Al^{3+} into FeOOH appears to stabilize this mineral, resulting in a lower solubility product.

However, different structural forms, despite identical chemical compositions, can have different solubility products. In the example given above, 10^{-34} is the solubility product for gibbsite, one crystalline form of $Al(OH)_3$. Other crystalline forms of $Al(OH)_3$, such as bayerite, have somewhat different values of K_{SO}. Noncrystalline forms of solids invariably are more soluble (i.e., have higher solubility products) than their crystalline counterparts. However, because the least soluble form is thermodynamically the most stable, the crystalline form is ultimately expected to determine the level of solubility.

In the case of $Al(OH)_3$ freshly precipitated from solution, there is a slow recrystallization from the amorphous to the crystalline form, with the result that the ion activity product, $(a_{Al^{3+}})\cdot(a_{OH^-})^3$, of the aqueous solution decreases over time, approaching the value for the crystalline mineral. Here, as in many chemical processes, activities of ions and molecules in solution are subject to control by reaction rates (kinetics) as well as equilibrium constants. The important role of kinetics in chemical reactions will be discussed later in this chapter.

Acidity and Basicity Constants. An acid is a chemical species that tends to lose or donate a proton, while a base is a species that tends to accept or add a proton. The strength of acidity for an acid, HA, can be measured from the equilibrium constant, K_a, of the dissociation reaction:

$$HA = H^+ + A^- \tag{1.43}$$

which is given by

$$K_a = \frac{(a_{H^+})(a_{A^-})}{a_{HA}} \tag{1.44}$$

Since protons in solution are actually hydrated, it is more accurate to write the reaction as the transfer of a proton from the acid to water:

$$HA + H_2O = H_3O^+ + A^- \tag{1.45}$$

However, the equilibrium expression for K_a is the same whether water is explicitly included in the reaction or not.

The strength of basicity for a base, B, is measured by the equilibrium constant, K_b, for the association reaction:

$$B + H_2O = BH^+ + OH^- \tag{1.46}$$

and expressed as

$$K_b = \frac{(a_{BH^+})(a_{OH^-})}{a_B} \tag{1.47}$$

From these definitions, hydrated Al^{3+} is seen to be a weak acid by virtue of the proton that it releases upon hydrolysis:

$$Al(H_2O)_6^{3+} = Al(H_2O)_5OH^{2+} + H^+ \qquad K_a = 10^{-4.97} \tag{1.48}$$

Similarly, carboxylic acids are weak to moderately strong acids, releasing protons to form the carboxylate anion:

$$R{-}COOH = R{-}COO^- + H^+ \qquad K_a \approx 10^{-3}\text{–}10^{-6} \tag{1.49}$$

The strength of acidity depends on the chemical nature of the organic group, R. These acidity and basicity constants are typically reported in negative logarithmic form, as $pK_a = -\log K_a$ and $pK_b = -\log K_b$. The pK_a of $Al(H_2O)_6^{3+}$ is then 4.97, for example.

These definitions describe the Lowry-Brønsted concept of acidity and basicity. The Lewis concept of acids and bases is more general, and may be useful for reactions in which protons are not involved. A *Lewis acid* is any substance that can accept electrons, and a *Lewis base* is any substance that can donate electrons. Small, high-charge metal ions such as Al^{3+} and Fe^{3+} are strong Lewis acids because they tend to complex with functional groups on molecules (Lewis bases) by accepting electrons from them:

$$Fe^{3+} + :NH_3 = Fe^{3+}-NH_3 \tag{1.50}$$

Lewis bases include H_2O, NH_3, and amines, as well as phenols and carboxylic acids in the dissociated ($R-O^-$, $R-COO^-$) form. All of these molecules and many others possess electron-donating coordinating groups, and are referred to collectively as *ligands.*

According to the Pearson terminology, Lewis acids and bases can be classified on a scale of "hard" to "soft." Soft bases (or ligands) are large molecules that are easily polarized; that is, they readily donate electrons to form covalent bonds. They selectively bond with soft acids, typically metal ions of relatively large radius and low charge. Hard bases tend to be small molecules that are not easily polarized and form less covalent and more ionic bonds, preferentially with hard acids—typically metal ions of small radius and high charge. In summary, then, the rule is:

Soft acids bond preferentially with soft bases and hard acids bond preferentially with hard bases.

A listing of the classification of acids and bases is given in Table 1.3. From this table it is apparent, for example, that Fe^{3+} (a hard acid) will tend to bond strongly with "hard" bases such as phosphate and carboxylate, but not with "softer" bases such as sulfide or aromatic amines. Fe^{2+}, a "softer" acid, bonds more strongly with these softer bases.

Complexation and Chelation Reactions. Bonding of metals with ligands in soil solution is important in determining the chemical behavior and toxicity of metals in

Table 1.3. Classification of Lewis Acids and Bases According to the Pearson Hardness Concept

	Acids	Bases
Hard	H^+, Li^+, Na^+, K^+ Mg^{2+}, Ca^{2+}, Sr^{2+} Ti^{3+}, Cr^{3+}, Mn^{2+}, Fe^{3+}, Co^{3+}, Al^{3+} Si^{4+}	NH_3, $R-NH_2$ (amines) H_2O, OH^-, O^{2-}, $R-OH$ (alcohols) CH_3COO^-, CO_3^{2-}, NO_3^-, PO_4^{3-}, SO_4^{2-}, F^-
Borderline	Fe^{2+}, Co^{2+}, Ni^{2+}, Cu^{2+}, $Zn^{2,+}$$Pb^{2+}$	$C_6H_5NH_2$ (aniline and other aromatic amines) C_5H_5N (pyridine), NO_2^-, SO_3^{2-}, Br^-
Soft	Cu^+, Ag^+, Cd^{2+}, Hg^+, Hg^{2+}	CN^-, CO, S^{2-}, $R-SH$ (sulfhydryl), $R-S^-$

soil. Depending on the composition of the solution, a large fraction of the soluble metal ions in soil solutions may actually be complexed with inorganic or organic ligands. For example, the toxic metal cadmium (Cd^{2+}) has a tendency to complex with as many as four Cl^- ions. The reactions can be written as four step-wise additions of one Cl^- ligand,[1] or as the overall (cumulative) complexation reactions of Cd^{2+}:

$$Cd^{2+} + Cl^- = CdCl^+ \qquad K_1 = 10^{1.98} \tag{1.51}$$

$$Cd^{2+} + 2\,Cl^- = CdCl_2^0 \qquad K_2 = 10^{2.60} \tag{1.52}$$

$$Cd^{2+} + 3\,Cl^- = CdCl_3^- \qquad K_3 = 10^{2.40} \tag{1.53}$$

$$Cd^{2+} + 4\,Cl^- = CdCl_4^{2-} \qquad K_4 = 10^{2.50} \tag{1.54}$$

Each equilibrium constant, K_n, has the general form

$$K_n = \frac{[CdCl_n^{(2-n)}]}{[Cd^{2+}][Cl^-]^n} \tag{1.55}$$

where the brackets denote concentrations of the complexed Cd, free Cd^{2+}, and free Cl^-, and activity corrections have been ignored. The total soluble Cd is given by the summed concentrations of all the molecular species:

$$\text{Total Cd} = [Cd^{2+}] + [CdCl^+] + [CdCl_2^0] + [CdCl_3^-] + [CdCl_4^{2-}] \tag{1.56}$$

Now, by substituting the appropriate expression for $[CdCl_n^{(2-n)}]$ from equation 1.55 into equation 1.56, total soluble Cd can be expressed in terms of $[Cd^{2+}]$, $[Cl^-]$ and the K_n values:

$$\text{Total Cd} = [Cd^{2+}] + K_1[Cd^{2+}][Cl^-] + K_2[Cd^{2+}][Cl^-]^2 + K_3[Cd^{2+}][Cl^-]^3 + K_4[Cd^{2+}][Cl^-]^4 \tag{1.57}$$

By factoring out $[Cd^{2+}]$ on the right-hand side of equation 1.57, a final equation is obtained for the activity of the free Cd^{2+} ion:

$$[Cd^{2+}] = \frac{\text{Total Cd}}{1 + K_1[Cl^-] + K_2[Cl^-]^2 + K_3[Cl^-]^3 + K_4[Cl^-]^4} \tag{1.58}$$

Now the effect of complexation on Cd^{2+} activity is apparent. The values of the equilibrium constants, K_1, K_2, K_3, and K_4, along with the activity of Cl^- in solution, determine the fraction of the total soluble Cd that is in the free Cd^{2+} (uncomplexed) form. High Cl^- activity would result in a small fraction of the soluble Cd being in the free cation form. For example, if the Cl^- concentration were 0.01 *M*, then equation 1.58 could be solved to show that 50 percent of the total soluble Cd is in the free Cd^{2+} form, while the rest is almost completely in the form of $CdCl^+$.

Metal complexation in solution can have the effect of increasing the apparent solubility of minerals. A case in point is the oxalic acid—Al hydroxide system. Oxalic acid, naturally occurring in soils as a product of biological activity, is a relatively strong dicarboxylic acid that dissociates readily:

1. These step-wise reactions can be obtained by subtracting two adjacent overall reactions, so that the step-wise K value is equal to (K_{n+1}/K_n).

$$HOOC-COOH = HOOC-COO^- + H^+ \qquad K_a^1 = 10^{-1.25} \tag{1.59}$$

$$HOOC-COO^- = {}^-OOC-COO^- + H^+ \qquad K_a^2 = 10^{-4.27} \tag{1.60}$$

The oxalate anion, $^-OOC-COO^-$, provides two coordinating groups for bonding with metals, and the resulting complex is termed a *chelate* because of the ringlike structure that is formed:

$$M(H_2O)_6^{n+} + {}^-OOC-COO^- = \left[(H_2O)_4M \begin{matrix} / OOC \\ \setminus OOC \end{matrix} \Big| \right]^{(n-2)+} + 2H_2O \tag{1.61}$$

The formation of chelates is favored because of the increase in entropy that results from the displacement of several water molecules from the coordination sites on the metal. In the reaction above, two molecules react to form three molecules, so that entropy (increased degrees of freedom of molecular motion) is generated by the reaction.

In general, then, any complexation reaction that has more moles of products than reactants will tend to have a positive entropy change, ΔS, and will proceed spontaneously unless an energy barrier exists. For the chelation of Al^{3+} by oxalate, the enthalpy of reaction, ΔH, is small because the replacement of water by carboxylate ligands (reaction 1.61) involves little change in bond energy. According to Table 1.3, both H_2O and carboxylate are "hard" Lewis bases and have comparable affinities for the hard acid Al^{3+}. The fact that acetate, CH_3COO^-, bonds only weakly with metals such as Al^{3+} despite its chemical similarity to the COO^- groups of oxalate is an indication of the dominant role of entropy in driving the chelation process.

Besides entropy, the other factors that determine the strength of chelation are the Lewis acid-base properties of the metal-ligand pair (determining the strength of the bond), and steric factors. The metal-oxalate chelate, for example, forms a five-member ring structure (see reaction 1.61) that is energetically favorable because it possesses little bond strain. Generally, five- and six-member ring structures have the least bond strain and are the most stable.

The stability constant for the Al^{3+}-oxalate complex is expressed as

$$K_s = \frac{(Al(COO^-)_2^+)}{(Al^{3+})((COO^-)_2^{2-})} = 10^{6.1} \tag{1.62}$$

where the parentheses denote activities. The value of K_s is determined by titration or other standard chemical methods. Now, the acid dissociation equations 1.59 and 1.60 can be combined with equation 1.62 to determine the degree of Al complexation at any particular pH. This is best demonstrated by first simplifying the chemical symbolism as follows:

$$\begin{aligned} [M] &= (Al^{3+}) \\ [L] &= ((COO)_2^{2-}) \\ [H] &= (H^+) \\ [HL] &= (HOOC-COO^-) \\ [H_2L] &= ((COOH)_2) \end{aligned} \tag{1.63}$$

where concentrations (square brackets) have been equated to activities (parentheses) on the assumption that the ionic strength of the solution is low. By convention, L refers to the chemical form of the organic ligand that actually complexes with the metal—in this case, the fully deprotonated oxalate anion. Now equations 1.59, 1.60, and 1.62 can be expressed in terms of this symbolism:

$$\frac{[ML]}{[M][L]} = 10^{6.1} \tag{1.64}$$

$$\frac{[H_2L]}{[HL][H]} = 10^{1.25} \tag{1.65}$$

$$\frac{[HL]}{[H][L]} = 10^{4.27} \tag{1.66}$$

Since there are now three equations with six variables ($[H_2L]$, [HL], [L], [H], [ML], [M]), the extent of metal complexation can be calculated only if more conditions are imposed on the system. For example, if the total concentration of soluble Al, oxalic acid, and the pH are specified, this fixes the extent of complexation. Consider the following example that uses these equations:

Example Problem: Ten grams of the mineral gibbsite are mixed into a liter of 10^{-3} *M* oxalic acid. The pH of the suspension is adjusted to 6.0 with NaOH. Calculate the concentrations of free Al^{3+} and complexed Al in solution at equilibrium, neglecting corrections for activity coefficients.

Solution: First, we recognize that the presence of the solid phase, gibbsite, controls the concentration (activity, to be more precise) of free Al^{3+} at a given pH (see equation 1.42):

$$[Al^{3+}][OH^-]^3 = 10^{-34} \tag{1.67}$$

The equilibrium constant, K_w, for the dissociation of water:

$$H_2O = H^+ + OH^- \qquad K_w = 10^{-14} \tag{1.68}$$

requires that $[H^+][OH^-] = 10^{-14}$, so that equation 1.67 can be reexpressed as

$$\frac{[Al^{3+}]}{[H^+]^3} = 10^8 \tag{1.69}$$

At pH 6.0, $[H^+] = 10^{-6}$, so that the concentration of free Al^{3+} is fixed at

$$[Al^{3+}] = 10^8[H^+]^3 = 10^{-10} \tag{1.70}$$

The presence of the Al-complexing organic, oxalate, has no effect on the activity of free Al^{3+} at equilibrium as long as excess gibbsite is present, but does increase the total solubility of Al by forming soluble Al-oxalate complexes. The extent of complex formation can be determined using equations 1.64, 1.65, and 1.66 and the fact that the concentrations of the various forms of oxalate (H_2L, HL, L, and ML) must sum up to the initial concentration of oxalic acid:

$$[H_2L] + [HL] + [L] + [ML] = 10^{-3} \tag{1.71}$$

Only four species have concentrations that are unknown, since [H] is fixed by the controlled pH of the suspension and [M] is fixed by the solubility product of gibbsite (equation 1.69). The solution to this problem can now be obtained since there are four equations (1.64, 1.65, 1.66, and 1.71) with four unknown concentrations. One way to solve these equations is to express each concentration in equation 1.71 in terms of [HL], producing the following expression:

$$10^{-4.75}\,[HL] + [HL] + 10^{1.73}\,[HL] + 10^{-2.17}\,[HL] = 10^{-3} \qquad (1.72)$$

This gives $[HL] = 1.83 \times 10^{-5}$. It is then a simple matter to solve for [L], $[H_2L]$, and [ML] using equations 1.64, 1.65, and 1.66 and the fact that both [M] and [HL] are known. The solutions are $[L] = 9.84 \times 10^{-4}$, $[H_2L] = 3.25 \times 10^{-10}$, and $[ML] = 1.24 \times 10^{-7}$. Note that [M] is much lower than [ML], so that complexed Al is the major form of dissolved Al.

It is clear that the presence of oxalate (and other complexing ligands) in solution can increase the total solubility of Al by forming soluble Al-ligand complexes. It is also evident that even this simple chemical system has a solution for one particular set of conditions that is obtained by rather laborious mathematical manipulation. Furthermore, this simple system would have had a much more difficult solution if certain assumptions had not been made. Not only were activity corrections ignored, but important chemical species were neglected for the sake of simplification. Al forms AlL_2 (1:2 type) and AlL_3 (1:3 type) complexes with oxalate, in addition to the 1:1 AlL complex specifically considered in the problem. A complete solution for the equilibrium concentrations of all the significant chemical species in the oxalic acid–gibbsite-water system is most readily achieved using a computer speciation program, which iteratively arrives at a solution to the set of simultaneous equations describing the system. This more exact solution is described in the section in Chapter 5 dealing with aluminum chemistry.

Even with numerous computer programs now available to calculate activities of ions and complexes in solution, it is important that the chemical principles behind these calculations be understood. Otherwise, chemically unreasonable solutions can be arrived at without the program user having the chemical "intuition" to recognize the error. For this reason the problem above was solved manually, to illustrate the origin of the equations in simple chemical principles.

1.2f. Electrochemistry

Half-Reactions and Electrode Potentials. Reduction-oxidation (redox) reactions involve electron donors (reductants or reducing agents) and electron acceptors (oxidants or oxidizing agents), analogous to the involvement of proton donors and proton acceptors in acid-base reactions. Redox reactions are characterized by the transfer of an integral number of electrons from the reductant to the oxidant molecule. The tendency of the redox reaction to proceed is determined by the *electromotive force* or *potential change* (E) for the particular reaction. Since there is always an electron donor that becomes oxidized by the transfer, and an electron acceptor that becomes reduced by the transfer, any redox reaction can be written as the sum of two half-reactions. For example, the redox reaction

$$Zn + 2\,H^+ \rightarrow Zn^{2+} + H_2 \qquad (1.73)$$

can be visualized as the sum of the half-reactions

$$Zn \rightarrow Zn^{2+} + 2e^- \qquad (1.74)$$

and

$$2H^+ + 2e^- \rightarrow H_2\,(g) \qquad (1.75)$$

where e^- symbolizes the electron. Reaction 1.75 is the process observed at a hydrogen electrode; that is, an inert conducting electrode (eg., platinum metal) immersed in water. If this electrode is immersed in an aqueous solution along with a second electrode composed of Zn metal, and the two electrodes are electrically connected, a *cell* is created in which reaction 1.73 can proceed *if* the potential (E) of the reaction as written is positive.

To determine the potential for the cell, the hydrogen electrode is by convention assigned a half-reaction potential (E_h) of 0.00 volt (V) under standard state conditions.[1] This means that if the H^+ ion activity and the H_2 gas activity are both 1.00, E_h for the hydrogen electrode is 0.00 volt. This convention can be symbolized as

$$H^+(1\,M) + e^- \rightarrow \tfrac{1}{2}H_2\,(1\text{ atm}) \qquad E_h^0 = 0.00\text{ V} \qquad (1.76)$$

In a similar way, the *standard state potentials* of all half-reactions, symbolized as E_h^0, are defined for standard state conditions of the reactants and products.

Now, since the cell reaction (1.73) proceeds spontaneously, the free energy change of the reaction, ΔG, must be negative (i.e., $\Delta G < 0$). The potential of the reaction is related to the free energy change by the equation

$$\Delta G = -nFE \qquad (1.77)$$

where n is the number (of moles) of electrons transferred by the redox reaction and F is Faraday's constant (F = 96,500 coulombs/mole of electrons). So, for standard state conditions,

$$\Delta G^0 = -nFE^0 \qquad (1.78)$$

Since under these conditions reaction 1.75 has $E_h^0 = 0.00$ volt, it is clear that reaction 1.74 must have $E_h^0 > 0$ for the whole cell reaction to proceed spontaneously (i.e., $\Delta G < 0$). In conclusion, then, several rules of electrochemistry become evident:

1. The sign of the electromotive force of a half-reaction or overall redox reaction depends on the direction in which the reaction is written.
2. Redox reactions in which $E > 0$ are spontaneous.
3. The number of moles or equivalents of reactants and products involved in a redox reaction does not affect the reaction's driving force E but does influence the free energy change ΔG.

The Nernst Equation. Because in real reactions the reactants and products are rarely at unit activity, standard state potentials, E^0, must be corrected to the relevant

1. Standard state conditions are defined as 1 atmosphere pressure and 25°C with activities of all chemical species equal to 1.0.

conditions. This is done using the Nernst relationship, which for any general redox reaction at 25°C between a moles of reactant A and b moles of B to produce c moles of C and d moles of D:

$$aA + bB = cC + dD \tag{1.79}$$

is

$$E = E^0 - \frac{0.059}{n} \log \frac{[C]^c[D]^d}{[A]^a[B]^b} \tag{1.80}$$

where the square brackets denote concentrations (or, more accurately, activities) of the reactants and products. This equation makes possible the calculation of the "driving force" for any redox reaction if the concentrations of all the reactants and products in solution are known, and if the standard-state potential E^0 of the reaction is known. The value of E^0 is simply calculated as the sum of the two potentials, E_h^0, of the half-reactions, which combine to give the overall reaction *as written*. These E_h^0 values are obtained from tables of standard electrode potentials, in which the most usual convention is to write half-reactions as reductions. This means that reaction 1.74 should be written in the opposite direction:

$$Zn^{2+} + 2e^- = Zn \qquad E_h^0 = -0.76 \text{ V} \tag{1.81}$$

The redox reaction (1.73) is then obtained by summing the two half-reactions:

$$\begin{array}{rcll} Zn & = & Zn^{2+} + 2e^- & E_h^0 = +0.76 \text{ V} \\ 2H^+ + 2e^- & = & H_2 (g) & E_h^0 = 0.00 \text{ V} \\ \hline Zn + 2H^+ & = & Zn^{2+} + H_2 (g) & E^0 = +0.76 \text{ V} \end{array} \tag{1.82}$$

While the electromotive (driving) force of a half-reaction obviously changes sign when the half-reaction is written in the reverse direction, it is considered desirable to define electrode potentials so that they are insensitive to the reaction direction being considered.[1] In this manner the standard electrode potential of the $Zn^{2+}-Zn$ couple is given as -0.76 volt, using the convention of writing the half-reaction as a reduction.

Table 1.4 lists some half-reactions that might be encountered in soil solutions. As the reduction potentials become more positive (from top to bottom), the tendency of the oxidized species to be reduced increases. The MnO_2-Mn^{2+} couple undergoes reduction more readily than the $Fe^{3+}-Fe^{2+}$ couple, for example.

The Concept of pε. The redox intensity, pε, is defined in a manner analogous to pH:

$$p\epsilon = -\log(e^-) \tag{1.83}$$

where pε expresses a hypothetical "electron activity" in solution (compared with proton activity in the case of pH). It measures the relative tendency of a solution to accept electrons. Reducing solutions have low pε values and tend to donate electrons

1. It is argued that an electrode potential measures real electrostatic charge on an electrode in an electrochemical cell, or electron activity in redox systems in general. These potentials, having physical reality, should not be subject to arbitrary convention.

Table 1.4. Reduction Half-Reactions and Their Standard Electrode Potentials (E_h^0) at 25°C

$O_2 + e^- = O_2^-$	−0.563 V
$Fe(OH)_3 + e^- = Fe(OH)_2 + OH^-$	−0.56 V
$2H^+ + 2e^- = H_2\,(g)$	0.00 V
$Cu^{2+} + e^- = Cu^+$	0.153 V
$Fe^{3+} + e^- = Fe^{2+}$	0.77 V
$O_2 + 4H^+ + 4e^- = 2H_2O$	1.229 V
$MnO_2 + 4H^+ + 2e^- = Mn^{2+} + 2H_2O$	1.29 V
$Mn^{3+} + e^- = Mn^{2+}$	1.51 V
$Co^{3+} + e^- = Co^{2+}$	1.808 V

to species placed in the solution; oxidizing solutions have high pϵ values and tend to accept electrons from species placed in the solution.

Consider the $Fe^{3+}-Fe^{2+}$ half-reaction:

$$Fe^{3+} + e^- = Fe^{2+} \qquad E_h^0 = +0.77 \text{ V} \tag{1.84}$$

The relationship between pϵ and E_h (at 25°C) is

$$p\epsilon = \frac{E_h(\text{volts})}{0.059} \tag{1.85}$$

The Nernst equation relates the activities of Fe^{3+} and Fe^{2+} to E_h:

$$E_h = E_h^0 - \frac{0.059}{n} \log \frac{(Fe^{2+})}{(Fe^{3+})} = 0.77 - 0.059 \log \frac{(Fe^{2+})}{(Fe^{3+})} \tag{1.86}$$

since the number of moles of electrons, n, transferred in reaction 1.84 is 1.0. If equation 1.86 is divided on both sides by 0.059, an equivalent expression results:

$$p\epsilon = \frac{E_h}{0.059} = 13.05 - \log \frac{(Fe^{2+})}{(Fe^{3+})} \tag{1.87}$$

Now if the solution were very acid and contained, say, 10^{-5} M Fe^{3+} and 10^{-3} M Fe^{2+}, equation 1.87 could be solved to give

$$p\epsilon = 13.05 - \log\left(\frac{10^{-3}}{10^{-5}}\right) = 11.05 \tag{1.88}$$

or for a solution with 10^{-5} M Fe^{3+} but only 10^{-7} M Fe^{2+}:

$$p\epsilon = 13.05 - \log\left(\frac{10^{-7}}{10^{-5}}\right) = 15.05 \tag{1.89}$$

This example confirms that more oxidized solutions have higher pϵ values.

1.2g. Kinetics

Reaction Rate Constants and Reaction Order. The principles of thermodynamics can be used to predict whether a particular chemical reaction can occur, but they fail to provide any information about the speed of the reaction. For example, it can be shown from thermodynamics that atmospheric nitrogen and oxygen should react spontaneously in water to form nitric acid:

$$2N_2 + 5O_2 + 2H_2O \rightarrow 4HNO_3 \tag{1.90}$$

because the free energy change, ΔG, of the reaction is negative. All life on earth owes its existence to the fact that this reaction is so slow as to be undetectable. Obviously, factors other than the net change in energy and entropy can limit the progress of a reaction.

For any chemical reaction, described in general by the equation

$$a\mathrm{A} + b\mathrm{B} \rightarrow p\mathrm{P} + q\mathrm{Q} \tag{1.91}$$

the rate of the reaction, R, is the same as the rate of disappearance of the reactants, or the rate of appearance of the products:

$$R = -\frac{1}{a}\frac{d[\mathrm{A}]}{dt} = -\frac{1}{b}\frac{d[\mathrm{B}]}{dt} = \frac{1}{p}\frac{d[\mathrm{P}]}{dt} = \frac{1}{q}\frac{d[\mathrm{Q}]}{dt} \tag{1.92}$$

where the square brackets indicate concentrations and t is time. Typically, then, R has units of moles liter^{-1} second^{-1}.

The dependence of R, the reaction rate, on the concentrations of reactants (or catalysts) has the general form

$$R = k[\mathrm{A}]^a[\mathrm{B}]^b \ldots \tag{1.93}$$

where k is the rate constant. The exponent, a, is called the *order of reaction* with respect to reactant A, and b is called the order of reaction with respect to B. More than two reactants can appear in equation 1.93, with the *total order of the reaction*, n, being the sum of all the exponents:

$$n = a + b + \cdots \tag{1.94}$$

For an overall first-order reaction, R takes the form

$$R = k[\mathrm{A}] \tag{1.95}$$

so that according to the definition of R (equation 1.92):

$$\frac{-d[\mathrm{A}]}{dt} = k[\mathrm{A}] \tag{1.96}$$

Integration of this equation over a time interval gives

$$\ln\frac{[\mathrm{A}]}{[\mathrm{A}]_0} = -kt \tag{1.97}$$

where $[\mathrm{A}]_0$ is the initial concentration of A and [A] is the concentration after time interval t. Thus, the rate constant for the reaction can be obtained by plotting ln[A] against time and measuring the slope of the straight line.

Although many reactions are not first-order overall, they can be treated as "pseudo first-order" reactions by controlling the reaction conditions. Consider, for example, the chemical reaction

$$\mathrm{A} + 2\mathrm{B} \rightarrow C \tag{1.98}$$

with a rate law of

$$R = \frac{d[C]}{dt} = k[A][B]^2 \qquad (1.99)$$

If the concentration of B is maintained at a high level such that [B] $\gg$ [A], then:

$$R \approx k'[A] \qquad (1.100)$$

and the reaction appears to be first-order. This allows equation 1.97 to be used to determine the value of k'.

The simplified analysis of kinetics given here is only valid if the back reaction can be neglected. For example, as reaction 1.98 proceeds, the product C accumulates and may begin to dissociate back to A and B. (Eventually, once the back reaction rate equals that of the forward reaction, steady-state or equilibrium is achieved.) For this reason, kinetic studies are typically done in the early stages of a reaction before back reactions begin to invalidate the definition of reaction rate as given by equation 1.92.

Temperature Effects on Reaction Rates. Chemical reactions typically have rate constants whose temperature dependence takes the mathematical form

$$k = Ae^{-E_a/RT} \qquad (1.101)$$

This is the Arrhenius equation, where A is a constant for any particular reaction and E_a is the *activation energy* for that reaction. This means that chemical reaction rates increase with temperature to an extent determined by the magnitude of the activation energy. Figure 1.6 illustrates that the temperature dependence of the forward reaction rate is a result of the "energy barrier," E_a, which must be overcome before the product(s) can form. Higher temperature provides the thermal energy to overcome this barrier. The figure also shows that the effective barrier to the reverse reaction can be much higher than that for the forward reaction, depending on the heat of reaction, ΔH. It is common, then, for chemical adsorption reactions on mineral surfaces to have fast forward reaction rates and slow reverse reaction rates. For example, the lead cation bonds to the surface of iron oxide:

$$Pb^{2+} + {>}Fe(III){-}OH \underset{k_2}{\overset{k_1}{\leftrightarrow}} {>}Fe(III){-}O{-}Pb^{2+} \qquad (1.102)$$

and the forward reaction rate, k_1, is much larger than the backward rate, k_2.

Diffusion Effects on Reaction Rates. Chemical reaction rates in solution and between solutions and solids, regardless of mechanism, are ultimately limited by the process of molecular diffusion. For example, if the chemical reaction

$$A + B \rightarrow \text{products} \qquad (1.103)$$

is so fast that, whenever A and B come into contact, they immediately react, then a concentration gradient will build until the rate of diffusion of A and B toward each other equals the reaction rate. Fast reactions of this kind are common and are said to be *diffusion limited.* The reaction rate constant, k, can never exceed the diffusion

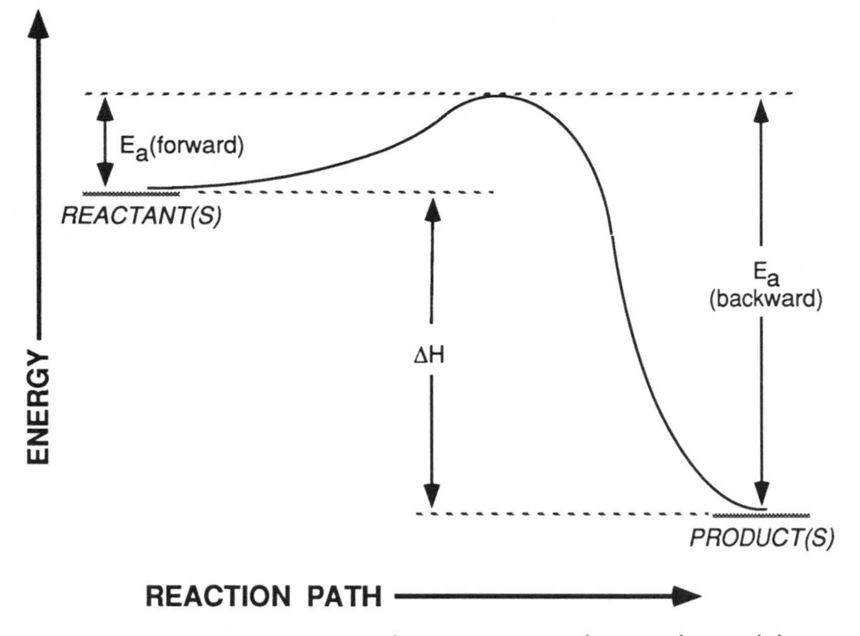

Figure 1.6. Energy diagram for an exothermic reaction, showing the much larger activation energy (E_a) for the backward relative to the forward process.

constant, k_D, no matter how rapidly A and B react once they come into contact. This limiting rate constant has the form

$$k_D = 4\pi(D_A + D_B)r\left(\frac{N_0}{1000}\right) \tag{1.104}$$

where D_A and D_B are the diffusion coefficients of A and B in water, r is the separation distance at which A and B are considered to be in contact, and N_0 is Avogadro's number. If A and B are ions of the same charge (e.g., A^+ and B^+, or A^- and B^-), the value of k_D predicted from equation 1.104 is higher than that observed; electrostatic repulsion in this case reduces the probability of direct encounter between A and B. High electrolyte (salt) concentrations in solution "shield" the individual ions from this repulsion, so that reaction 1.103 has a rate constant that *increases* with increasing electrolyte concentration. If, on the other hand, A and B have opposite charge, the intrinsic rate constant should be higher but will *decrease* with increasing electrolyte concentration. For reactions between neutral molecules, the rate constant should be fairly insensitive to electrolyte concentration.

Writing the diffusion rate as dw/dt, the number of grams of molecules transferred across a boundary per second, the equation describing Fick's law of diffusion defines a diffusion coefficient, D:

$$\frac{dw}{dt} = -Da\left(\frac{dc}{dx}\right) \tag{1.105}$$

where dc/dx is the concentration gradient and a is the cross-sectional area of the boundary. The diffusion coefficient can be interpreted in terms of the equation

$$D = \frac{RT}{6\pi r N_0 \eta} \tag{1.106}$$

where R is the gas constant, r is the molecular radius, and η is the solvent viscosity. From this equation it is seen that for a given molecule in water, temperature is the only variable, and equation 1.105 predicts a greater diffusion rate of molecules at higher temperature. Consequently, the rates of diffusion-limited solution reactions should increase *in proportion to temperature.*

For reactions at solid surfaces, then, the rate-limiting process can be:

1. Diffusion of reactant and product molecules or ions through the solution "film" that surrounds the particle *(film diffusion)*
2. Diffusion within the particle *(particle diffusion)*
3. The chemical reaction itself and its associated activation energy

Diffusion of molecules or ions through crystalline solids is extremely slow at 25°C, so that the term *solid diffusion* should generally be interpreted to mean transfer through micropores, faults, or interfaces of the solid rather than through the lattice itself. Only at very high temperatures does true diffusion of ions and molecules into a crystal lattice become significant.

References

Huheey, J. E. 1972. *Inorganic Chemistry.* New York: Harper & Row.

Suggested Additional Reading

Brown, T. L., H. E. LeMay, Jr., and B. E. Bursten, 1991.*Chemistry. The Central Science.* 5th ed. Englewood Cliffs, N.J.: Prentice-Hall.

Drever, J. I. 1988. *The Geochemistry of Natural Waters.* 2nd ed. Englewood Cliffs, N.J.: Prentice-Hall, chap. 2.

Stumm, W., and J. J. Morgan, 1981. *Aquatic Chemistry.* 2nd ed. New York: Wiley, chaps. 1–3.

Questions

1. Given that the density of solid sodium chloride is 2.165 g/cm^3, what are the dimensions of a cube that contains 1 mole of solid NaCl? If the distance between centers of adjacent Na^+ and Cl^- ions is 2.8 Å, how many Na^+ and Cl^- ions lie along each edge of the cube? Calculate Avogadro's number from these data.

2. Calculate the lattice energy (in kcal/mole) for NaCl if the Madelung constant for this solid is 1.75 and the value of n is 9. The radii of Na^+ and Cl^- are 1.02 and 1.81 Å, respectively. If KCl has about the same Madelung constant as NaCl, how does its lattice energy compare with that of NaCl?

3. When considering a pair of ions of charge z^+ and z^-, the attractive electrostatic potential between them is given by $\frac{-z^+z^-e^2}{r}$, where r is the distance between the ions. There is also a shorter range repulsive potential, of general form B/r^n, where B and n are constants. Thus, the total potential energy E has the form

$$E = \frac{B}{r^n} - \frac{z^+ z^- e^2}{r}$$

(a) Find the value of r, r_0, for which the potential energy is at a minimum. This is the stable separation distance.
(b) From the solution in (a), obtain an expression for the constant B.
(c) Substitute the expression for B into the equation for total potential energy and obtain the equation for the potential energy minimum, E_0. Compare the form of this equation with the Born-Lande equation. What is the physical significance of the value of n?

4. The water molecule has a dipole moment, μ, of 1.84×10^{-18} esu-cm (esu = electrostatic unit). If water dipoles in ice are separated by 2.76 Å, estimate the potential intermolecular attraction energy (in kcal/mole) due to oriented dipole-dipole interaction. Compare this quantity with the enthalpy of sublimation of ice (11.3 kcal/mole), which represents the amount of energy needed to break these bonds. Why might this calculation underestimate the actual energy of water molecule interaction?

5. Calculate the energy of ion-dipole interaction for a water molecule in the inner hydration sphere of both Na^+ and Al^{3+}. Use the dipole moment for water given in question 4. The radii of Na^+, Al^{3+}, and H_2O are 1.02, 0.53, and 1.35 Å, respectively.

   If there are six water molecules in the inner-sphere hydration shell of Na^+ and Al^{3+}, how much of the free energy of hydration estimated from equation 1.21 can be accounted for by inner-sphere hydration? Is this a true estimate of the ion-dipole interaction energy? Explain.

6. Calculate the activity coefficients of Na^+ and Al^{3+} ions dissolved in 0.01 M $CaCl_2$. Assume that the Na^+ and Al^{3+} ions do not contribute significantly to the total ionic strength of the solution.

7. As seen from Figure 1.4, Cu^{2+} and Mg^{2+} have similar ionic potentials, yet Cu^{2+} tends to hydrolyze in water to a much greater extent than Mg^{2+}. Explain this behavior.

8. For the chemical reaction in water

$$CaCl_2 + Na_2CO_3 = CaCO_3 + 2NaCl$$

determine the standard free energy of reaction, ΔG^0, using the standard free energies of formation given below. Calculate the value of the equilibrium constant K for the reaction. Is this reaction as written spontaneous under standard state conditions?

Standard Free Energies of Formation	ΔG_f^0 (kcal/mole)
Na^+	−62.59
Ca^{2+}	−132.52
Cl^-	−31.371
CO_3^{2-}	−126.18
$CaCO_3$ (calcite)	−270.18

9. Calculate the molar concentration of H^+, HAc, and Ac^- in a solution that is prepared from 0.150 mole HCl, 0.100 mole acetic acid (HAc), and enough water to make 1.00 liter of solution. What is the pH of this solution? (The dissociation constant for HAc is 1.85×10^{-5}, and HCl is completely dissociated in aqueous solution.)

10. The solubility product of magnesium hydroxide, $Mg(OH)_2$, is 1.8×10^{-11}. What is the solubility (g/liter) of $Mg(OH)_2$ in pure water? What is the pH of this saturated solution?

11. The Mn^{2+} ion complexes with phthalic acid in aqueous solution, forming a bidentate complex with the phthalate anion. If this anion is given the symbol L^{2-}, then the complexation reaction can be written

$$Mn^{2+} + L^{2-} = MnL^0 \qquad \log K(20°C) = 2.74$$

The association reactions of phthalate with protons are given by

$$H^+ + L^{2-} = HL^- \qquad \log K = 5.41$$
$$H^+ + HL^- = H_2L \qquad \log K = 2.95$$

If a solution is prepared in which the initial concentration of $MnCl_2$ is 5×10^{-4} M and that of phthalic acid is 5×10^{-3} M, and the pH is then adjusted to 7, calculate the fraction of soluble Mn that is complexed with the organic acid at equilibrium. (Ignore activity corrections, which are small in this solution of low ionic strength.)

12. The activity of Fe^{3+} in aqueous solution is often controlled by the precipitation of $Fe(OH)_3$:

$$Fe(OH)_3\,(s) = Fe^{3+} + 3OH^- \qquad \log K_{so} = -39.1$$

Use this fact to reexpress equation 1.86 so that the potential, E_h, is explicitly defined in terms of pH.

Calculate the E_h of the $Fe(OH)_3$–Fe^{2+} half-cell at pH 4, 5, 6, and 7 for $(Fe^{2+}) = 10^{-6}$ M.

Calculate the E_h of the O_2-H_2O half-cell also at pH 4, 5, 6, and 7 for the atmospheric level ($P_{O_2} = 0.2$ atmospheres) of oxygen gas.

Determine the potential (E) and free energy (ΔG) for the oxidation of 10^{-6} M Fe^{2+} at these four pH values. What general statement can you make about the effect of pH on Fe^{2+} oxidation in water?

13. For a second-order reaction with $R = k[A]^2$, derive the equation that must be used to obtain the rate constant from a straight-line plot.

14. How would you recognize a reaction that has zero-order kinetics?

2

Soil Solids: Composition and Structure

2.1. ELEMENTAL, CHEMICAL, AND PHYSICAL COMPOSITION OF SOIL

Soils are complex materials, reflecting the variability of the parent rock material and organic residues from which they form. Nevertheless, their elemental composition, particle size, and mineralogy can be related more or less systematically to the nature of the parent material and the degree to which this material has been altered by weathering (a subject that will be expanded on in Chapter 6 of this book). Figure 2.1 compares the average elemental composition of the granitic rock of the earth's crust with that of two soils having very different age and origin—a relatively young soil of glacial origin and a highly weathered soil of the tropics. The latter soil has developed with the loss of much of the silica (a process termed *desilication*) and most of the basic[1] elements (Ca, Mg, K, Na) that were initially present in the parent material. Thus, this soil is very different from its parent material in elemental makeup. Conversely, the composition of the younger soil reflects the elemental composition of its parent rock. However, even in this case the soil has developed physical and mineralogical properties that are fundamentally different from those of the parent material. Processes of physical weathering have reduced the particle size of the mineral grains, increasing the exposed surface area. Simultaneously, chemical weathering processes have produced even more profound changes, dissolving or altering the rock-forming minerals (termed *primary* minerals) to ultimately produce layer silicate and oxide minerals. These mineral products of weathering, collectively referred to as *secondary* minerals, are typically clay sized ($< 2\ \mu m$ in diameter). Because they possess very high surface areas, they contribute substantially, along with the decomposed organic matter *(humus)*, to the chemical reactivity of soils. Consequently, the structure and reactivity of soil clays and humus will be a prominent topic in this book.

Any mineral with a particle size of $< 2\ \mu m$ is, by definition, part of the *clay* fraction of a soil. However, the term *clay mineral* has a different connotation historically, usually referring to the secondary layer silicates, which are the dominant inor-

1. "Basic" refers here to the fact that these four elements form relatively strongly alkaline hydroxide salts.

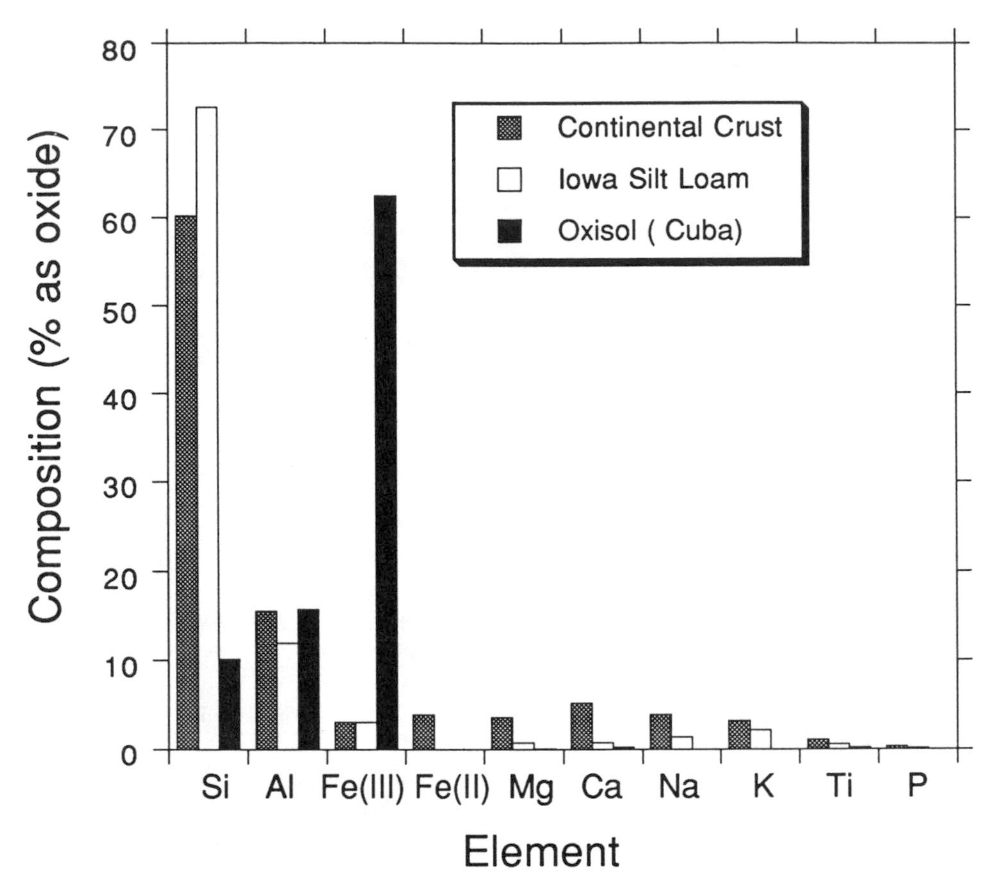

Figure 2.1. Elemental composition of an intensely weathered soil (oxisol) and a less strongly weathered soil, expressed as percent by weight of the oxide form. Also shown is the average elemental composition of the continental crust for comparison.

ganic constituents in the fine fraction of mineral soils throughout the temperate regions of the world. This term should be broadened to include allophanes and the oxidic minerals of Fe and Al; the latter minerals are common components of the clay fraction in soils of the humid tropics and subtropics. Conventions in terminology can produce confusion, because clay-sized particles of primary minerals commonly found in glaciated soils (e.g., quartz), are clays but are not classified as clay minerals.

In Figure 2.2, the relationship between soil particle size and mineralogy is illustrated qualitatively, demonstrating the prevalence of secondary minerals in the clay fraction. The primary minerals, which originally formed under conditions of high temperature and pressure, are unstable in the soil environment. Consequently, primary minerals, when they are reduced to small particle size by physical weathering, tend to chemically decompose (weather) rapidly. Nevertheless, clay-sized quartz particles persist in soils because of the resistance of the quartz structure to chemical decomposition, as will be discussed later in this chapter, but even this mineral eventually dissolves as silica is leached from the soil. Thus we see that reaction rates

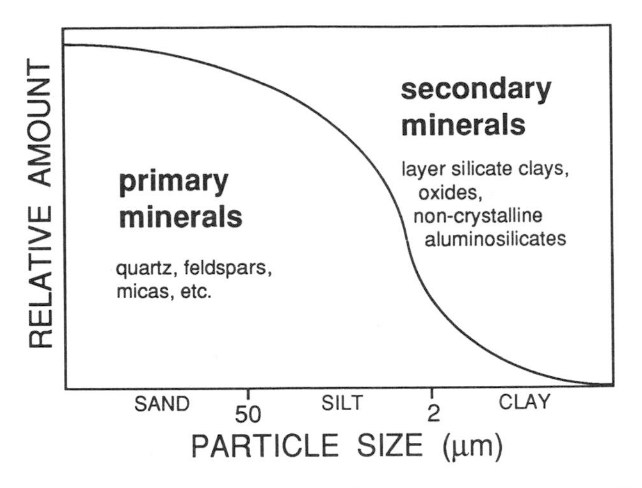

Figure 2.2. Typical abundance of primary and secondary minerals in different size fractions of the soil.

(determined by principles of chemical kinetics) are often more important than the inherent stability of minerals (determined from thermodynamic properties) in deciding the mineralogical composition of soils. The issue of slow reaction rates will be a recurring theme throughout the discussion of soil behavior in this text, as it will in most cases be very difficult to distinguish soil systems that are at chemical equilibrium from those that are undergoing gradual change toward a new equilibrium. Most soils are not at equilibrium or even steady state, but mineralogical changes are not likely to be detectable except on geological time scales.

2.2. MINERALS

A description of the minerals occurring most commonly in soils will be given in this section. Before this can be done, however, some basic concepts used to describe ionic solids need to be presented. The reader with no background in mineralogy, and unfamiliar with terms such as unit cells, lattice points, and lattice planes, should read the appendix of this chapter first.

2.2a. Principles of Ionic Solid Structures

The oxyanion, O^{2-}, is the most common elemental form in the earth's crust, making up 46.6 percent by weight and 93.8 percent by volume of the earth's surface. Most of the mineral structures are formed by bonding between this large anion and the generally much smaller metallic cations (principally Si^{4+}, Al^{3+}, Fe^{3+}, Fe^{2+}, and Mg^{2+}). The metal-oxygen bonds are largely ionic in nature, so that the "hard sphere" model of atom packing can be used to predict likely atomic arrangements. In fact, for structures with pure ionic bonding, the oxyanions pack together in the most space-efficient manner possible (termed the *close-packed arrangement*), and the metal cations fit into the interstices of these anions.

There are two distinguishable close-packed arrangements, hexagonal and cubic. They are equally efficient in packing, and both are formed by stacking planar layers of close-packed atoms (shown in Figure 2.3a) in such a way that the second layer has its atoms centered over the interstices created by three adjacent atoms in the first layer (as shown in Figure 2.3b). However, in hexagonal closest packing (HCP), the atoms of the added third layer are aligned directly above the atoms of the first layer, so that the layer arrangement is labeled ABABABAB, repeating every two layers. In cubic closest packing (CCP), the atoms of the third layer are positioned above interstices of the second layer, but are *not* directly above the atoms of the first layer, so that the layer arrangement is labeled ABCABCABC. These two nonequivalent arrangements of atoms (or O^{2-} ions in the specific case of silicate and oxide minerals) are compared in Figure 2.3c and d.

Metal cations occupy the holes created in the HCP and CCP arrangement of oxyanions, and for **n** anions there are **n** *octahedral* holes and 2**n** *tetrahedral* holes. The

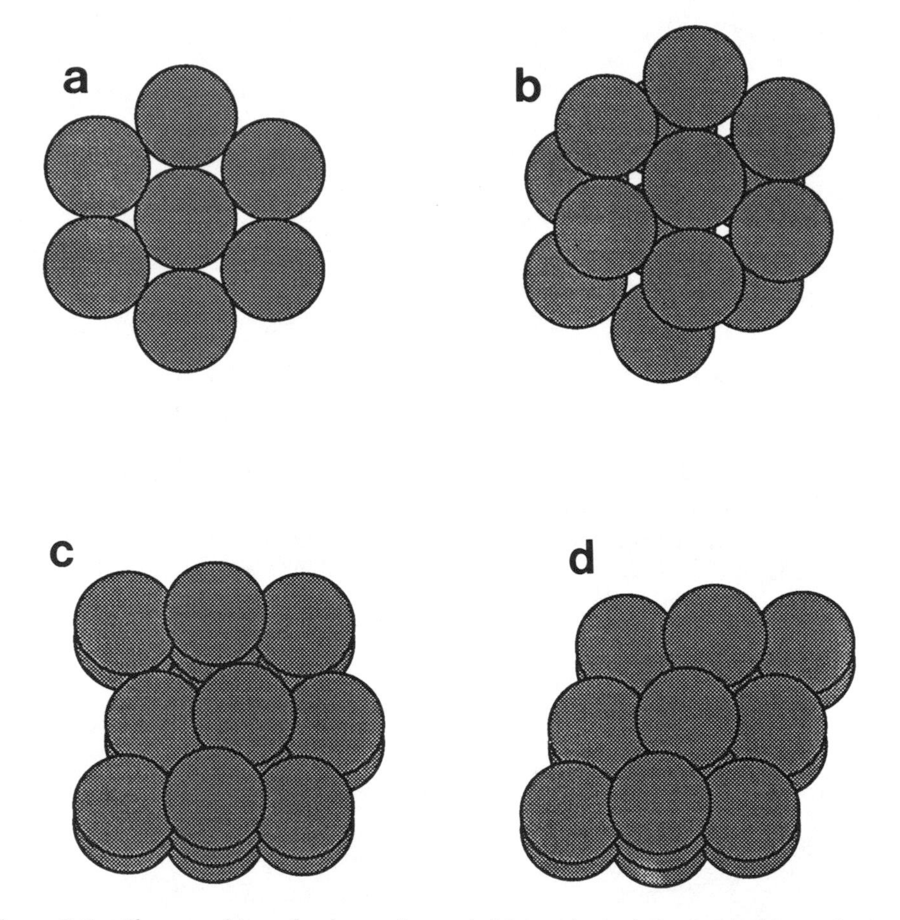

Figure 2.3. Close-packing of spheres (atoms): (a) one layer (top view), (b) two layers (top view), (c) three layers, hexagonal close packed (side view), and (d) three layers, cubic close packed (side view).

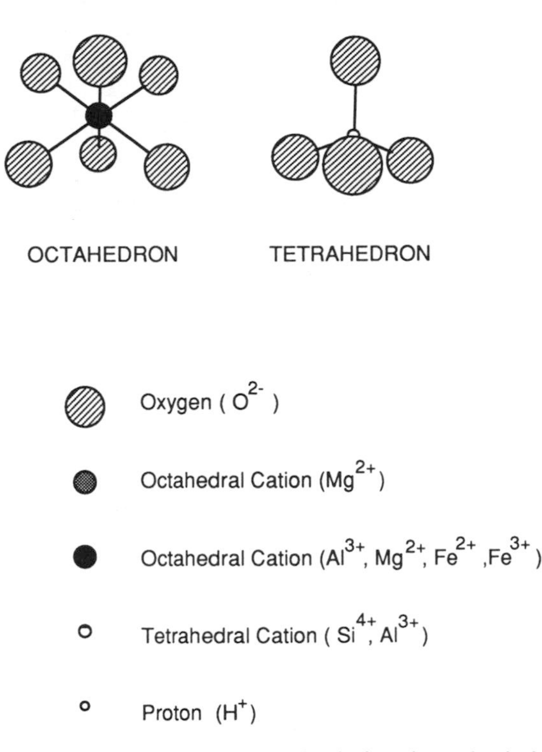

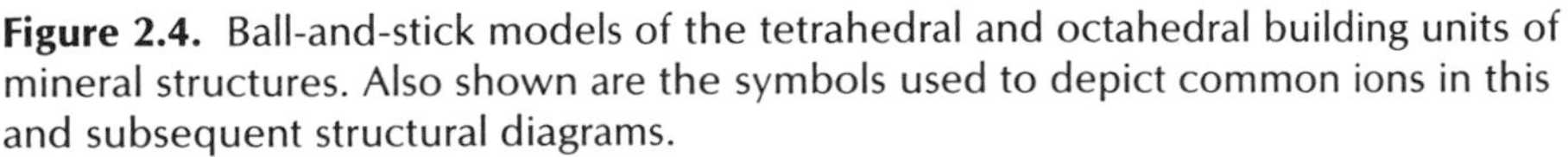

Figure 2.4. Ball-and-stick models of the tetrahedral and octahedral building units of mineral structures. Also shown are the symbols used to depict common ions in this and subsequent structural diagrams.

larger octahedral holes are created by six adjacent close-packed O^{2-} anions, while the smaller tetrahedral holes are formed by four adjacent close-packed O^{2-} anions. These two anion arrangements are depicted in Figure 2.4. The size of metal cation that can stably occupy these different holes is dictated by the *radius ratio,* r_+/r_-, depending on the radius of the cation, r_+, and anion, r_-. For example, considering a cross-sectional view of the octahedral hole, shown in Figure 2.5, four oxyanions in a square arrangement define the size of the hole. Simple trigonometry can then be used to prove the relationships below:

$$\frac{2r_-}{2r_- + 2r_+} = \cos 45° = 0.707 \tag{2.1}$$

$$\frac{r_+}{r_-} = 0.414 \tag{2.2}$$

The value of 0.414 is termed the *limiting radius ratio* for octahedral coordination. The cation is stable in the octahedral hole if it is large enough to keep the anions from "touching," thereby preventing strong repulsive forces from destabilizing the structure. In effect, the limiting radius ratio sets a limit on how *small* (rather than how large) the cation can be relative to the anion. The octahedral arrangement represents a *coordination number* of 6; that is, the cation is surrounded by six equidistant

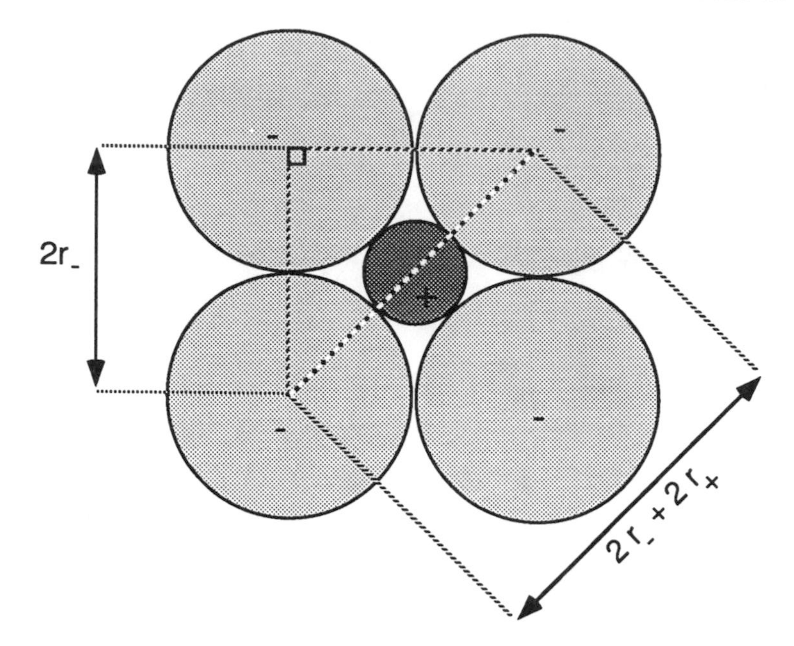

Figure 2.5. Cross-sectional view of the octahedral site, with dimensions given in terms of the cation radius, r_+, and anion radius r_-.

anions. This coordination type should be stable for radius ratios of 0.414 or higher, although much higher radius ratios cause higher coordination numbers to be favored. Thus, as shown in Table 2.1, radius ratios greater than 0.732 favor 8-coordination over 6-coordination. Furthermore, radius ratios greater than 0.225 and lower than 0.414 favor 4-coordination (tetrahedral geometry).

The most common elements in the earth's crust (after oxygen) are Si and Al, with valences of +4 and +3, respectively. The radius ratios of Si^{4+} and Al^{3+} coordinated to O^{2-} are 0.29 and 0.36, respectively. Thus, Si^{4+} can reside only in tetrahedral holes. The Al^{3+} ion is also predicted to occupy tetrahedral holes, and does so in many primary minerals, but the radius ratio is sufficiently close to the limiting value of octahedral coordination that 6-coordination of Al^{3+} is common. In fact, under the low temperature and pressure conditions that prevail during secondary mineral formation, 6-coordination of aluminum is more common than 4-coordination. Tetrahe-

Table 2.1. Limiting Radius Ratios for Coordination Numbers Commonly Found in Minerals

Coordination Number	Geometry	Limiting Radius Ratio
4	Tetrahedral	0.225
6	Octahedral	0.414
8	Cubic	0.732
12	Dodecahedral	1.00

drally coordinated Al in soil minerals is often inherited from primary minerals that have been altered by weathering processes.

The complication of covalent bonding between the cation and anion can reduce the predictive accuracy of the radius ratio. For example, the sulfide mineral, CdS, has $r_+/r_- = 0.53$ and a predicted Cd^{2+} coordination number of 6. Yet Cd^{2+} in this mineral is found in 4-coordination because this coordination geometry maximizes the degree of covalent bonding between Cd^{2+} and S^{2-}. Clearly, the limiting radius ratio is a parameter having utility in ionic solids only.

Mineral structures are composed in an ordered fashion from silica and alumina tetrahedra (MO_4), octahedra of Al, Fe, and Mg (MO_6), and various other polyhedra (where M symbolizes the metal cation). They form according to certain rules, deduced by Pauling, that minimize electrostatic repulsion forces and maximize attractive forces. These *Pauling rules* have been stated as follows:

1. A coordinated polyhedron of anions is found about each cation, the cation-anion distance being determined by the radius sum and the coordination number of the cation by the radius ratio.
2. In a stable coordination structure, the total strength of the valency bonds that reach an anion from all the neighboring cations is equal to the charge of the anion.
3. The existence of edges, and particularly of faces, common to two anion polyhedra in a coordinated structure decreases its stability; this effect is large for cations with high valency and small coordination number, and is especially large when the radius ratio approaches the lower limit of stability of the polyhedron.
4. In a crystal containing different cations, those of high valency and small coordination number tend not to share polyhedron elements[1] with each other.
5. The number of essentially different kinds of constituents in a crystal tends to be small.

Many of the structures formed, especially the oxides, are based on the closest packing of oxyanions. Others, such as the silicates, are not wholly close packed because of the overriding influence of Si—O covalent bonding in determining atomic positions.

2.2b. Important Primary Mineral Structures

The principal *primary mineral* groups are the silica minerals (including quartz), feldspars, feldspathoids, olivines, pyroxenes, amphiboles, and micas. All are silicates and can be classified structurally based on the arrangement of connected silica (SiO_4) tetrahedra. Table 2.2 summarizes the classification system for these common rock-forming minerals.

The mineral structures will be described briefly, progressing from the least to most complex arrangement of silica tetrahedra.

1. The corners, edges, and faces imagined to be formed by the positions of the oxygen atoms in the MO_4 and MO_6 units are defined as "elements," while "constituents" are assemblies of these units that combine to form the structure.

Table 2.2. Structural Classification of Silicates

Classification	Structural Arrangement	Si/O Ratio	Examples
Nesosilicates (island)	Independent tetrahedra	1:4	Olivine $(Mg,Fe)_2SiO_4$
Inosilicates (chain)	Continuous single chains of tetrahedra, each sharing two oxygens	1:3	Pyroxenes (e.g., $MgSiO_3$)
	Continuous double chains of tetrahedra sharing alternately two and three oxygens	4:11	Amphiboles [e.g., $Mg_7(Si_4O_{11})_2(OH)_2$]
Phyllosilicates (sheet)	Continuous sheets of tetrahedra, each sharing three oxygens	2:5	Micas [e.g., phlogopite $KMg_3(AlSi_3O_{10})(OH)_2$]
Tectosilicates (framework)	Continuous framework of tetrahedra, each sharing all four oxygens	1:2	Silica (e.g., quartz, SiO_2), feldspars [e.g., orthoclase $K(Si_3AlO_8)$], feldspathoids

Olivine. This mineral consists of isolated SiO_4^{4-} units held together by octahedrally coordinated Mg^{2+} or Fe^{2+}, and is therefore classified as an island silicate. The O^{2-} anions are approximately hexagonally close-packed in this structure; consequently, the rule of **n** octahedral holes and **2n** tetrahedral holes for every **n** oxygen anions applies. The formula of olivine is $(Mg,Fe)_2SiO_4$, so that one-eighth of the tetrahedral sites (holes) must be occupied by Si^{4+}, while one-half of the octahedral sites must be filled by Mg^{2+} and Fe^{2+}. The arrangement of the atoms and the tetrahedral and octahedral units is depicted in Figure 2.6. This structure is seen to obey Pauling's rule

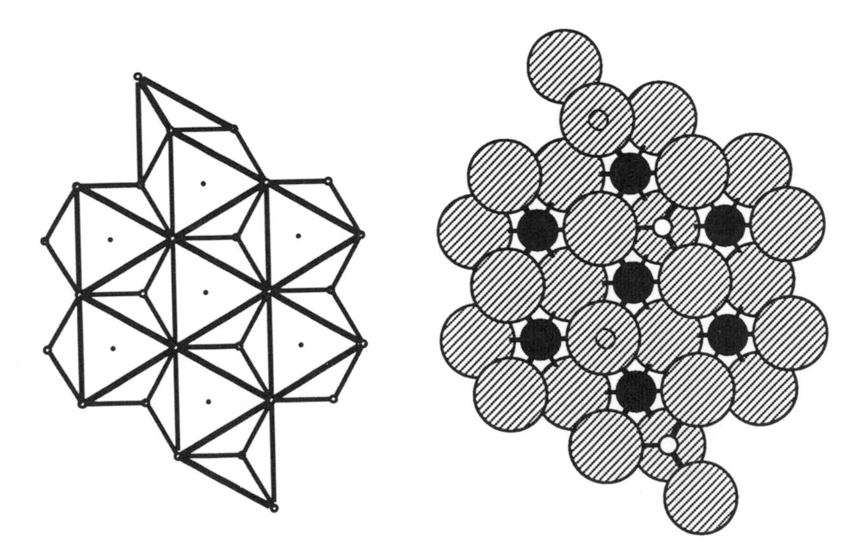

Figure 2.6. Polyhedral and ball-and-stick diagrams of a section of the olivine structure (two layers of O^{2-} atoms thick).

number 4, since the SiO_4 tetrahedra do not share corners, edges, or faces with one another. The MgO_6 (or FeO_6) octahedra share edges with one another, but share faces with SiO_4 tetrahedra. According to rule number 3, this sharing of faces is the least stable structural feature of olivine, and would only be expected for a mineral formed under high temperature and pressure conditions.

Pyroxenes. These minerals are chain silicates, in which the SiO_4 units are polymerized into very long chains with the formula $(SiO_3^{2-})_n$, as shown in Figure 2.7a. The negative charge of these chains is balanced by Mg^{2+} in 6-coordination and the larger

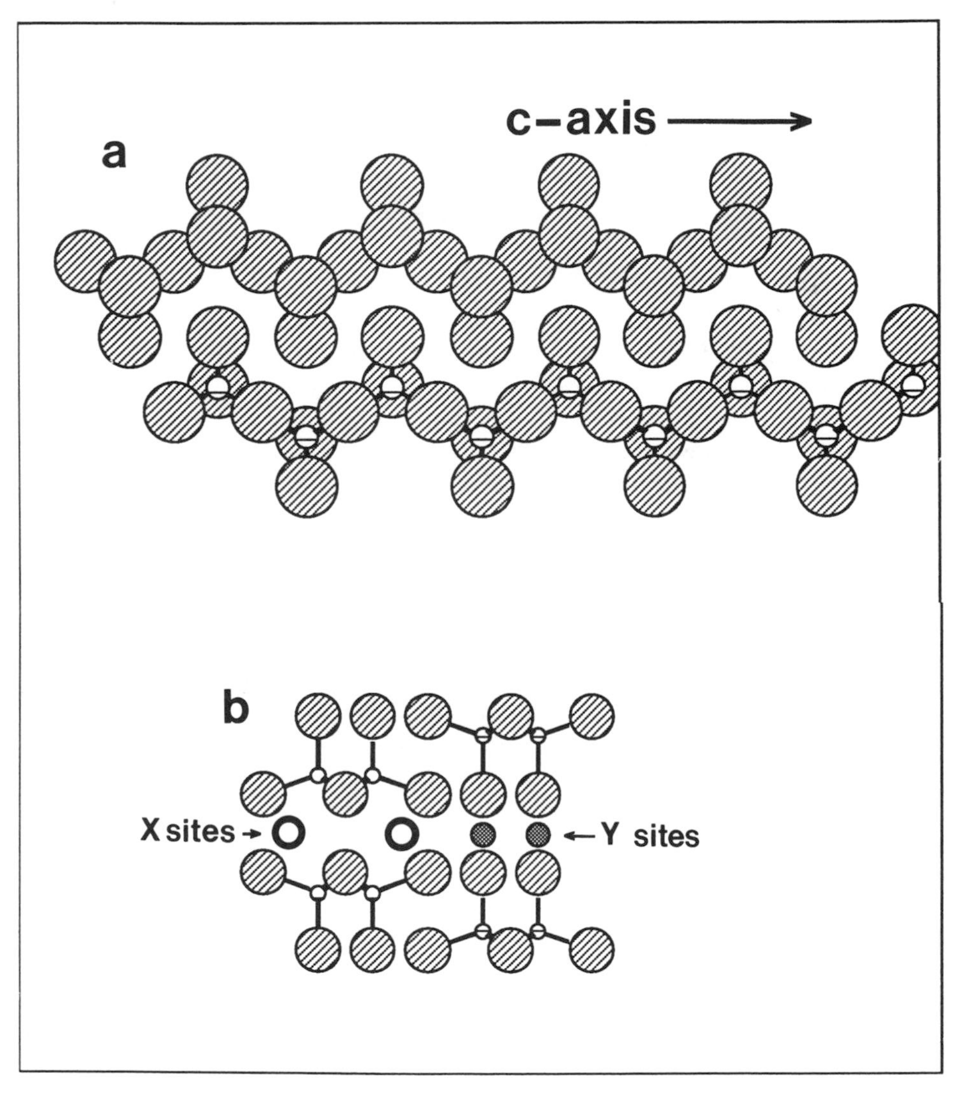

Figure 2.7. View of the single chains of pyroxene (a) from above, (b) end-on. The c-axis is parallel to the length of the chains.

Ca^{2+} ion in 8-coordination, residing in sites between the silica chains. A view along the length of the chains (the c-axis[1]) reveals the location of these two types of coordination sites (Figure 2.7b). Pyroxenes have the chemical formula $XYSi_2O_6$, where Y is a small cation (e.g., Mg^{2+}, Fe^{2+}, Li^+), but X may be larger (e.g., Ca^{2+}, Na^+), and the sum of the charge of X and Y is 4+. Example pyroxenes are $MgSiO_3$ (enstatite), $CaMgSi_2O_6$ (diopside), and $NaAlSi_2O_6$ (jadeite).

Amphiboles. These structures can be visualized to form by the cross-linking of two pyroxene chains, creating a wider, infinitely long, planar chain with the formula $(Si_4O_{11}^{6-})_n$. A single chain of this type is shown in Figure 2.8a. The chains are stacked together in much the same way as with pyroxene, as revealed in Figure 2.8b from an edge view, but the wider chains result in a different ratio of X-type and Y-type sites for large and small cations linking the chains together. The amphibole formula is $X_2Y_5Si_8O_{22}(OH)_2$, where the larger cations occupy the X sites. A common amphibole has the formula $Ca_2Mg_5Si_8O_{22}(OH)_2$, but many different cations including certain trace metals can occupy the X and Y sites. Amphiboles also contain structural OH^-, located in the holes created by forming the double silica chain, as is evident in Figure 2.8a.

Micas. If the double-chain amphibole structure diagrammed in Figure 2.8a is extended in two dimensions by the bonding of all three basal O^{2-} atoms of each tetrahedron with Si^{4+} atoms of other tetrahedra, a sheet silicate (phyllosilicate) is formed with the structure shown in Figure 2.9a. This polymer, extended infinitely in two dimensions, has the formula $(Si_4O_{10}^{4-})_n$, and is the basis of the mica structure (as well as the layer silicate clays, as will be discussed later in this chapter).

The structural variations in different micas arise from the degree of isomorphous substitution of Al^{3+} for Si^{4+} in the silica sheet, which generates negative charge in the layers, and from the identity of the octahedrally coordinated cations that bind these silica sheets together. Common micas include muscovite and biotite, in which one of four tetrahedral sites is occupied by Al^{3+} instead of Si^{4+}, generating a negative structural charge that is balanced by K^+ ions residing between layers. In Figure 2.9b, an edge view of mica layers reveals the position of these K^+ ions, as well as the position of octahedral cations, each of which is coordinated to four apical O^{2-} atoms of the silica layers and two structural OH^- groups. The K^+ ions reside in the adjacent hexagonal holes created by the hexagonal array of silica tetrahedra, and are therefore positioned directly above and below structural OH^- groups. In muscovite, the octahedral cations are Al^{3+}, and the mica is classified as *dioctahedral* because only two out of three of the available octahedral sites in the half-unit cell[2] are occupied. In biotite, the octahedral cations are Mg^{2+} and Fe^{2+}, and the mica is *trioctahedral,* with

1. The pyroxenes have orthorhombic symmetry and the crystallographic axis system is chosen so that the three mutually perpendicular axes (a, b, c) are oriented with the c-axis along the length of the chain. The repeat distance, or c-dimension, along the chain is about 5.3 Å.

2. The unit cell is the smallest structural unit from which the entire mineral structure can be built. For micas, this requires the equivalent of two Si_4O_{10} units. However, for convenience, the chemical formula is commonly based on a single Si_4O_{10} unit or half-unit cell.

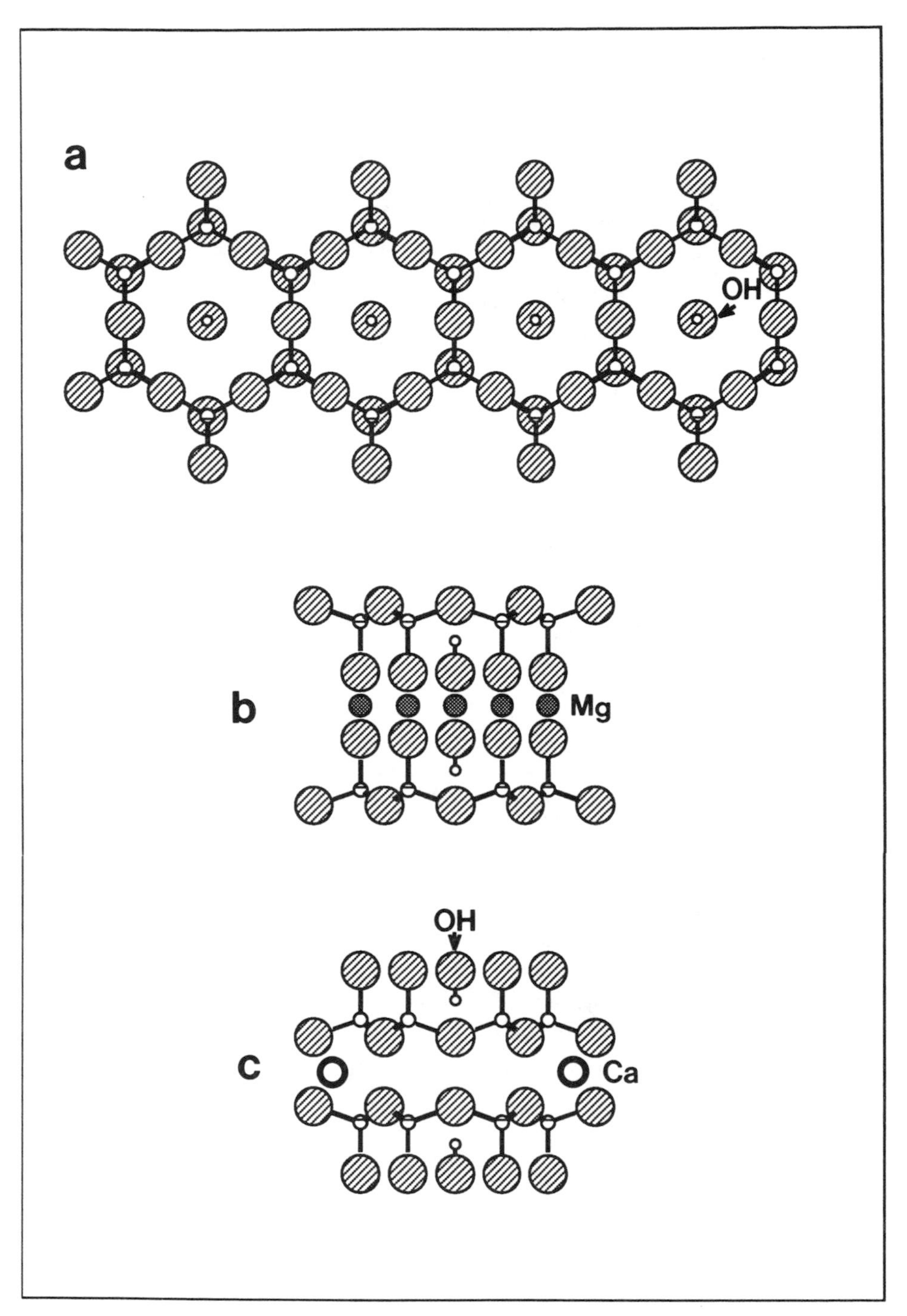

Figure 2.8. View of the double chains of amphibole (a) from above, (b) end-on showing Mg sites, and (c) end-on showing Ca sites. The OH^- in all hydrous minerals is depicted as an O^{2-} ion attached to a proton.

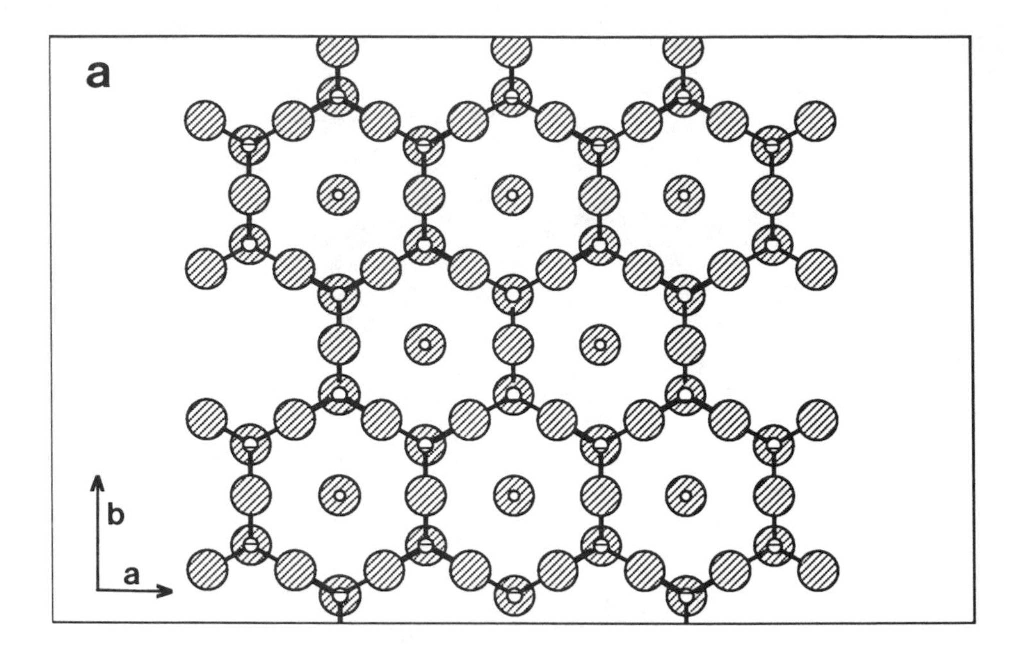

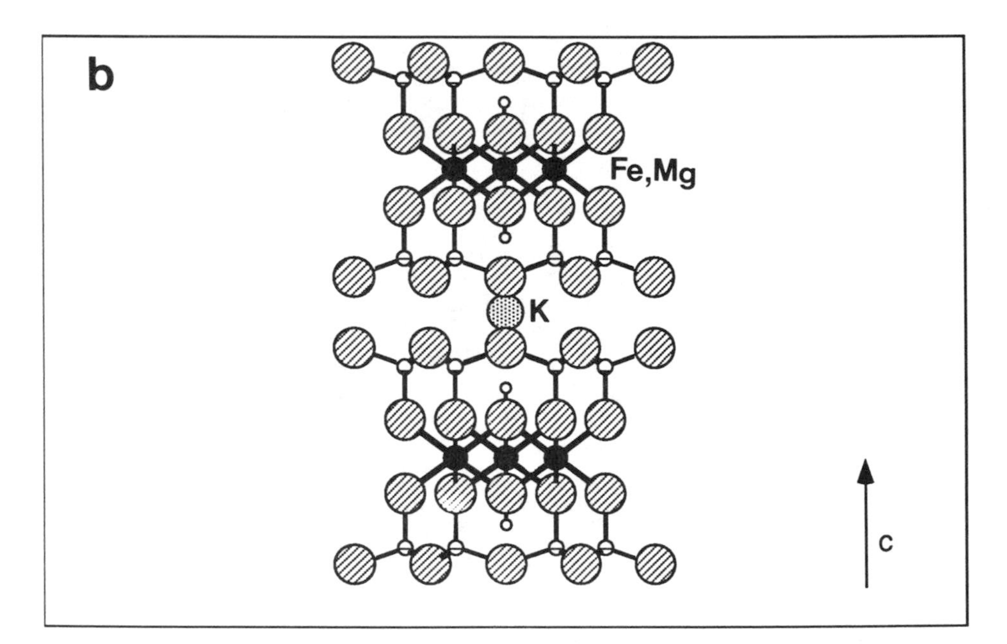

Figure 2.9. View of the phyllosilicate sheets of mica (a) from above, and (b) end-on, showing the location of Mg or Fe in the trioctahedral form of mica. The c-axis in phyllosilicates is perpendicular to the plane of the sheets.

all three of the octahedral sites in the half-unit cell occupied, as illustrated in Figure 2.9b. The half-unit cell formulae for these two micas are

$$KAl_2(Si_3Al)O_{10}(OH)_2 \qquad K(Mg,Fe^{2+})_3(Si_3Al)O_{10}(OH)_2$$

muscovite *biotite*

The repeating dimensions of the mica crystal along the a-, b-, and c-axes, which are nearly perpendicular to one another as shown in Figures 2.9a and 2.9b, are about a = 5.3 Å, b = 9.2 Å, and c = 10.2 Å. The c-dimension reflects the fact that the individual mica sheets are 9.6 Å thick, and the embedded K^+ ions prevent these sheets from stacking together in the most space-efficient manner. Consequently, the 10.2-Å spacing (as determined by x-ray diffraction of the mineral) is diagnostic for mica and micalike clay minerals.

Silica Minerals (Quartz). Although silicas can occur in nature as several distinct minerals, quartz is the most common and is a major component of many igneous rocks. With the formula SiO_2, quartz is classified as a framework silicate because the Si—O—Si bonding (sharing of tetrahedral oxygens) extends into all three dimensions, and is in fact maximized. This is illustrated by the very regular structure of cristobalite (a high-temperature form of SiO_2) in Figure 2.10. The quartz structure is similarly constructed from the three-dimensional linkage of SiO_4 tetrahedra, but is a less open structure with more distortion in the individual tetrahedra. These framework structures are not close packed and the more open ones have space for ionic impurities to be "stuffed" into them. The three-dimensional Si—O—Si bonding causes these structures to possess no inherent weak points as well as very sluggish dissolution kinetics. Consequently, quartz is particularly long-lived in soils despite

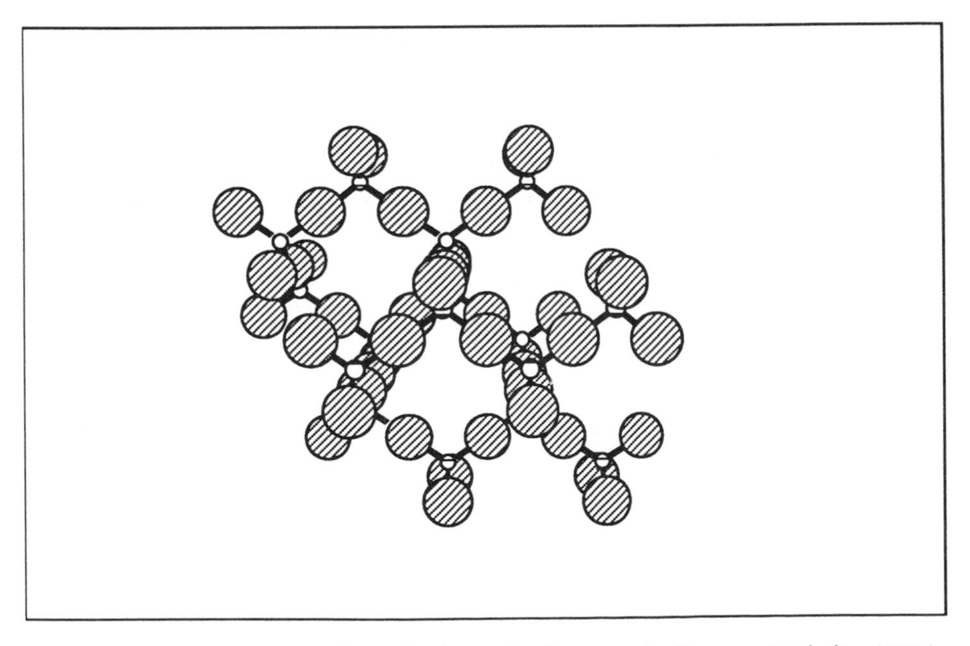

Figure 2.10. Arrangement of tetrahedra in the framework silicate, cristobalite (SiO_2).

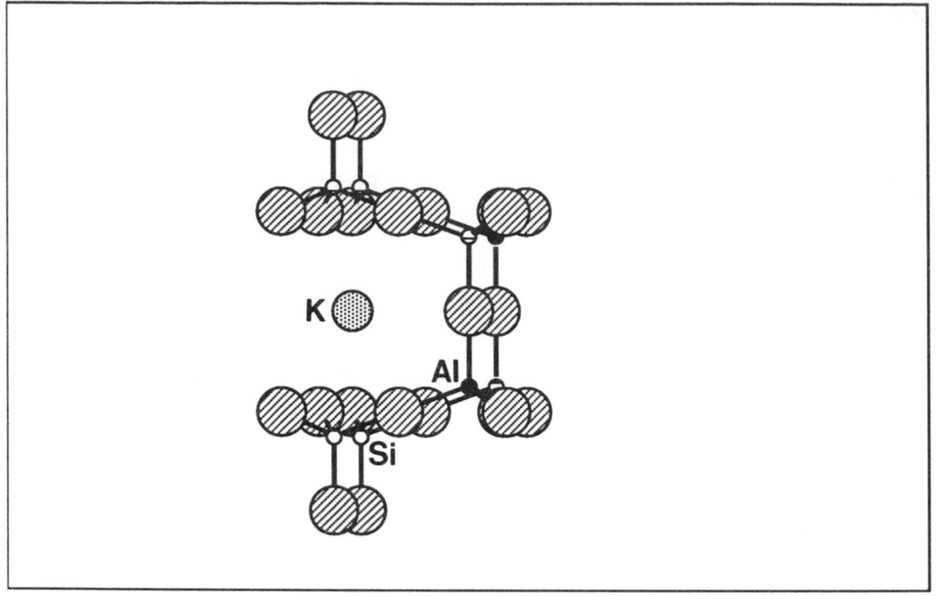

Figure 2.11. Arrangement of tetrahedra in the framework silicate, feldspar, showing the "cage" surrounding the K^+ ion.

the fact that, from a thermodynamic standpoint, it is expected to dissolve as secondary aluminosilicate minerals are formed.

Feldspars. Like the silica minerals, feldspars are framework silicates with all tetrahedral corners shared. By replacing one in four of the Si^{4+} atoms by Al^{3+}, the anionic framework $Si_3AlO_8^-$ is obtained from SiO_2. If the framework charge is balanced by structural K^+, the formula of potassium feldspar, KSi_3AlO_8, is obtained. This feldspar has two common structural *polymorphs* (different structures with the same chemical formula), orthoclase and microcline. If the charge is balanced by structural Na^+, the feldspar obtained is albite, $NaSi_3AlO_8$.

It is also possible for half of the tetrahedra to contain Al^{3+} in place of Si^{4+}, producing the $Si_2Al_2O_8^{2-}$ anionic framework. The charge must then be balanced by a divalent structural cation. If this cation is Ca^{2+}, the mineral name of the feldspar is anorthite. Mixed compositions are also possible, with one of the most common being the solid solution[1] between albite and anorthite, referred to as *plagioclase feldspars.*

The exact spatial arrangement of tetrahedra in feldspars is variable, depending on the size of the metal cation balancing the framework charge. This cation resides in a "cage" created by four-membered and eight-membered rings of connected tetrahedra. The approximate (idealized) geometry of one side of this cage in K-feldspar is diagrammed in Figure 2.11. The cage is sufficiently large that it must be occupied by

1. A solid solution is a homogeneous mineral of variable composition. The end members of this solid solution do not necessarily have the same structure.

large cations such as Ba^{2+} to prevent distortion. Smaller cations such as Ca^{2+} or Na^{+} electrostatically pull the tetrahedra together to form an elongated cage that is less symmetrical than that shown in Figure 2.11.

2.2c. Clays

Layer Silicates. Although the common primary minerals include island, chain, sheet, and framework silicates, the most stable and persistent silicates, which occur as weathering products (secondary minerals) in the clay fraction of soils, are sheet silicates. Figure 2.9a depicts the structure of the tetrahedral sheet in these minerals, which is comparable to the tetrahedral structure of mica. For the layer silicate clays, however, numerous structural combinations of the tetrahedral sheet with octahedrally coordinated metal cations are possible.

First, there are the 1:1 layer silicates, so named because each individual layer is constructed from one tetrahedral (silica) sheet and one octahedral (commonly alumina) sheet. The sheets are bonded together by the sharing of O^{2-} ions between the octahedral cation (Al^{3+}) and Si^{4+}, as shown in the structural diagram of Figure 2.12. The individual layers stack up to form a crystal of the 1:1 mineral. If this mineral is dioctahedral it has the name *kaolinite,* with the ideal chemical formula $Si_4Al_4O_{10}(OH)_8$. The ideal structure has no charge, and the layers are held together rather tenaciously by hydrogen bonding between the OH^- groups of one layer and the O^{2-} ions of the adjacent layer. For this reason, kaolinites do not swell in water (the interlayer surfaces have no tendency to separate except in extremely polar solvents) and have low surface areas and cation exchange capacities (see Chapter 3,

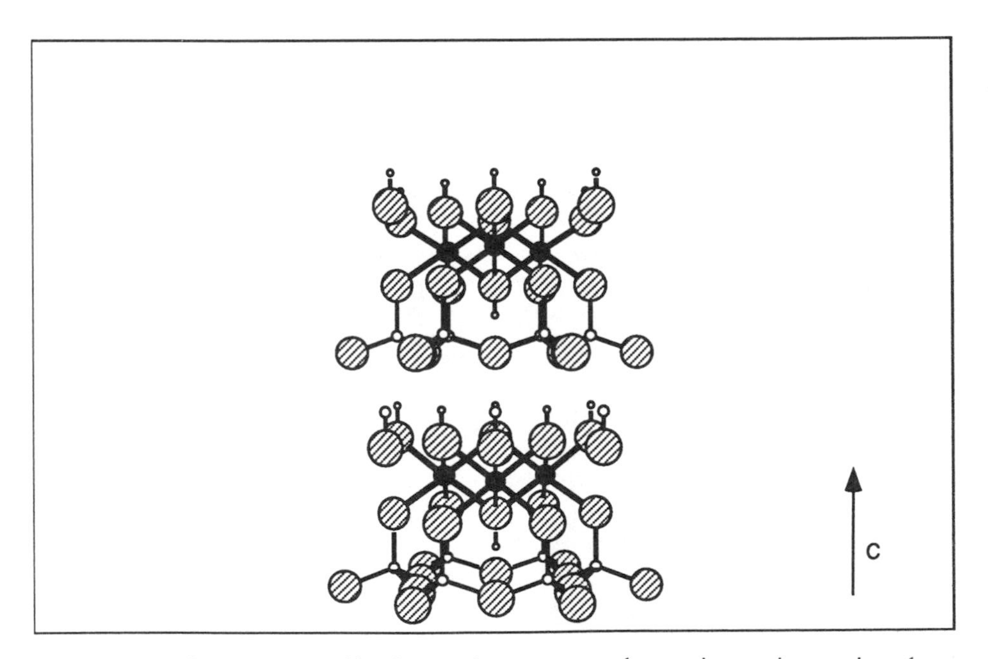

Figure 2.12. The structure of kaolinite, showing two of many layers that stack to form an hexagonal crystal.

section 3.2). The low cation exchange capacity that is observed in pure kaolinites ($<$ 1 centimole/kg) can be attributed to very limited isomorphous substitution[1] of Mg^{2+} for Al^{3+} or Al^{3+} for Si^{4+} in the structure.

Individual kaolinite layers are 7 Å thick, so that the stacked layers produce a repeat distance (c-axis spacing) of 7 Å, a fact used to identify kaolinite in soils by x-ray diffraction. The mineral halloysite, which has the same 1:1 layer structure as kaolinite, differs from kaolinite in the fact that it has a single sheet of water molecules between the layers. This increases the c-spacing to 10.1 Å. Mild heating of halloysite will dehydrate and collapse it irreversibly to the kaolinite 7-Å spacing.

Many of the layer silicate clays common in soils are based on the mica structure (shown in Figure 2.9b) in which two tetrahedral sheets sandwich a single sheet of octahedrally coordinated cations. Consequently, they are termed 2:1 layer silicates. Conceptually, it is useful to start with the neutral framework of the talc and pyrophyllite structures, representing the trioctahedral (Mg^{2+} in the octahedral sheet) and dioctahedral (Al^{3+} in the octahedral sheet) members of the 2:1 group. These have the ideal formulae given below:

$$Mg_3Si_4O_{10}(OH)_2 \qquad Al_2Si_4O_{10}(OH)_2$$

talc *pyrophyllite*

While these minerals are not common in soils, they can be used as prototypes for the many variations on the 2:1 structure that occur in soils. Common ones that will be described here are: (a) smectite, (b) vermiculite, (c) illite, and (d) chlorite. Each of these mineral names represents a clay group and a reasonably well-defined range of chemical compositions.

Smectites, which are based on either the trioctahedral 2:1 (talc) or dioctahedral 2:1 (pyrophyllite) structure, differ from these neutral structures by the presence of isomorphous substitution in the octahedral or tetrahedral sheet. For example, the dioctahedral smectite, montmorillonite, has the general formula

$$M^+_{x+y}(Al_{2-x}Mg_x)(Si_{4-y}Al_y)O_{10}(OH)_2$$

where $x \approx 0.4$ and $x > y$. The individual layers carry negative charge as a result of the isomorphous substitution, and most of this charge is localized in the octahedral sheet of montmorillonite.[2] In the above formula, the charge can be balanced by monovalent exchangeable cations, M^+, residing between the layers, as shown in Figure 2.13. These interlayer cations are generally hydrated and readily displaced into solution by other cations in the soil, a process termed *cation exchange* (see Chapter 3).

Smectites vary in chemical composition and location of structural charge, but they have in common a relatively low layer charge (compared with mica), which allows the individual layers to separate to large dimensions in water, conferring unique and dramatic swelling properties to this particular mineral group. In Table 2.3, some measured c-axis (sometimes called basal) spacings are listed for montmo-

1. Isomorphous substitution is the replacement of one structural ion by another of equal or different charge, without fundamentally altering the structure.

2. Layer charge can also arise from vacancies in the octahedral sheet, that is, a deficit of octahedrally coordinated cations.

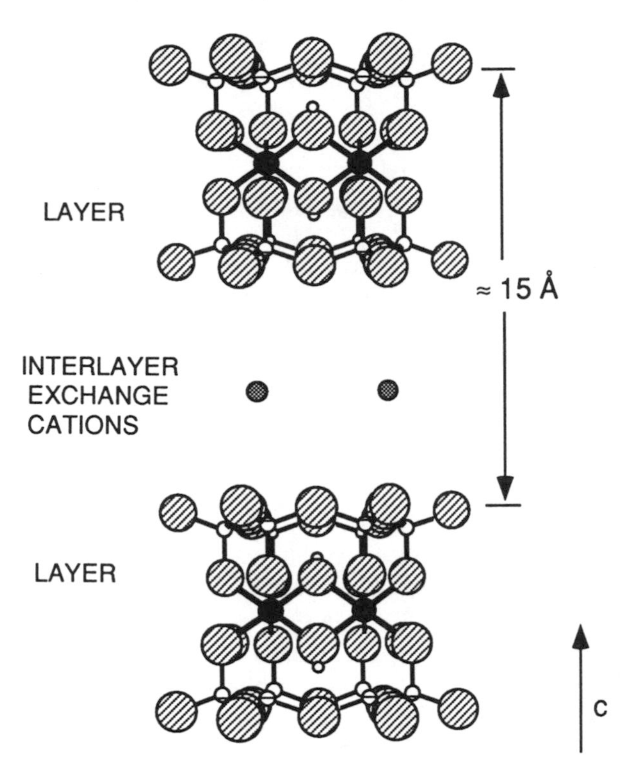

Figure 2.13. The structure of a dioctahedral smectite, seen edge-on. The c-axis spacing is about 15 Å in this diagram.

rillonite with different exchangeable cations in the interlayers. These spacings point out the important role of the exchangeable cations in clay swelling. Of the exchangeable cations commonly encountered in soils, Na^+ generates the greatest degree of expansion when water is added to smectites. A more complete explanation of clay swelling and the related phenomenon of clay dispersion is provided in Chapter 8, section 8.4.

Table 2.3. Measured c-Axis Spacings for Common Vermiculite and Montmorillonite Minerals Saturated with Exchangeable K^+, Na^+, and Mg^{2+} Cations

		c-Spacing (Å)		
Mineral	Exchange Ion	Air-Dry	Wet	Glycerol Added
Montmorillonite	K^+	11.6	≫20	17
	Na^+	12.4	≫20	17
	Mg^{2+}	14.7	19.5	17
Vermiculite	K^+	10.0	10.0	10.0
	Na^+	12.0	14.7	12.8
	Mg^{2+}	14.2	14.2	14.2

Vermiculites, like smectites, have structures based on the 2:1 sheet silicate structure of pyrophyllite or talc, with both the dioctahedral and trioctahedral forms occurring in soil. Dioctahedral vermiculite is common in the clay fraction of soils, while trioctahedral vermiculite often occurs as large particles, being derived from the weathering of the structurally analogous primary minerals, namely trioctahedral micas or chlorites. Vermiculites have a higher layer charge (and consequently a higher cation exchange capacity) than smectites, with about 0.6 to 0.9 unit of charge per half-unit cell. This is, however, less layer charge than that of micas from which they are often derived. Micas ideally have one in four of their tetrahedral sites occupied by Al^{3+}, which is equivalent to a layer charge of 1.0 unit per half-unit cell. As micas are weathered to vermiculite, the structural K^+ is replaced in the interlayer region by hydrated cations such as Mg^{2+}. Details of mica weathering are given in Chapter 6.

The formula for a typical trioctahedral vermiculite is

$$Mg_{0.33}(Mg,Al,Fe^{3+})_3(Si_3Al)O_{10}(OH)_2$$

where it is seen that part of the negative charge in the tetrahedral sheet is compensated by excess positive charge in the octahedral sheet. This is one way that the net layer charge of vermiculite could become less than that of its mica "parent."

Vermiculites do not swell in water as extensively as smectites (see Table 2.3). The greater layer charge of vermiculite, localized in the tetrahedral sheet, appears to cause the exchangeable cations to electrostatically pull the layers together more energetically and limit the extent of layer separation. Typically, vermiculite has only two molecular sheets of water between the layers, producing a characteristic c-axis spacing of 14 Å (Figure 2.13 illustrates this approximate spacing for 2:1 layer silicates). K^+ (and other large weakly hydrating cations such as Cs^+ and Ba^{2+}) readily displace the hydrated cations such as $Mg(H_2O)_6^{2+}$ from these interlayer regions of vermiculite, causing the layers to collapse together and "fixing" the K^+ into a micalike structure.

The separation between vermiculites and smectites is ambiguous, despite the disparity in swelling behavior of "ideal" members of these clay groups. This ambiguity is evident in Figure 2.14, which reveals a transition zone between the free swelling of "typical" smectites and the limited (two water-layer) expansion of "typical" vermiculites. Clays having intermediate magnitudes of tetrahedral and octahedral layer charge fall into this transition category, and could be termed high-charge smectites or low-charge vermiculites. Tetrahedral layer charge is more influential than an equivalent amount of octahedral layer charge in shifting 2:1 clays into the region of limited expansion.

Illite is the name of a rather poorly defined secondary mineral that could be described as a clay-sized hydrous mica. Although its structure is approximated by that of the primary mineral, mica (see Figure 2.9), it is much less crystalline than mica, containing less K^+ and more structural water[1] than expected for the ideal mica structure. The 10-Å c-spacing allows for straightforward identification of illite in soils since the spacing is unchanging under different conditions, unlike those of smectite and vermiculite. The micalike structure of illite means that the K^+, trapped (fixed)

1. Structural water is a term referring to the structural OH^- groups, as opposed to the adsorbed water, of a mineral. The ideal mica has two OH^- groups for each half-unit cell.

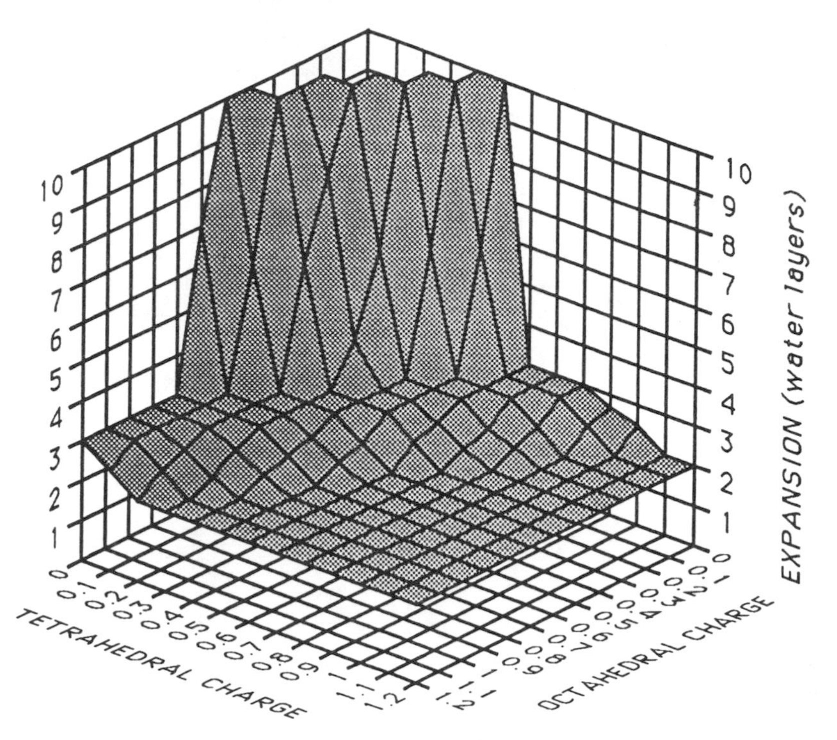

Figure 2.14. Dependence of 2:1 layer silicate clay expansion in water on structural layer charge.

between silicate layers, prevents clay swelling. As a result, illite has a comparatively low cation exchange capacity and specific surface area (see Chapter 3, section 3.2), with little interlayer surface accessible for cation exchange or water adsorption.

Chlorites are classified as a separate layer silicate mineral group, but in soils, may actually have a common origin with the vermiculites. Like vermiculites, chlorites have tetrahedrally charged 2:1 layers that can be trioctahedral or dioctahedral in nature, with the dioctahedral chlorites usually found in the clay-sized fraction of soils. Unlike vermiculites, chlorites balance at least part of this charge by a positively charged metal hydroxide sheet sandwiched between the 2:1 silicate layers. This structure, termed a 2:1:1 layer silicate, is diagrammed in Figure 2.15. A generalized structural formula is

$$\underbrace{[(M^{2+},M^{3+})_3(Si,Al)_4O_{10}(OH)_2]^{x-}}_{\textit{2:1 layer}}\ \underbrace{[(M^{2+},M^{3+})_3(OH)_6]^{x+}}_{\textit{interlayer hydroxide sheet}}$$

The primary mineral, chlorite, which occurs in rocks as large crystals, possesses an interlayer sheet composed largely of $Mg(OH)_2$. Since the mineral brucite is composed of magnesia sheets with the same basic structure, the single interlayer sheet in chlorite is termed the *brucite layer*. Isomorphous substitution of part of the Mg^{2+} by Al^{3+} produces a positively charged hydroxide sheet (see formula above) that props the 2:1 layers apart at a c-spacing of 14 Å. This rigid interlayer contrasts with the hydrated interlayer of vermiculite, and even though chlorite and vermiculite have similar c-

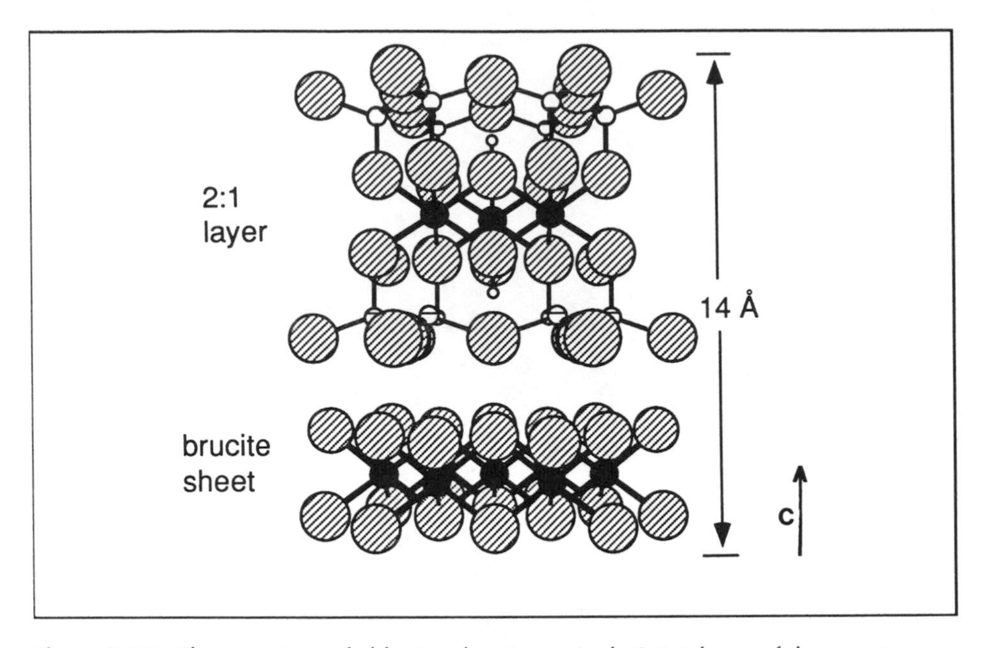

Figure 2.15. The structure of chlorite, showing a single 2:1:1 layer of the structure (≈14-Å c-axis dimension).

spacings under conditions found in soil, they are distinguished by the fact the chlorite spacing is very heat stable while vermiculite dehydrates and collapses on heating. Consequently, chlorites are identified in soils by the existence of stable 14-Å x-ray diffraction peaks, but this test does not identify the nature of the 2:1 layer or the hydroxide interlayer. In many cases, particularly in acid soils, the hydroxide interlayer is likely to be gibbsitelike (composed largely of $Al(OH)_3$) rather than brucitelike.

Chlorites have rather low specific surface areas and cation exchange capacities (see Chapter 3, section 3.2), a result of the blockage of the interlayer regions by hydroxy sheets, and do not expand at all in water. Exchangeable cations are likely to be found only on external surfaces and edges of chlorite particles.

A diagram summarizing the structural features of the important layer silicate clays is shown in Figure 2.16. All of these clays possess a platy morphology to some extent, but it is most notable in the smectite group. These clays have particles with dimensions perpendicular to the c-axis (parallel to the silicate sheet) on the order of microns (10^{-6} m), and dimensions along the c-axis on the order of nanometers (10^{-9} m). In fact, in a fully expanded smectite, each particle consists of a single 2:1 layer only 10 Å thick. Particles of kaolinite, on the other hand, have much larger c-axis dimensions because they are formed from hundreds of 1:1 layers bonded together. The platy morphology of layer silicate clays contributes to many of their unique and interesting physical properties: shrink-swell behavior, lubricating ability associated with their slippery feel when wet, and ability to seal up and prevent water flow in porous media such as soils. Reasons for, and consequences of, some of these properties will be discussed in more detail in Chapters 3 and 8.

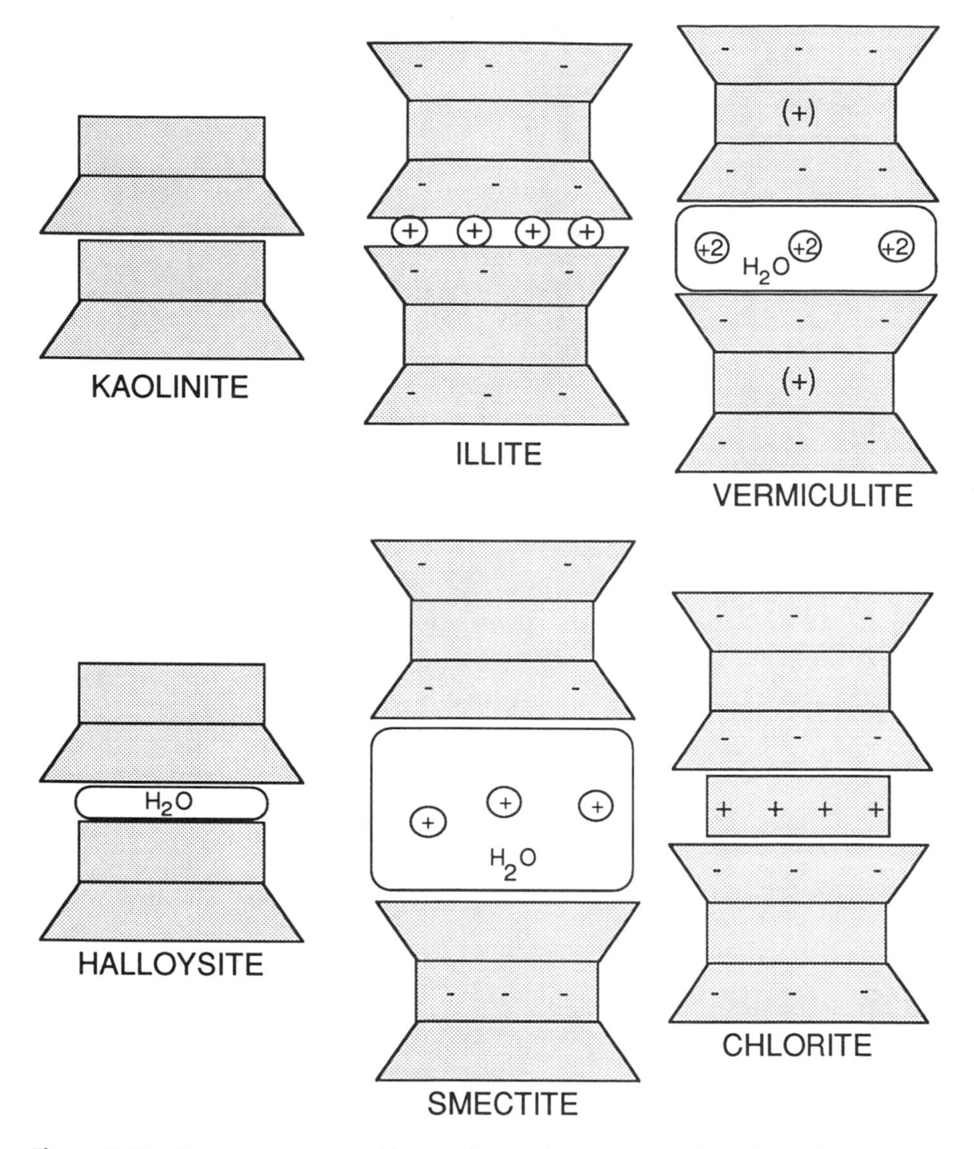

Figure 2.16. Common groups of layer silicate clay structures found in soils, pictured in terms of their tetrahedral (▲) and octahedral (■) sheets. The usual locations of structural charge and exchange cations are indicated by − and + signs.

Oxides. The impression should not be left that the layer silicates are the dominant clay minerals in all soils. While this may be true in many regions of the world where the soils have not been subjected to intense or prolonged weathering (e.g., soils in glaciated or arid regions), large areas of the earth's surface are characterized by ancient soils that formed from the parent rock material over hundreds of thousands, even millions, of years. The mineral fraction of these soils is typically composed largely of nonsilicate minerals, namely Fe and Al oxides and hydroxides. Because these areas tend to be found in the tropics or subtropics, and much of the early work in soil mineralogy and chemistry was done on soils of more temperate climates, it is only more recently that the oxide minerals (a collective term meant here to include the oxides, hydroxides, and oxyhydroxides) have received the attention they deserve. Manganese oxides have some properties similar to those of the more abundant Fe and Al oxides, but also have unique properties discussed in Chapter 7.

In some respects, the oxides are structurally simpler than the layer silicates, consisting of hexagonal or cubic close-packed O^{2-} and/or OH^- anions with Fe^{3+}, Al^{3+}, Mn^{4+} or Mn^{3+} residing in octahedral sites. A summary of properties of common Fe and Al oxide minerals is given in Table 2.4.

The arrangement of the AlO_6 and FeO_6 octahedral units, and the degree to which they share corners (one shared oxygen), edges (two shared oxygens), or faces (three shared oxygens), distinguishes the individual oxide minerals. These different sharing arrangements are diagrammed in Figure 2.17. The structure of three of the most common crystalline oxides found in soils—gibbsite, goethite, and hematite—are depicted in Figures 2.18, 2.19, and 2.20.

Unlike layer silicate clays, the oxides of Fe and Al are not inclined to develop structural charge as a result of isomorphous substitution. Consequently they have very low cation exchange capacities despite sometimes possessing impressively large surface areas. The surfaces do, however, develop limited charge (negative or positive) in response to the pH of the surrounding solution, and this process will be discussed

Table 2.4. Oxides of Fe and Al Found in Soils

Formula	Name	Oxyanion Packing[a]	Comments
α-FeOOH	Goethite	hcp	Common in temperate-region soils. Gives soils characteristic yellow-brown color
α-AlOOH	Diaspore	hcp	Uncommon in soils. Isostructural with goethite
γ-FeOOH	Lepidocrocite	ccp	Uncommon except in wet soils. Color similar to goethite.
γ-AlOOH	Boehmite	ccp	Uncommon in soils. Isostructural with lepidocrocite.
$Fe_2O_3 \cdot nH_2O$	Ferrihydrite	Disordered	Common in temperate-region soils. Reddish-brown to yellow-brown.
α-Fe_2O_3	Hematite	hcp	Common in soils of hot climates, both humid and dry. Reddish-brown color.
γ-$Al(OH)_3$	Gibbsite	hcp (open packed)	Common in humid tropics. Not common in cool temperate-region soils.
γ-Fe_2O_3	Maghemite	ccp	Widespread but minor. May be product of oxidation-reduction cycles in soil. Magnetic.
Fe_3O_4	Magnetite	ccp	Widespread but minor. Black, magnetic. May have origin similar to maghemite.

[a]Hcp and ccp refer to hexagonal and cubic closest packing.

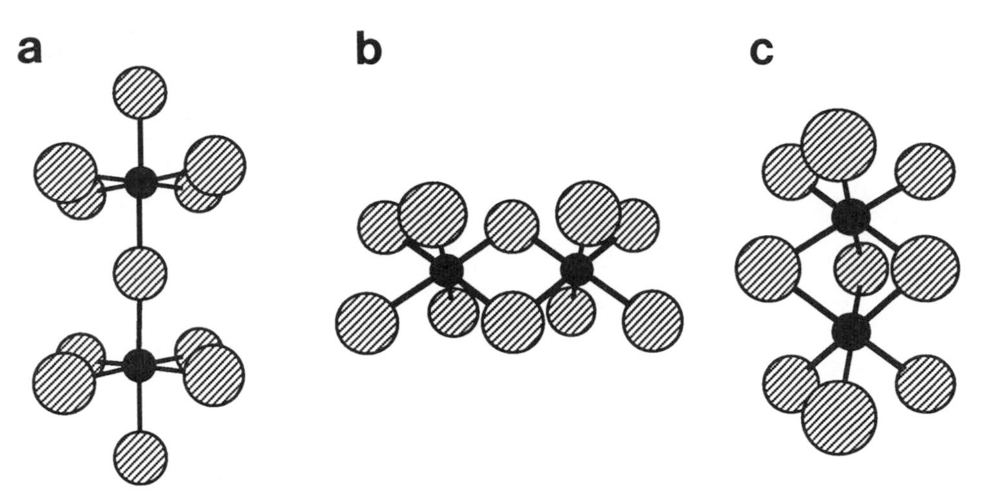

Figure 2.17. The three ways in which octahedra can share "elements" (see Pauling's rule number 4). (a) Corner sharing. (b) Edge-sharing. (c) Face-sharing.

in Chapter 3. Furthermore, the surfaces have considerable capacity to chemisorb metal ions as well as inorganic and organic anions, and this important property of oxides will be discussed in Chapter 4.

Noncrystalline Minerals. It is not unusual for soils developed from recently deposited parent materials (e.g., volcanic ash, glacial till) to contain poorly crystallized aluminosilicate or oxide minerals. Soils formed from volcanic ash commonly contain *allophane,* a noncrystalline hydrous aluminosilicate. Allophanes have also been identified in the clay fraction of spodosols formed from glacial tills, and in soils of the tropics formed by intense weathering of basic igneous rocks. Although the structure of allophane is variable, it appears to consist of extremely small spherical aluminosilicate particles in which Al^{3+} occupies both octahedral and tetrahedral sites, the latter increasing as the Si/Al ratio in the allophane increases. Tetrahedral Al^{3+} would be expected to generate permanent negative charge in the structure, yet there is no indication that allophanes possess permanent charge. Instead, the cation exchange capacity of allophane is determined by the pH of the soil solution (see Chapter 3). Allophane can develop much more charge per unit of weight than aluminum or iron oxides.

One mineral related to allophane, but having a higher degree of order in its structure, is *imogolite.* It has the ideal formula $SiO_2 \cdot Al_2O_3 \cdot 2H_2O$ and consists of a single gibbsitelike sheet curved around on itself to form the outer surface of a tube. It therefore has no tetrahedral Al. The inside surface has isolated Si^{4+} atoms bonded to three O^{2-} ions of the sheet, with the fourth coordination position of Si^{4+} occupied by an OH^- directed inward toward the center of the tube. The ideal structure is uncharged, but charge develops on the surface in response to the pH in soil solution (see Chapter 3). This mineral is unusual in that it seems to be able to trap ions inside its tubular structure.

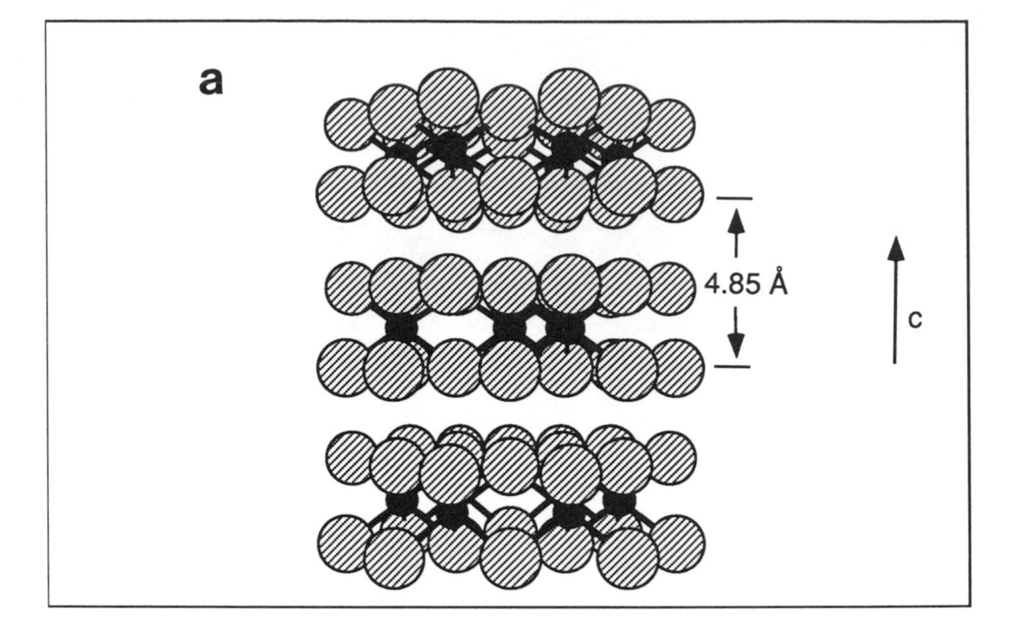

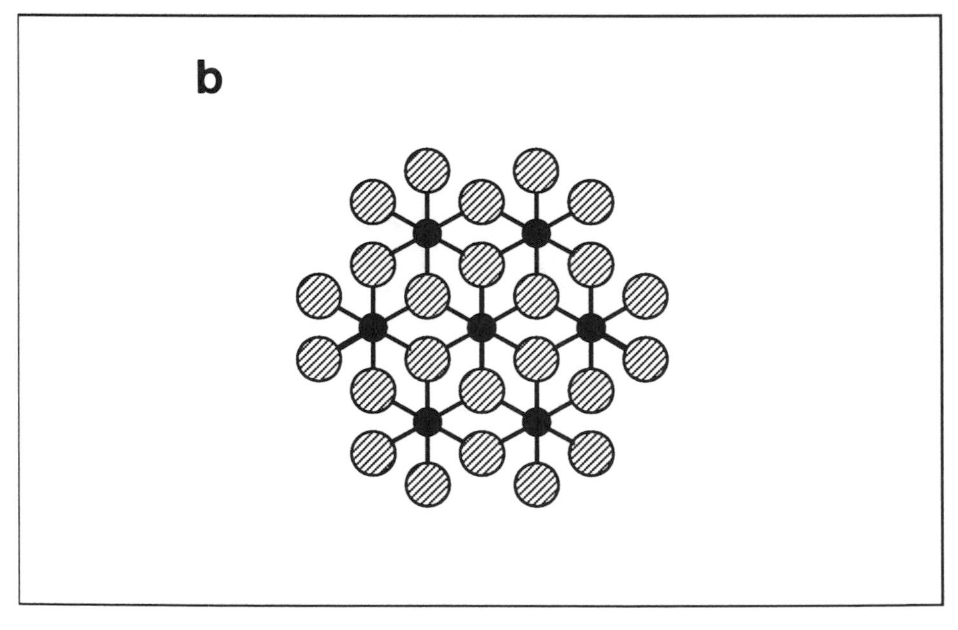

Figure 2.18. Structure of gibbsite showing (a) edge view and (b) top view of three sheets.

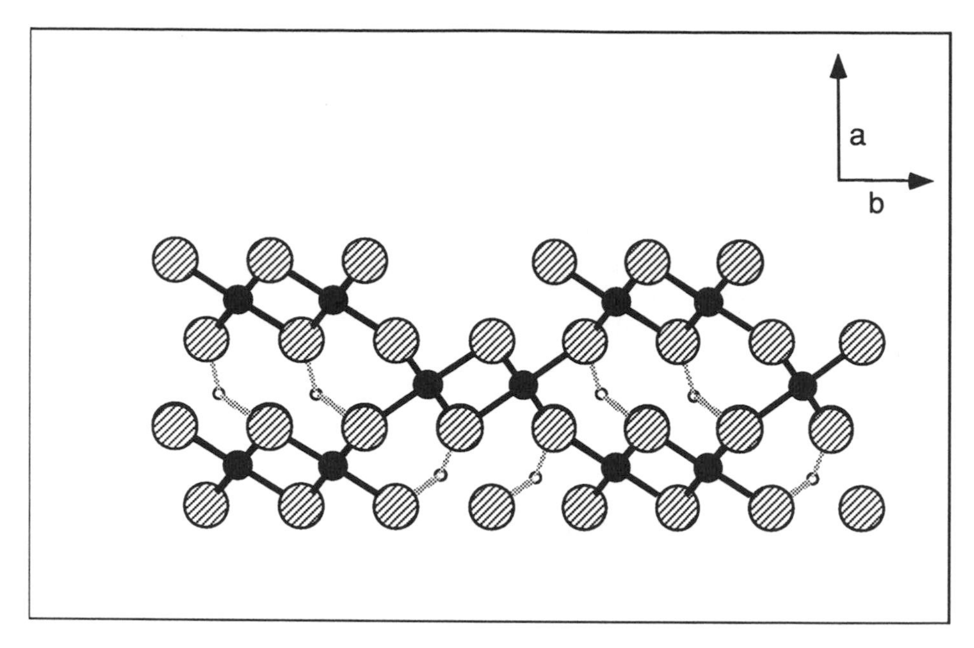

Figure 2.19. Structure of goethite showing end-on view of double rows of FeO_6 octahedra. Approximate positions of the protons and H-bonds (nonsolid lines) are shown.

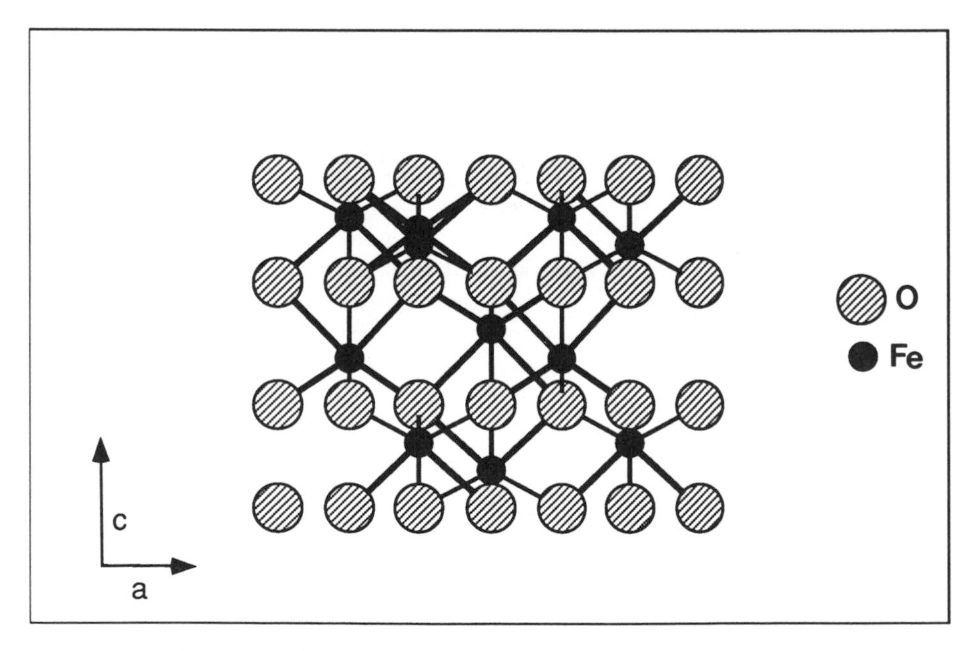

Figure 2.20. Structure of hematite showing displacement of Fe^{3+} ions from ideal octahedral positions.

Imogolite can be a major component of clays in soils developed from volcanic ash, but has also been discovered as a minor constituent of spodosols formed in northern temperate climates. The fibrous appearance of imogolite under high magnification, with tubules that are as much as several microns in length but only 21 Å in diameter, distinguishes it from allophane. Its formation is favored over that of allophane in acidic environments.

Noncrystalline oxides, particularly *ferrihydrite,* are common in soils because the presence of soluble silica and organic matter tends to inhibit crystallization into more stable, better ordered oxides of Fe. Even so, ferrihydrite is considered to be unstable, gradually transforming to hematite in tropical or subtropical climates or to goethite in humid temperate climates.

2.3. ORGANIC MATTER

In partnership with the clay fraction, organic matter has an extremely important influence on the chemical and physical properties of soils. Critical and beneficial functions of organic matter include:

1. Maintenance of good pore structure accompanied by improved water retention
2. Retention of nutrients (e.g., Ca^{2+}, Mg^{2+}, K^{+}, NH_4^{+}, Mn^{2+}, Fe^{3+}, Cu^{2+}) by cation exchange
3. Release of nitrogen, phosphorus, sulfur, and trace elements by mineralization[1]
4. Adsorption of potentially toxic organics (pesticides, industrial wastes, etc.)

Much of this activity can probably be attributed to *humus,* a high surface area material that is the product of decomposition of dead plant and animal matter in soil.

2.3a. Classification of the Components of Organic Matter

Humus, which has lost all the visible features of the organic residues from which it formed, is subdivided into amorphous brown-colored polymers according to the scheme depicted in Figure 2.21. The polymers, termed *humic substances,* can be further separated (operationally, if not intrinsically) into *humic acid, fulvic acid,* and *humin,* according to their solubility in strong acid and base. Aqueous sodium hydroxide extracts humic and fulvic acid from soils, leaving the humin unextracted. Acidification of this dark-colored extract then causes the humic acid to precipitate, while the fulvic acid remains soluble.

Although this scheme of separation appears to be somewhat crude and arbitrary, it does achieve a degree of segregation of polymeric materials as gauged by their more important chemical properties. Table 2.5 summarizes some of these properties, showing that the fulvic acid-humic acid-humin sequence represents a continuum of

1. Mineralization is the microbial process by which organic compounds are decomposed and carbon dioxide is released.

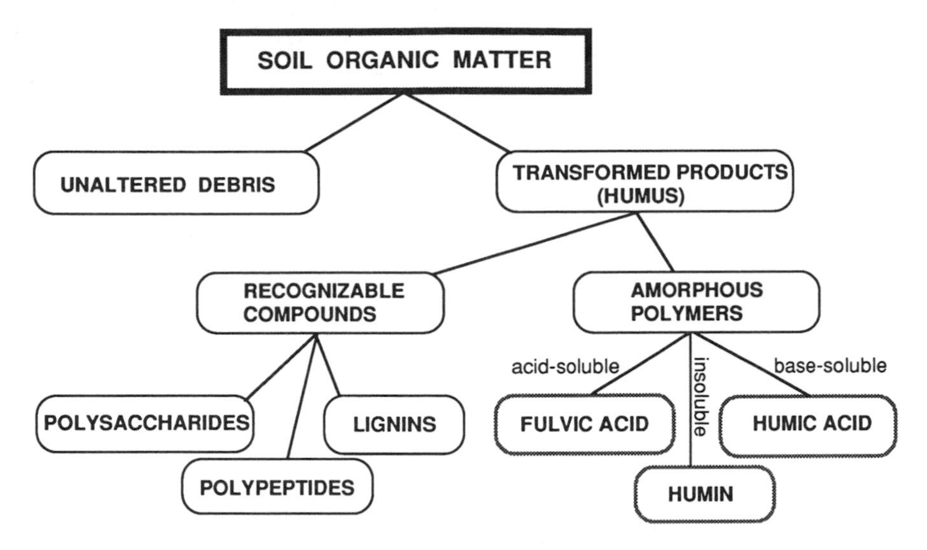

Figure 2.21. A classification scheme for soil organic matter. (After M.H.B. Hayes and R. S. Swift. 1978. The chemistry of soil organic colloids. In D. J. Greenland and M.H.B. Hayes (eds.), *The Chemistry of Soil Constituents.* New York: Wiley.)

Table 2.5. Some Fundamental Properties of Humic Substances

	Fulvic Acid	Humic Acid	Humin[a]
Molecular wt.	1000–5000	10,000–100,000	>100,000
% C	42–47	51–62	>62
% O	45–50	31–36	<30
% N	2.0–4.1	3.6–5.5	>5
Acid content (moles/kg)[b]	14	5	<5

[a]Values for humin are uncertain because of difficulty in separating this fraction from the mineral particles for elemental analysis.

[b]The acid content is equivalent to the potential cation exchange capacity once the acidity is neutralized by alkali.

properties with increasing resemblance to lignin (a complex aromatic polymer) suggested in the higher molecular weight fractions.[1]

Appendix: Basic Concepts of Mineralogy

Crystals of all minerals are conceptualized in terms of an array of points in space, called a *lattice.* The lattice is useful because it represents the periodic nature of the mineral structure regardless of the specific atoms involved. The *unit cell* is the repeat unit by which the crystal lattice can be built. The lattice parameters used to describe

1. Recent studies using solid-state NMR reveal that some of the decomposition-resistant polymers in soils may have long-chain alkane groups.

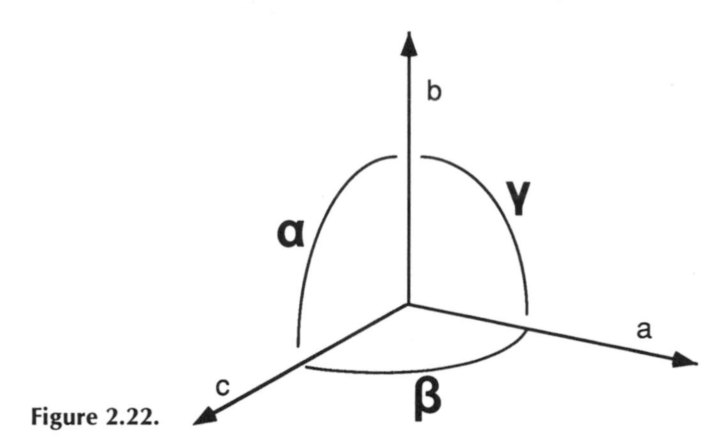

Figure 2.22.

the unit cell are the *crystallographic axes* **a**, **b**, **c** (of length *a*, *b*, and *c* units) and the angles α, β, γ between them, as shown in Figure 2.22.

The unit cell that is chosen to describe a particular mineral structure is the one that is the most convenient to work with. There are seven different kinds of cells necessary to describe all possible lattices, and these cells are the basis for the seven different crystal systems listed in the table.

Crystal System	Lattice Types[a]	Axial Ratios	Axial Angles
Triclinic	P	a ≠ b ≠ c	$\alpha \neq \beta \neq \gamma$
Monoclinic	P,C	a ≠ b ≠ c	$\alpha = \gamma = 90° \neq \beta$
Orthorhombic	P,C,I,F	a ≠ b ≠ c	$\alpha = \beta = \gamma = 90°$
Tetragonal	P,I	a = b ≠ c	$\alpha = \beta = \gamma = 90°$
Cubic	P,I,F	a = b = c	$\alpha = \beta = \gamma = 90°$
Trigonal	R	a = b = c	$\alpha = \beta = \gamma \neq 90°, <120°$
Hexagonal	P	a = b ≠ c	$\alpha = \beta = 90°, \gamma = 120°$

[a]P = primitive, I = body-centered, F = face-centered, C = side-centered, R = rhombohedral.

There are only fourteen different arrangements of points in space that satisfy the definition of a lattice. These are known as the Bravais lattices, listed above under lattice types. Diagrams of these space lattices are found in introductory mineralogy texts.

In a lattice, one can specify *lattice points, lattice directions,* and *lattice planes.* A lattice point is located by its coordinates, that is, its position in relation to the crystallographic axes. A lattice direction for any line in a lattice can be described by first drawing a line through the origin parallel to the given line, then listing the coordinates, (**u**,**v**,**w**), of any point on the line. Then [**uvw**] are the indices of the direction of that line and any line parallel to it. The values of **u**, **v**, **w** are always converted to a set of smallest integers by multiplication or division. For example, [½½1], [112], and [224] all represent the same direction, but [112] is the preferred form. Negative indices are written with a bar over the number (e.g., [$11\bar{2}$]). Lattice planes are specified by their *Miller indices,* which are the reciprocals of the fractional intercepts that the

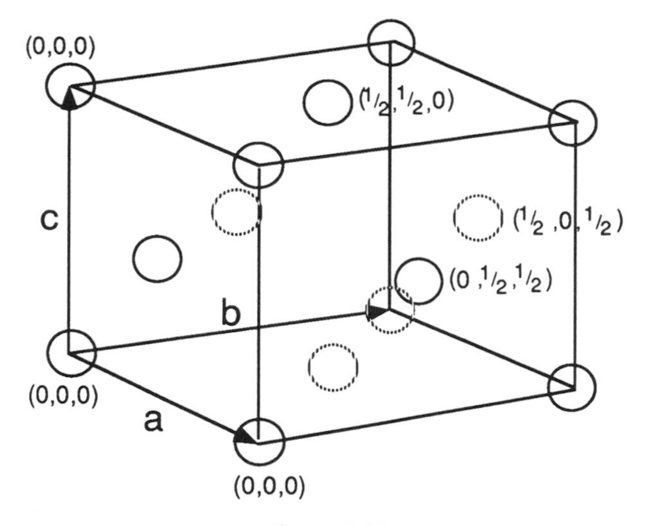

Figure 2.23.

plane makes with the crystallographic axes. The Miller indices of a plane are written as (**hkl**), and negative indices are again written with a bar over the number.

These definitions can be explained using a face-centered lattice, shown in Figure 2.23. The origin of the unit cell, labeled (0,0,0), can be chosen at any corner, since the lattice continues to infinity in all directions. The lattice point coordinates are then obtained by tracing the path to each point along the three axes. For example, the center-face points at the front and back of a cell can be reached by a translation of ½ the unit cell distance along axis **a**, no distance along **b**, and ½ the unit cell distance along axis **c**. These lattice points are then labeled (½,0,½). The other face-centered atoms, located by the same procedure, are found at lattice positions (½,½,0) and (0,½,½). These four lattice points, (0,0,0), (½,0,½), (½,½,0) and (0,½,½), completely describe the face-centered lattice. Note that this determination of lattice points is independent of the angles between axes.

Continuing to use the same unit cell described above, lattice directions can be specified. For example, a diagonal line drawn from (0,0,0) to (1,1,1) has the lattice direction [111], as depicted in Figure 2.24. A line from the origin to halfway between the basal atoms on the opposite side of the cell has the lattice direction [1 ½ 0], or preferably [210].

Lattice planes can be defined on the same unit cell. Consider, for example, a plane drawn normal to the **a** axis. This plane can never intercept the **b** or **c** axes, so that its interception points on the **a**, **b**, and **c** axes can be considered to be 1, ∞, and ∞, respectively. The reciprocals of 1, ∞, ∞ are 1,0,0 so that the Miller indices of this plane are (**100**). This plane, drawn on the unit cell in Figure 2.25 is often called the A-face of the crystal lattice, being perpendicular to the **a** axis. B and C faces are defined analogously.

A plane that is parallel to the (**100**) plane and intercepts the unit cell at a/2, has the Miller indices (**200**).

The atoms of a crystal are set in space either on the points of a Bravais lattice or

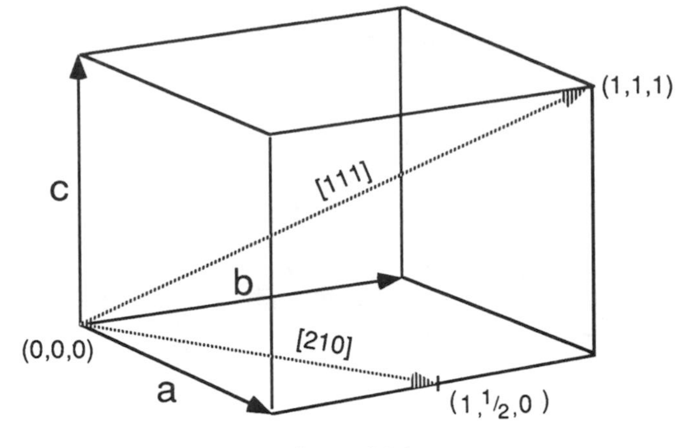

Figure 2.24.

in some fixed relation to those points. There may not be an atom at every lattice point, so that lattice and structure are *not* interchangeable concepts.

In crystalline minerals, planes of atoms exist parallel to the lattice planes. Because x-rays are coherently scattered from atoms, a beam of x-rays appears to be reflected from the parallel planes. Diffracted beams are observed when the reflected x-rays from parallel planes of atoms constructively interfere. The condition for appearance of diffraction beams is given by the Bragg law:

$$n\lambda = 2d \sin \theta \qquad n = 1,2,3,\ldots$$

which defines the angle of incidence, θ, of the x-ray beam on the crystal plane that causes diffraction from planes of atoms separated by the distance d. In this equation, λ is the wavelength of the x-rays and n is an integer. Thus the parallel planes of atoms along **(100)** can, in principle, generate multiple diffraction beams as the angle of incidence is increased, the first-order diffraction ($n = 1$), second-order diffraction ($n =$

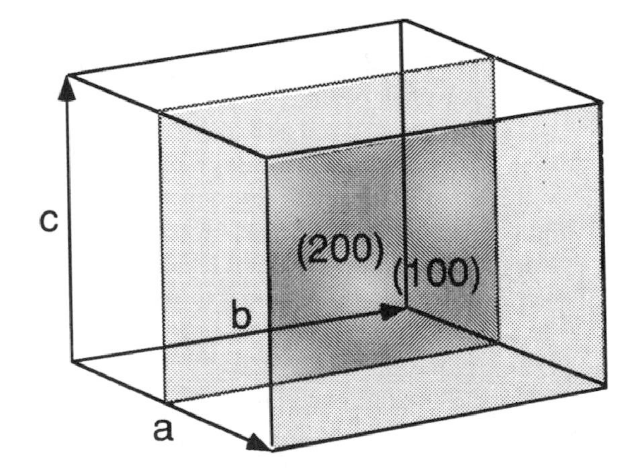

Figure 2.25.

2), and so on. However, a more detailed analysis is necessary to determine the intensity of diffraction from individual lattice planes. Some lattice planes produce no observable diffraction beams because the beam intensity is at or near zero.

References

Hayes, M.H.B. and R. S. Swift. 1978. The chemistry of soil organic colloids. In D. J. Greenland and M.H.B. Hayes (eds.) *The Chemistry of Soil Constituents.* New York: Wiley, pp. 179–320.

Suggested Additional Reading

D. G. Schulze. 1989. An introduction to soil mineralogy. J. B. Dixon and S. B. Weed (eds.), In *Minerals in Soil Environments.* Madison, Wis.: Soil Science Society of America, pp. 1–34.

G. Brown, A.C.D. Newman, J. H. Rayner, and A. H. Weir. 1978. The structures and chemistry of soil clay minerals. In Greenland and Hayes (eds.), *The Chemistry of Soil Constituents,* pp. 29–178.

Questions

1. From the list of ionic radii given below:

Ion	Radius (Å)
Li^+	0.78
Na^+	0.98
K^+	1.33
Mg^{2+}	0.78
Ca^{2+}	1.06
Sr^{2+}	1.27
Ba^{2+}	1.43
Al^{3+}	0.57
Si^{4+}	0.39
Fe^{2+}	0.83
Fe^{3+}	0.67
O^{2-}	1.32

   (a) Determine which metal ions are likely to substitute into (1) the octahedral sheet and (2) the tetrahedral sheet of layer silicates.

   (b) Calculate from geometric principles the limiting radius ratio for 8-coordination of metal ions with oxyanions. Which metals of the above list are expected to enter sites of 8-coordination? Find an example of 8-coordination in the primary minerals.

2. Apply Pauling rules to the structure of gibbsite, $Al(OH)_3$. Are all of these rules obeyed by this structure?

3. The hematite structure appears to violate Pauling's third rule. Explain how this is done without excessively destabilizing the structure.

4. Applying Pauling's second rule to the surface of goethite (see Figure 2.19), position protons on the surface oxygen atoms to create a charge-balanced structure.

5. The Si—O hexagonal array of layer silicates, shown in Figure 2.9a, defines a rectangular unit cell of dimensions *a*, *b*, and *c*, where *a* and *b* are the repeat distances in the plane of the sheet. Calculate these repeat distances in terms of bond lengths or structural units, and show that $b/a = \sqrt{3}$.

6. Show that the smallest repeat unit of a layer silicate sheet has the formula $Si_4O_{10}^{4-}$.

7. The humic and fulvic acid fractions of a soil, analyzed for elemental composition, are found to contain the following (data from Hayes and Swift, 1978):

	%C	%O	%H	%N
Humic acid	57.6	35.3	5.2	4.8
Fulvic acid	42.6	44.6	5.0	4.1

(a) Calculate the molar C/O/H ratio in the humic and fulvic acids and compare them with those of simple aliphatic and aromatic acids. What does this suggest about the nature of humic and fulvic acids?

(b) Estimate a maximum potential cation exchange capacity (negative charge) for the humic and fulvic acid in units of millimoles per kilogram, assuming that most of the oxygen content can be assigned to carboxylic acids.

3

Ion Exchange

3.1. CONCEPT AND SOURCE OF CATION EXCHANGE CAPACITY IN SOILS

Ion exchange reactions were first recognized in the mid-nineteenth century to be important to nutrient dynamics in soils. In fact, the ability of soil colloids to reversibly adsorb cations from solution was understood as an ion exchange reaction long before the chemical structures of the clay minerals and organic matter were sufficiently resolved to suggest an origin for the soil's negative charge. The previous chapter discussed the structural origin of much of this charge. The present chapter deals with the specific details of surface charge and exchange between cations (and anions), and outlines the concepts behind the equations used to quantitatively describe ion exchange.

CEC (cation exchange capacity) is the quantity of cations reversibly adsorbed (expressed as moles of positive charge) per unit weight of mineral. Conventional units for CEC are centimoles per kilogram (cmoles/kg), although millimoles per kilogram are more acceptable according to the International System of Units.

The structural (permanent) negative charge of layer silicate clays, as shown in Table 3.1, ranges from near zero in minerals possessing little or no isomorphous substitution to over 150 cmoles/kg in vermiculites. However, the surface density of charge is much less varied because the specific surface area (area per unit weight of clay) tends to increase in proportion to the structural charge or CEC, as the data in Table 3.2 show. An obvious exception is seen in the micas, where the structural charge is very high yet the surface area and CEC are low. In this particular case, K^+ ions fixed in the interlayer region prevent the internal surfaces of the mica from becoming accessible to solution and cation exchange processes. It is necessary to conclude that structural charge and CEC are not always equal in magnitude, because some portion of the charge sites may be inaccessible.

The cation exchange properties of silicate clay minerals with high permanent charge densities (i.e., smectites and vermiculites) are determined largely by the cation-bonding characteristics of permanent octahedral and tetrahedral charge sites, described in Table 3.3. On these clays, while "dangling bonds" on the silicate particle edges can generate pH-dependent charge, the relative magnitude of this charge is low and of little importance for cation exchange.

TABLE 3.1 Chemical Composition and Charge Characteristics of Representative Layer-Silicate Minerals

Mineral	Chemical formula	Charge Per Half Unit Cell		Structural Charge, cmol(−)kg^{-1}
		Tetrahedral	Octahedral	
	2:1 Dioctahedral			
Pyrophyllite	$Si_4Al_2O_{10}(OH)_2$	0	0	0
Montmorillonite	$Ca_{0.165}Si_4(Al_{1.67}Mg_{0.33})O_{10}(OH)_2$	0	−0.33	92
Beidellite	$Ca_{0.25}(Si_{3.5}Al_{0.5})Al_2O_{10}(OH)_2$	−0.5	0	135
Nontronite	$Na_{0.50}(Si_{3.5}Al_{0.5})Fe_2^{3+}O_{10}(OH)_2$	−0.5	0	117
Muscovite	$K_{0.94}(Si_{3.11}Al_{0.89})(Al_{1.95}Mg_{0.05})O_{10}(OH)_2$	−0.89	−0.05	237
	2:1 Trioctahedral			
Talc	$Si_4Mg_3O_{10}(OH)_2$	0	0	0
Hectorite	$Na_{0.33}Si_4(Mg_{2.67}Li_{0.33})O_{10}(OH)_2$	0	−0.33	89
Saponite	$Na_{0.33}(Si_{3.67}Al_{0.33})Mg_3O_{10}(OH)_2$	−0.33	0	87
Vermiculite	$Mg_{0.31}(Si_{3.15}Al_{0.85})(Mg_{2.69}Fe_{0.23}^{3+}Fe_{0.08}^{2+})O_{10}(OH)_2$	−0.85	+0.23	157
Phlogopite	$K(Si_3Al)Mg_3O_{10}(OH)_2$	−1.0	0	240
	1:1 Dioctahedral			
Kaolinite	$Si_2Al_2O_5(OH)_4$	0	0	0[a]
	1:1 Trioctahedral			
Serpentine	$Si_2Mg_3O_5(OH)_4$	0	0	0

[a]There is evidence that isomorphous substitution in kaolinite generates ≈ 1.0 cmol (−) kg^{-1} of permanent charge.

Source: M. B. McBride. 1989. "Surface Chemistry of Soil Minerals." In J. B. Dixon and S. B. Weed (eds.), *Minerals in Soil Environments.* Madison, Wis.: Soil Science Society of America.

Table 3.2. Surface Areas and CEC Values of Some Common Clay Minerals

	Specific Surface Area (m^2/g)	CEC (cmoles/kg)
Kaolinite[a]	5–20	1–15
Illite	80–150	10–40
Vermiculite	300–500	100–150
Smectite	700–800	70–120

[a]There is evidence that the higher values of CEC and surface area are due to mineral impurities in kaolinite.

Source: Modified from L. D. Baver, 1948. *Soil Physics.* New York: Wiley.

On the other hand, silicate clays possessing little or no permanent charge (such as kaolinite), as well as oxides and allophanic minerals, have much lower CEC values. For these minerals, the "edge" sites, described in Table 3.3, assume much greater importance. The *silanol* group is the most likely edge site to adsorb cations at typical soil pH values (< 7), because the dissociation of this group:

$$\equiv Si-OH = \equiv Si-O^- + H^+ \tag{3.1}$$

is believed to occur to some extent even at low pH. This group attracts metal ions electrostatically, forming a bond according to the reaction

$$\equiv Si-O^- + M(H_2O)_y^{n+} = \equiv Si-O^- \cdot\cdot\cdot\cdot M(H_2O)_y^{n+} \tag{3.2}$$

Metal ions on these sites are exchangeable.

Other common edge sites are associated with dangling $-Fe-OH_2$ and $-Al-OH_2$ groups, which are weakly acidic. For these sites, dissociation to form cation exchange sites occurs only at high pH:

$$>Fe-OH_2^{+1/2} = >Fe-OH^{-1/2} + H^+ \tag{3.3}$$

and it is unlikely that these sites contribute significantly to the cation exchange properties of most soils. Nevertheless, the sites are able to complex metals inclined toward covalent bonding:

$$>Fe-OH^{-1/2} + M(H_2O)_y^{n+} = >Fe-O-M(H_2O)_{y-1}^{(n-3/2)+} + H^+ \tag{3.4}$$

This is termed *inner-sphere* or *coordination* bonding because the metal ion is bonded directly to the surface group without any intervening water molecules. Bonding of this type has features quite different from that of electrostatic bonding to permanent charge sites; it is more specific (depends on the identity of the metal ion and the surface group), directional, and less reversible. In other words, it has the characteristics of covalent bonding. This type of bonding will not be further considered in this chapter because it is not a cation exchange process, but rather falls under the broad category of "strong" adsorption (chemisorption), which will be dealt with in Chapter 4.

Table 3.3. Classification of Mineral Surface Adsorption Sites (M^{3+} = Al, Fe; M^{2+} = Mg, Fe)

Structure	Type	Affinity for H^+	Metal Bonding	Examples
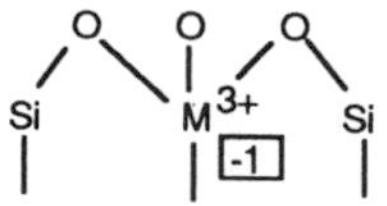	Tetrahedral, permanent	Low	Medium-range, electrostatic, fairly nonspecific	Some smectites, all vermiculites
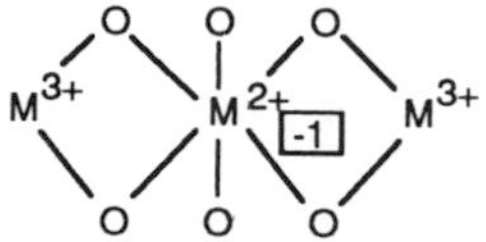	Octahedral, permanent	Low	Long-range, electrostatic, nonspecific	Some smectites
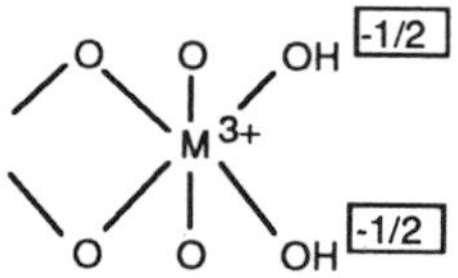	"Edge," pH-dependent (terminal OH)	High	Partially covalent, very specific	Fe oxides, Al oxides, layer silicate edges
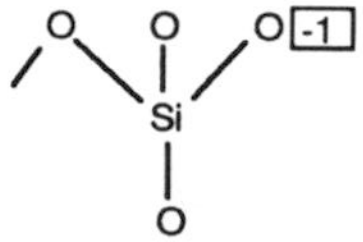	"Edge," pH-dependent (terminal O)	Moderate	Largely electrostatic, fairly specific	Silica, allophane, layer silicate edges

3.2. CATION EXCHANGE ON CLAYS WITH PERMANENT CHARGE

3.2a. Theory of Exchange Equations and Ion Selectivity

Although ion exchange is not a chemical reaction in the usual sense—the bonds broken and formed are long-range electrostatic bonds of low energy—the exchange process is usually written in the formalism of a chemical reaction. For example, the exchange of Mg^{2+} ions from a layer silicate clay surface by $CaCl_2$ can be expressed as

$$CaCl_2\,(aq) + MgX_2\,(s) = MgCl_2\,(aq) + CaX_2\,(s) \tag{3.5}$$

where (aq) and (s) refer to the aqueous and solid (exchanger) phases, respectively, and Cl^- and X^- denote the aqueous chloride anion and negatively charged exchange site. The ion exchange equilibrium constant, K_{eq}, can then be expressed as

$$K_{eq} = \frac{(MgCl_2) \cdot (CaX_2)}{(CaCl_2) \cdot (MgX_2)} \tag{3.6}$$

where the round brackets denote chemical activities (see Chapter 1 for a discussion of activity). If the solid phases MgX_2 and CaX_2 are treated as pure solids, their chemical activities can be assigned values of unity by convention. This would simplify the equilibrium expression to

$$K_{eq} = \frac{(MgCl_2)}{(CaCl_2)} \tag{3.7}$$

with the result that, as long as <u>any</u> Mg^{2+} or Ca^{2+} ions were present on exchange sites, the ratio of $MgCl_2$ and $CaCl_2$ activities in solution would remain constant. Early experiments with clays revealed that this was clearly not the case, but that the activity, or "active mass," of MgX_2 and CaX_2 depended on the quantity of Mg^{2+} and Ca^{2+} adsorbed on the clay. Thus, equation 3.6 was seen to be the appropriate form for the exchange equation, but the question that immediately arose was how to mathematically describe and measure the "active masses" of the Ca^{2+} and Mg^{2+} cations associated with the exchanger. In other words, how does one quantify (MgX_2) and (CaX_2) in equation 3.6?

One possible solution would be to regard the mixture of two ions, Ca^{2+} and Mg^{2+}, adsorbed on the clay as the two-dimensional equivalent of a homogeneous mixture of two gases, A and B. For gases the activity is expressed by the vapor pressure, P; if the gas mixture is *ideal,*[1] then the partial pressure, P_A and P_B, of the two gaseous components can be shown to be proportional to the mole fractions, x_A and x_B, of the components in the mixture:

$$P_A = x_A P_A^0 \qquad P_B = x_B P_B^0 \tag{3.8}$$

This is the expression of Raoult's law for ideal gases, where P_A^0 and P_B^0 are the vapor pressures of the pure (unmixed) gases (see Chapter 1).

A similar type of expression is found for the activities of ions in an ideal *solid*

1. Ideal gas mixtures are ones in which the molecules do not interact sufficiently to produce significant deviation in the ideal gas law ($PV = nRT$). Ideal aqueous solutions are ones in which the dissolved ions interact so weakly that ionic concentrations can be equated to ionic activities.

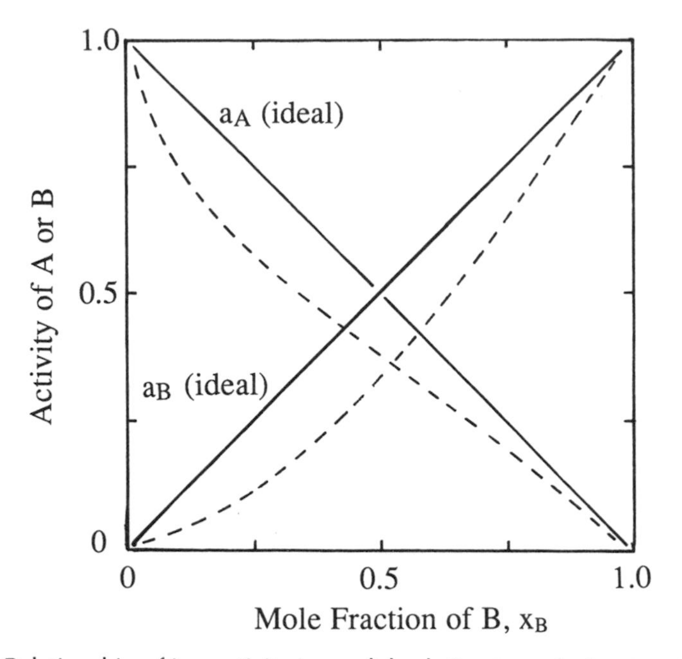

Figure 3.1. Relationship of ion activity in a solid solution to mole fraction. Solid lines represent the ideal case; broken lines depict activities resulting from nonideal behavior.

solution. Thus the activities, a_i, of cations A and B in the solid of composition $A_{1-x}B_xY$ (where Y is an anion and $x \leq 1$) are given by their mole fraction in the solid[1]:

$$a_A = 1 - x \qquad a_B = x \tag{3.9}$$

The relationship of ion activity to mole fraction is shown in Figure 3.1.

In summary, then, either the gas or the solid ideal solution model applied to a mixture of exchange cations would convert equation 3.6 to an ion exchange equation of the form

$$K_S = \frac{(MgCl_2) \cdot M_{Ca}}{(CaCl_2) \cdot M_{Mg}} \tag{3.10}$$

where M_i symbolizes mole fractions of exchange cations, defined in terms of molar quantities of adsorbed ions, $[CaX_2]$ and $[MgX_2]$:

$$M_{Ca} = \frac{[CaX_2]}{[MgX_2] + [CaX_2]} \tag{3.11}$$

1. In ideal solid mixtures, the linear proportionality of activity to mole fraction is a direct consequence of the fact that the entropy of mixing, S_M, is given by an expression of the form: $S_M = x \ln x + (1 - x) \ln (1 - x)$. This equation reflects the number of ways of incorporating x moles of ion B and $(1 - x)$ moles of ion A into the solid matrix, that is, the degree of disorder of the component ions in the solid.

$$M_{Mg} = \frac{[MgX_2]}{[MgX_2] + [CaX_2]} \tag{3.12}$$

This form of exchange equation, which equates the "active mass" of adsorbed cations to the mole fraction of these cations on the exchange sites, is often referred to in the soil chemistry literature as the *Vanselow equation.* Ion exchange equations of this form are the mathematical expression of the important hypothesis of ion exchange, that:

> *The active mass, or activity of exchangeable ions is measured, not by the* quantity *of these ions in the adsorbed state, but rather by the fraction of exchange sites that they occupy.*[1]

The soluble anion is usually not explicitly considered in cation exchange equations, because the activities of electrolytes can be expressed in terms of the activities of their constituent ions, that is,

$$(MgCl_2) = (Mg^{2+})(Cl^-)^2 \tag{3.13}$$

$$(CaCl_2) = (Ca^{2+})(Cl^-)^2 \tag{3.14}$$

When these expressions are substituted into equation 3.10, the activity of Cl^- in solution is seen to cancel out.

The simple model of exchange equilibrium described by equation 3.10 provides a reasonably accurate description of exchange between ions of equal charge (i.e., $A^{n+}-B^{n+}$ exchange where $n = 1, 2,$ or 3). Even in these cases, however, some nonideal behavior is usually detected from the nonconstancy of the value of K_S in equation 3.10. The value of K_S tends to vary as a function of the fraction of exchange sites occupied by ion A or B. This behavior is equivalent, in the solid solution analogy, to the appearance of additional entropy or energy terms that arise because of the inability of ions A and B to completely substitute for one another in the solid without introducing some structural strain or site selectivity. Only for ions of closely matched radius and charge can the resulting solid solution be expected to display ideal behavior. Consequently, most real solid solutions have ion activity functions more like the broken lines than the solid lines shown in Figure 3.1. In a similar way, few cation exchange systems are truly ideal if adherence to equation 3.10 is used as the criterion of ideality. Nonideal behavior is then accounted for by defining the activities of adsorbed cations in terms of *corrected* mole fractions:

$$(CaX_2) = f_{Ca} M_{Ca} \qquad (MgX_2) = f_{Mg} M_{Mg} \tag{3.15}$$

where the correction factors, f_i, can be considered somewhat analogous to the activity coefficients used to correct ionic concentrations to activities in nonideal solutions. The K_S value of equation 3.10 is then recognized to be a *selectivity coefficient* that is not strictly constant for different values of M_{Ca}, but introduction of the correction factors produces the expression

$$K_E = \frac{(MgCl_2) \cdot f_{Ca} M_{Ca}}{(CaCl_2) \cdot f_{Mg} M_{Mg}} = K_S \cdot \frac{f_{Ca}}{f_{Mg}} \tag{3.16}$$

1. One consequence of this hypothesis is that ion exchange equilibrium should be insensitive to the concentration of exchanger (clay) in suspension. This has not been clearly established by experiment but is widely assumed.

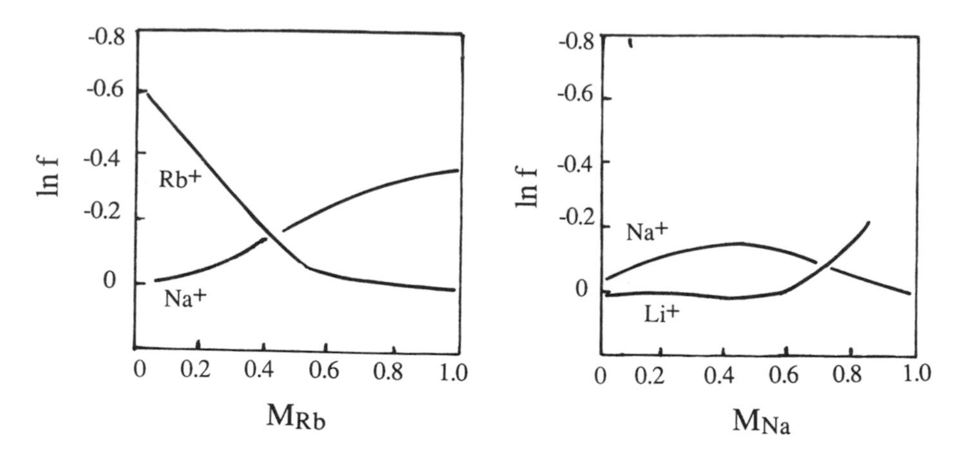

Figure 3.2. Dependence of *f*-values of adsorbed cations on the mole fraction of exchange sites occupied by Rb^+ and Na^+ in the Na^+/Rb^+ and Li^+/Na^+ exchange reaction on smectite. (Adapted from R. G. Gast. 1969. Standard free energies of exchange for alkali metal cations on Wyoming bentonite. *Soil Sci. Soc. Am. Proc.* 33:37–41; 1972. Alkali metal cation exchange on Chambers montmorillonite. *Soil Sci. Soc. Am. Proc.* 36:14–19.)

so that, by definition, K_E is a constant, equivalent to K_{eq} in equation 3.6. The nonideality is expressed within the ratio of correction factors, f_{Ca}/f_{Mg}.[1]

In general, the f_i values are a function of the cation composition of the exchanger. From the point of view of the ideal solid solution model, variation in f_i quantifies factors that contribute to nonideal behavior. These might be electrostatic cation-cation repulsion, cation segregation (see discussion later in this chapter), or any other energy or entropy contribution not allowed for in the description of ideal solutions. Intuitively, one might expect that cations with similar properties would display near-ideal exchange behavior, while cations with very different properties would deviate from ideality. This expectation is borne out, as shown in Figure 3.2 and 3.3, as the f_i values and selectivity coefficient K_S are nearly constant and close to unity for Li^+-Na^+ exchange on smectite over a range of exchange cation composition. Conversely, Na^+-Rb^+ exchange on the same clay has a larger value of K_S, indicating strong preference for Rb^+ over Na^+ and a clear tendency for this preference to diminish at higher levels of Rb^+ on exchange sites. The *f*-values, plotted as functions of the exchange ion composition in Figure 3.2, show that the *f*-value of Rb^+ adsorbed on the clay exchange sites at low levels ($M_{Rb} < 0.1$) is much less than 1.0, increasing toward 1.0 at higher levels. A similar pattern is seen for the *f*-value of adsorbed Na^+. The fact that both f_{Rb} and f_{Na} approach 1.0 for the Rb^+-saturated and Na^+-saturated clays is not coincidence; it stems from the arbitrary choice of these saturated end members as reference states with *f*-values assigned the value of 1.0. Without this convention, the selectivity coefficient could not be interpreted in terms of variability in f_{Rb} and f_{Na} independently, but only in terms of the activity *ratio* f_{Rb}/f_{Na}. Therefore, the absolute *f*-values plotted in Figure 3.2 are arbitrary. Nevertheless, the relative variation and asymmetry of these *f*-value functions provide an instructive interpretation of the variation

1. Experimentally, f_{Ca} and f_{Mg} cannot be separately quantified without making arbitrary assumptions about the activities of reference states of adsorbed cations.

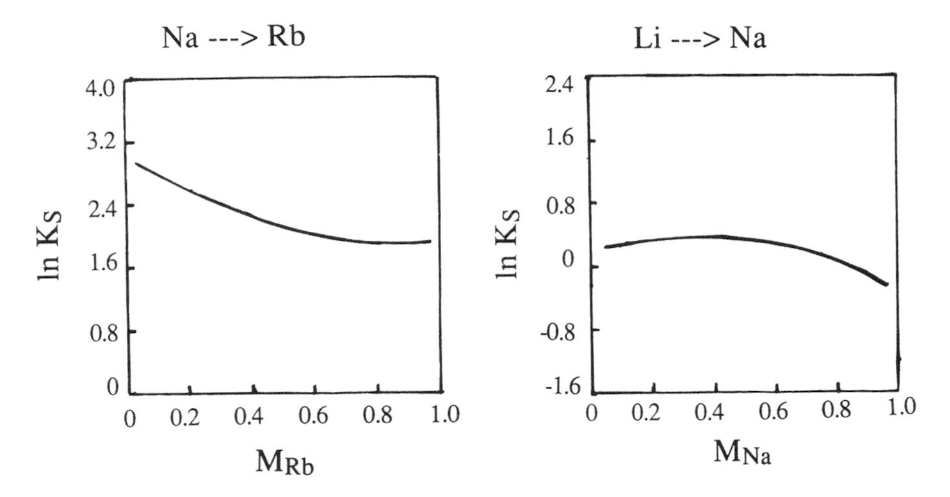

Figure 3.3. Dependence of the selectivity coefficient, K_S, on the mole fraction of exchange sites occupied by Rb^+ and Na^+ in the Na^+/Rb^+ and Li^+/Na^+ exchange reaction on smectite. (Adapted from R. G. Gast. 1969. Standard free energies of exchange for alkali metal cations on Wyoming bentonite. *Soil Sci. Soc. Am. Proc.* 33:37–41; 1972. Alkali metal cation exchange on Chambers montmorillonite. *Soil Sci. Soc. Am. Proc.* 36:14–19.)

in K_S that is apparent in Figure 3.3. The higher selectivity at low Rb^+ levels on the exchange sites favors Rb^+ adsorption, and is interpreted as a lowered "activity coefficient" of the adsorbed Rb^+ when these ions occupy a small fraction of the sites. The actual mechanism producing this nonideal behavior is not known, but could involve a nonideal entropy of mixing between Rb^+ and Na^+, two ions with very different radii and hydration energies.

Na^+-Li^+ exchange, on the other hand, reveals more nearly ideal behavior, with a K_S value close to unity that varies relatively little with exchange ion composition. As a result, the f_{Na} and f_{Li} values calculated from equation 3.16 are near unity and fairly insensitive to changes in the exchanger (clay surface) composition. Because Na^+ and Li^+ are small ions with fairly high hydration energies, their near-ideal exchange behavior may reflect their similarity as well as their tendency to form weak (outer-sphere) electrostatic bonds with the silicate surface.

This discussion points to a second important hypothesis of ion exchange, that qualifies the first one:

> *The active mass of strongly adsorbed exchangeable cations on permanent-charge clays is lower than that estimated from the mole fraction of these cations on the exchange sites when ideal behavior is assumed.*

This principle becomes important when thinking about the factors that limit uptake of metal ions by plant roots or leaching of these ions through soils.

Generally, it is the cations with the largest ionic radii and lowest hydration energies that adsorb most strongly on permanent charge sites of clay minerals. This principle is clearly demonstrated by the exchange results of Table 3.4. Preference for the

Table 3.4. Cation Exchange Data for the Replacement of Na^+ by Other Alkali Metals on Smectite

Replacing Ion (M^+)	Radius (Å)	Hydration Energy (kcal/mol)	K_S^a	Reaction Heat (ΔH) (kcal/mol)
Li^+	0.60	124	0.71	+0.019
K^+	1.33	77	2.9	−1.16
Rb^+	1.48	72	7.1	−1.92
Cs^+	1.69	66	29	−2.65

[a] K_S here has the mathematical form $K_S = [(Na^+) \cdot M_{M+}]/[(M^+) \cdot M_{Na+}]$, and the value reported is for a fixed exchanger composition of $M_{Na+} = 0.5$.

Source: R. G. Gast. 1969. Standard free energies of exchange for alkali metal cations on Wyoming bentonite. *Soil Sci. Soc. Am. Proc.*33:37–41; 1972. Alkali metal cation exchange on Chambers montmorillonite *Soil Sci. Soc. Am. Proc.* 36:14–19.

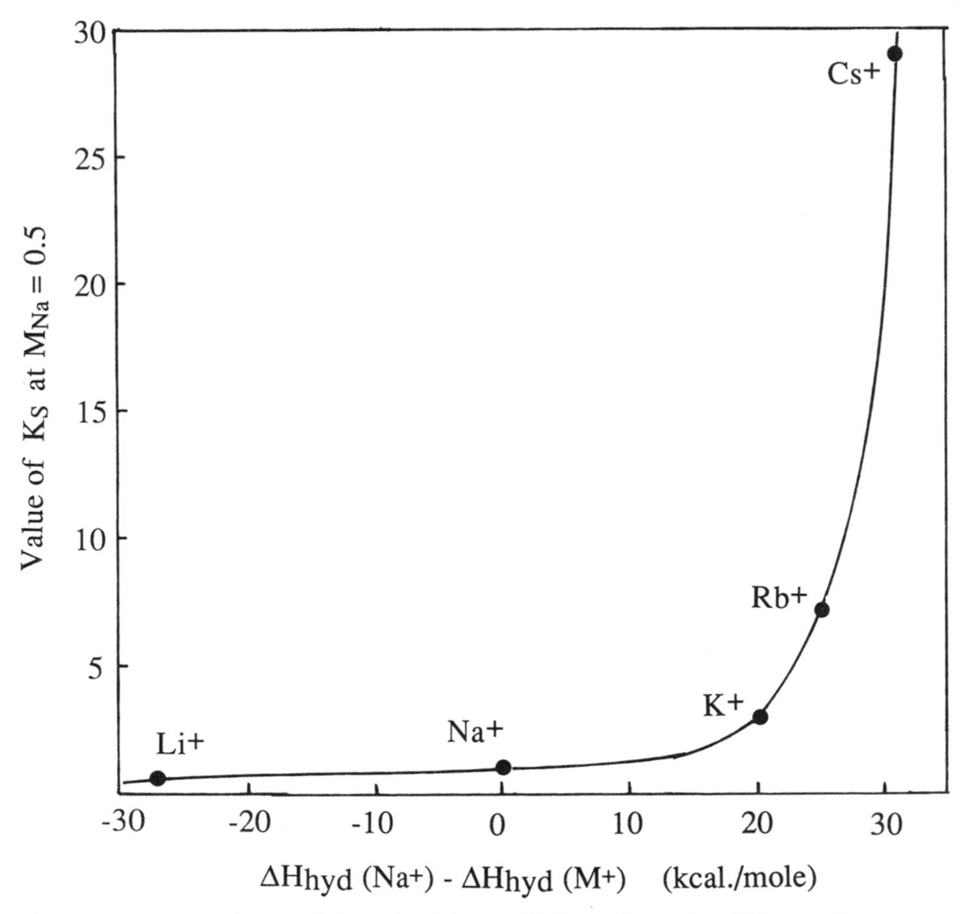

Figure 3.4. Dependence of the selectivity coefficient, K_S, on the difference in hydration energy of the two alkali metal cations undergoing exchange.

adsorbing cation becomes particularly strong when the difference in the hydration energies of the two exchanging cations is more than 20 kcal/mol. This point is illustrated by the relationship shown in Figure 3.4.

3.2b. Limitations in Extending the Ideal Exchange Equation

As long as the ions involved in exchange have the same charge, an exchange equation similar to equation 3.10, which describes ideal mass-action behavior, is adequate. For example, nearly ideal behavior is observed for $Ca^{2+}-Mg^{2+}$ ion exchange on smectite, as might be expected from the similar chemical properties of these two ions. In other words, equation 3.10 has a nearly constant value of K_S close to unity regardless of the fraction of exchange sites occupied by Ca^{2+} or Mg^{2+}. This behavior is expected in all cases of exchange between ions of equal charge if the ions are fairly closely matched in radius. However, care must be taken in extending such a simple equation to describing the behavior of clays having limited expandability. Cations such as K^+, Cs^+, or NH_{4+} may become trapped in illites and vermiculites, so that apparent K_S values are far from unity. Collapse of the interlayer regions on these weakly hydrating ions can produce high selectivities favoring their adsorption.[1] The K^+ ion fits snugly into the hexagonal cavity that is built into layer silicate surfaces and becomes trapped or "fixed" in a nonexchangeable form. Ion exchange equations cannot adequately quantify this kind of behavior.

Even $Mg^{2+}-Ca^{2+}$ exchange on vermiculite is anomalous relative to "normal" ion exchange. The K_S value, near unity at low levels of Mg^{2+} on Ca^{2+}-vermiculite, increases markedly at high levels of Mg^{2+} adsorption ($M_{Mg} > 0.5$). The strong preference for Mg^{2+} may be related to the different c-axis spacings of the wet Ca^{2+} and Mg^{2+}-saturated vermiculites, which are 15.4 and 14.6 Å, respectively. When ion exchange sites have dimensions that are roughly equal to the size of the hydrated exchanging ions, ions may be excluded on the basis of their hydrated size, and extreme values of the selectivity coefficient can result.

3.2c. The Eisenman Energy Model of Cation Exchange

A descriptive picture of cation exchange, attributed to Eisenman (1961), can suggest a reason for the different behavior of ions of different radius. The electrostatic attraction energy, E_{att}, between an adsorbed cation, A^+, and the surface charge site is inversely proportional to the sum of r_s, the effective radius of the charge site, and r_A, the radius of cation A^+:

$$E_{att} \propto \frac{e^2}{(r_s + r_A)} \tag{3.17}$$

where e is the electronic charge unit. The exchange cation at the charge site is depicted in Figure 3.5a, but the direct contact shown here would require that energy

1. However, if insufficient K^+ is present in solution to occupy a significant fraction of exchange sites and initiate this collapse, low affinity of K^+ for expanded vermiculite may be observed.

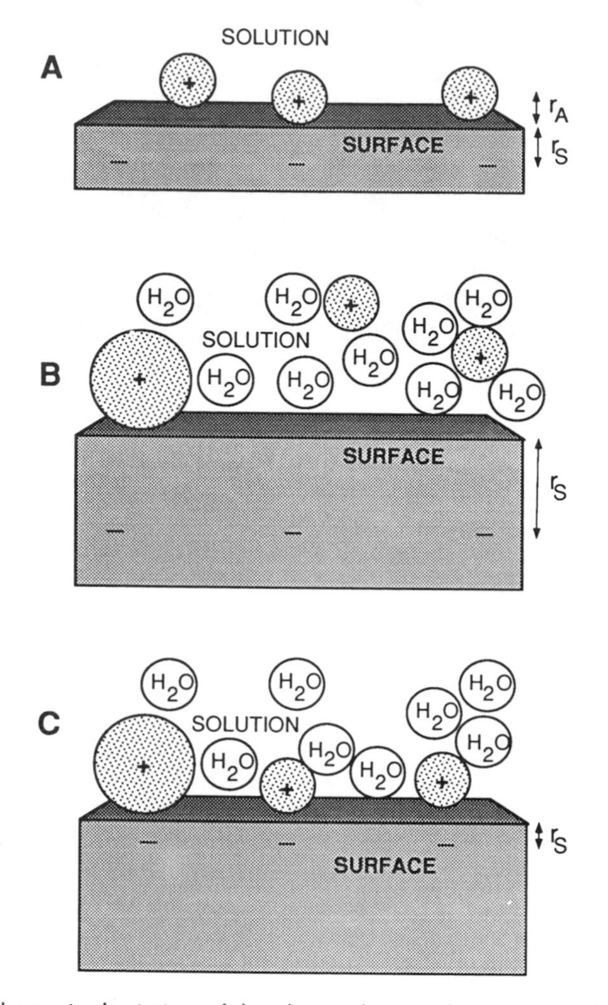

Figure 3.5. Schematic depiction of the clay surface-exchange cation interaction in the case of (a) no water molecules present, (b) water present on a "weak field" exchanger, and (c) water present on a "strong field" exchanger.

be expended to "push" aside the water molecules separating the cation from the surface. Since this water is retained by hydration of the site and the cation, the energy that must be expended should be proportional to the energy of hydration of the site, E_s, and of the cation, E_A. Therefore the total energy change, E_{tot}, resulting from the movement of ion A^+ from solution into contact with the surface is given by:

$$E_{tot} \propto \left\{ -\left(\frac{e^2}{r_s + r_A} \right) + E_s + E_A \right\} \tag{3.18}$$

The adsorption of ion A^+ must be accompanied by the desorption of ion B^+, so that the overall change in energy for the exchange of ion B^+ by ion A^+ is

$$\Delta E_{\text{tot}} \propto \left\{ \left(\frac{e^2}{r_s + r_B} - \frac{e^2}{r_s + r_A} \right) - (E_B - E_A) \right\} \tag{3.19}$$

where r_B and E_B are the radius and hydration energy of ion B.

From this simple electrostatic model, two limiting cases become evident. If the structural charge is located well beneath the surface (as is the case for layer silicate clays with isomorphous substitution in the octahedral layer), then the distance between the exchange cation and the site of negative charge ($r_s + r_A$ or $r_s + r_B$) must be large, so that the electrostatic term in equation 3.19 becomes small. This means that the energy of exchange, ΔE_{tot}, is determined largely by the difference in hydration energies of ions A^+ and B^+. This is the *weak field* case, depicted in Figure 3.5b, in which strongly hydrating cations such as Li^+, Ca^{2+}, and Mg^{2+} are not attracted strongly enough to the surface to overcome the forces of hydration. Only weakly hydrating cations such as K^+ and Cs^+ are expected to come into direct contact with the surface. Such a picture of cation adsorption is appropriate for smectites, with the result that the energy of the exchange reaction is dependent on the hydration properties of the exchanging cations. This dependence is clear from Table 3.4 and Figure 3.4, where the weakly hydrating cations spontaneously displace strongly hydrating cations from the clay, and energy is released in the process.[1]

The other possible limiting case for ion exchange occurs if the structural charge is at or near the solid surface, so that r_s is small relative to r_A and r_B. This means that the electrostatic term of equation 3.19 is *not* negligible and may contribute more to ΔE_{tot} than the hydration term. This is called the *strong field* case, diagrammed in Figure 3.5c, in which all cations are in direct contact (inner-sphere association) with the clay surface. Although no permanent-charge clays are known to possess strong-field behavior, studies of Fe oxide minerals, which develop negative charge by the dissociation of bound H_2O molecules:

$$> \text{Fe}{-}\text{OH}_2 = > \text{Fe}{-}\text{OH}^- + \text{H}^+ \tag{3.20}$$

have revealed a preference for smaller cations in the order

$$\text{Li}^+ > \text{K}^+ > \text{Cs}^+$$

which is consistent with strong field exchange behavior. This order of selectivity is the opposite of that observed on permanent-charge layer silicate clays, and may reflect the small effective radius (r_s) of the charged hydroxyl group at the oxide surface (further discussion of surface charge and ion exchange on oxidic surfaces is found later in this chapter).

The Eisenman model, while conceptually useful, is incomplete because it does not take into account entropy (disorder) changes associated with exchange, which contribute to the free energy of the reaction. For ions of equal charge, these entropy changes are often quite small, but for exchange between ions of different charge, the change in entropy may be substantial, driving the exchange in one direction or the other. The sources of increasing entropy in an exchange reaction are changes in disorder of water molecules and cations, both adsorbed and in solution.

1. At constant temperature and pressure, spontaneous processes occur with a decrease in the free energy, ΔG, of the closed exchange system.

3.2d. Exchange Between Ions of Unequal Charge—the Concentration-Charge Rule

The previous section discussed the special case of exchange between cations of equal charge. However, the general form of the cation exchange equation allows for exchange between A^{n+} and B^{m+}, where $n+$ and $m+$ may not be equal:

$$nA^{m+}X_m\,(s) + mB^{n+}Cl_n\,(aq) = mB^{n+}X_n\,(s) + nA^{m+}Cl_m\,(aq) \qquad (3.21)$$

where X denotes one mole of exchange sites and Cl symbolizes the chloride anion chosen arbitrarily as the nonadsorbing counterion. The selectivity coefficient is then expressed by the equation

$$K_S = \frac{N_B^m}{N_A^n} \cdot \frac{(ACl_m)^n}{(BCl_n)^m} \qquad (3.22)$$

where N_A and N_B symbolize the fraction of exchange sites occupied by cations A and B.[1] By definition in a two-cation exchange system, $N_A + N_B = 1$. For the common examples of divalent-monovalent ion exchange, equation 3.22 becomes

$$K_S = \frac{N_{B2+}}{N_{A+}^2} \cdot \frac{(ACl)^2}{(BCl_2)} \qquad (3.23)$$

If the mixture of exchange cations on the clay surface behaves ideally, K_S should remain essentially constant over a range of BCl_2 and ACl electrolyte concentrations. If the electrolyte concentration is lowered by dilution of the clay suspension with pure water, then the activity ratio in equation 3.23, $(ACl)^2/(BCl_2)$, will decrease, and N_{B2+}/N_{A+}^2 must increase to maintain a constant K_S. This shift in favor of adsorption of the higher-charge cation at lower electrolyte concentrations is called the *concentration-charge effect,* and is a necessary consequence of any exchange process in which the total number of cations on the solid is changed.[2]

The concentration-charge effect can be derived from ion exchange equation 3.23. The equation is first simplified by the principle that electrolyte activities can be expressed in terms of products of the activities of the constituent ions. Thus, $(ACl) = (A) \cdot (Cl)$ and $(BCl_2) = (B) \cdot (Cl)^2$. The solution activity ratio, $(ACl)^2/(BCl_2)$, is then equated to $(A)^2/(B)$, and equation 3.23 can be rewritten without the explicit inclusion of the solution anion:

$$K_S = \frac{N_{B2+}}{N_{A+}^2} \cdot \frac{(A^+)^2}{(B^{2+})} \qquad (3.24)$$

The mole fractions of A^+ and B^{2+} in solution are

$$M_A = \frac{m_A}{m_A + m_B} = \frac{m_A}{m_T} \qquad (3.25)$$

1. N_A and N_B are equivalent rather than mole fractions. The choice of equivalent or mole fraction for use in the exchange equation is an arbitrary one, but it affects the values of K_S in exchange systems.

2. Replacement of one B^{2+} by 2 A^+ cations has a probability proportional to the chance of finding two A^+ ions in the vicinity of this B^{2+} cation. This probability is given by $[A^+]^2$.

$$M_B = \frac{m_B}{m_A + m_B} = \frac{m_B}{m_T} \tag{3.26}$$

where m_A and m_B are the number of moles of A^+ and B^{2+} in the solution phase, and m_T is the sum of moles of A^+ plus B^{2+} in solution. But, if solution concentrations of ions can be approximated to activities, then $(A^+) = [A^+] = m_A/V$, and $(B^{2+}) = [B^{2+}] = m_B/V$, where V is the volume of the solution phase and square brackets denote concentrations (moles/liter). Substituting these expressions into equation 3.24 we obtain

$$K_S = \frac{N_{B2+}}{N^2_{A+}} \cdot \frac{(m_A/V)^2}{(m_B/V)} \tag{3.27}$$

Now, by applying the definitions of m_A and m_B from equations 3.25 and 3.26, equation 3.27 is rewritten:

$$K_S = \frac{N_{B2+}}{N^2_{A+}} \cdot \frac{M^2_A}{M_B} \cdot \frac{m_T}{V} \tag{3.28}$$

But since $m_T/V = M_T$, where M_T is the total molarity of the solution, the selectivity expression becomes

$$K_S = \frac{N_{B2+}}{N^2_{A+}} \cdot \frac{M^2_A}{M_B} \cdot M_T \tag{3.29}$$

This equations reveals that, for a constant molar ratio of A^+ and B^{2+} in solution, as the molarity of the solution phase (M_T) decreases, the fraction of exchange sites occupied by B^{2+} increases. This is the mathematical expression of the concentration-charge rule, which holds true as long as K_S is fairly constant over a range of cation composition on the exchanger surface. The equation also shows that, for exchange between ions of unequal charge, the occupancy of the exchange sites is not uniquely determined by the mole *ratio* of cations in solution; the total molarity (or normality) of the solution must also be specified.

From the general nature of this derivation, it is also apparent that the concentration-charge rule applies to all cation pairs involved in exchange as long as their ionic charges are different. The higher charge cation will always be the one that is preferentially adsorbed as the cations in solution become more dilute.

In Figure 3.6, the predicted adsorption of Ca^{2+} on a Na^+-saturated clay at three different total concentrations of electrolyte, calculated from equation 3.23, is plotted. Although no intrinsic selectivity of the exchange sites for Ca^{2+} was assumed in generating these exchange isotherms (K_S was assigned a value of 1), it is clear that low salt concentrations produce nearly Ca^{2+}-saturated exchange sites even when Na^+ is the predominant cation in solution. The consequence of this concentration-charge effect in soils is that, in humid climates that produce frequent leaching of salts from soils such that the electrolyte concentration remains low, the cations with higher charge (Ca^{2+}, Mg^{2+}, Al^{3+}) dominate the exchange sites. This principle is verified by actual data for $Ca^{2+}-Na^+$ exchange on soil, presented in Table 3.5.

In soils of arid climates, higher electrolyte concentrations favor Na^+ and K^+ adsorption on exchange sites. Regardless of climate, periodic cycles of soil wetting and drying are likely to produce wide fluctuations in electrolyte concentrations, with the result that the cationic composition of the soil's exchange sites is not constant.

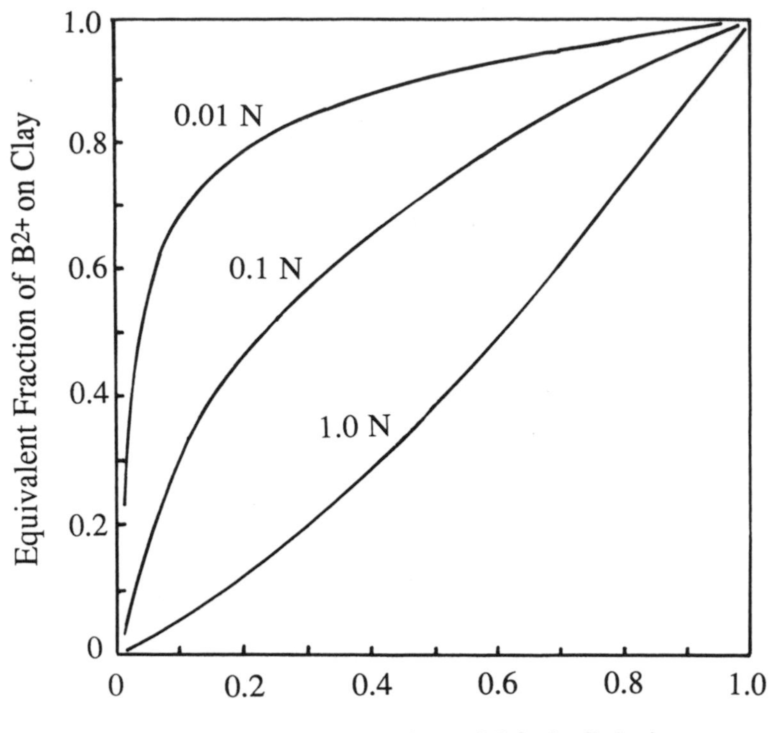

Figure 3.6. Calculated isotherms for divalent cation (B^{2+}) adsorption on a smectite initially saturated with monovalent cations at three electrolyte concentrations in solution.

Table 3.5. Ca^{2+} Exchange by Na^{+} Under Conditions of Fixed Quantities of Soil (100 g) and NaCl (100 mmoles), but Variable Solution Volume

Volume of NaCl Solution (liters)	Ca^{2+}Exchanged(mmoles(+)/liter)
1.0	16.5
2.5	14.5
5.0	12.9
10.0	11.1

Source: W. P. Kelley. 1948. *Cation Exchange in Soils.* Reinhold, N.Y.: American Chemical Society.

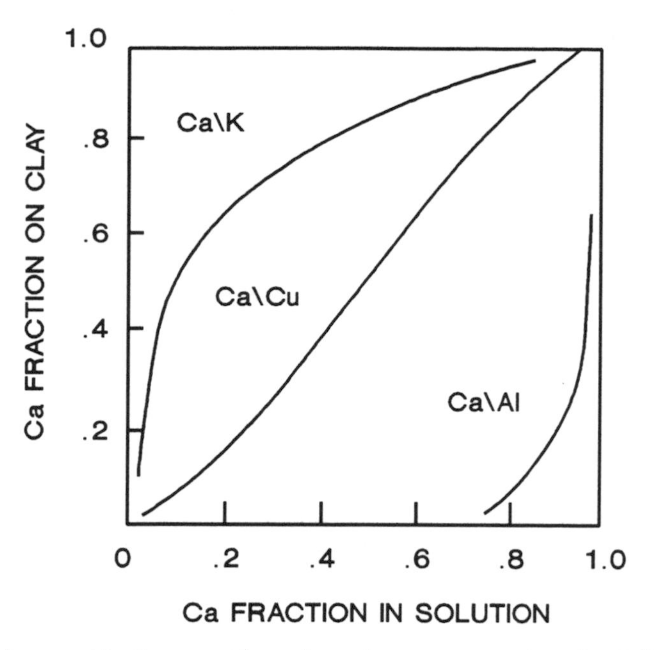

Figure 3.7. Measured isotherms at low electrolyte concentrations for Ca^{2+} adsorption on K^{+}-, Cu^{2+}-, and Al^{3+}-saturated smectites.

3.2e. Exchange Between Ions of Unequal Charge—Thermodynamic Principles

At low salt concentrations, Ca^{2+} is an effective competitor for clay exchange sites occupied by monovalent ions but competes less well for sites occupied by high-charge cations such as Al^{3+}. This fact is illustrated by the exchange results for smectite shown in Figure 3.7. A tempting conclusion from results of this kind is that electrostatic forces control the cation exchange process, explaining the commonly observed order of metal (M) cation preference on clays:

$$M^{3+} > M^{2+} > M^{+}$$

However, studies of the thermodynamic properties of ion exchange have revealed an unexpected result:

> *The spontaneous exchange of low-charge by high-charge cations on clays is an endothermic*[1] *process.*

This principle is confirmed by the data reported in Table 3.6. In these exchange reactions on smectite, NH_4^{+} or K^{+} exchange by divalent metal ions is driven by an increase in entropy, *not* by an energy decrease in the clay-solution exchange system.

1. An endothermic reaction consumes energy, so that the enthalpy change of the reaction, ΔH, is defined to be positive.

Table 3.6. Thermodynamic Data for the Exchange of NH_4+ and K^+ on Smectite by Alkaline Earth Metals

Adsorbing Ion	Displaced Ion	ΔG^0(kcal/mol)	ΔH^0(kcal/mol)	ΔS^0(eu)[a]
Mg^{2+}	$2NH_4^+$	+1.02	+2.79	+5.92
Mn^{2+}	$2NH_4^+$	+1.00	+2.74	+5.83
Ca^{2+}	$2NH_4^+$	+1.00	+2.70	+5.72
Sr^{2+}	$2NH_4^+$	+0.99	+2.20	+4.13
Ba^{2+}	$2NH_4^+$	+0.94	+1.92	+3.30
Ca^{2+}	$2\,K^+$	+1.86	+3.87	+6.7

[a]Entropy units (eu) are cal/degree-mole.

Source: M. Gilbert and R. van Bladel. 1970. Thermodynamics and thermochemistry of the exchange reaction between NH_4+ and Mn^{2+} in a montmorillonite clay. *J. Soil Sci.* 21:38–49; A. T. Hutcheon. 1966. Thermodynamics of cation exchange on clay:Ca-K-montmorillonite. *J. Soil Sci.* 17:339–355.

The endothermic nature of this exchange process on smectites has been verified from the effect of temperature on the equilibrium constant of exchange, K_E. The van't Hoff equation:

$$\frac{d \ln K}{dT} = \frac{\Delta H^0}{RT^2} \tag{3.30}$$

is derived from thermodynamics, predicting that for an endothermic reaction (positive ΔH^0), the equilibrium shifts further in the direction that the reaction is written as the temperature is raised. This is in fact the case for exchange involving cations of unequal charge, with the equilibrium shifting in favor of adsorption of the higher charge ion at higher temperature.

Further confirmation of the endothermic nature of divalent ion exchange of monovalent ions on smectites has resulted from direct measurement of heat uptake or release during exchange.

The increase in entropy, which results when divalent cations displace monovalent cations from the surface into solution, could arise from increased disorder of either cations or water molecules in the exchange system. The more strongly hydrating divalent ions may have greater rotational mobility at the clay surface (more disorder) than the monovalent ions, the latter having some tendency to form ion pairs with the exchange sites. Furthermore, it can be argued that the replacement of monovalent by divalent cations on the clay must produce a gain of entropy because there are more energy-equivalent ways by which the divalent cations could occupy the array of singly-charged exchange sites. This contribution to the free energy of exchange is called *configurational entropy.*

An important contribution to the entropy of exchange is associated, not with the clay, but with the solution phase. The entropy of high-charge cations in aqueous solution is much more negative than that of low-charge cations because of the order imposed on water molecules by strongly hydrating ions. Since the adsorption of high-charge cations from solution involves the simultaneous release of low-charge cations to solution, a large net *increase* in the entropy of the solution phase results. This may be counterbalanced at least in part by increased order of the water retained on the clay as the exchange sites become more fully occupied by high-charge cations.

The interesting conclusion that high-charge cation adsorption occurs favorably as a result of an entropy gain and not a lowering of the energy in the clay-water exchange system appears more reasonable when the geometry of multivalent cations arrayed at the layer silicate clay surface is considered. In Figure 3.8a, the possible arrangement of Ca^{2+} and Na^{+} ions on a single clay platelet is drawn schematically. While Na^{+} ions can approach the individual charge sites fairly closely, Ca^{2+} ions cannot because of their strongly held hydration shells and the fact that smectite clay charge sites are typically separated by 10 to 15 Å. As a result, from the standpoint of electrostatic bonding energy, the Na^{+} form of clay may actually be more stable than the Ca^{2+} form. This would explain the endothermic nature of Na^{+} exchange by Ca^{2+}.

If the clay platelets were to become aligned into packets (called quasi-crystals), the geometry might favor divalent and multivalent cations as suggested by the diagram in Figure 3.8b. In these packets, Ca^{2+} ions are found to occupy the interlayer regions, while Na^{+} ions segregate from Ca^{2+} ions and reside on the external surfaces.

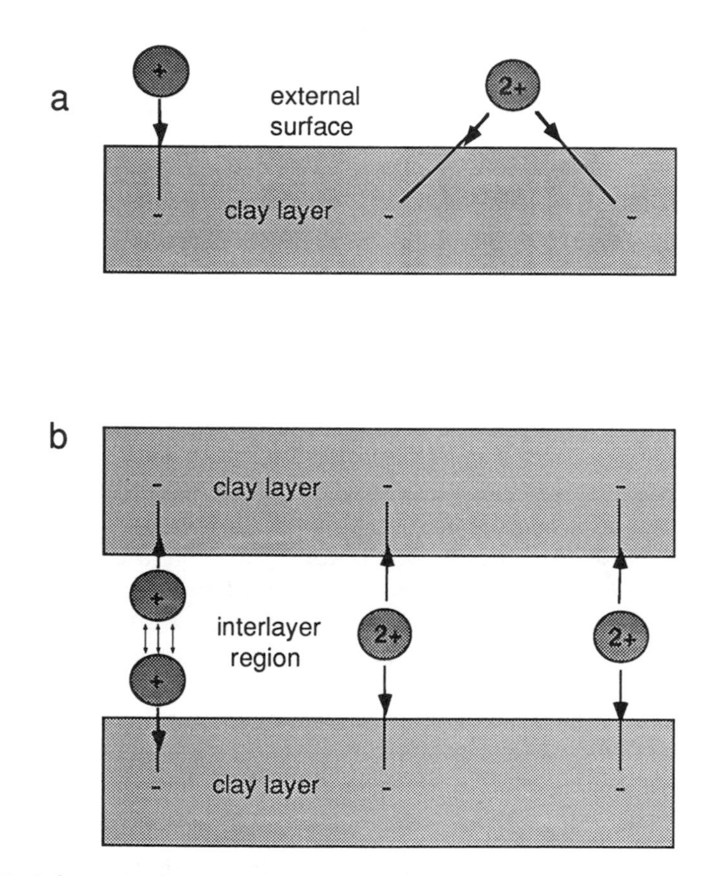

Figure 3.8. Schematic diagram of monovalent (e.g., Na^{+}) and divalent (e.g., Ca^{2+}) exchange cations retained electrostatically on a smectite with octahedral layer charge, showing (a) a single platelet edge-on, and (b) two platelets edge-on, forming an interlayer region.

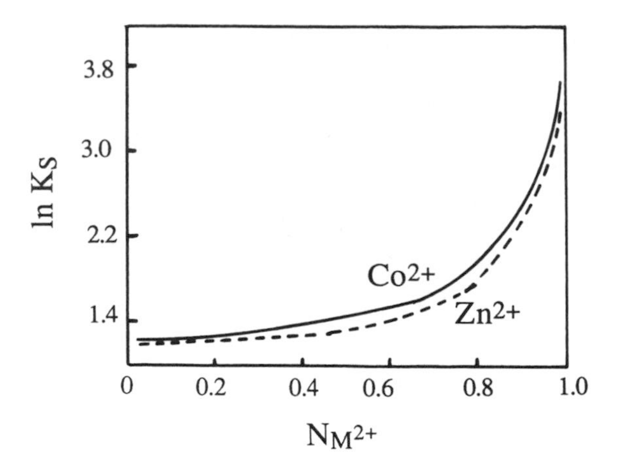

Figure 3.9. Selectivity coefficients for Co^{2+}/Na^{+} and Zn^{2+}/Na^{+} exchange on smectite, plotted as a function of $N_{M^{2+}}$, the fraction of exchange sites occupied by Co^{2+} or Zn^{2+}. (Adapted from Maes et al., 1975. Thermodynamics of transition metal ion exchange in montmorillonite. In S. W. Bailey (ed.), *Proceedings of the International Clay Conference,* Mexico City.)

As the diagram suggests, the fact that the population of monovalent ions needed to occupy the interlayer region and satisfy the CEC is double that of the divalent ions may further discourage the entry of Na^{+} into the more compressed interlayer of the Ca^{2+} clay; electrostatic repulsion between closely spaced Na^{+} ions could be significant.

A common observation with smectites, that the selectivity for divalent cations relative to monovalent cations (as measured by K_S) increases as the divalent cation occupies a greater fraction of the exchange sites, may be a result of the different selectivity displayed by external and interlayer exchange sites toward divalent and monovalent cations. The exchange results of Figure 3.9 reveal that, as Co^{2+} and Zn^{2+} displace more and more Na^{+} from exchange sites, creating packets of platelets from the individual separated platelets of the Na^{+}-saturated clay, the selectivity coefficient shifts more in favor of these divalent ions. It seems that the restricted interlayer thickness of about 10 Å (depicted in Figure 3.10), created by the displacement of Na^{+} from the clay, provides a favorable site for divalent and multivalent cations. This causes exchange cations of different charge to segregate ("demix") to some extent on layer silicate clays, as shown in Figure 3.11. One consequence of quasicrystal formation is that exchange between monovalent and divalent or polyvalent cations is not perfectly reversible, at least on smectites. These clays retain some "memory" of their exchange history in the extent of quasicrystal formation.

3.2f. A Statistical Mechanical Approach to Cation Exchange

The degree of success of the "ideal ion mixture" model of ion exchange, which allows for nonideal behavior by introducing activity coefficients for individual ions in much the same way that the activities of ions in aqueous solutions are adjusted for nonideality,

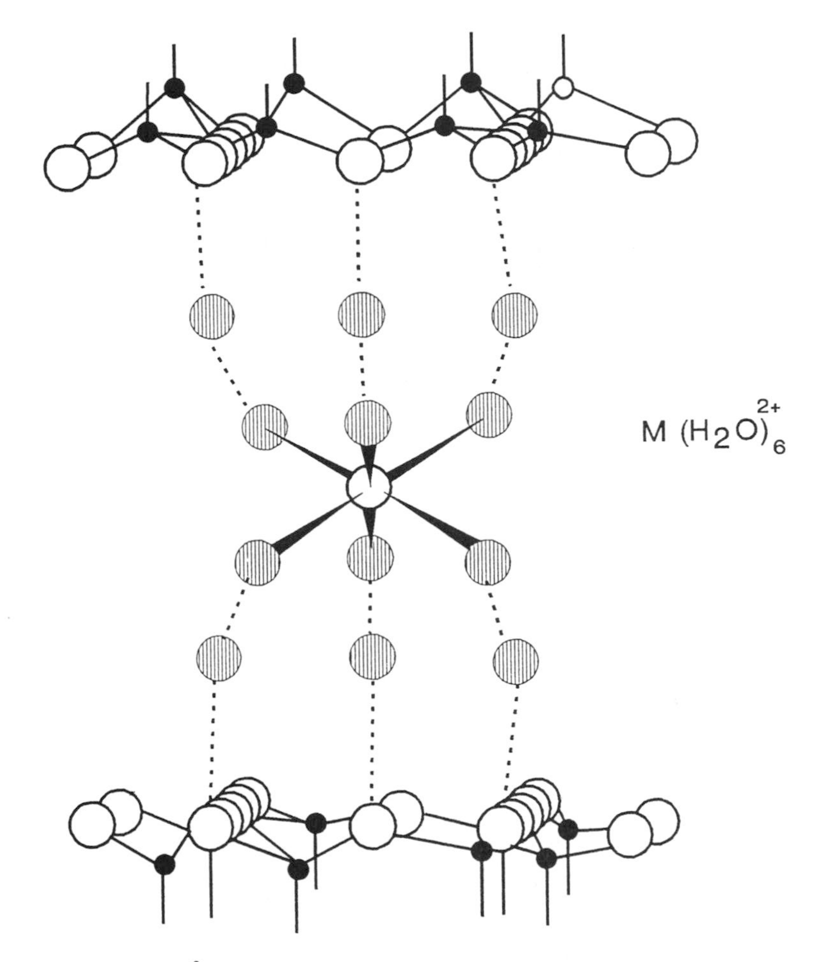

Figure 3.10. The ≈10-Å interlayer of smectite created by the adsorption of hydrated divalent metal ions, $M(H_2O)_6^{2+}$. Shaded circles represent water molecules; open circles depict silicate oxygens.

depends on the extent to which adsorbed ions can be likened to ions in homogeneous solutions or solid matrices. Adsorbed cations in close proximity to clay surfaces experience a direction-dependent (anisotropic) electrostatic force. The "ideal mixture" model, while providing a formalism for expressing cation exchange data in terms of "activity" coefficients (f-values), gives no insight into the physical cause of this apparent deviation from nonideality.

A different approach to ion exchange, using principles of statistical thermodynamics, affords a more penetrating analysis of "ideal" exchange behavior. To the extent that the exchange ions are bound closely to the surface, the entropy decrease associated with this localization and ordering of cations can be factored into the exchange model. This entropy term, referred to as *configurational entropy,* can be calculated for the extreme cases of molecules either confined to discrete sites or possessing free lateral mobility along the surface. The equations given by Adamson (1976), which describe the differential molar configurational entropy for these two cases, are

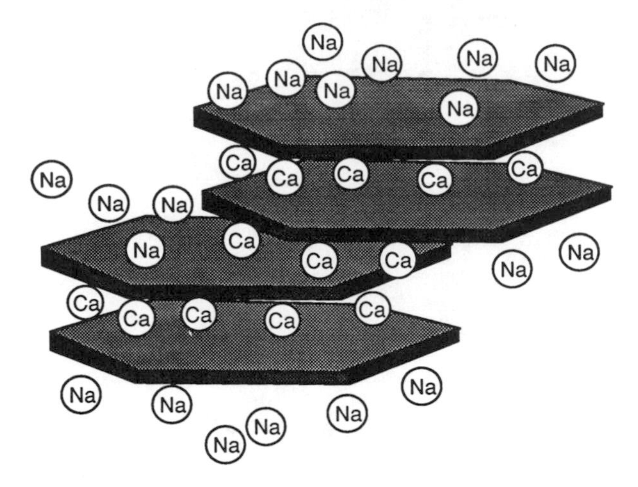

Figure 3.11. Depiction of Na^+/Ca^{2+} cation segregation between internal and external exchange sites of layer silicate clay particles.

$$S_l = -R \ln \left(\frac{\Theta_i}{1 - \Theta_i} \right) \tag{3.31}$$

$$S_m = -R \ln \Theta_i \tag{3.32}$$

where m and l symbolize "mobile" and "localized" ions, R is the gas constant, and Θ_i is the fraction of exchange sites occupied by cation i. These equations apply strictly only for exchange between monovalent ions because localized multivalent ions may be able to occupy exchange sites in more than one way, complicating the description of configurational entropy.

The derivation of equation 3.31 for differential entropy is as follow. Consider S exchange sites on which N_A ions of A^+ and N_B ions of B^+ are localized. In a two-ion exchange system, it is necessary that $N_A + N_B = S$. Now, the total number of different ways that S ions can be arranged on S sites is $S!$ However, of these arrangements, many correspond to interchanges of A^+ ions with A^+ ions and B^+ ions with B^+ ions. These are not distinguishable and must not be counted, and since there are $N_A!$ such arrangements of A^+ and $N_B!$ such arrangements of B^+, the number of distinguishable ways, W, of arranging all ions on the exchange sites is

$$W = \frac{S!}{N_A!\, N_B!} \tag{3.33}$$

Since the entropy of mixing (configurational entropy) is a function of W:

$$\Delta S_{\text{mix}} = k \ln W \tag{3.34}$$

where k is the Boltzmann constant, then

$$\Delta S_{\text{mix}} = k \ln \left\{ \frac{S!}{N_A!\, N_B!} \right\} = k \ln S! - k \ln N_A! - k \ln N_B! \tag{3.35}$$

For large numbers, Stirling's approximation ($\ln x! = x \ln x - x$) can be used to eliminate factorials from the equation:

$$\Delta S_{\text{mix}} = kS \ln S - kS - (kN_A \ln N_A - kN_A) - (kN_B \ln N_B - kN_B) \tag{3.36}$$

Recalling that $S = N_A + N_B$, this expression can be simplified to

$$\Delta S_{mix} = kS \ln S - kN_A \ln N_A - kN_B \ln N_B \tag{3.37}$$

or

$$\begin{aligned}\Delta S_{mix} &= k(N_A + N_B) \ln S - kN_A \ln N_A - kN_B \ln N_B \\ &= -kN_A \ln \left\{\frac{N_A}{S}\right\} - kN_B \ln \left\{\frac{N_B}{S}\right\} \\ &= \frac{-(kS)N_A}{S} \ln \left\{\frac{N_A}{S}\right\} - \frac{(kS)N_B}{S} \ln \left\{\frac{N_B}{S}\right\}\end{aligned} \tag{3.38}$$

If 1 mole of exchange sites is being considered, then S is equal to Avogadro's number and the gas constant, R, can replace kS to give the molar entropy of mixing:

$$\Delta S_{mix} = -R\,\Theta_A \ln \Theta_A - R\,\Theta_B \ln \Theta_B \tag{3.39}$$

where Θ_i symbolizes the mole fractions of adsorbed A^+ and B^+ (i.e., $\Theta_A = N_A/S$, $\Theta_B = N_B/S$). The *differential* entropy of mixing is then obtained by differentiating equation 3.39 with respect to Θ_A or Θ_B:

$$\frac{\partial(\Delta S_{mix})}{\partial \Theta_A} = -R \ln \left\{\frac{\Theta_A}{1 - \Theta_A}\right\} \tag{3.40}$$

This is the origin of equation 3.31, showing how the configurational entropy changes as the ionic composition on the surface changes.

If exchange ions are not localized on the surface, the derivation of configurational entropy would be somewhat different. In this case, the placement of one ion on an exchange site does not reduce the sites available to subsequent ions, and each of the N_A and N_B ions can be placed in S ways. Consequently:

$$W = \frac{S^{N_A} S^{N_B}}{N_A!\, N_B!} \tag{3.41}$$

$$= \frac{S^S}{N_A!\, N_B!} \tag{3.42}$$

Using the same mathematical procedure as that above, the entropy of mixing is given by

$$\Delta S_{mix} = R - R\,\Theta_A \ln \Theta_A - R\,\Theta_B \ln \Theta_B \tag{3.43}$$

which is equivalent to equation 3.39 for localized ions except for the added entropy, R, a measure of the greater degree of motional freedom for delocalized ions. However, the *differential* entropy for delocalized ions, $\partial(\Delta S_{mix})/\partial\Theta_A$, is identical to that given in equation 3.40 for localized ions. This equation differs from the more general equation 3.32, applicable to molecular adsorption, because ion exchange is a special case of adsorption, requiring that all adsorption sites be occupied. It appears, then, that the degree of cation delocalization on the surface does not influence the ion exchange equilibrium *if both ionic species have equal degrees of surface mobility.*

3.2g. A General Statistical Mechanical Model of Cation Exchange

Consider the simple case of $Na^+ - K^+$ exchange on a permanent-charge clay to see how energy and entropy influence the exchange process and cation selectivity. Based on the "ideal mixture" model, the equilibrium exchange expression is

$$K_E = \frac{M_{K+} \cdot (Na^+)}{M_{Na+} \cdot (K^+)} \tag{3.44}$$

where M_i symbolizes the molar quantities of adsorbed ions. The form of this equation can be confirmed as correct from the following derivation. The differential entropy term for K^+ ions localized on exchange sites is given by

$$S_{K^+} = -R \ln \left\{ \frac{M_{K+}}{1 - M_{K^+}} \right\} \tag{3.45}$$

(see equation 3.40 and the preceding derivation for the statistical mechanical origin of equation 3.45). This entropy change is referred to as the *entropy of mixing* because it is the mixing together of K^+ and Na^+ on the exchange sites that generates the entropy described by equation 3.45 and increases disorder in the system.

If the bonding energy change per mole of K^+ adsorbed (differential molar energy) is given the symbol E_{K^+}, then the change in free energy of the clay per mole of K^+ adsorbed, G_c, is

$$G_c = E_{K^+} - TS_{K^+} = E_{K^+} + RT \ln \left\{ \frac{M_{K^+}}{1 - M_{K^+}} \right\} \tag{3.46}$$

An expression for the free energy change in the solution phase, G_s, as a result of the adsorption of K^+ is obtained by recognizing that the free energy change per mole of dissolved electrolyte is the definition of the chemical potential, μ, of the electrolyte. For example, for KCl:

$$G_{KCl} = \mu_{KCl} = \mu^0_{KCl} + RT \ln (KCl) \tag{3.47}$$

In this equation, $\mu^0{}_{KCl}$ is the chemical potential of dissolved KCl in the standard state. The activities of the two electrolytes in solution (NaCl, KCl) change as K^+ exchanges Na^+, and the activity of NaCl increases while that of KCl decreases in solution. The overall free energy change in solution per mole of K^+ adsorption is

$$G_s = \mu_{NaCl} - \mu_{KCl} \tag{3.48}$$

By applying equation 3.47 and the principle that electrolyte activities can be expressed as the product of activities of the component ions, equation 3.48 can be reexpressed in a form independent of the nonadsorbing Cl^- anion:

$$G_s = \mu_{Na^+} - \mu_{K^+} \tag{3.49}$$

At equilibrium, the change in free energy of the solution phase must be exactly balanced by the change in free energy of the clay phase; that is, the free energy change of the entire system must be zero:

$$G_c + G_s = 0 \tag{3.50}$$

Substituting the expressions for G_c and G_s (equations 3.46 and 3.49) into equation 3.50, the exchange equilibrium condition is obtained:

$$\mu_{K^+} - \mu_{Na^+} = E_{K+} + RT \ln \left\{ \frac{M_{K^+}}{1 - M_{K^+}} \right\} \tag{3.51}$$

By extending the definition of chemical potentials (equation 3.47) to cations ($\mu_{A^+} = \mu^0_{A^+} + RT \ln (A^+)$), and converting equation 3.51 from logarithmic to exponential form, the equilibrium expression becomes:

$$e^{\Delta\mu^0/RT}\, e^{-E_K/RT} = \frac{M_{K^+} \cdot (Na^+)}{(1 - M_{K^+}) \cdot (K^+)} \tag{3.52}$$

where $\Delta\mu^0$ is equal to $\mu^0_{K^+} - \mu^0_{Na^+}$, and the left side of equation 3.52 is seen to be a constant for any particular clay exchange system at a controlled temperature. This, then, is exactly equivalent to exchange equation 3.44, which is derived from the "ideal mixture" model. Now, however, the nature of the equilibrium constant, K_E, is revealed in terms of standard state chemical potentials of the cations in solution ($\mu^0_{K^+} - \mu^0_{Na^+}$) and the difference in surface bonding energy, E_{K^+}, between K^+ and Na^+. This energy term is negative when K^+ replaces Na^+ on exchange sites. Equation 3.52 reveals that, in general, a large energy term would result in a selectivity coefficient much greater than unity.

It now becomes clear that this most commonly assumed form of ion exchange equation results only for an ideal mixture of ions at the surface, with the ideal entropy of mixing being the consequence of the two exchange cations being equally localized at sites. In such a symmetrical (and uncommon!) exchange process, the differential molar entropy and energy are symmetrical functions of M_{A^+}, where A^+ is the preferred cation, as shown in Figure 3.12 using the example of K^+-Na^+ exchange. This figure reveals that, while the energy gained by exchange would tend to drive the reaction all the way toward exchange site saturation by K^+ ($M_{K^+} = 1$), the mixing entropy is decreasing (differential entropy is negative) once M_{K^+} exceeds 0.5. The net result of these two opposing factors is that *complete* exchange site saturation by one cation, even a strongly preferred cation such as K^+, cannot easily be achieved.

3.2h. The Reality of Nonideal Exchange Behavior

It is common practice in ion exchange experiments to maintain the concentration of one of the exchanging cations at a sufficiently high level that its solution activity does not vary significantly during the exchange reaction. Considering again the Na^+-K^+ exchange system, if the activity in solution of one of these cations, say Na^+, is maintained nearly constant by a high background concentration of electrolyte (e.g., NaCl), then equation 3.44 simplifies to

$$(K^+) = A \cdot \left(\frac{M_{K^+}}{M_{Na^+}}\right) \tag{3.53}$$

where A is a constant for any given concentration of electrolyte. But $M_{Na^+} = 1 - M_{K^+}$ in this two-cation system, assuming a fixed quantity of exchange sites. Equation 3.53 is therefore equivalent to

$$(K^+) = A \cdot \frac{M_{K^+}}{(1 - M_{K^+})} \tag{3.54}$$

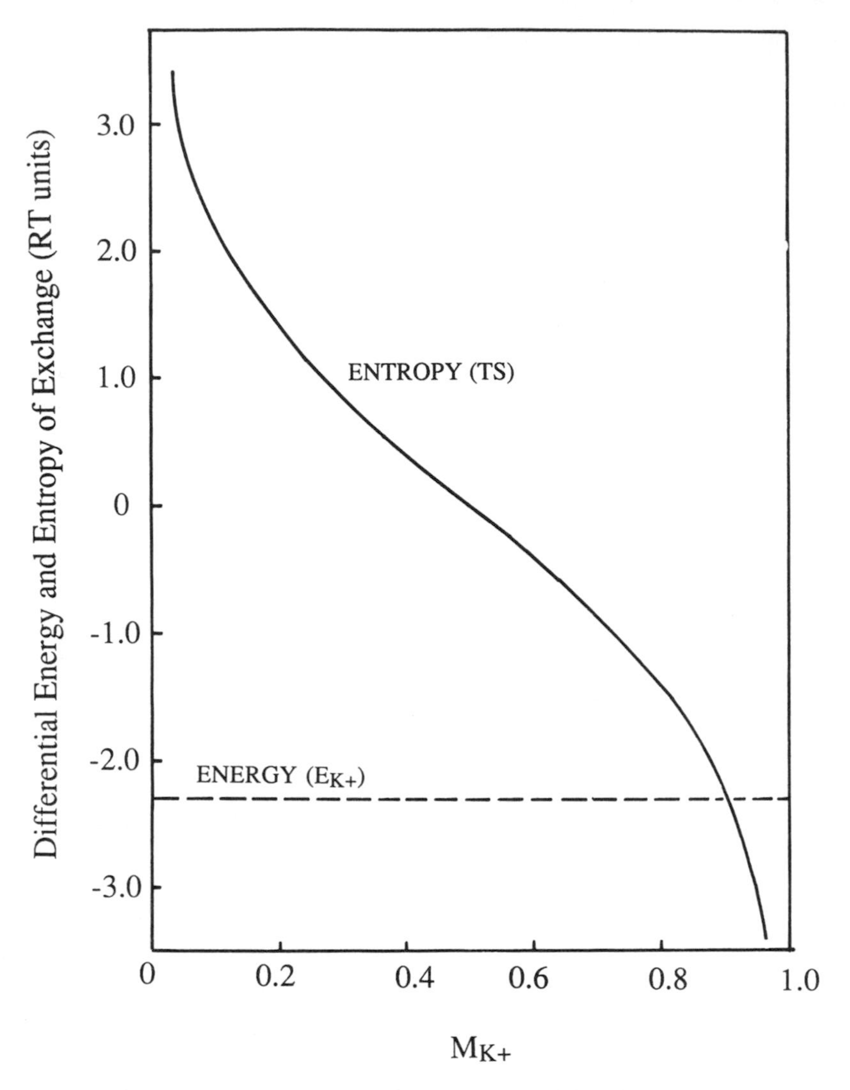

Figure 3.12. Change in entropy (TS) and energy(E_{K+}) of the clay phase per mole of K^+ adsorbed, plotted as a function of the mole fraction of K^+ on exchange sites, M_{K+}. One RT unit at room temperature is 2.5 kiloJoules (0.6 kilocalories) of free energy.

Under conditions where one of the exchange cations is being held essentially constant by a "bathing" electrolyte, equation 3.54 is just as valid as 3.44 in predicting the equilibrium state. This equation has the form of the Langmuir adsorption equation, which is widely used to describe chemical adsorption processes (see Chapter 4 for a description of the Langmuir equation). However, adherence of cation exchange behavior to equation 3.54 is not proof that the "ideal mixture" model of adsorbed ions is adequate. A true test of equation 3.44 and constancy of the exchange coefficient (K_E) requires that both the Na^+ and K^+ activities in solution be widely varied.

Such tests have shown exchange between ions of the same charge and similar radius to conform fairly closely to equation 3.44 (see Figure 3.3).

Certain patterns of nonideal behavior become apparent when cations whose properties are *not* closely matched are involved in ion exchange. One of these patterns or rules is:

> *The* K_S *value that defines the degree of selectivity for one monovalent cation over another tends to decrease as the preferred ion is increasingly adsorbed. This decrease is most notable when the two cations have greatly different radii.*

It seems likely that this "deviant" behavior is linked to the reduced motional freedom (lower entropy) of weakly hydrating cations at the exchanger surface.

When exchange between cations of *different* charge is considered, a more serious failure of the "ideal mixture" model (equation 3.22) is encountered. Although this model is applied widely in soil chemistry, its failure to predict actual exchange behavior over a wide range of electrolyte concentrations, as illustrated in Figure 3.13 by $Cd^{2+}-Na^{+}$ exchange on smectite, has led to selectivity coefficients being defined for

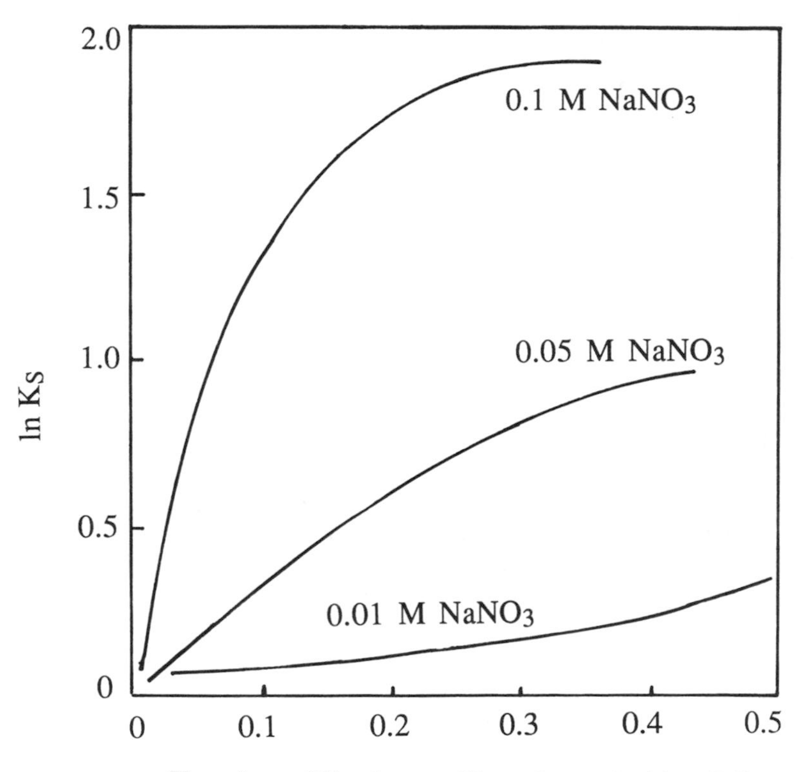

Figure 3.13. Variation of the selectivity coefficient K_s for Cd^{2+}/Na^{+} exchange on smectite as a function of the adsorption level of Cd^{2+} and the electrolyte concentration.

each specific electrolyte concentration. Needless to say, this creates a very awkward situation for applying equation 3.22 to exchange in the real world. However, a general pattern of behavior can be summarized as:

> *The exchange of high-charge cations from permanent-charge clays by increasing concentrations of lower-charge cations is overestimated by the "ideal-mixture" equation.*

Thus, in Figure 3.13 we see that the K_S value from the "ideal-mixture" equation (which measures the preference for Cd^{2+} on the clay) increases in magnitude as more $NaNO_3$ is introduced into solution. In a similar way, if one uses the K_S value measured by exchange at low electrolyte concentration to predict behavior at higher electrolyte concentration, the exchange equation will overestimate the ability of cations like Na^+ and K^+ to replace higher charge metals (e.g., Ni^{2+}, Al^{3+}) from the clay. A particularly troublesome problem has been the overestimation of Al^{3+} exchange from clays by added Ca^{2+}.

The source of nonideal exchange behavior in the case of cations having different charge could be: mixing entropy functions (multivalent cations can occupy exchange sites in different ways), cation-cation repulsion energy, or changes in the clay site geometry. The latter factor is a likely source of nonideal behavior in smectites, as higher electrolyte concentrations create a condition favoring adsorption of multivalent cations over monovalent cations by collapsing the interparticle spacing of the clay and forming quasicrystals (see section 3.2e). The dependence of Cd^{2+} selectivity on electrolyte concentration, shown in Figure 3.13, may be one manifestation of this.

3.2i. Rates of Exchange Reactions

Ion exchange reactions at surface sites exposed to solution are extremely fast. In fact, the kinetics of ion exchange have not generally been accessible to measurement by conventional methods. Cation exchange on clays without narrow interlayer regions (e.g., kaolinites, fully dispersed smectites) appears to be instantaneous. On the other hand, exchange on clays such as vermiculites and partially collapsed smectites can be very slow, limited by the rate of cation diffusion through the narrow interlayer region. In principle, vigorous stirring of the clay-solution system should eliminate film diffusion as a rate-limiting process, leaving particle diffusion as the controlling process (see Chapter 1). Particle diffusion becomes a predominant factor with larger clay particles because the diffusion pathway to the exchange site is longer. As the temperature is raised, diffusion-limited exchange becomes more rapid because the diffusion process is driven by the random thermal motion of ions and molecules, which becomes more vigorous at higher temperature.

It is not surprising that exchange on vermiculite and micaceous minerals should be very slow, particularly when cations able to fully collapse the interlayer space are involved (e.g., K^+, NH_4^+, Cs^+). However, under certain conditions, even reactions such as $Ca^{2+}-Na^+$ exchange on smectites can be slow. A glimpse at the clay quasicrystal structure, shown in Figure 3.11, that is organized upon the partial replacement of monovalent cations by divalent cations suggests that a barrier to cation diffusion may be raised by the narrow interlayer space. This suggestion seems to be confirmed by cation exchange data of the type shown in Figure 3.14, revealing

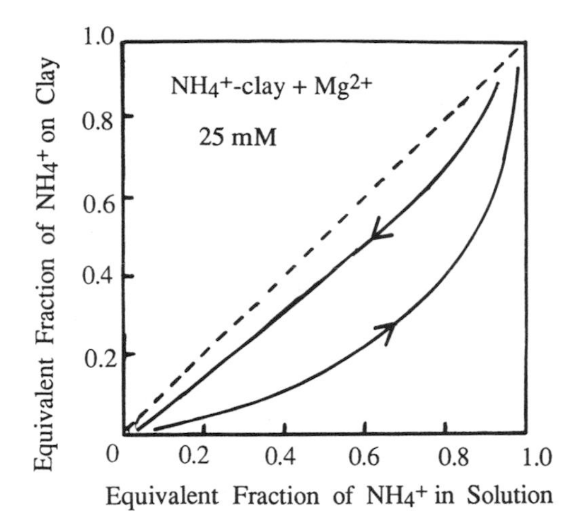

Figure 3.14. Adsorption and desorption isotherms for the exchange reaction between NH_4^+–smectite and Mg^{2+} at constant normality (25 m*M* of cation charge in solution). (Adapted from R. van Bladel and H. Laudelout, 1967. Apparent irreversibility of ion-exchange reactions in clay suspensions. *Soil Sci.* 104:134–137.)

$NH_4^+ - Mg^{2+}$ exchange on smectite to be pathway dependent. Adsorption of NH_4^+ on Mg^{2+}-saturated smectite produces an isotherm described by a particular value of the selectivity coefficient, K_S. The reverse reaction, Mg^{2+} adsorption on NH_4^+-saturated smectite, is described by a different value of K_S. In fact, the K_S value indicates a greater preference for Mg^{2+} when starting from the Mg^{2+}-saturated clay than when starting from the NH_4^+-saturated clay. This apparent existence of two different equilibrium states, both seemingly stable, is termed *hysteresis*. It seems that the exchange reaction is more sluggish in one direction than the other because of the different physical arrangement of the clay particles in the Mg^{2+} and NH_4^+ forms. The Mg^{2+} form, with its quasicrystals and diffusion-limiting narrow interlayers, may inhibit exchange of Mg^{2+} by NH_4^+, creating a higher preference for Mg^{2+}. If this explanation is the correct one, the cation exchange is not truly irreversible. In practical terms, however, it is necessary to treat this exchange reaction as having different "equilibrium" states depending on the reaction direction because the time required for true equilibrium is long relative to the duration of interest in clays and soils.

3.2j. Proton Involvement in Cation Exchange

The simple models of cation exchange presented here have not yet considered H^+ as a competing cation, even though it is always present in clay-water systems. H^+ cations are produced by water dissociation or, more importantly (because carbonic acid is more acidic than water), by the dissociation of carbonic acid formed from the dissolution of CO_2 in water:

$$CO_2 + H_2O = H_2CO_3 \tag{3.55}$$

$$\underset{\textit{carbonic acid}}{H_2CO_3} = H^+ + HCO_3^- \tag{3.56}$$

Equilibrium expressions for these two reactions are

$$K_H = \frac{[H_2CO_3^*]}{P_{CO2}} = 10^{-1.41} \tag{3.57}$$

$$K_1 = \frac{[H^+][HCO_3^-]}{[H_2CO_3^*]} = 10^{-6.3} \tag{3.58}$$

In these equations the asterisk denotes that the concentration of H_2CO_3 includes dissolved and hydrated CO_2 (i.e., $CO_2 \cdot H_2O$) in addition to true carbonic acid. P_{CO2} is the partial gas pressure of CO_2 in the gas phase in equilibrium with the solution.

If we reconsider K^+-Na^+ exchange on smectite, nearly ideal two-ion exchange behavior is seen when the process is conducted in 0.01 *M* NaCl; that is, the selectivity coefficient K_S is fairly constant. On the other hand, if the electrolyte concentration is very low, Na^+ cations, and to a lesser extent K^+ cations, are exchanged from the clay by protons (more precisely, H_3O^+ ions) generated from carbonic acid dissociation. As a result, the *apparent* value of K_S, if based on the incorrect assumption that only K^+ and Na^+ are competing for exchange sites, can vary greatly depending on the degree of proton participation. For example, addition of 1 g of K^+-saturated smectite to 1 liter of distilled water (exposed to air) is followed by about 5 to 10 percent displacement of the exchangeable K^+ from the clay. For Na^+-saturated clays, an even greater displacement of exchange cations follows the immersion of the clays in water. The reaction causing this exchange is

$$H_2CO_3 + Na^+\text{-smectite} = Na^+ + HCO_3^- + H^+\text{-smectite} \tag{3.59}$$

This process, historically referred to as *hydrolysis*[1] because it can proceed even from the weak dissociation reaction of water molecules, causes a marked rise in the solution pH. To avoid confusion with more appropriate usage of the term "hydrolysis," exchange reactions involving protons generated from water or carbonic acid and producing alkalinity in solution will be referred to as *hydrolytic exchange.*

Unless proton occupancy of exchange sites is taken into account, hydrolytic exchange will seem to create nonstoichiometric exchange between cations. An example is shown in Figure 3.15, where the adsorption of K^+ on a Na^+-smectite appears to release more Na^+ into solution than mole-for-mole exchange could produce.

Because the extent of hydrolytic exchange depends on the ease with which H^+ can displace metal ions from exchange sites (see reaction 3.59), only monovalent exchange cations such as Na^+ are substantially subjected to hydrolytic exchange. Clays and soils saturated with Ca^{2+}, Mg^{2+}, or Al^{3+} show little hydrolytic exchange when immersed in water. However, soils of arid climates, which can possess significant exchangeable Na^+, rise in pH as they are diluted in water. Such soils often have pH values in excess of 9, due at least in part to the phenomenon of hydrolytic exchange. The properties of these alkaline soils will be discussed further in Chapter 8.

1. This term is, unfortunately, used to describe several quite different chemical reactions. Hydrolysis as described here should not be confused with the reactions of metal cations with water to form hydroxy-metal complexes.

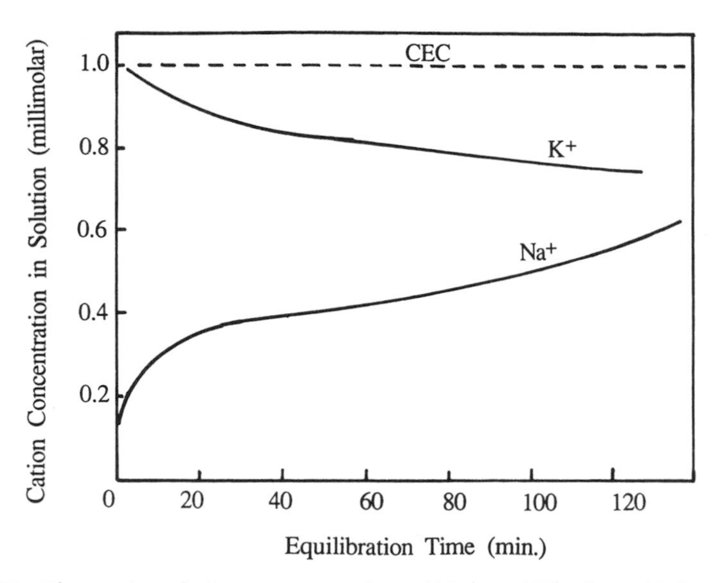

Figure 3.15. Change in solution concentration of Na^+ and K^+ after a Na^+-saturated smectite suspension is immersed in a 1.0-m*M* KCl solution. The solution phase is separated from the clay by an ion-permeable membrane. The broken line depicts the concentration of Na^+ that would be in solution if all of the Na^+ were exchanged from the clay.

Problem: Consider a calculation of the pH and extent of hydrolytic exchange that results when 1.0 g of Na^+-smectite is placed into 1.0 liter of distilled water that is open to the atmosphere. The smectite has a CEC of 0.9 mmole/g. Two reactions involving protons (3.56 and 3.59) must be considered. The first, which describes the dissociation of carbonic acid to form bicarbonate and a proton, has an equilibrium condition given by equation 3.58. The second is the exchange of Na^+ from the clay by H^+:

$$Na^+\text{-smectite} + H^+ = H^+\text{-smectite} + Na^+ \qquad (3.60)$$

with an equilibrium described approximately by the equation

$$K_{ex} = \frac{[Na^+] \cdot M_{H+}}{[H^+] \cdot M_{Na+}} \qquad (3.61)$$

where M_i symbolizes molar quantities on exchange sites, as before. The affinity of protons for permanent charge sites is low because the proton is a very strongly hydrated cation, and evidence suggests that the value of K_{ex} for $H^+ - Na^+$ exchange is about 1.0.

For water exposed to the atmosphere, the concentration of dissolved CO_2, $[H_2CO_3^*]$, is determined by the partial pressure of CO_2 (P_{CO2}) in the atmosphere according to Henry's law:

$$[H_2CO_3^*] = K_H \, P_{CO2} = 10^{-1.41} \times 0.0003 \approx 10^{-5}\, M \qquad (3.62)$$

Here it is assumed that the partial pressure of CO_2 in the air is about 0.0003 atmospheres, although it would be much higher in the *soil's* atmosphere because of biological activity. When this value for $[H_2CO_3^*]$ is inserted into equation 3.58, the relationship

$$[H^+]\,[HCO_3^-] = 10^{-6.3} \times 10^{-5} = 10^{-11.3} \tag{3.63}$$

is obtained. This combined with the exchange requirement that

$$\frac{[Na^+] \cdot M_{H^+}}{[H^+] \cdot M_{Na^+}} \approx 1.0 \tag{3.64}$$

permits the equilibrium condition to be calculated.

Consider the initial (before hydrolysis) and equilibrium quantities of Na^+ in solution, Na^+ on the clay, and H^+ on the clay, summarized below:

	M_{Na^+} (mmoles)	M_{H^+} (mmoles)	$[Na^+]$(mM)
Initial	0.9	0.0	0.0
Equilibrium	$0.9 - x$	x	x

Here x is the unknown quantity of Na^+ exchanged from the clay by the hydrolytic exchange mechanism. When the equilibrium values from this table are substituted into equation 3.64, the expression

$$\frac{[Na^+]}{[H^+]} \cdot \frac{M_{H^+}}{M_{Na^+}} = \frac{(x)(10^{-3})}{[H^+]} \cdot \frac{x}{0.9 - x} = 1.0 \tag{3.65}$$

results, where the 10^{-3} factor converts millimolarity to molarity units. The H^+ ion concentration in solution (activity corrections are ignored here) can then be expressed in terms of x:

$$[H^+] = \frac{(x^2)\,(10^{-3})}{0.9 - x} \tag{3.66}$$

But $[H^+]$ is subject to control by carbonic acid dissociation according to equation 3.63. In addition, the dissociation of HCO_3^- to form CO_3^{2-} and release a proton can also influence the equilibrium, but this reaction can be neglected as long as the equilibrium pH does not rise to very alkaline values. The rule of electroneutrality (charge conservation) requires that the HCO_3^- anions in solution be balanced by Na^+ and H^+. That is:

$$[HCO_3^-] = [Na^+] + [H^+] = 10^{-3}\,x + [H^+] \tag{3.67}$$

Inserting this expression for $[HCO_3^-]$ into equation 3.63, the equation is obtained:

$$[H^+] = \frac{10^{-11.3}}{10^{-3}\,x + [H^+]} \tag{3.68}$$

The expression for $[H^+]$ given by equation 3.66 can now be substituted into equation 3.68, producing the equation:

$$\frac{x^4}{0.81 - 1.8x + x^2} + \frac{x^3}{0.9 - x} = 10^{-5.3} \tag{3.69}$$

The solution to this equation, obtained by trial and error, is $x \approx 0.017$. Since x is the quantity of Na^+ hydrolytically exchanged, then 0.017/0.9, or about 2 percent, of the Na^+

on exchange sites is predicted to be displaced into solution. The resulting pH is about 6.5, calculated by substituting the value of x into equation 3.66 or 3.68. (Note that in the absence of the clay, the pH of the water would be determined by carbonic acid dissociation, and equation 3.63 would be sufficient to calculate the pH. Since, in this simpler case, no cation other than H^+ is in solution, $[H^+] = [HCO_3^-]$, and equation 3.63 predicts a pH of 5.65). The pH is higher with the clay present because of the adsorption of H^+ and $NaHCO_3$ accumulation in solution. If a greater concentration of clay had been assumed, the final pH would have been even higher than 6.5.

The same problem can now be solved assuming that 10^{-2} *M* NaCl, instead of distilled water, is the initial solution into which the Na^+-smectite is immersed. Activity corrections are again ignored here, although they could be significant at this higher ionic strength (see Chapter 1). The initial and equilibrium conditions are again described in terms of x, the quantity of Na^+ exchanged by H^+:

	M_{Na^+} (mmoles)	M_{H^+} (mmoles)	$[Na^+]$(mmoles/l)
Initial	0.9	0.0	10.0
Equilibrium	$0.9 - x$	x	$10 + x$

The exchange equation (3.64) requires that

$$\frac{[Na^+]}{[H^+]} \cdot \frac{M_{H^+}}{M_{Na^+}} = \frac{(10 + x)(10^{-3})}{[H^+]} \cdot \frac{x}{0.9 - x} = 1.0 \quad (3.70)$$

Rearranging gives the expression for $[H^+]$:

$$[H^+] = (10 + x)(10^{-3}) \frac{x}{(0.9 - x)} \quad (3.71)$$

Again, as in the last problem, the charge of any HCO_3^- in solution must be balanced by H^+ or by Na^+ released by hydrolytic exchange. Consequently, equation 3.68 is still valid when 10^{-2} *M* NaCl is present. By substituting the expression for $[H^+]$ (equation 3.71) into equation 3.68, a new relationship for x is found:

$$\frac{100x^2 + 20x^3 + x^4}{0.81 - 1.8x + x^2} + \frac{10x^2 + x^3}{0.9 - x} = 10^{-5.3} \quad (3.72)$$

By trial and error, the solution to this equation is found to be $x \approx 0.0002$. It is clear, then, that almost no hydrolytic exchange occurs *if* the initial concentration of electrolyte in solution is high; that is, Na^+ exchange by H^+ is suppressed by high Na^+ activity. The value of x, substituted into equation 3.71, gives a solution for $[H^+]$, and reveals that the pH in this system would be 5.65, the same as the pH in aqueous solution when no clay is present.

Significant replacement of monovalent metal cations on layer silicate clay surfaces by protons can occur if the electrolyte concentration is very low. The long-term result, beyond hydrolytic exchange, is acidic decomposition of the clay structure in part, and release of structural Al^{3+} or Mg^{2+} to solution. These multivalent cations may then readsorb onto exchange sites, influencing the rheological properties of clays in very dilute salts. Some of the anomalous behavior of Na^+-smectites suspended in dilute solutions of electrolytes may be attributed to these processes that arise from hydrolytic exchange and subsequent decomposition reactions. It is important to remember that, as a larger and larger fraction of exchange sites becomes H^+ - occupied from hydrolytic exchange, most clay structures become increasingly unstable. Thus, reaction 3.59 cannot proceed very far in clays before dissolution reactions start to consume the acidity building up at the surface.

The mechanism of hydrolytic exchange, because it generates acidity on the colloids while forming alkalinity in solution, can lead to the spatial separation of acidity on soil solids from alkalinity in solution. The result is an important rule of soil chemistry:

> *Leaching by pure water of soils containing negatively charged colloids is necessarily a soil-acidifying process.*

It will be seen in Chapter 6 that this principle helps to explain the fact that soils of humid climates become acidified by natural processes. In Chapter 8, hydrolytic exchange will be found to generate alkalinity under environmental conditions that prevent significant leaching.

3.3. ION EXCHANGE ON OXIDES AND OTHER VARIABLE-CHARGE MINERALS

3.3a. Introduction to the PZC Concept for Amphoteric (Variable-Charge) Minerals

Oxides and hydroxides of Al, Fe, Mn, and Si possess little or no permanent surface charge, but generate cation and anion exchange capacity (CEC and AEC) as a result of the adsorption of protons and hydroxyl ions:

$$\underset{(CEC)}{>Fe{-}OH]^{-1/2}} \underset{+\,OH^-}{\overset{+\,H^+}{=}} \underset{(AEC)}{>Fe{-}OH_2]^{+1/2}} \qquad (3.73)$$

Earlier models of oxide surfaces had assumed a two-step process of charge buildup, based on the premise that a given surface metal–hydroxyl group could exist in one of three chemical forms:

$$Fe{-}O^- \overset{+\,OH^-}{\longleftarrow} Fe{-}OH^0 \overset{+\,H^+}{\longrightarrow} Fe{-}OH_2^+ \qquad (3.74)$$

However, the $Fe{-}O^-$ form is unlikely in all but highly alkaline solutions because $Fe{-}OH^0$ is a very weak acid. It is now believed that the OH groups responsible for surface charge (i.e., most reactive groups) are terminal groups, bonded to a single metal (e.g., Fe) ion. The application of Pauling's valence rule shows that the actual charge on this type of group is $-\frac{1}{2}$ or $+\frac{1}{2}$, as shown by equation 3.73, depending on whether the hydroxyl group is protonated or not. Consequently, the most realistic scheme of oxide charge development is one in which the two chemical forms described by equation 3.73 coexist at the surface in relative populations determined by the solution pH. An electrically neutral surface is one with an equal population of the two groups.

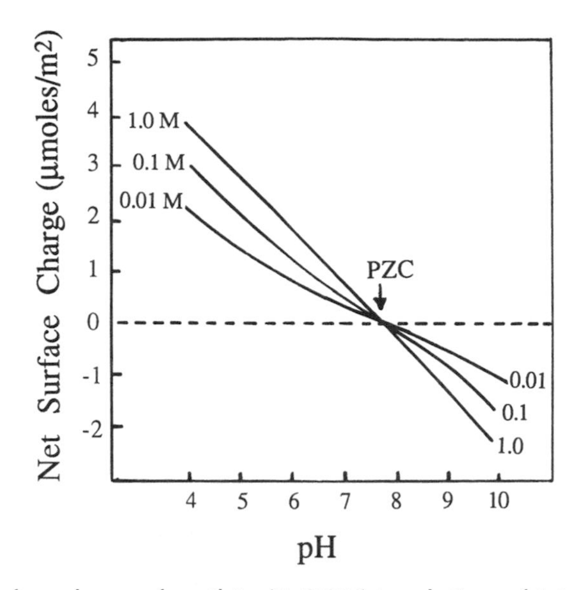

Figure 3.16. Surface charge of goethite (FeOOH) in solutions of 1.0, 0.1, and 0.01 *M* NaCl, estimated from the difference in proton and hydroxyl uptake ($H^+ - OH^-$) by the oxide during acid-base titration. PZC, point of zero charge. (Adapted from F. J. Hingston, 1970. Specific adsorption of anions on goethite and gibbsite. Ph.D. dissertation. University of Western Australia, Perth.)

The fact that these mineral surfaces are amphoteric (able to act as both acid and base) is apparent from titration curves such as those shown in Figure 3.16, which indicate that H^+ or OH^- adsorption on oxides is both a function of solution pH and electrolyte concentration. Besides oxides, the noncrystalline aluminosilicates (allophane and imogolite) also possess this form of pH-dependent charge, but these minerals have such high specific surface areas that their CEC or AEC can be much higher than that of the oxides. Figure 3.17 illustrates this fact for an allophane, which can adsorb upward of 0.8 mmole/g of exchangeable anions at low pH, and as many cations at high pH. Comparison of the ion exchange capacity of a permanent-charge clay (montmorillonite) with that of allophane in Figure 3.17 emphasizes the greater importance of pH as a variable controlling ion adsorption on the amphoteric minerals.

The measure of relative affinity of H^+ and OH^- for amphoteric mineral surfaces is given by the *point of zero charge* (PZC), which is the pH at which surface negative charge (CEC) equals surface positive charge (AEC). For pure amphoteric solids, this PZC is commonly estimated from the crossover point of acid-base titration curves obtained at several electrolyte concentrations (see Figure 3.16).[1] This method assumes that each H^+ (or OH^-) ion consumed by the solid generates one positive (or negative) charge at the solid surface. However, at extreme pH values, H^+ or OH^-

1. This type of PZC is called the *point of zero salt effect* because it is the unique pH at which the addition of electrolyte does not produce a net gain in either H^+ or OH^- adsorption.

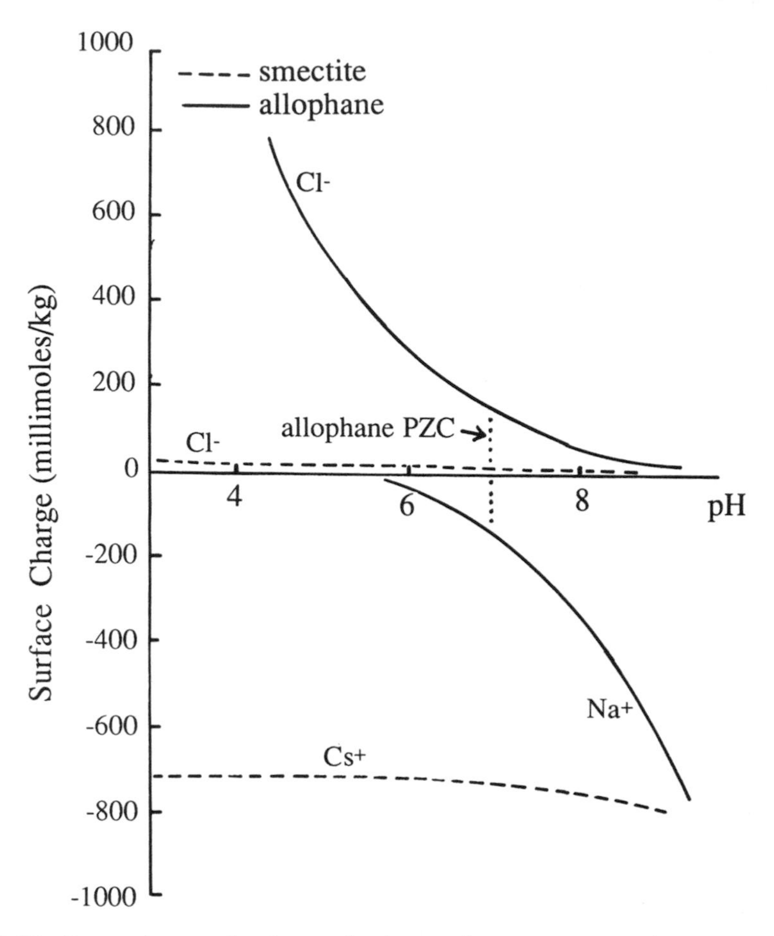

Figure 3.17. Dependence of cation and anion exchange capacity of smectite and allophane on solution pH, measured by the adsorption of cations (Cs^+ for smectite, Na^+ for allophane) and the chloride anion. PZC, point of zero charge.

may be consumed by mineral dissolution. For example, in the titration of aluminum hydroxide, reactions such as

$$Al(OH)_3 + 3\ H^+ = Al^{3+} + 3\ H_2O \tag{3.75}$$

and

$$Al(OH)_3 + OH^- = Al(OH)_4^- \tag{3.76}$$

are possible at low and high pH, respectively. The assumptions of this method for PZC determination then become untenable. This means that the extreme PZC values (< 4, > 9) often reported for some minerals using the titration method should be regarded with caution.

A more reliable method for estimating PZC is the direct measurement of AEC (e.g., Cl^- adsorption) and CEC (e.g., Na^+ or K^+ adsorption) at several pH values.

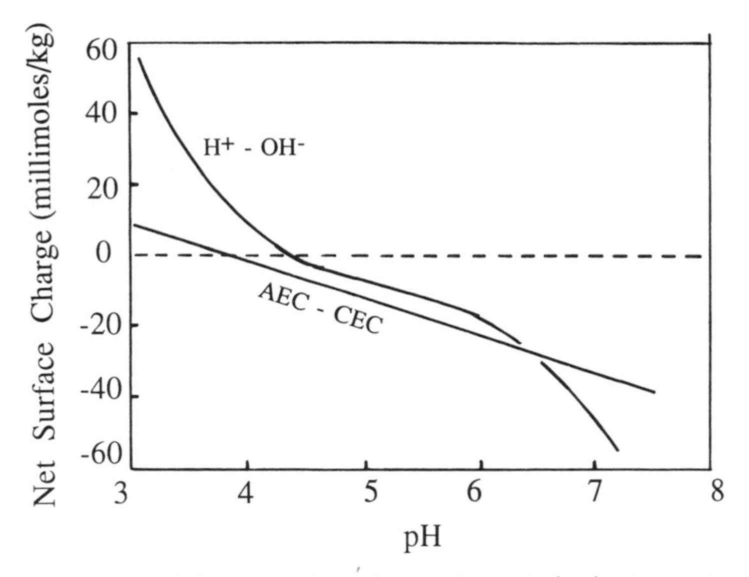

Figure 3.18. Estimates of the net surface charge of an oxisol A-horizon using the method of anion and cation adsorption (AEC − CEC) and the method of acid-base titration (H^+–OH^-). The point of zero charge (PZC), based on the AEC − CEC curve, is 3.86. (Adapted from E. Marcano-Martinez and M. B. McBride, 1989. Comparison of the titration and ion adsorption methods for surface charge measurement in oxisols. *Soil Sci. Soc. Am. J.* 53:1040–1045).

This type of PZC[1] is simply the pH at which AEC = CEC. In Figure 3.17, for example, the PZC defined this way for allophane is about 7, but the PZC for montmorillonite is not defined because CEC ≫ AEC at all pH values. The same method applied to soils gives a PZC that is the net product of all of the ion-adsorbing surfaces in the soil. As such, soil PZC values are not easy to interpret. The net surface charge of a highly weathered soil (oxisol), measured as the difference between anion and cation exchange capacity, AEC-CEC, is shown in Figure 3.18 to be positive below pH 3.9 and negative above this pH. The PZC of the soil is therefore 3.9, the pH at which AEC − CEC = 0. For comparison, the net H^+ consumption (H^+ consumed minus OH^- consumed), measured from titration data, is plotted for the same soil in Figure 3.18. The net H^+ consumption appears to give quite different estimates of soil charge above pH 6.5 and below pH 4. The difference is attributed to reactions such as 3.75 and 3.76, which consume H^+ or OH^- without generating surface charge. Therefore, the PZC estimated from AEC-CEC is taken to be the most reliable value.

In any determination of PZC, the ions that occupy the exchange sites and are subjected to measurement, called the *index ions,* are chosen based on their tendency to adsorb by essentially nonselective electrostatic bonds. These ions may not be in direct contact with the surface, being readily exchangeable by other ions of like charge. Salts of these ions (for example, NaCl, KNO_3) are referred to as "indifferent

1. This is often referred to as the *point of zero net charge.*

electrolytes" because the ions appear to adsorb passively, attracted to the surface positive or negative charge generated by H^+ or OH^- reaction. In this simplified view, the H^+ and OH^- ions are seen as special because they alone control the extent of surface charge buildup, and are referred to as *potential-determining ions* (PDI) for reasons that will be discussed later.

Cations and anions that adsorb by forming short directional bonds with the surface cannot be considered to be indifferent in character. These ions actually alter the surface charge by the very process of adsorption, and their bonding is classified as *chemical adsorption* (or chemisorption) in this text. Examples of chemisorption include copper and phosphate adsorption on iron oxides:

$$>Fe{-}OH]^{-1/2} + Cu^{2+} \rightarrow >Fe{-}O{-}Cu]^{+1/2} + H^+ \tag{3.77}$$

$$>Fe{-}OH_2]^{+1/2} + HPO_4^{2-} \rightarrow >Fe{-}O{-}PO_3H]^{-3/2} + H_2O \tag{3.78}$$

Chemisorbed cations increase surface positive charge and shift the PZC higher, while chemisorbed anions increase surface negative charge and shift the PZC lower. Adsorption of this type, which often controls the solubility of environmentally important ions such as sulfate, phosphate, and heavy metals, will be discussed further in Chapter 4.

In the absence of chemisorbing ions (other than H^+ and OH^-, of course), the charge of the metal oxide or hydroxide mineral surface is determined by the relative affinity of the metal-hydroxy surface group for protons, that is, by the reversible reaction:

$$>M{-}OH]^{-1/2} \underset{OH^-}{\overset{H^+}{\leftrightarrow}} >M{-}OH_2]^{+1/2} \tag{3.79}$$

The valence and radius of the structural metal ion, M, controls the strength of the M—O bond. Consequently, oxides or hydroxides of metals having high valence and small radius (e.g., SiO_2, MnO_2) have more acidic $M{-}OH_2$ surface groups than oxides of metals with lower valence (e.g., Fe and Al oxides). The strong electrical field of metals such as Si^{4+} and Mn^{4+} has a powerful polarizing effect on the bonding electrons, strengthening the M—O bond and weakening the affinity of the oxygen atom for protons. This means that the commonest oxides and hydroxides in soil clays, those of Fe^{3+} and Al^{3+}, have a relatively high affinity for protons, and reaction 3.79 leans toward the right side. For these common soil oxides, the pH at which reaction 3.79 favors neither the forward nor backward direction is usually somewhere between 7.5 and 9. This pH is the PZC of the oxide. Since the pH in most soils does not exceed 8, it is apparent that pure oxides should not be important contributors to CEC. Only AEC is expected to be significant (albeit small) in soils containing Fe and Al oxides and hydroxides. On the other hand, allophanes, with PZC values below 7 and high surface areas, possess quite high pH-dependent cation and anion exchange capacities (see Figure 3.17). Therefore, soils rich in allophane, which are often formed from volcanic ash deposits, display high capacities to adsorb cations and anions, although the ratio of CEC to AEC depends on the pH. Organic anions, important constituents of soil organic matter, bond selectively on the anion exchange sites, suppressing adsorption of "indifferent" anions such as chloride. Consequently, surface soils typically have lower anion exchange capacities than the corresponding subsoils.

Besides these important relations between PZC and ion adsorption by soils, the physical properties of soils may also be sensitive to the PZC. It is a general principle of colloid chemistry that colloidal particles dispersed in water tend to maintain their dispersed state if the particles possess relatively high surface charge, either positive or negative. Conversely, aggregation of these individual particles (i.e., flocculation) is most favored when the surface charge is low. This leads to the rule that:

The most flocculated state of a colloidal suspension composed of a single mineral occurs at the pH equal to the mineral's PZC.

At the PZC, not only are the surface negative (CEC) and positive (AEC) charges equal in magnitude, but the ability of the surface to adsorb ions of either charge is at a minimum. Consequently, interparticle electrostatic repulsion and osmotic swelling forces reach a minimum at pH = PZC, and aggregated structures are favored.

Whether this simple principle relating dispersibility to PZC is widely useful for soils under real conditions is not yet certain. Oxide minerals in soils can be intimately associated with layer silicate clays, or may have organic anions, phosphate, silicate, and metals adsorbed on their surfaces. Their "effective" PZC values may be very different from the PZCs reported for pure oxide minerals. Indeed, reported PZCs for highly weathered soils with mainly Fe and Al oxide mineralogy are in the range of 3.5 to 4.5, much lower than the 7 to 9 range of the pure oxides. The permanent charge of silicate clays, especially kaolinite, often found in these soils, and the presence of organic matter, may be at least partly responsible for shifting the PZC to such low values.[1] A question is then raised about the pH of maximum dispersibility of the soil particles. Is it determined by the PZC of the individual minerals or by the overall PZC of the soil? It is logical that the dispersibility of a mineral phase in the soil should be a function of the charge of that mineral, and the overall PZC may be of little use in predicting the dispersing behavior of soil components.

Water-dispersible clay, which is widely used as a measure of a soil's susceptibility to erosion, crusting, and other undesirable degradation processes, increases upon liming some highly weathered soils. Liming removes the aggregating effects of Al^{3+} on clays and organic matter and increases the negative charge on these colloids. The result is increased interparticle repulsion and dispersion. Low levels of organic matter in these highly weathered soils actually seem to exacerbate the problem of dispersibility. A possible mechanism explaining this behavior is the bonding of organic acids to low-charge minerals such as oxides and kaolinite, thereby increasing negative charge at the mineral surfaces. It is perhaps ironic that organic matter can have deleterious effects, while Al^{3+} has beneficial effects on soil structure, in contradiction to organic matter and Al^{3+} effects on plant growth.

The importance of colloidal dispersion in degrading the physical properties and agricultural productivity of soils is further discussed in Chapter 8.

The PZC in its most simple definition is the pH at which the net surface charge is zero. In soils, this concept of PZC is complicated by the presence of permanent in addition to pH-dependent charge sites. As shown in Figure 3.19a, an increase in ionic

1. Destruction of the organic matter in these soils shifts the PZC to higher pH.

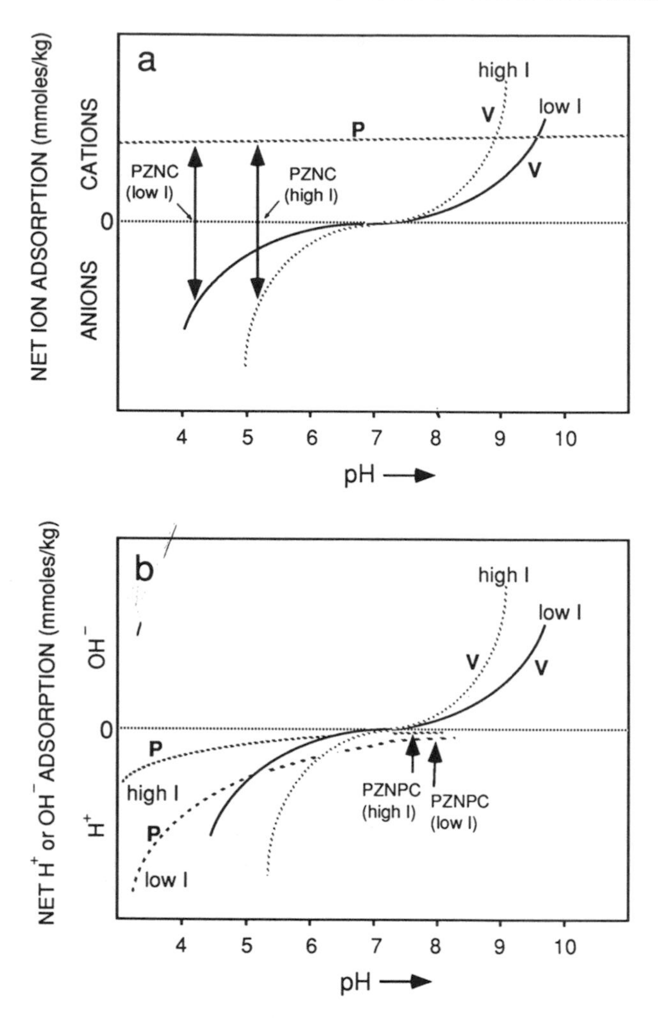

Figure 3.19. Effect of ionic strength, *I*, on the point of zero net charge (PZNC) (a) and the point of zero net proton charge (PZNPC) (b) of a soil containing both permanent (*P*) and variable (*V*) charge minerals.

strength, I, would affect only the pH-dependent component of the total charge, and the PZC as measured by the point of zero net charge (PZNC) would shift to *higher* pH. On the other hand, the point of zero net proton charge (PZNPC), defined as the pH at which charge due to an excess of surface-bonded protons (or OH^- ions) is zero, may actually shift toward *lower* pH at higher ionic strength. This effect, shown in Figure 3.19b, results because indifferent salts suppress H^+ adsorption on permanent charge sites while promoting H^+ adsorption on variable-charge sites. In these mixed-mineral systems, unique crossover points in titration curves (see Figure 3.16), termed the *points of zero salt effect* (PZSE), may not exist. In any event, the relationship of the PZSE to soil surface charge is not altogether straightforward.

3.3b. Model for Charge Development on Amphoteric (Variable-Charge) Mineral Surfaces

A simple chemical model can be used to show how the charge of oxides and other variable-charge minerals depends on pH and ionic strength. Consider the chemisorption of an acid HX:

$$\begin{matrix} >\text{Fe}-\text{OH}\,|^0 \\ >\text{Fe}-\text{OH}_2| \end{matrix} + \text{HX} = \begin{matrix} >\text{Fe}-\text{OH}_2|^+ \\ >\text{Fe}-\text{OH}_2| \end{matrix} \cdots \text{X}^- \tag{3.80}$$

or base, MOH:

$$\begin{matrix} >\text{Fe}-\text{OH}\,|^0 \\ >\text{Fe}-\text{OH}_2| \end{matrix} + \text{MOH} = \begin{matrix} >\text{Fe}-\text{OH}|^- \\ >\text{Fe}-\text{OH}| \\ +\,\text{H}_2\text{O} \end{matrix} \cdots \text{M}^+ \tag{3.81}$$

These reactions generate anion and cation exchange capacity on an iron oxide in this example. X^- and M^+ symbolize the anion and cation of the acid and base, where MX is an indifferent electrolyte. The chemical identity of the acid and base is lost upon reaction with the surface, but the exact chemical nature of the surface-bound ions need not be known to describe thermodynamically the equilibrium state of the oxide-electrolyte system.

The Langmuir equation (derived in Chapter 4), which is applicable to the chemisorption of H^+ or OH^- on oxides, has the general form:

$$\Theta_i = \frac{Ba_i}{1 + Ba_i} \tag{3.82}$$

where Θ_i is the fraction of adsorption sites occupied by the adsorbate; a_i is the activity of the adsorbate, i, in solution; and B is a bonding constant that depends on the energy of the adsorbate-adsorbent bond formed. Since the adsorbates in this case are HX and MOH, then Θ_+ and Θ_- can be defined as the fraction of reactive surface sites ($-$Fe$-$OH groups) that have reacted with HX and MOH to form positive and negative surface charge, respectively:

$$\Theta_+ = \frac{B_{HX}\, a_{HX}}{1 + B_{HX}\, a_{HX}} \qquad \Theta_- = \frac{B_{MOH}\, a_{MOH}}{1 + B_{MOH}\, a_{MOH}} \tag{3.83}$$

Since the activities of HX and MOH in solution can be expressed as the product of activities of ionic species:

$$a_{HX} = a_{H+} \cdot a_{X-} \qquad a_{MOH} = a_{M+} \cdot a_{OH-} \tag{3.84}$$

and since the dissociation constant of water requires that $a_{H+} \cdot a_{OH-} = 10^{-14}$, then equations 3.83 become

$$\Theta_+ = \frac{B_{HX}\, a_{H+} \cdot a_{X-}}{1 + B_{HX}\, a_{H+} \cdot a_{X-}} \qquad \Theta_- = \frac{10^{-14}\, B_{MOH}\, (a_{M+}/a_{H+})}{1 + 10^{-14}\, B_{MOH}\, (a_{M+}/a_{H+})} \tag{3.85}$$

It is likely that surface OH^- groups bonded to only one Fe atom (terminal groups) are responsible for both the negative and positive charge development described by reactions 3.80 and 3.81. If the number of such groups is taken to be S μmoles per gram of oxide, then the positive or negative surface charge, Γ_+ or Γ_-, can be expressed in terms of fraction of sites reacted and S, which limits the capacity to develop charge to $S > \Gamma_{+,-} > 0$:

$$\Gamma_+ = S\Theta_+ \qquad \Gamma_- = S\Theta_- \tag{3.86}$$

Anion and cation adsorption is believed to be competitive on amphoteric surfaces, with the same sites serving to adsorb both cations and anions. Furthermore, simultaneous adsorption of cations (MOH) and anions (HX) is possible at separate sites. It is necessary then to correct equations 3.83 for the effect of competition for limited sites[1]:

$$\Theta_+ = \frac{B_{HX}\, a_{H+} \cdot a_{X-}}{1 + B_{HX}\, a_{H+} \cdot a_{X-} + 10^{-14}\, B_{MOH}\, (a_{M+}/a_{H+})} \tag{3.87}$$

$$\Theta_- = \frac{10^{-14}\, B_{MOH}\, (a_{M+}/a_{H+})}{1 + B_{HX}\, a_{H+} \cdot a_{X-} + 10^{-14}\, B_{MOH}\, (a_{M+}/a_{H+})} \tag{3.88}$$

Equations 3.87 and 3.88 define the fraction of surface sites that are positively and negatively charged. Since pH by definition is the negative logarithm of the H^+ ion activity in solution (pH $= -\log a_{H+}$), these equations show that surface charge on amphoteric minerals must be a function of pH. Furthermore, because the electrolyte concentration determines the activities of M^+ and X^- in solution, a_{M+} and a_{X-}, it is clear from these equations that surface charge depends on the electrolyte level. B_{HX} and B_{MOH} are bonding constants whose values are not initially known for any particular oxide. However, the measured PZC of the oxide, which is by definition the value of pH at which $\Theta_+ = \Theta_-$, estimates the *relative* bonding strength of HX and MOH on the oxide. Knowledge of this PZC can be used to reduce B_{HX} and B_{MOH} in equations 3.87 and 3.88 to a single unknown parameter, B.

The *net* surface charge of the oxide, σ_o, is given by

$$\sigma_o = \Gamma_+ - \Gamma_- = S(\Theta_+ - \Theta_-) \tag{3.89}$$

The predicted net surface charge of an hypothetical oxide would then depend on pH and electrolyte concentration as shown in Figure 3.20, where a PZC of 8 is assumed and the number of surface sites, S, and the bonding constant are assigned arbitrary values. These charge curves resemble actual curves obtained by titration of iron oxides (see Figure 3.16), correctly predicting, at least qualitatively, increased surface charge at higher electrolyte concentrations.

This model for surface charge has only two adjustable fitting constants once the PZC of the amphoteric mineral is known. The bonding constant, B, could in principle be determined from theory, but is best treated as a fitting parameter. The quantity of sites, S, that potentially could develop positive or negative charge can be estimated in principle by

1 The form of the competitive Langmuir adsorption isotherm will not be derived here, but it has been derived by Adamson (1976) for two competing gases on a surface.

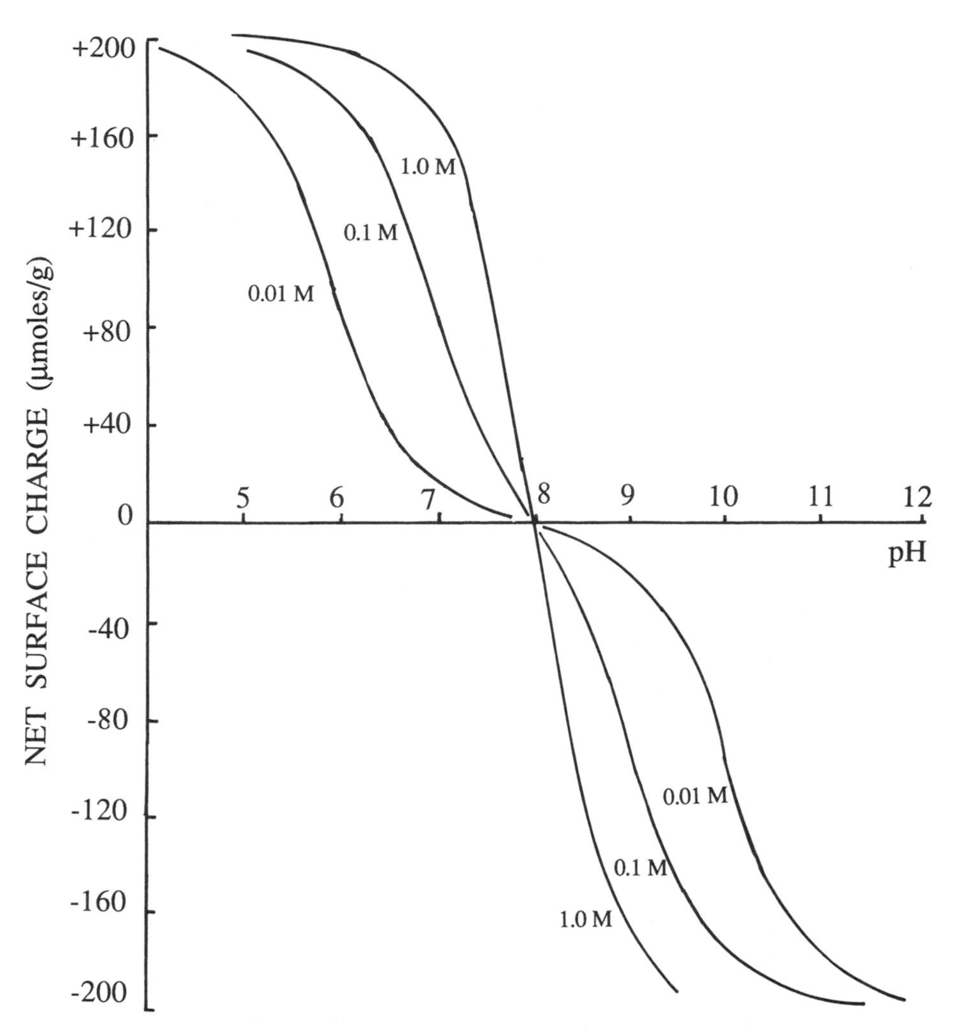

Figure 3.20. Theoretical net surface charge of an hypothetical amphoteric oxide with PZC (point of zero charge) = 8, calculated at three concentrations (0.01, 0.1, and 1.0 *M*) of indifferent electrolyte.

spectroscopic techniques and a knowledge of the surface area of the mineral. In practice, however, this estimate is difficult to make. Estimates of adsorption densities of protons on oxide surfaces range as high as one excess H^+ per 34 $Å^2$, measured for hematite at pH 4 in 1 *M* KCl. The model limits surface charge to a maximum regardless of pH, and is physically more reasonable than the diffuse double-layer model (discussed in the appendix of this chapter) that predicts that the magnitude of charge should increase exponentially as the solution pH is adjusted away from the PZC.

The model provides a chemical explanation for charge buildup on variable-charge surfaces as salt concentrations are increased in solution. The explanation is that the activities of both base, MOH, and acid, HX, are increased by the addition of electrolyte, MX, to solution. From equation 3.84 it can be seen that adding electrolyte to solution at low pH, by increasing the anion activity a_{X^-}, necessarily increases the activity of the acid in solution and forces reaction 3.80 further to the right. The surface charge becomes more positive as a result. Conversely, at high pH the addition of electrolyte increases the chemical activity of the base, and reaction 3.81 is forced further to the right with the formation of more negative surface charge. Since reaction 3.80 consumes acid from solution, while reaction 3.81 consumes base, the general principle is found that:

> *Addition of neutral salts to suspensions of variable-charge minerals (or soils containing these minerals predominantly) shifts the pH of the suspensions toward the PZC.*

3.3c. The Variable-Charge Model and PZC in Soil Chemistry

The model developed in the last section shows that, when the solution pH equals the PZC of a variable-charge mineral, addition of electrolyte increases the positive and negative surface charge equally, so that the pH does not change and the *net* charge (AEC-CEC) remains at zero. This unique pH is called the *point of zero salt effect* (PZSE) for obvious reasons, and is easily measured by adjusting the pH of soil or mineral suspensions over a range using strong acid and base, then adding neutral salt (indifferent electrolyte such as NaCl) at these different pH values until the pH that is not shifted by salt addition is found.

The usefulness of the variable-charge and PZC concepts is limited in soils with signficant quantities of permanent-charge clays. Permanent-charge sites complicate the physical interpretation of measured PZC values and may cause the soil to have no PZNC; that is, the soil retains net negative charge at all pH values (see Figure 3.19a). Such soils respond to neutral salt addition by a lowered pH, which means that permanent-charge clays and soils containing these clays do not usually possess a PZSE, at least in the pH range relevant to the soil environment.

The quantity of charge developed on the amphoteric (variable-charge) minerals is likely to be insignificant compared with that on permanent-charge clays. There are important exceptions to this generalization, however. Allophanic soils of volcanic origin can have high CEC or AEC values, depending on the pH. Highly weathered soils of tropical and subtropical climates typically are composed of kaolinites and oxides of Al and Fe, so that permanent charge is a minor contribution to the total surface charge. As Figure 3.18 demonstrates, the CEC and AEC values of these soils are low, and AEC can exceed CEC under natural pH conditions. Much of the CEC in highly weathered soils is provided by organic matter rather than the mineral fraction.

Soils composed of variable-charge minerals generally have higher anion exchange capacities than soils composed predominantly of permanent-charge minerals. This produces chemical behavior that contradicts certain generalizations about soils that are based on experience with soils of northern temperate climates. For example, the assumption that anions such as NO_3^- are completely mobile in soils is invalid in some acid soils of the tropics, which have appreciable anion exchange capacities. Both chloride, Cl^-, and NO_3^- behave "indifferently" on anion exchange sites; that is, they bond electrostatically, show little relative selectivity for sites, and are readily displaced by other anions. The more strongly adsorbed (chemisorbed) anions, such as phosphate, form specific bonds with variable-charge minerals; their behavior is a special case that will be discussed in Chapter 4.

3.4. PH-DEPENDENT CHARGE OF ORGANIC MATTER

3.4a. Dissociation of Weak Organic Acids

The predominant functional groups of soil organic matter contain oxygen atoms, with the carboxylic and phenolic groups accounting for most of the titratable acidity in humus. These organic acids dissociate by reaction with base:

$$\mathrm{R{-}C(OH){=}O} + \mathrm{NaOH} = \mathrm{R{-}C(O^-Na^+){=}O} + \mathrm{H_2O} \tag{3.90}$$

$$\mathrm{R{-}C_6H_4{-}OH} + \mathrm{NaOH} = \mathrm{R{-}C_6H_4{-}O^-Na^+} + \mathrm{H_2O} \tag{3.91}$$

creating negative charge at the organic surface that must then be balanced by cations (such as the Na^+ shown in the reaction above). However, because carboxylic acids are generally more strongly acidic than phenols, carboxylate groups are mainly responsible for creating the CEC attributed to organic matter in soils.

A simple "monoprotic" acid dissociation model can be used to describe pH-dependent charge on humus, where the organic acid, HA, forms the anion, A^-:

$$\mathrm{HA} = \mathrm{H^+} + \mathrm{A^-} \tag{3.92}$$

and the acid dissociation constant, K_a, is defined by

$$K_a = \frac{[\mathrm{H^+}][\mathrm{A^-}]}{[\mathrm{HA}]} \tag{3.93}$$

Equation 3.93, expressed in terms of concentrations (square brackets) can be rewritten in the logarithmic form:

$$\mathrm{p}K_a = -\log K_a = \mathrm{pH} - \log\left\{\frac{[\mathrm{A^-}]}{[\mathrm{HA}]}\right\} \tag{3.94}$$

If α is defined as the degree of dissociation of the acidic functional groups, then:

$$\alpha = \frac{[\mathrm{A^-}]}{[\mathrm{HA}] + [\mathrm{A^-}]} \tag{3.95}$$

and equations 3.94 and 3.95 are combined to give

$$\mathrm{pH} = \mathrm{p}K_a + \log\left\{\frac{\alpha}{(1-\alpha)}\right\} \tag{3.96}$$

This equation correctly describes the titration curve of *soluble* monoprotic acids such as acetic acid (see Figure 3.21). Soil organic acids are, however, much more

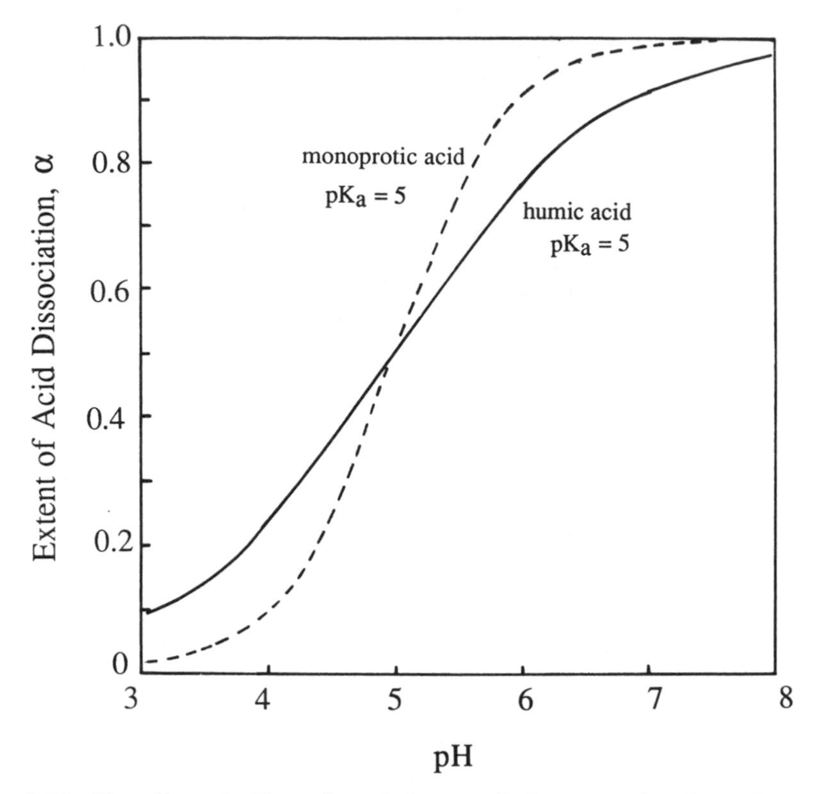

Figure 3.21. The effect of pH on dissociation, α, of a humic acid and simple monoprotic organic acid in solution, both with pK_a of 5, calculated from the Henderson–Hasselbach equation.

complex. They are often in solid or colloidal form, not dissolved in water. They are heterogeneous, with an array of different functional groups possessing different pK_a values. Finally, they are polymeric and probably polyprotic (having more than one acid group on each molecule). For these reasons at least, equation 3.96 cannot be expected to describe the dissociation behavior of humic substances. Nevertheless, a generalized version of this equation, termed the Henderson–Hasselbach equation, describes fairly satisfactorily the dissociation (titration curves) of humic materials. This equation:

$$pH = pK_a + n \log \left\{ \frac{\alpha}{(1 - \alpha)} \right\} \tag{3.97}$$

ostensibly differs from equation 3.96 only in the introduction of an empirical constant, n. However, the pK_a in this equation is defined as the "apparent pK_a" for the humic substance, and is not a true acid dissociation constant.

The meaning of the pK_a parameter in equation 3.97 should be interpreted with caution. Like the true pK_a of a soluble monoprotic acid, it is numerically equal to the pH at which 50% of the organic acid is dissociated (see Figure 3.21). However, at pH values higher than this pK_a value, electrostatic models predict that the buildup of charge on the organic polymer during dissociation would inhibit further dissociation. Probability-based models similarly predict that the successive removal of protons from a single molecule would become increasingly unlikely. This prediction has gained credibility from titrations of synthetic polycarboxylic acids in which all acidic groups are chemically identical. Even in these homogeneous weak acids, the dissociation of a fraction of the acidic groups weakens the apparent acidity of the remaining groups. This phenomenon explains, at least in part, the lower dissociation of humic acid at high pH compared with dissociation of a simple monoprotic acid (see Figure 3.21).

The result of this "charging" effect on inhibiting dissociation is that the "apparent pK_a" of polyprotic organic acids can only be compared directly with the pK_a of monoprotic organic acids when the polyprotic acid is undissociated ($\alpha = 0$). Thus, the pK_a value at $\alpha = 0$ is called the intrinsic pK_a of the humic material. The relationship of the apparent pK_a to the intrinsic pK_a is given by

$$pK_a\,(\text{apparent}) = pK_a\,(\text{intrinsic}) + W\,N\,\alpha \qquad (3.98)$$

In this equation, N is the number of carboxylic acid groups per molecule, and W is a constant attributed to electrostatic interaction energy.

Even when the "charging" effect is considered, the pK_a of soil organic matter, unlike the pK_a of a monoprotic acid, tends to decrease as the ionic strength (salt concentration) of the soil solution is increased. This means that soil organic matter seems to be more acidic in more concentrated salt solutions. Synthetic polycarboxylic acids also show this behavior. It becomes clear that the application of solution concepts such as pK_a to complex charged polymeric solids has numerous pitfalls.

The two parameters, pK_a and n, can be varied to fit the titration curves of soil organic matter to equation 3.97. When this is done, the value of n is usually found to be about 2. The two parameters are determined by plotting pH as a function of $\log\{\alpha/(1-\alpha)\}$. A straight-line relation should be obtained with slope n and intercept pK_a. The relationship between solution pH and degree of dissociation of the acidic groups is plotted in Figure 3.21, comparing a humic acid having a pK_a of 5.0 with a simple monoprotic acid with a pK_a of 5.0. The humic acid clearly dissociates to a lesser degree and in a more nearly linear fashion in response to a pH increase than does the simple acid. In fact, the organic matter found in soils dissociates more or less in linear proportion to the pH. Since dissociation creates exchange sites, α is related to the CEC of organic matter by the simple equation

$$\alpha = \frac{\text{CEC (actual)}}{\text{CEC (potential)}} \qquad (3.99)$$

where CEC (potential) is the sum total of the acidic functional groups in the soil organic matter. Accordingly, an α-pH relationship such as that described in Figure 3.21 can be used to estimate the CEC of organic matter at any pH. For example, at pH 7, the CEC is given by

$$\text{CEC (pH 7)} = \alpha\,(\text{pH 7}) \times \{\text{total organic acid content}\} \qquad (3.100)$$

The organic acid content of the organic matter can be measured by titration with base (estimated values of acid content for fulvic and humic acids are reported in Table 2.5). In soils, however, titrations neutralize mineral as well as organic acidity, so that it is not possible to obtain a reliable estimate of the organic acid content. Nevertheless, an estimation of the contribution of soil organic matter to CEC has been made from the empirical relationship

$$\text{CEC (mmoles/kg)} = -600 + 500\ \text{pH} \tag{3.101}$$

For soils near neutral pH, this means that a "rule of thumb" can be applied: each percent (by weight) of organic matter contributes about 30 mmoles/kg to the CEC.

Soil organic matter becomes more soluble at higher pH as dissociation increases the surface negative charge of the individual organic particles or molecules. Intermolecular and intramolecular electrostatic repulsion is believed to be responsible for this increased solubility, although charge site hydration and osmotic forces are probably important as well. In the "random coil" model of humic substances, the repulsive forces cause an unfolding and expansion that are associated with dispersion and dissolution of the organics. The flocculated and dispersed forms of organic polymers are shown schematically in Figure 3.22. Soluble organics can elevate the solubility of natural and pollutant metal cations bonded to the organic molecules, and may also serve to mobilize organic pollutants adsorbed on humic substances.

3.4b. Metal Cation Exchange on Organic Matter

Organic matter, as will be discussed further in Chapter 4, limits the solubility of certain metal cations in soils by forming coordination complexes with them. These complexes typically involve some covalency, and the strength of the bond formed is sensitive to the chemical properties of the metal as well as the organic coordinating group. However, the negatively charged surface of organic matter also functions as a "weak-field" exchanger; that is, adsorption can occur by nonselective electrostatic forces. For this type of adsorption, the charge and radius of the metal ion are the most critical factors controlling selectivity. In general, large cations preferentially displace small cations, with the order of selectivity for alkali and alkaline earth metals being

$$Cs^+ > Rb^+ > K^+ > Na^+ > Li^+$$

$$Ba^{2+} > Sr^{2+} > Ca^{2+} > Mg^{2+}$$

This order is evidence that the strongly hydrating cations such as Li^+ and Mg^{2+} retain their hydration shells when adsorbed at the organic sites.

The trivalent metal cations, Fe^{3+} and Al^{3+}, form highly stable complexes with organic matter, each metal bonding to two or more functional groups. Some Fe^{3+}-humic complexes are so stable (kinetically or thermodynamically) that they resist dissociation over the pH range of 3 to 10. In effect, then, Fe^{3+} and Al^{3+} block potential cation exchange sites in organic matter, reducing the CEC. It is not uncommon for the stable organic matter found in very acid mineral soils, where Al^{3+} is soluble, to contribute little to the soil's CEC.

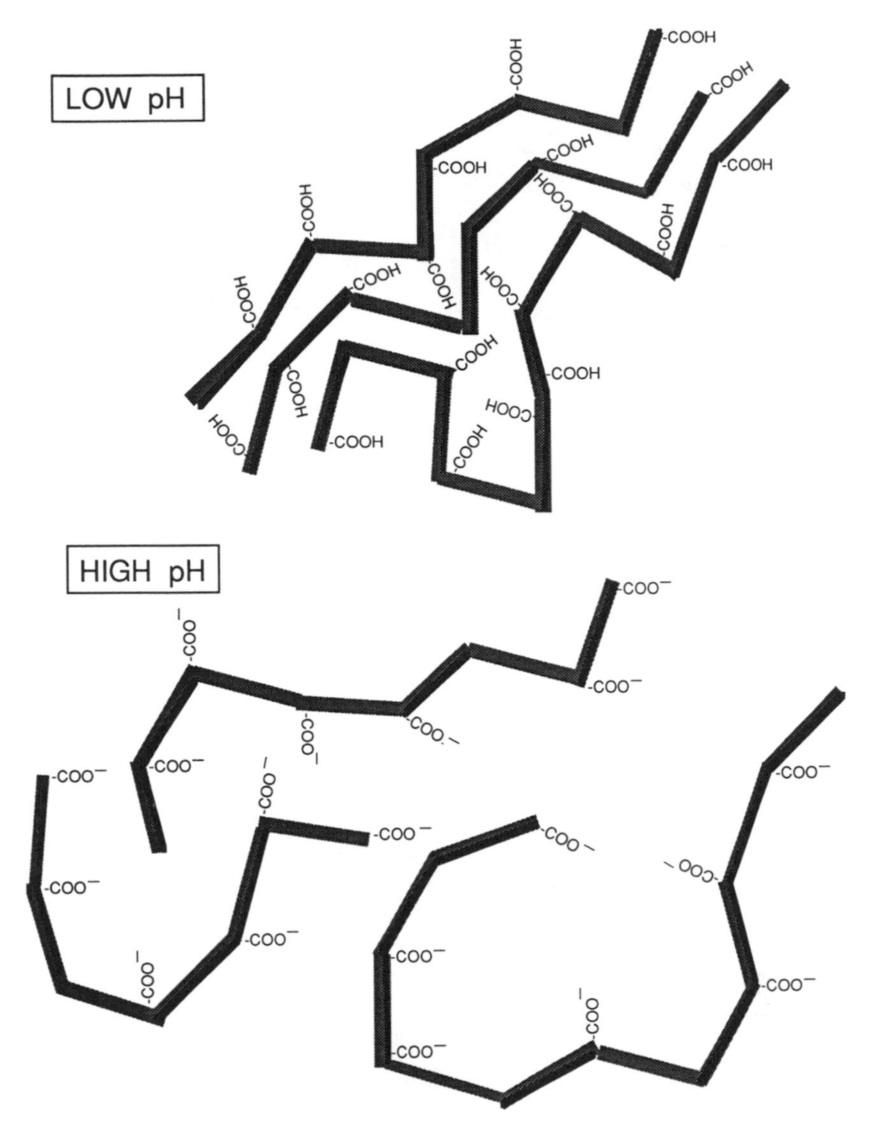

Figure 3.22. "Random coil" model of humic substances in soil, showing the flocculated (low pH) and dispersed (high pH) forms of the organic polymers.

3.5. THE RETARDATION OF ION MOVEMENT IN SOIL BY EXCHANGE

Soils are often relied on to attenuate the migration of potentially hazardous metals into ground or surface waters. For cationic metals, this retardation of movement is easily understood as due to exchange on negatively charged soil colloids. Suppose we consider the downward leaching of a radioactive pollutant, $^{137}Cs^{+}$, from a nuclear disposal site. If it were not adsorbed on exchange sites, it would move at the same

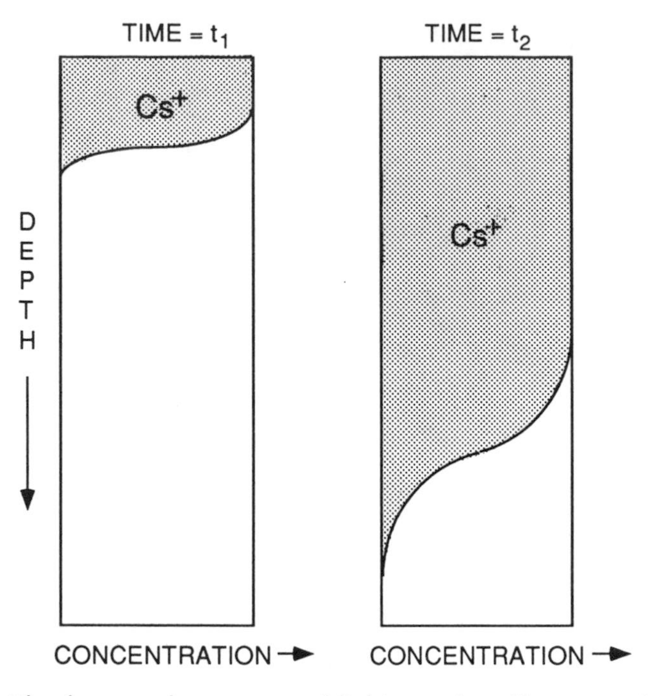

Figure 3.23. The downward movement of Cs^+ in a soil profile as a result of leaching by water. The shaded areas denote the presence of Cs^+.

velocity, v, as the leaching water. Because it is adsorbed, its velocity, v_{Cs}, is less than v. Imagining that the Cs^+ moves down on a front as diagrammed in Figure 3.23, we can estimate the rate of downward penetration of this front from the equation

$$v_{Cs} = \frac{v}{1 + (\rho_B/\phi)K_d} \tag{3.102}$$

where ρ_B is the bulk density of the soil, ϕ is the porosity of the soil, and K_d is the distribution coefficient for Cs^+ ion adsorption on the soil.

K_d is related to the previously discussed K_s of cation exchange. The general form of equation for Cs^+ adsorption, assuming that the soil exchange sites are occupied mostly by Ca^{2+}, is

$$K_s = \frac{(N_{Cs})^2 \cdot [Ca^{2+}]}{(N_{Ca}) \cdot [Cs^+]^2} \tag{3.103}$$

Now, since N_{Cs} and N_{Ca} represent the fractions of exchange sites occupied by Cs^+ and Ca^{2+}, they can be represented as m_{Cs}/CEC and m_{Ca}/CEC, the quantities of exchange sites occupied by Cs^+ and Ca^{2+} divided by the total quantity of exchange sites. But in this two-cation system, m_{Ca} = CEC − m_{Cs}, so that equation 3.103 becomes

$$K_S = \frac{(m_{Cs}/\text{CEC})^2 \cdot [Ca^{2+}]}{\left(\frac{(\text{CEC} - m_{Cs})}{\text{CEC}}\right) \cdot [Cs^+]^2} \tag{3.104}$$

Assuming that the predominant cation in soil solution is Ca^{2+}, $[Ca^{2+}]$ can be replaced in equation 3.103 by $N - [Cs^+]$, where N is the total normality of the solution. But Cs^+ is assumed to be a minor ion in both solution and on exchange sites, so that $N - [Cs^+] \approx N$, and CEC $- m_{Cs} \approx$ CEC. Equation 3.104 then simplifies to

$$K_S = \frac{(m_{Cs})^2 \cdot N}{(\text{CEC})^2 \cdot [Cs^+]^2} \tag{3.105}$$

Rearranging to obtain an expression for the quantity of adsorbed Cs^+, the result is

$$m_{Cs} = \text{CEC}\ \sqrt{K_S/N}\,[Cs^+] \tag{3.106}$$

By defining a distribution coefficient K_d such that $K_d = \text{CEC}\ \sqrt{K_S/N}$, we finally obtain a simple relationship between the quantity of Cs^+ adsorbed on the soil solids (mmoles/kg) and the concentration of Cs^+ in solution (mmoles/kg):

$$m_{Cs} = K_d\,[Cs^+] \tag{3.107}$$

This derivation reveals that K_d depends on the CEC of the soil, the selectivity of exchange sites, and the concentration of salts in soil solution. K_d is a property of the particular soil *and soil solution* that exists at the site of metal movement, and must be measured experimentally.

The magnitude of $1 + (\rho_B/\phi)K_d$ is the *retardation factor,* estimating the degree to which the cation's movement is reduced. K_d is simply the ratio of moles of Cs^+ adsorbed to moles in solution, and must be determined by an adsorption experiment using the soil material from the site of interest. The K_d can have a value of 100 or more for Cs^+.

The use of equation 3.102 is limited to situations in which: (1) K_d is fairly constant over the range of adsorption considered, (2) the ion exchange reaction is rapid and reversible, and (3) the concentration of Cs^+ (or other metal pollutant) in soil solution is low relative to that of the other cations. For soils containing vermiculites and other micaceous minerals, the requirement of exchange reversibility is unlikely to be met. These minerals fix Cs^+ between layers, so that desorption is very much slower than adsorption.

Large retardation factors can mean that the leaching of metal cations from the soil surface into the subsoil is slow, even assuming that metal adsorption is by exchange processes only. As will be explained in Chapter 4, many metals adsorb in addition by strong forces, and this form of metal retention in soils is likely to be practically irreversible, leading to long-term immobilization.

Appendix: The Diffuse Double-Layer Model of Exchangeable Cations

Little has been said in this chapter about the location of exchangeable cations in the vicinity of charged clay surfaces. The spatial distribution of these "counterions" (ions with charge opposite that of the surface) cannot easily be determined experimentally, and evidence supporting different models of counterion distribution at the mineral-solution interface is usually indirect. The colloidal properties of clays, such as dispersion, flocculation, and swelling, are influenced by the counterion distribution. *Electrophoretic mobility* of colloidal particles, determined by applying an electrical field to dispersed particles and measuring the direction and velocity of particle migration, is believed to give some indication of the average position of exchange cations at the particle surface. If the counterions are on average separated by some distance from the surface, then the particle charge is not compensated close to the surface, and the particle moves in the applied field toward the electrode with charge opposite that of the particle.

A commonly used model for describing counterion distribution at a charged surface is based on the Gouy-Chapman diffuse double-layer (DDL) theory. This model assumes that the surface can be visualized as a structurally featureless plane with evenly distributed charge, while the counterions are considered point charges in a uniform liquid continuum. In this simplified picture, the equilibrium distribution of counterions is described by the Boltzmann equation:

$$n(x) = n_o\, e^{-ze\psi(x)/kT} \tag{3.108}$$

where $n(x)$ and $\psi(x)$ are (respectively) the local concentration of counterions and the electrical potential at a distance, x, from the surface; n_o is the concentration of counterions in bulk solution (i.e., at $x = \infty$); e is the unit of electronic charge; k is the Boltzmann constant; T is temperature (K); z is the charge of the counterion.

Equation 3.108 predicts a higher local concentration of cations near a negatively charged clay surface than in bulk solution, and a lower concentration of anions near the surface than in solution. Figure 3.24 shows this predicted distribution of monovalent cations and anions near the clay surface for two different concentrations of electrolyte in solution. More modern statistical mechanical models of this clay interfacial region have predicted that ion-ion correlation (electrostatic) effects should cause deviations from this classical picture, such as the positive adsorption of anions at intermediate distances from the surface when the cation is divalent or multivalent.

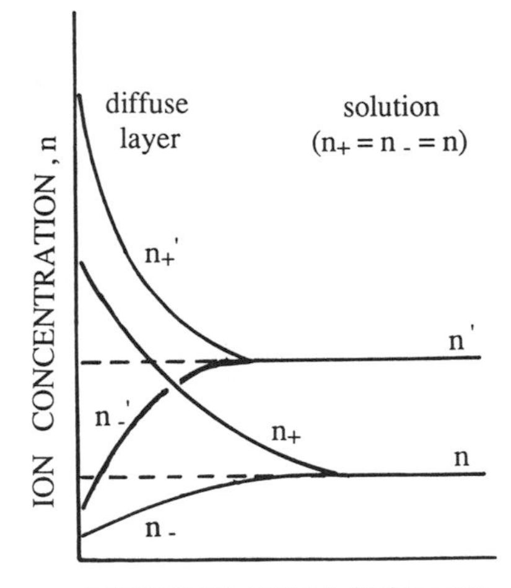

Figure 3.24. Diffuse double-layer model of cation and anion distribution near a permanent-charge clay surface. (Adapted from H. van Olphen, 1977. *An Introduction to Clay Colloid Chemistry* 2nd ed. New York: Wiley.)

These models reveal that the classical picture of counterion distribution is probably valid for monovalent ions only.

According to the classical diffuse double-layer model, the electrical potential, $\psi(x)$, develops as the thermal energy of the counterions causes them to diffuse away from the surface to some extent. The degree of diffuseness of the double layer increases at higher temperature. By combining the Boltzmann equation with other classical (continuum) equations that establish interrelationships among electrical potentials, local charge imbalances, and surface charge density, an equation can be obtained that describes the decrease of electrical potential with distance from the charged surface. For clay surfaces with low charge densities, and assuming that the electrolyte in solution consists of monovalent cations and anions ($z_+ = z_- = 1$), an approximate solution is valid:

$$\psi(x) = \psi_o e^{-\kappa x} \tag{3.109}$$

where ψ_o is the potential at the surface ($x = 0$). The parameter κ, can be thought of as the reciprocal of the double-layer thickness, $1/\kappa$, that is, the extent of counterion distribution away from the surface. This double-layer thickness is controlled, according to the theory, by certain variables in the clay-solution system as described by the equation

$$\kappa = \mathrm{A} \cdot z \left(\frac{n_o}{\epsilon kT} \right)^{1/2} \tag{3.110}$$

where A is a constant, z is the counterion charge, n_o is the electrolyte concentration, and ϵ is the dielectric constant of the solvent. Accordingly, exchange cations with low charge (e.g., Na^+), solutions with low salt concentrations, and solvents with high dielectric constants (i.e., water) should expand the double layer of negatively charged clays. An expanded double layer means that the negative electrical potential extends further from the surface, and electrostatic particle-particle repulsion becomes more probable, increasing the tendency of the particles to remain suspended in water as a stable colloidal dispersion.

Although the DDL theory was initially applied to clays of permanent charge, it was later used to describe the electrical potential at variable-charge (e.g., oxide) surfaces. This application required that the Nernst equation be first used to determine the surface electrical potential, ψ_o, as a function of the solution pH:

$$\psi_o = -2.303 \frac{RT}{F} (\mathrm{pH} - \mathrm{PZC}) \tag{3.111}$$

Here, F is the Faraday constant and R is the ideal gas constant. However, this model of charge development on surfaces predicts that surface charge should increase exponentially as the pH is adjusted away from the point of zero charge, PZC.

There are good reasons to believe that the application of the Nernst equation and the DDL model to oxides and other variable-charge mineral surfaces is inappropriate. To begin with, cations and anions may adsorb on oxides by direct coordination to the charged surface group. Even for monovalent cations and anions, a high percentage of the bonds with oxide surfaces are believed to be of the inner-sphere type.

Consequently, the selectivity of metal ions for Fe oxides follows the order: $Li^+ > Na^+ > K^+$. This order is the opposite of that expected for exchangeable cations that are separated from the surface by hydration water, suggesting that the counterions are very near the surface on average. A description of more localized ion bonding appears to be needed.

Further objection to the Nernst-DDL description of variable-charge surfaces follows from the assumption used that the activity of H^+ at the surface does not change with pH. Since the H^+ ion chemically adsorbs on oxides, this assumption is undoubtedly fallacious, and oxides are likely to behave in a non-Nernstian manner.

References

Adamson, A. W. 1976. *Physical Chemistry of Surfaces,* 3rd ed. New York: Wiley.

Baver, L. D. 1948. *Soil Physics.* New York: Wiley.

Eisenman, G. 1961. On the elementary origin of equilibrium ionic specificity. In A. Kleinzeller and A. Kotyk (eds.), *Symposium on Membrane Transport and Metabolism.* New York: Academic Press, pp. 163–179.

Gast, R. G. 1969. Standard free energies of exchange for alkali metal cations on Wyoming bentonite. *Soil Sci. Soc. Am. Proc.* 33:37–41.

Gast, R. G. 1972. Alkali metal cation exchange on Chambers montmorillonite. *Soil Sci. Soc. Am. Proc.* 36:14–19.

Gilbert, M. and R. van Bladel. 1970. Thermodynamics and thermochemistry of the exchange reaction between NH_4^+ and Mn^{2+} in a montmorillonite clay. *J. Soil Sci.* 21:38–49.

Hingston, F. J. 1970. Specific adsorption of anions on goethite and gibbsite. Ph.D. dissertation, University of Western Australia, Perth.

Hutcheon, A. T. 1966. Thermodynamics of cation exchange on clay: Ca-K-montmorillonite. *J. Soil Sci.* 17:339–355.

Kelley, W. P. 1948. *Cation Exchange in Soils.* Reinhold, N.Y.: American Chemical Society, Monograph No. 109.

Maes, A., P. Peigneur, and A. Cremers, 1975. Thermodynamics of transition metal ion exchange in montmorillonite. In S. W. Bailey (ed.), Proceedings of the International Clay Conference, Mexico City.

Marcano-Martinez, E. and M. B. McBride. 1989. Comparison of the titration and ion adsorption methods for surface charge measurement in oxisols. *Soil Sci. Soc. Am. J.* 53:1040–1045.

McBride, M. B. 1989. Surface chemistry of soil minerals. In J. B. Dixon and S. B. Weed (eds.), *Minerals in Soil Environments.* Madison, Wis.: Soil Science Society of America, pp. 35–88.

van Bladel, R. and H. Laudelout. 1967. Apparent irreversibility of ion-exchange reactions in clay suspensions. *Soil Sci.* 104:134–137.

van Olphen, H. 1977. *An Introduction to Clay Colloid Chemistry.* 2nd ed. New York: Wiley.

Suggested Additional Reading

Bolt, G. H. 1967. Cation-exchange equations used in soil science—a review. *Netherlands J. Agric. Sci.,* pp. 81–103.

Helfferich, F. 1962. *Ion Exchange.* New York: McGraw-Hill.

Reichenberg, D. 1966. Ion-exchange selectivity. In J. A. Marinsky (ed.), *Ion Exchange: A Series of Advances.* Vol. 1. New York: Marcel Dekker, pp. 227–276.

Sposito, G. 1981. *The Thermodynamics of Soil Solutions.* New York: Oxford University Press, Chaps. 5 and 6.

Thomas, G. W. 1977. Historical developments in soil chemistry. *Soil Sci. Soc. Am. J.* 41:230–237.

Questions

1. The exchange equation for the reaction

$$B^{2+} + 2\ A^{+}\text{-clay} = B^{2+}\text{-clay} + 2\ A^{+}$$

has the form

$$K_S = \frac{N_{B2+} \cdot [A^{+}]^2}{N_{A+}^2 \cdot [B^{2+}]}$$

If $K_S = 1$ is assumed (indicating nonpreferential adsorption of the divalent cation):

(a) What is the fraction of exchange sites occupied by A^+ and B^{2+} if $[B^{2+}] = 1.0$ molar and $[A^+] = 1.0$ molar?

(b) Suppose the exchange system of (a) is diluted 1000 times with pure water, keeping the ratio of B^{2+} to A^+ in solution essentially constant. Now what fractions of exchange sites are occupied by A^+ and B^{2+}?

2. Suppose a soil, whose mineral fraction is dominantly composed of smectite, has a CEC of 200 mmoles (−)/kg, balanced by 180 mmoles (+)/kg of exchangeable Ca^{2+} and 20 mmoles (+)/kg of exchangeable Na^+. Measured concentrations of Ca^{2+} and Na^+ in the soil solution, which occupies 30 percent of the soil's volume, are 0.025 and 0.04 *M*, respectively. Answer the following questions.

(a) What is the value of the selectivity coefficient, K_S, for Ca^{2+} exchange of Na^+ on this soil?

(b) What would be the expected effect on the composition of exchange sites if rainfall increased the soil's water content? What would be the effect if evaporation reduced the water content?

(c) What would be the expected effect of adding water with the composition of 0.025 *M* Ca^{2+} and 0.04 *M* Na^+ to the soil?

(d) Suppose irrigation water with 0.05 *M* Ca^{2+} and 0.01 *M* Na^+ were applied to the soil over a period of years. Estimate the new composition of cations on the exchange sites.

3. For the adsorption of Ca^{2+} on a Na^+-saturated smectite, what do you expect to be:

(a) the approximate magnitude and sign of the reaction enthalpy, ΔH^0?

(b) the approximate magnitude and sign of the reaction entropy, ΔS^0?

(c) the effect of raising the temperature?

4. A 10-g sample of Na^+-saturated montmorillonite (CEC = 1.0 mmole/g) is mixed with 1.0 liter of water containing 10 ppm Na^+ and 10 ppm K^+ as the only cations. If the exchange equation

$$K_S = \frac{[Na^{+}] \cdot M_{K+}}{[K^{+}] \cdot M_{Na+}} = 2.0$$

can be accepted as valid for this system, calculate the equilibrium concentrations of Na^+ and K^+ in solution.

5. Several 0.100-g samples of K^+-saturated smectite (CEC = 920 mmoles(−)/kg) were placed in containers with 100 ml of NaCl solutions of different concentrations. After equi-

librium was reached, the solutions were analyzed for K^+ and Na^+. The results are given below.

Initial Concentrations (M)		Final Concentrations (M)	
(Na^+)	(K^+)	(Na^+)	(K^+)
10^{-3}	0	7.80×10^{-4}	2.46×10^{-4}
2.5×10^{-3}	0	2.18×10^{-3}	3.43×10^{-4}
5.0×10^{-3}	0	4.71×10^{-3}	4.19×10^{-4}
10^{-2}	0	9.70×10^{-3}	5.01×10^{-4}

(a) Calculate the fractions of exchange sites occupied by Na^+ and K^+ at equilibrium for each NaCl level.
(b) Calculate the selectivity coefficient at each NaCl level, ignoring any correction for non-ideal behavior of the ions in aqueous solution.
(c) Provide a possible explanation for any variation in K_S that is noted.
(d) Using equation 3.52, and assuming that the standard state chemical potential of K^+ and Na^+ in solution are about equal (i.e., $\mu^0_{K+} = \mu^0_{Na+}$), find the relationship between K_S and the energy of the exchange reaction. How does this compare with the energy of a chemical reaction?
(e) From the relationship derived in (d), what should be the effect on selectivity of changing the temperature?

6. Assuming 90 cm of annual rainfall, estimate the maximum quantity of base cation (moles/hectare) that could be exchanged and leached out of an acid soil in one year by:
(a) the acidity in pristine rainfall.
(b) acid rain with a pH of 4.
(c) the acidity generated in the soil by biological processes, if P_{CO2} in the soil air is 0.03 atmospheres.

7. For the adsorption of divalent trace metals, M^{2+}, (e.g., Zn^{2+}, Co^{2+}, Ni^{2+}, Cu^{2+}, Mn^{2+}), on permanent-charge clays with exchange sites largely occupied by Ca^{2+}, the expression

$$K_S = \frac{N_{M2+} \cdot [Ca^{2+}]}{N_{Ca2+} \cdot [M^{2+}]} = 1.0$$

provides a reasonable description of exchange as long as the pH is low enough that hydrolysis of the trace metals does not become a complicating factor.
(a) Derive a simplified approximate Ca^{2+}–M^{2+} exchange expression based on the assumption that the trace metal occupies a very small fraction of the exchange sites.
(b) Use the expression from (a) to calculate the fraction of cation exchange sites that would be occupied by Zn^{2+} if the solution contained 10^{-3} *M* Ca^{2+} and 10^{-7} *M* Zn^{2+} at equilibrium.

8. A sample of pure goethite, FeOOH, with a PZC of 7.5, is suspended in 10^{-3} *M* NaCl. The pH is adjusted to 5.0 with acid. What would be the effect of increasing the NaCl concentration to 10^{-2} *M* on:
(a) solution pH?
(b) CEC and AEC?
(c) net surface charge?

9. The following data (Marcano-Martinez and McBride, 1989) were obtained for K^+ and Cl^- adsorption on the A and B horizons of an oxisol (mineralogy of both horizons is very similar, dominantly kaolinite and Fe and Al oxides).

Horizon	pH	Cl^- adsorbed (mmoles/kg)	K^+ adsorbed(mmoles/kg)
A	3.0	15.2	6.5
	3.2	11.4	5.8
	3.8	7.8	7.9
	4.1	5.8	6.8
	4.7	2.7	11.9
	6.3	0.74	20.9
	6.7	0.76	28.9
B	3.2	25.9	7.6
	3.4	22.8	8.1
	3.75	16.5	6.4
	4.1	13.5	7.1
	4.85	6.9	13.4
	6.4	0.55	24.2
	6.7	0.52	33.1

(a) Graph CEC, AEC, and net charge as a function of pH for each horizon. Determine the PZNC of the A and B horizon soils.
(b) Is there any indication that these soils have some permanent charge? How would permanent charge affect the PZNC of a variable-charge soil?
(c) Propose a mechanism to explain the difference in AEC for the A and B horizons.
(d) If NO_3^- adsorbs as an "indifferent" anion like Cl^-, calculate the quantity of nitrate (in units of kilograms of elemental N per hectare) that the A - horizon could prevent from leaching at pH 4.5. Assume a soil depth of 30 cm and a bulk density of 1.5 g/cm^3.

10. Suspensions of clay-sized FeOOH in 0.001 M NaCl were adjusted over a range of pH with NaOH and HCl, then shaken and allowed to settle in test tubes. One day later, light transmittance through the undisturbed supernatants was measured. The results were as follows:

pH	4.18	4.64	5.06	5.88	6.88	7.28	7.75	8.54	8.74
transmittance (%)	9.4	21.4	41.9	70.2	91.5	88.0	59.8	57.6	49.5

(a) Explain the results on the basis of surface charge.
(b) What is the PZC of the oxide based on this experiment? Could CO_2 dissolved in the water have influenced the PZC?
(c) How would adsorbed Cu^{2+}, phosphate, or fulvic acid affect the results of this experiment?

11. Calculate the relationship between the two bonding constants, B_{HX} and B_{MOH} for an oxide with PZC = 9 (hint: at PZC, $\Gamma_+ = \Gamma_-$). How many adjustable parameters does this leave in the surface charge model for this particular oxide?

12. The organic fraction of a soil is found to have a CEC of 2500 mmoles/kg at pH = 5 and a CEC of 3500 mmoles/kg at pH = 6. If the Henderson–Hasselbach equation (with n =

2) can be used to describe the dissociation of acidic functional groups in this complex material, calculate the following:

(a) the total quantity of acid functional groups (mmoles/kg).

(b) the apparent pK_a of the functional groups.

(c) the CEC at pH 7.

4

Chemisorption and Precipitation of Inorganic Ions

Soils are remarkable for their ability to remove ions from water by sorption reactions. Sorption is defined broadly here as the transfer of ions from the solution phase to the solid phase. In Chapter 3 it was shown that those elements that take the form of cations are retained in soils by cation exchange on clays and humus. However, more selective and less reversible sorption reactions such as complexation with organic functional groups and bonding on variable-charge minerals (e.g., oxides, allophane) can also retain and even immobilize metal cations. Elements having the form of anions in solution are retained in soils primarily by selective bonding (chemisorption) processes at variable-charge mineral surfaces and layer silicate particle edges. These types of cation and anion adsorption are collectively referred to as *specific adsorption* to distinguish them from ion exchange. In addition, chemical precipitation may participate in the removal of ions from water and is treated in this chapter as a part of sorption.

This chapter discusses the important mechanisms by which ions, whether present naturally in the soil or introduced by pollution, are sorbed by specific adsorption and precipitation, processes that are inherently less reversible than ion exchange. The chapter is intended to provide a fundamental basis for understanding inorganic cation and anion adsorption and precipitation, the reactions that limit solubility of many of the elements.

4.1. SORPTION OF METAL CATIONS

4.1a. Cation Exchange Reactions

The permanent charge sites of layer silicate clays retain cations by nonspecific electrostatic forces, so that, unless the pH is high enough to favor metal hydrolysis, cation selectivity on these clays follows the rules of cation exchange (see Chapter 3). This is not to say that ion exchange processes cannot result in strong apparent preference for one ion over another; for example, multivalent cations effectively displace monovalent cations from clay exchange sites at low ionic strength according to the concentration-charge rule. Even so, the preference does not originate from a specific bonding process, but is a consequence of the mathematical form of the ion exchange equation.

Consider, for example, Cu^{2+}-Na^+ exchange:

$$Cu^{2+} + 2Na^+\text{-clay} = Cu^{2+}\text{-clay} + 2Na^+ \qquad (4.1)$$

The selectivity coefficient can be written

$$K_S = \frac{[Na]^2 N_{Cu}}{[Cu]\,(N_{Na})^2} \qquad (4.2)$$

where [Na] and [Cu] represent the molarity of the metals in solution, and N_{Na} and N_{Cu} symbolize the fraction of clay exchange sites occupied by Na^+ and Cu^{2+}. In general, if equation 4.2 is even approximately correct, Cu^{2+} will show a strong tendency (relative to Na^+) to concentrate on clay surfaces from solutions of low ionic strength. Despite this, it is unlikely that cation exchange limits the solubility of Cu and other trace metals, except perhaps in metal-polluted soils. Because of the low concentrations of most trace metals in unpolluted soil, specific adsorption mechanisms usually explain their very low solubilities in soil solution. Cation exchange can, however, greatly retard metal ion movement through soils. This is demonstrated in Chapters 3 (subsection 3.5) and 9 (subsection 9.2b) using the examples of Cd^{2+} and Cs^+.

4.1b. Features of Chemisorption on Aluminosilicates and Oxides

The transition and heavy metals, referred to hereafter as trace metals, are important to plants and animals as both micronutrients and toxic elements. Many of them occur in the soil environment in cation form. As naturally occurring elements, some of these cations are incorporated into primary and secondary mineral structures and may be very unavailable. Schemes for complete extraction of these metals from soils require extreme treatments, including dissolution of certain minerals. As pollutants, the metals may enter the soil in organically complexed form or as metal salts. In the latter case, the metal cations then adsorb on mineral and organic surfaces.

Noncrystalline aluminosilicates (allophanes), oxides, and hydroxides of Fe, Al, and Mn, and even the edges of layer silicate clays to a lesser extent, provide surface sites for the chemisorption of transition and heavy metals. All of these minerals present a similar type of adsorptive site to the soil solution: a valence-unsatisfied OH^- or H_2O ligand bound to a metal ion (usually Fe^{3+}, Al^{3+}, or $Mn^{3+,4+}$). For example, on iron oxides, a trace metal, M, may bind according to the reaction

$$>Fe-OH]^{-1/2} + M(H_2O)_6^{n+} \rightarrow\ >Fe-O-M(H_2O)_5]^{(n-3/2)+} + H_3O^+ \qquad (4.3)$$

This reaction has at least four features that distinguish it from cation exchange:

1. Release of as many as n H^+ ions for each M^{n+} cation adsorbed
2. A high degree of specificity shown by particular minerals for particular trace metals
3. Tendency toward irreversibility, or at least a desorption rate that is orders of magnitude slower than the adsorption rate
4. A change in the measured surface charge toward a more positive value.

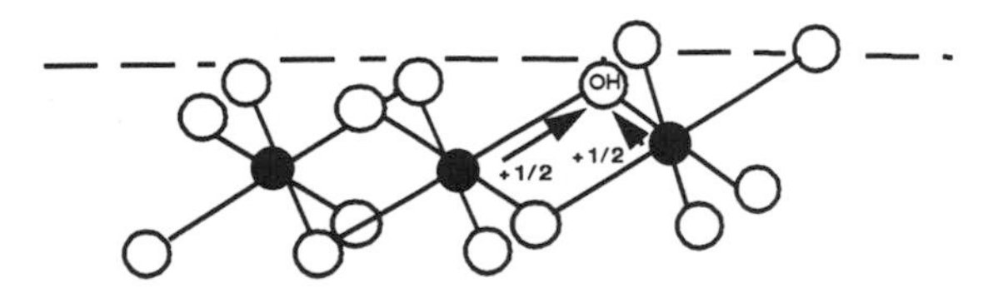

Figure 4.1. Cross section of the planar (001) surface of gibbsite depicting bridging (charge-balanced) OH^- groups. (From M. B. McBride. 1989. Reactions controlling heavy metal solubility in soils. In B. A. Stewart (ed.), *Advances in Soil Science* 10:1–56.)

This last features implies that the adsorbed metal and its charge become part of the mineral surface, thereby shifting the PZC to higher pH.

Some minerals are much more reactive than others on a weight basis. Reactivity is limited by the type and number of valence-unsatisfied (terminal) groups at the surface. For example, gibbsite has no such groups on its (001) crystal faces because each OH^- group is coordinated to two Al ions, and is valence satisfied according to Pauling rules (see Figure 4.1). Since (001) surfaces constitute most of the surface area of the platelike gibbsite crystals, this mineral, even in the microcrystalline form, chemisorbs very small quantities of metals such as Cu^{2+}. Bonding probably occurs at edges that possess OH^- or H_2O groups coordinated to a single Al^{3+} ion. In contrast, noncrystalline oxides and allophanes possess larger numbers of valence-unsatisfied groups because of their structural disorder, and can chemisorb more trace metals.[1]

Figure 4.2 suggests a rationale for the chemical behavior of mineral surface bonding groups, based on the electron-withdrawing power of the cation(s) coordinated to the surface oxyanion. The degree of electron polarization away from the bonding lone-pairs of the O atom, estimated from the electronegativity of the coordinated H^+ and Al^{3+}, provides a basis for predicting the tendency of different sites on oxides and silicates to bond covalently with metals. For example, an O atom bonded to one Al^{3+} and one H^+ (Figure 4.2b) is more prone to bond with metals than O bonded to two Al^{3+} and one H^+ (Figure 4.2a). The OH^- group of Figure 4.2a is termed a *bridging* OH^-, while that of Figure 4.2b is termed a *terminal* OH^-. This latter OH^- is more likely to accept an additional proton in acid solution, thus forming a positively charged $Al{-}OH_2^+$ site. Conversely, this same OH^- is thought to be less acidic than the bridging OH^-, resisting dissociation to the anionic ($Al{-}O^-$) form. It may seem surprising that the group that deprotonates with difficulty also bonds most strongly to metals, as the proton must be displaced for the metal to form a surface bond. However, the terminal OH^- group, once deprotonated, is a strong Lewis base that bonds favorably with trace metals.

As will be seen later in this chapter, the protonated terminal site is susceptible to ligand exchange by phosphate and other strongly bonding anions. These anions exchange H_2O readily from the terminal group in the $Al{-}OH_2^+$ form, as exchange of H_2O is easier than

1. It is possible to estimate the number of reactive sites by adding the reagent NaF to the mineral. The F^- anion displaces singly coordinated OH^- groups into solution, raising the pH.

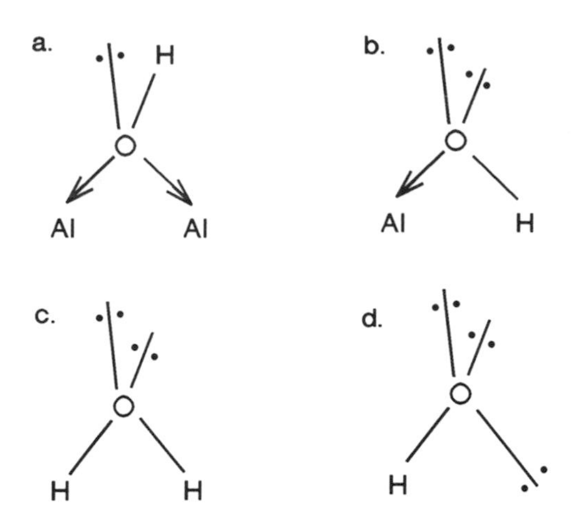

Figure 4.2. Schematic diagram of the oxygen atom in the bonding environment of (a) a bridging hydroxyl of an oxide, (b) a terminal hydroxyl of an oxide, (c) a water molecule, and (d) a free hydroxyl (OH^-) ion. (From M. B. McBride. 1989. Reactions controlling heavy metal solubility in soils. In B. A. Stewart (ed.), *Advances in Soil Science* 10:1–56.)

exchange of OH^-. It seems that the same terminal OH^- groups on minerals can adsorb *both* metal cations and anions, but by different mechanisms. Competition between cations and anions for adsorption is then possible, because valence-unsatisfied OH^- groups are common to both mechanisms. These reactive groups have the highest density per unit weight on poorly crystalline oxide and aluminosilicate surfaces, so that minerals with high metal adsorption capacities also tend to have high anion adsorption capacities. However, recent studies of Fe oxides suggest that anions and cations may prefer different site geometries, because metal cations bond at edges of Fe octahedra whereas oxyanions like phosphate bond to two adjacent octahedral corners, that is, to O atoms bonded to different Fe atoms (Manceau et al., 1992).

Since the adsorbing metal ion bonds directly, via an *inner-sphere mechanism,* to O atoms at the surface, the properties of the surface, and therefore the nature of the metal constituting the adsorption site, should dictate to some extent the tendency for adsorption. For example, if oxides of different metals are compared, it is found that the valence and coordination number of the constituent metal are correlated with reactivity. Comparing silica with alumina, $>Si-OH$ groups are found to be more acidic than $>Al-OH$ groups. The fact that Si has a valence of +4, a charge shared more or less equally by four coordinated O atoms, means that a surface OH group (terminal) on silica "feels" a charge of $+4/4 = +1$ from Si. By the same reasoning, OH (terminal) on alumina "feels" $+3/6 = +1/2$ charge from Al^{3+}. This means that electron density is drawn more strongly away from the OH of silica, thereby creating a stronger acid on the silica surface. Furthermore, the dissociated $-Si-O^-$ group is a weak Lewis base because of this polarization effect, and chemisorption of trace metals is not very favorable on silica.

In general, then, the ratio of metal valence to coordination number, referred to

Table 4.1. Acid Dissociation Constants of Oxide Surface OH Groups

Group	pK_1[a]	pK_2[b]	Valence/Coordination No.
>Si−OH	<2	6–7	$+4/4 = 1$
>Ti−OH	3–4	7–9	$+4/6 = 0.67$
>Fe−OH	6.5	9	$+3/6 = 0.5$
>Al−OH	5–7.5	8–10	$+3/6 = 0.5$

[a] $K_1 = \{S{-}OH\}[H^+]/\{S{-}OH_2{+}\}$ (S is the metal ion of the particular surface group).
[b] $K_2 = \{S{-}O^-\}[H^+]/\{S{-}OH\}$.
Source: Adapted from P. W. Schindler and W. Stumm. 1987. The surface chemistry of oxides, hydroxides and oxide minerals. In W. Stumm (ed.), *Aquatic Surface Chemistry.* New York: Wiley.

hereafter as "shared charge," seems to be a useful parameter in predicting surface acidity and reactivity of oxides, and the basic concept outlined here for silica and alumina can be applied to other metal oxides. The approximate surface acid dissociation constants for several oxides, compared with their ratios of valence to coordination number, are listed in Table 4.1.

Extending the argument above, which is based on simple electrostatics, the tendency of O atoms at surface sites to chemisorb metals should vary depending on the valence and coordination number of the metal constituting the adsorbent. That is, the adsorbing groups listed toward the bottom of Table 4.1 should chemisorb metals more effectively than those toward the top. For groups with very similar values of valence/coordination number, say Al−OH and Fe(III)−OH, the electronegativity of the constituent metal also has to be considered in explaining differences in trace metal adsorption behavior.

Electronegativity is an important factor in determining which of the trace metals chemisorb with the highest preference. The more electronegative metals should form the strongest covalent bonds with O atoms on any particular mineral surface. For some commonly studied divalent metals, the predicted order of bonding preference would be

$$Cu > Ni > Co > Pb > Cd > Zn > Mg > Sr$$

On the other hand, on the basis of electrostatics, the strongest bond should be formed by the metal with the greatest charge-to-radius ratio. This would produce a different order of preference for the same metals:

$$Ni > Mg > Cu > Co > Zn > Cd > Sr > Pb$$

and would also predict that trivalent trace metals such as Cr^{3+} and Fe^{3+} would chemisorb in preference to all of the divalent metals listed above.

Mn oxides show particularly high selectivity for Cu^{2+}, Ni^{2+}, Co^{2+}, and Pb^{2+}, perhaps indicating an important contribution from covalent bonding to adsorption. On the other hand, Fe and Al oxides as well as silica adsorb Pb^{2+} and Cu^{2+} most strongly of the divalent metals listed above, suggesting that neither a purely electrostatic nor a covalent model of surface bonding is adequate. Because Pb^{2+} and Cu^{2+} happen to be the two most easily hydrolyzed of the listed metals, it might be suggested that adsorption and hydrolysis are correlated in some way. Perhaps the metal adsorbs in a hydrolyzed form, for example, by the reaction

$$>S-OH + M^{n+} = >S-O-M-OH^{(n-2)+} + 2H^{+} \tag{4.4}$$

where S is the metal of the adsorbing surface. Or perhaps the metal-surface bonding reaction is favored by the same metal properties that favor the hydrolysis reaction—namely, a combination of high charge, small radius, and "softness" (polarizability). After all, there are chemical parallels between bonding of a metal ion with a surface O^{2-} or OH^-, and coordination of the same metal ion with OH^- in solution. This is made apparent by writing the chemisorption and hydrolysis reaction for the metal, M^{n+}:

$$\underset{\text{chemisorption}}{>S-OH + M^{n+} = >S-O-M^{(n-1)+} + H^{+}} \tag{4.5}$$

$$\underset{\text{hydrolysis}}{H-OH + M^{n+} = H-O-M^{(n-1)+} + H^{+}} \tag{4.6}$$

The tendency for easily hydrolyzed metals to appear to be strongly adsorbed on soil minerals may be caused in part by the manner in which adsorption is traditionally measured. Typically, the pH is adjusted over a range, and the amount of metal removed from solution is calculated from the change in solution concentration. As the pH is raised from an acid value, chemisorption is initially favored, but even before adsorption sites become saturated, metal ions cluster into metal oxide or

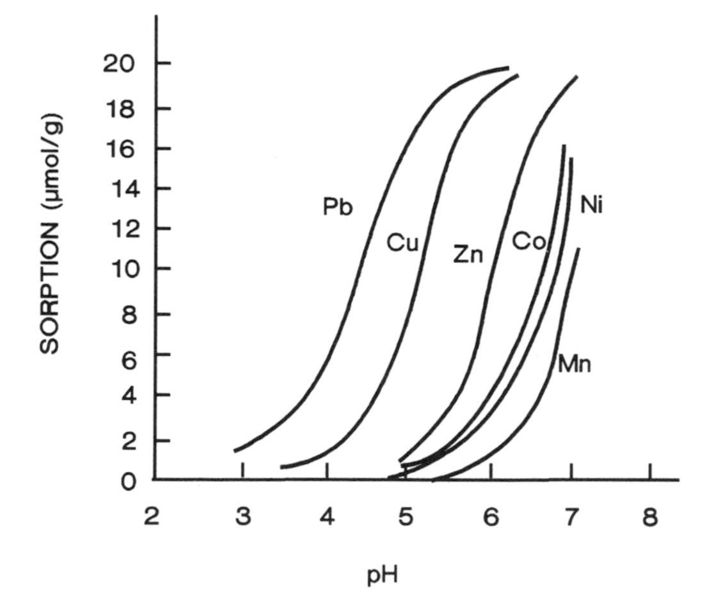

Figure 4.3. Metal cation adsorption versus pH on hematite. (Adapted from R. M. McKenzie. 1980. The adsorption of lead and other heavy metals on oxides of manganese and iron. *Aust. J. Soil Res.* 18:61–73.)

hydroxide nuclei at the surface. Ultimately, precipitation as a separate phase of the metal oxide or hydroxide ensues. This sequence of processes usually appears as a "sorption continuum," described by a smooth sorption isotherm such as that shown in Figure 4.3. Because the onset of metal chemisorption and the beginning of metal hydroxide precipitation are not often separated by a wide margin of pH, the chemisorption/nucleation/precipitation sequence is rarely resolved into discrete processes by clear discontinuities in the sorption curves.

Since precipitation ultimately removes all of the strongly hydrolyzing metals from solution as the pH is raised, isotherms of the type depicted in Figure 4.3 necessarily reach 100 percent "sorption" at high pH regardless of the nature and quantity of adsorbing surfaces. It is not possible to tell from the isotherm how much of the reported "sorption" is chemisorption and how much is precipitation; this depends on the relative quantities of reactive surface groups and adsorptive metals. Sorption of relatively large quantities of metals on a solid with limited bonding sites is likely to favor precipitation (usually as metal hydroxides or oxides) as the dominant process. In that event, the correlation between sorption and tendency for the metal to hydrolyze is expected to be high. The importance of precipitation as a mechanism of metal removal from solution will be discussed in a later section of this chapter.

4.1c. A Mathematical Model of Chemisorption

Surface hydroxyl groups of minerals protonate at low pH:

$$>S-OH + H^+ = >S-OH_2^+ \tag{4.7}$$

or dissociate (deprotonate) at high pH

$$>S-OH = >S-O^- + H^+ \tag{4.8}$$

The equilibrium constants for these two reactions, expressed as dissociations, are:

$$K_1 = \frac{\{S-OH\}[H^+]}{\{S-OH_2^+\}} \tag{4.9}$$

$$K_2 = \frac{\{S-O^-\}[H^+]}{\{S-OH\}} \tag{4.10}$$

where { } symbolize the concentrations of bracketed surface species in the adsorbing solid (absorbate).[1]

Once the surface hydroxyl groups dissociate, they serve as Lewis bases toward metal cations. The metals complex with one, or possibly two, of the deprotonated groups as follows:

$$>S-OH + M^{n+} = >S-OM^{(n-1)+} + H^+ \tag{4.11}$$

$$2 >S-OH + M^{n+} = >(S-O)_2M^{(n-2)+} + 2H^+ \tag{4.12}$$

1. To write equations 4.9 and 4.10 it is necessary to assume that activities of the bracketed surface species can be approximated by their molar concentrations.

The equilibrium "constants" for these two complexation reactions are

$$K_1^M = \frac{\{S-OM^{(n-1)+}\}[H^+]}{\{S-OH\}[M^{n+}]} \tag{4.13}$$

$$K_2^M = \frac{\{(S-O)_2M^{(n-2)+}\}[H^+]^2}{\{S-OH\}^2[M^{n+}]} \tag{4.14}$$

Consider now an actual adsorbing solid in aqueous suspension that has a finite concentration of reactive surface hydroxyl groups, symbolized by $\{S_T\}$. At any particular pH, these groups are partitioned into the protonated, uncharged, and deprotonated forms such that

$$\{S_T\} = \{S-OH_2^+\} + \{S-OH\} + \{S-O^-\} \tag{4.15}$$

Suppose that a metal ion, M^{n+}, is introduced into this suspension, but for the sake of simplicity it will be assumed that it adsorbs by reaction 4.11 only (no bidentate bonding as described by reaction 4.12). Now equation 4.15 must be modified to include the metal-occupied sites:

$$\{S_T\} = \{S-OH_2^+\} + \{S-OH\} + \{S-O^-\} + \{S-OM\} \tag{4.16}$$

By combining equations 4.9 and 4.13, the expression

$$K_1^M = \frac{\{S-OM\}[H^+]}{K_1(\{S-OH_2^+\}/[H^+])[M^{n+}]} \tag{4.17}$$

is obtained, and an expression for $\{S-OH_2^+\}$ from equation 4.16 can then be substituted into equation 4.17, giving

$$K_1^M = \frac{\{S-OM\}[H^+]^2}{K_1(\{S_T\}-\{S-OH\}-\{S-O^-\}-\{S-OM\})[M^{n+}]} \tag{4.18}$$

Substituting for $\{S-OH\}$ and $\{S-O^-\}$ from equations 4.13 and 4.10, respectively, and rearranging gives

$$\frac{\{S-OM\}}{\{S_T\}} = \frac{K_1^M[M^{n+}]}{\left(\frac{[H^+]^2}{K_1}\right) + [H^+] + K_2 + K_1^M\,[M^{n+}]} \tag{4.19}$$

which is the general solution for the fraction of surface sites occupied by metal ion, $\{S-OM\}/\{S_T\}$. For a given quantity of solid adsorbent, this fraction is directly proportional to the amount of metal ions adsorbed from solution.

There are several ways that equation 4.19 could be used in describing chemisorption; these will be discussed as three separate situations.

Situation 1: *Adsorption of the metal from solution at constant pH, as a function of metal ion concentration,* $[M^{n+}]$

This experiment is commonly done by buffering the solution or using a pH controller to prevent the adsorption process itself from lowering the pH. It pertains to soil environments because of the strong buffering effect of soils on solution pH.

At constant pH, equation 4.19 takes the form:

$$\frac{\{S-OM\}}{\{S_T\}} = \frac{K_1^M[M^{n+}]}{B + K_1^M[M^{n+}]} = \frac{[M^{n+}]}{B/K_1^M + [M^{n+}]} \tag{4.20}$$

where the constant, B, is given by

$$B = \frac{[H^+]^2}{K_1} + [H^+] + K_2 \tag{4.21}$$

Equation 4.20 is a Langmuir function that, if plotted, takes different shapes depending on the nature of the adsorbing surface and the pH, as shown in Figure 4.4. At low pH, B is relatively large, which both reduces chemisorption and alters the shape of the adsorption isotherm. The values of K_1 and K_2, properties of the surface, change the adsorption function by altering B. A larger K_1 decreases B (favors chemisorption); K_1 measures the tendency of protonated surface sites to dissociate (protonated sites cannot bond with metal ions). A larger K_2 increases B and would seem to disfavor adsorption, contrary to expectation, because K_2 is a measure of the tendency of sites

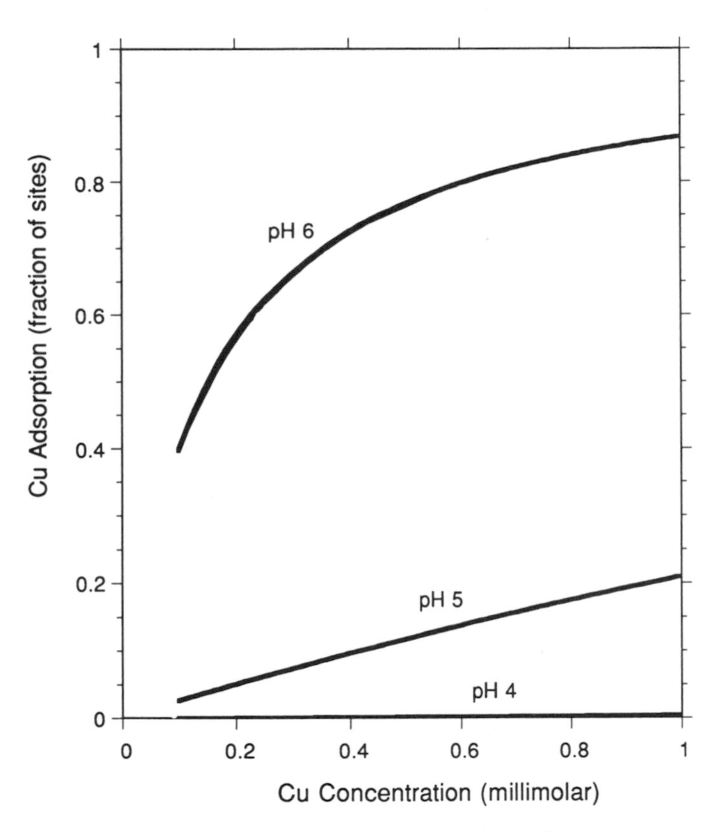

Figure 4.4. Predicted adsorption of Cu^{2+} as a function of Cu^{2+} concentration on an Al hydroxide having acid dissociation constants of $K_1 = 10^{-5.3}$ and $K_2 = 10^{-8.1}$, and a complexation constant of $K_1^{Cu} = 10^{-2.1}$. Surface bonding by the bidentate mechanism is neglected for this calculation.

to dissociate to the anionic form that actually bonds with metals. It is necessary, however, to view the metal bonding reaction (4.11) as a two-step process

$$\underset{\text{dissociation}}{>\text{S}-\text{OH} = >\text{S}-\text{O}^- + \text{H}^+} \tag{4.22}$$

$$\underset{\text{complexation}}{>\text{S}-\text{O}^- + \text{M}^{n+} = >\text{S}-\text{OM}^{(n-1)+}} \tag{4.23}$$

Only the first of these two reactions is defined by the dissociation constant K_2. The second reaction is defined by the intrinsic bonding constant K_1^B. The two reactions summed together give the metal bonding reaction, so that the previously defined constant K_1^M actually equals the product of two constants, $K_2K_1^B$. On the basis of this more complete analysis, it can be shown that the B/K_1^M term in equation 4.20 becomes

$$\frac{B}{K_1^M} = \frac{[\text{H}^+]^2}{K_1K_2K_1^B} + \frac{[\text{H}^+]}{K_2K_1^B} + \frac{1}{K_1^B} \tag{4.24}$$

Now it is clear that a larger K_2 actually *decreases* the magnitude of B/K_1^M, which, according to equation 4.20, should increase metal adsorption. That is, more acidic surface (S—OH) groups should adsorb metals more readily. Similarly, a larger K_1^B, reflecting a stronger surface oxygen-metal bond, should favor adsorption. Stronger bonds form at surface $\text{S}-\text{O}^-$ groups that are strong Lewis bases. However, surface groups that are more acidic tend at the same time to be weaker Lewis bases, and the net effect of these two opposing trends on metal chemisorption is not easily predicted. In any event, the predictions of the chemisorption model (equation 4.19), are in qualitative agreement with results of metal adsorption experiments.

Situation 2: *Adsorption of the metal as a function of pH at constant metal ion concentration,* $[M^{n+}]$, *in solution*

This experiment is not commonly done because it is difficult to maintain a measurable concentration of the soluble metal at higher pH. It could conceivably pertain to situations in soils where cation exchange on silicate clays "buffers" the metal ion concentration at the same time that adsorption on variable-charge minerals occurs. It is an unlikely situation that will not be analyzed further.

Situation 3: *Adsorption of the metal as a function of pH at uncontrolled metal ion concentration,* $[M^{n+}]$, *in solution*

In this most common type of experiment, the total quantity of metal in the system, M_T, is held constant while the pH is raised. It pertains to the quite feasible situation in soils where a particular quantity of metal pollutant has to be made less soluble by raising the pH with liming materials. Initially, at very low pH, all of the metal is in solution, but as the pH is adjusted higher, part of the metal is adsorbed. The mole quantity of metal adsorbed is $\{\text{S}-\text{OM}\} \times W$, where $\{\text{S}-\text{OM}\}$ has units of moles of metal bonded per unit weight of adsorbent and W is the weight of the adsorbent in the suspension. The mole quantity of metal in solution is $[\text{M}^{n+}] \times V$, where $[\text{M}^{n+}]$ has units of moles per unit volume and V is the volume of solution in the suspension. Mass balance then requires that

$$M_T = \text{constant} = \{S-OM\} \cdot W + [M^{n+}] \cdot V \tag{4.25}$$

This relationship allows $[M^{n+}]$ to be expressed in terms of controlled experimental variables of interest, causing the general adsorption equation 4.19 to take the form

$$\frac{\{S-OM\}}{\{S_T\}} = \frac{(K_1^M/V)(M_T - \{S-OM\} \cdot W)}{[H^+]^2/K_1 + [H^+] + K_2 + (K_1^M/V)(M_T - \{S-OM\} \cdot W)} \tag{4.26}$$

This is a quadratic equation in terms of the variable of interest, $\{S-OM\}$, the amount of metal adsorbed, as is demonstrated by rearranging equation 4.26:

$$\{S-OM\}^2 \frac{K_1^M W}{V} - \{S-OM\} \left(\frac{\{S_T\} K_1^M W}{V} + \frac{[H^+]^2}{K_1} + [H^+] + K_2 + \frac{K_1^M M_T}{V} \right) + \frac{\{S_T\} K_1^M M_T}{V} = 0 \tag{4.27}$$

To solve this equation for $\{S-OM\}$, the "constants" K_1, K_2, and K_1^M must be known, and the system variables $\{S_T\}$, W, V, M_T, and $[H^+]$ need to be specified. This means that the weight of the adsorbent, the volume of the suspension, the total quantity of metal adsorbate present, and the pH all have to be specified before the amount of adsorbed metal is determined.

This chemical model does not deal with the observed fact that the "constants" K_1, K_2, and K_1^M are not true constants, especially as a high level of metal adsorption is reached. Specifically, the measured value of the metal binding constant, K_1^M, decreases at high adsorption levels. This may result from the presence of numerous different types of metal adsorption sites (and mechanisms) at mineral surfaces, or it may reflect the inadequacy of assumptions in this simple model of surface dissociation and complexation that is based on principles of solution coordination chemistry. The latter explanation is usually invoked, since it is believed that a small part of the surface positive and negative charge, developed at low and high pH, respectively, may not be balanced by ions in direct coordination with the surface. Instead, a fraction of these ions are thought to occupy a "diffuse layer" near the surface, causing the mineral particle surface to develop an electrical potential. This potential would explain the electrophoretic mobility of oxide and other mineral particles that are devoid of permanent charge. The electrostatic field associated with the "diffuse layer" could either promote or suppress ionic adsorption. For example, as metal cations adsorb, their contribution of positive charge to the surface might repel other metal cations. How important the "diffuse layer" really is on variable-charge minerals of soils remains to be established. For some mineral surfaces, notably sulfides, effects of a "diffuse layer" on ion adsorption appears to be negligible.

Some adsorption models attempt to adjust for the nonconstant nature of the equilibrium "constants." The *constant-capacitance model* uses K values for dissociation and metal complexation that are adjusted for the effect of the electrical potential at the surface, employing the empirical reaction

$$\log K \text{ (adjusted)} = \log K_{int} - C \cdot \{S-OM\} \tag{4.28}$$

where K_{int} is the intrinsic K value for the surface groups and C is a constant. This equation produces an effective K value that decreases as more metal cations are adsorbed.

Example Problem: Consider aluminum hydroxide $Al(OH)_3$, with 100 mmoles/kg of reactive surface Al—OH groups, 1.0 g of which is suspended in 100 ml of 5×10^{-4} *M* $CuCl_2$ solution. For this system, the controlled experimental variables are

$$\{S_T\} = 0.1 \times 10^{-3} \text{ moles/g}$$
$$W = 1.0 \text{ g}$$
$$V = 0.1 \text{ liter}$$
$$M_T = 0.05 \times 10^{-3} \text{ moles } (50\ \mu\text{moles}) \text{ Cu}$$

The dissociation and Cu^{2+} complexation constants for $Al(OH)_3$ are estimated to be

$$K_1 = 10^{-5.3} \text{ moles/liter}$$
$$K_2 = 10^{-8.1} \text{ moles/liter}$$
$$K_1^{Cu} = 10^{-2.1} \text{ (unitless)}$$

and for this hypothetical system, metal bonding via two surface groups (bidentate) is ignored. Determine how the chemical model predicts the metal adsorption level, {S—OM}, to increase as the solution pH is raised.

Solution: The values (listed above) for the variables and constants in this system are entered into equation 4.27, and the pH is set to several fixed values between 4 and 7. From these solutions for the amounts of adsorbed metal, {S—OM}, adsorption curves, plotted in Figure 4.5 for the Cu level of 50 μmoles, are formed. Most of the Cu^{2+} chemisorption is predicted to occur in the pH range 4.5 to 6.5.

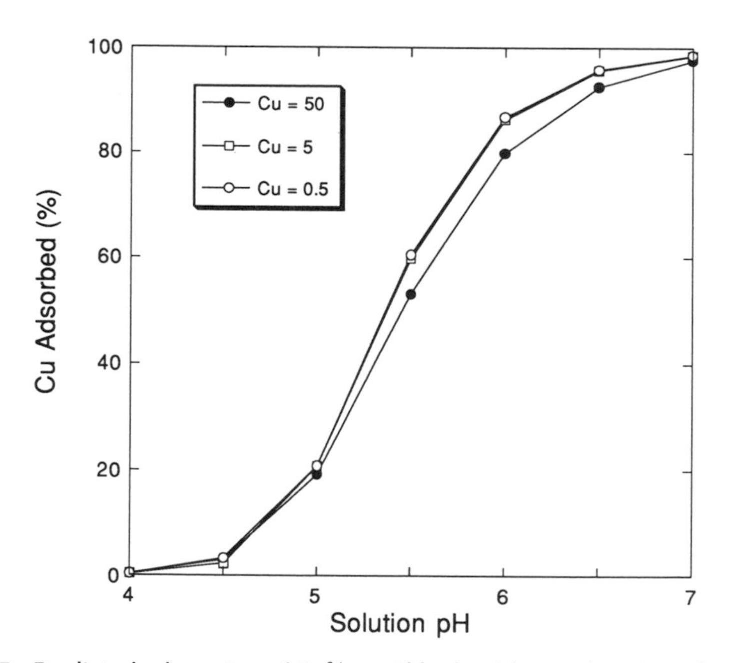

Figure 4.5. Predicted adsorption of Cu^{2+} on Al hydroxide as a function of pH. Three quantities of Cu^{2+}—50, 5, and 0.5 μmoles—are compared. The constants and assumptions used in this calculation are the same as those for Figure 4.4.

The above problem can be solved for smaller quantities of Cu^{2+} introduced into the $Al(OH)_3$ suspension. The result, as shown by a comparison of the curves for 50, 5, and 0.5 μmoles of Cu (Figure 4.5), reveals that the total quantity of Cu^{2+} present in the adsorption system, M_{Cu}, has an effect on chemisorption, but pH is the overriding factor *unless* sufficient Cu is present to saturate the surface bonding sites. As bonding sites approach saturation, the model predicts that adsorption curves shift toward higher pH, as is noted for the 50-μmole level of Cu in Figure 4.5 compared with the lower Cu levels. This arises from the effect of a high degree of occupancy of adsorption sites in lowering the tendency for further chemisorption.

The pH-dependent chemisorption seen in Figure 4.5 provides an explanation for the general observation in soil clays that specific adsorption of trace metal cations increases with pH. That is:

> *Adsorbed trace metals become increasingly nonexchangeable as the pH is raised.*

This is particularly true for strongly hydrolyzing metals.

4.1d. Selectivity and pH in Trace Metal Cation Sorption

The importance of metal/adsorbent ratio has been confirmed experimentally in the Zn^{2+}-goethite and Cd^{2+}-goethite systems, where reducing the quantity of goethite while increasing the less reactive adsorbent, kaolinite, to maintain the same total quantity of adsorbent shifts the S-shaped adsorption curves of trace quantities of Zn^{2+} and Cd^{2+} to higher pH (Tiller et al., 1984). The same general tendency is seen for trace metal adsorption in soils, but the explanation may lie partly in the heterogeneous nature of soil surfaces and the fact that the first-occupied adsorption sites form stronger bonds with the metal. Furthermore, soil clays possess an ability to adsorb trace levels of metals at much lower pH than expected from the variable-charge mineral model developed in section 4.1c. There is some indication that this low-pH adsorption is due to cation exchange on permanent charge clays or organic matter, even though most experiments that attempt to measure specific adsorption employ a background electrolyte such as 10^{-2} M $CaCl_2$, thereby inhibiting exchange adsorption of the trace metal cation by competition. It must be kept in mind, however, that only a very small fraction of exchange sites need be occupied to adsorb a significant fraction of low-concentration trace metal in solution, so that excess Ca^{2+} in solution may not fully inactivate ion exchange as a mechanism of trace metal adsorption. In any event, trace metal sorption in soil clays containing layer silicates tends to begin at lower pH and increase more gradually with pH than the S-shaped curves of Figure 4.5 predict.

The critical importance of metal (adsorbate) to soil (adsorbent) ratio can be stated in a basic rule:

> *The preference of adsorbing surfaces for trace metal cations (relative to the prevalent cation in soil solution, usually Ca^{2+}) diminishes with increasing adsorption level.*

Preference or affinity is measured by a selectivity or distribution coefficient (see section 3.5). The reduction of this selectivity with increased adsorption is most pronounced in systems favorable to chemisorption; that is, in cases where the affinity of the trace metal for the soil surfaces is high at low adsorption levels. For example, strongly hydrolyzing metals added to soils with pH 6 or higher produce high selectivity coefficients that diminish markedly at higher metal loadings.

In severely metal-polluted soils, hydrolysis and precipitation can remove hydrolysis-prone metals from solution as the pH approaches neutrality, so that experimental "sorption" curves, which include both chemisorption and precipitation, tend to be more abrupt than the one shown in Figure 4.5. For example, as the pH of the Cu/$Al(OH)_3$ system is adjusted upward, copper hydroxide can precipitate if insufficient adsorption has occurred to keep the $(Cu^{2+})(OH^-)^2$ activity product below the solubility product of $Cu(OH)_2$. In the absence of adsorption, 10^{-5}, 10^{-4}, and 10^{-3} M Cu^{2+} would begin to be removed from solution as $Cu(OH)_2$ at pH 6.8, 6.3, and 5.8, respectively. This means that, in contrast to metal adsorption curves, metal hydroxide and oxide precipitation curves shift to lower pH as the total metal in the system increases.

Each trace metal has its own characteristic S-shaped "sorption" curve, as exemplified by the Cu^{2+} sorption curves of Figure 4.5. In general, as was pointed out earlier, tendency toward sorption seems to be correlated with ease of metal hydrolysis. Consequently, the sorption curve for a strongly hydrolyzing metal such as Fe^{3+} is centered at lower pH than that shown in Figure 4.5, while the curve for a weakly hydrolyzing metal such as Cd^{2+} is found at higher pH. Sorption curves for a number of metals with different hydrolyzing tendencies are shown in Figure 4.3.

4.1e. Desorption of Chemisorbed Metals

Specific adsorption of metal cations on soil clays is slower than cation exchange, often increasing gradually over several days. Trace metal cations adsorbed on these clays gradually lose much of their initial lability (as measured by diminishing self-exchange rates) over a period of days. Consequently, trace metal adsorption on soils and the mineral components of soils is considered to be highly nonreversible.

The model of chemisorption developed in subsection 4.1c, based as it is on the assumption of equilibrium reactions with defined K values, gives no hint that the chemisorption of metals might be nonreversible. And yet, nonreversibility seems to be the rule rather than the exception. In other words, the curves shown in Figure 4.5 might only be valid for describing the forward (adsorption) reaction as the pH is raised. If the pH is subsequently lowered, the desorption curve is likely to be shifted to lower pH relative to the adsorption curve; that is, less Cu^{2+} desorbs at each particular pH than expected (assuming that the forward adsorption curve represents equilibrium).

Generally, sorption of metals seems to be more nearly reversible at low than at high pH. This may arise from the fact that the monodentate complexation reaction (4.11) should give way to the bidentate reaction (4.12) at higher pH. The latter reaction, involving two metal-surface bonds, is expected to have a very slow rate in the reverse direction (desorption). Studies of heavy metal bonding on pure oxides have indicated that the adsorption reaction step is fast and probably diffusion controlled ($k_f \approx 10^5$ moles^{-1} liter sec^{-1}), whereas the desorption reaction step has a rate constant that may be as much as three orders of magnitude slower.

Adsorption may or may not require a significant activation energy, E_a, depending on whether a strong bond must be broken to allow the metal ion to coordinate to the surface. The metal may have to lose one or more H_2O molecules of its hydration shell, for example. In contrast, assuming that the adsorption reaction is energetically favorable (exothermic), desorption always requires an activation energy to at least overcome the adsorption energy (see Chapter 1, section 1.2g). Consequently, many chemisorption reactions have a much higher activation energy in the reverse than the forward direction. The Arrhenius equation (1.101) then predicts that the adsorption reaction should be faster than the desorption reaction. The commonly observed "nonreversibility" in metal adsorption may be the result of the long time period that would be required for desorption to be complete. This would be particularly noticeable in the case of bidentate metal bonding, presumed to be an energetically very favorable reaction. The large activation energy for desorption would derive from the need to break two metal-surface bonds.

4.2. SORPTION OF ANIONS

Anions are known to attach to the oxide and silicate mineral fraction of soils. However, certain anions are able to bond to soil organic matter as well. Borate, $B(OH)_4^-$, is notable in this regard, bonding according to a reaction of the type:

$$\begin{matrix} >C-OH \\ >C-OH \end{matrix} + B(OH)_4^- = \begin{matrix} >C-O \\ >C-O \end{matrix} B \begin{matrix} OH \\ OH \end{matrix} \qquad (4.29)$$

where the involved organic groups can be aliphatic or aromatic. Some anions may bond indirectly to organic groups through a bridging metal ion such as Al^{3+} or Fe^{3+}. Nevertheless, most anions adsorb very little in humus, and it is safe to say that anion bonding at mineral surfaces accounts for most of the anion retention in soils. It is this latter process that will be discussed in detail in this section.

4.2a. Chemisorption of Anions on Variable-Charge Minerals

As with metal cations, anion chemisorption occurs on soil minerals that possess surface hydroxyl groups. The most important minerals in this regard are noncrystalline aluminosilicates (allophanes); oxides and hydroxides of Fe, Al, and Mn; and layer silicate clays (edge sites only). It is the H_2O or valence-unsatisfied OH^- ligands bound to surface metal ions (usually Fe, Al, or Mn) that are the sites of chemisorption. In general terms, the surface reaction can be written

$$>S-OH + A^{n-} = >S-A^{(n-1)-} + OH^- \qquad K_A^1 \qquad (4.30)$$

or, for the binuclear reaction:

$$2 >S-OH + A^{n-} = \begin{matrix} >S \\ >S \end{matrix} A^{(n-2)-} + 2OH^- \qquad K_A^2 \qquad (4.31)$$

where A^{n-} is an anion[1] of charge $-n$, and $>S-OH$ is a reactive metal hydroxyl group. Reactions 4.30 and 4.31 are termed *ligand exchange reactions* because the anion displaces OH^- or H_2O from coordination positions of a metal ion at the surface. These reactions are favored by low pH, as is evident from the release of OH^- into solution. Low pH causes surface OH^- groups to accept protons, and since H_2O is an easier ligand to displace from metal bonding sites than OH^-, this facilitates the ligand exchange. Therefore, from the point of view of kinetics as well as equilibrium thermodynamics, low pH promotes anion adsorption.

As was pointed out in Chapter 3, anions may adsorb on positively charged mineral surfaces by anion exchange as well, a process involving nonspecific electrostatic forces. Ligand exchange is distinguished from anion exchange based on the following characteristics:

1. Release of OH^- into solution
2. A high degree of specificity shown toward particular anions
3. A tendency to be nonreversible, or at least for desorption to be much slower than adsorption
4. A change in the measured surface charge to a more negative value

The rules that describe reactivity of minerals toward anions are much the same as those outlined for cation adsorption. Specifically, the most reactive groups are valence-unsatisfied surface hydroxyl ions, found at layer silicate edges, on oxides of Fe, Al, and Mn and on noncrystalline aluminosilicates. Reactive groups are most numerous on noncrystalline minerals such as ferrihydrite, allophane, and imogolite. Terminal OH^- groups, such as that depicted in Figure 4.2b, readily protonate according to reaction 4.7, and it is thought to be this protonated form that actually enters into the ligand exchange with anions.

The reasons for selectivity shown by particular anions for particular mineral surfaces are not very well understood. There is a generally valid rule that

> *If an anion shows a tendency to bond strongly with a particular metal ion in solution, it will show a comparably strong affinity for a surface M−OH site composed of the same metal.*

Often, strong anion-metal ion attraction is manifested as a tendency to form insoluble precipitates. Consider, for example, the contrasting degree to which anions such as phosphate, sulfate, and nitrate complex with Fe^{3+}. Phosphate pairs strongly with Fe^{3+}, resulting in the precipitation of the very insoluble solid, $FePO_4$. Sulfate pairs less strongly, and ferric sulfate salts are correspondingly more soluble. Nitrate does not pair in solution with Fe^{3+}, and ferric nitrate is a very soluble salt. Not coincidentally, chemisorption of phosphate, sulfate, and nitrate to ferric oxides follows this same order, with phosphate being the most strongly adsorbed anion.

The reason for comparing solution reactions to surface reactions here is simple: in both cases, an $Fe^{3+}-$anion bond (or bonds) is formed. It is reasonable to expect that the bond energy should be about the same in solution and at the surface; consequently, a tendency for anions to form ion pairs (or precipitate) transfers to a sim-

1. Only inorganic anions are being considered in this chapter, although organic anions bond by similar mechanisms, as discussed in Chapter 10.

Table 4.2. Chemical Characteristics of Important Anions in Soil Chemistry[a]

Oxyanion	Formula	"Shared Charge"	Electronegativity
Borate	$B(OH)_4^-$	3/4 = 0.75	2.04
Silicate	SiO_4^{4-}	4/4 = 1.0	1.90
Hydroxyl	OH^-	1/1 = 1.0	2.2
Phosphate	PO_4^{3-}	5/4 = 1.25	2.19
Arsenate	AsO_4^{3-}	5/4 = 1.25	—
Selenite	SeO_3^{2-}	4/3 = 1.33	—
Carbonate	CO_3^{2-}	4/3 = 1.33	2.55
Molybdate	MoO_4^{2-}	6/4 = 1.5	2.35
Chromate	CrO_4^{2-}	6/4 = 1.5	—
Sulfate	SO_4^{2-}	6/4 = 1.5	2.58
Selenate	SeO_4^{2-}	6/4 = 1.5	—
Nitrate	NO_3^-	5/3 = 1.67	3.04
Perchlorate	ClO_4^-	7/4 = 1.75	3.16
Halide			
Fluoride	F^-	—	3.98
Chloride	Cl^-	—	3.16
Bromide	Br^-	—	2.96
Iodide	I^-	—	2.66

[a]The shared charge and the electronegativity (Pauling) are properties of the oxygen atom (Lewis base) and central atom of the oxyanion, respectively. For the halide anions, shared charge is not a meaningful concept.

ilar tendency to chemisorb on metal oxides and other minerals. A complication of these interrelated tendencies is that chemisorption and precipitation of anions are almost inseparable participating processes in "sorption," because conditions that favor one generally favor the other.

Many elements of environmental concern, Cr, As, and Se, for example, occur as oxyanions in soils, some adsorbing much more strongly than others on oxides and other variable-charge minerals. Some anions of interest in soil chemistry are listed in Table 4.2, ranked according to a measure of the positive charge that the oxyanion's central atom shares formally with each bonded O atom. This "shared charge" is determined by dividing the valence of the central atom by the number of bonded O atoms.[1] Assuming that it is the deprotonated O atoms of oxyanions that actually bond with metals such as Fe^{3+} and Al^{3+} on mineral surfaces, then the smaller the "shared charge," the greater the effective negative charge residing on each O atom, and the stronger the metal-oxyanion ionic bond. The concept, demonstrated for phosphate in Figure 4.6, seems to work fairly well because the ranking of anions in Table 4.2 according to "shared charge" follows approximately the order of anion affinity for oxides and alumino-silicates. In cases of oxyanions with comparable values of "shared charge," further chemical factors such as electronegativity have to be

1. For weak acid oxyanions that contain H atoms, such as $B(OH)_4^-$, an argument can be made to modify the "shared charge" definition to include the +1 charge shared by the proton with the O atom, but this has not been done in Table 4.2.

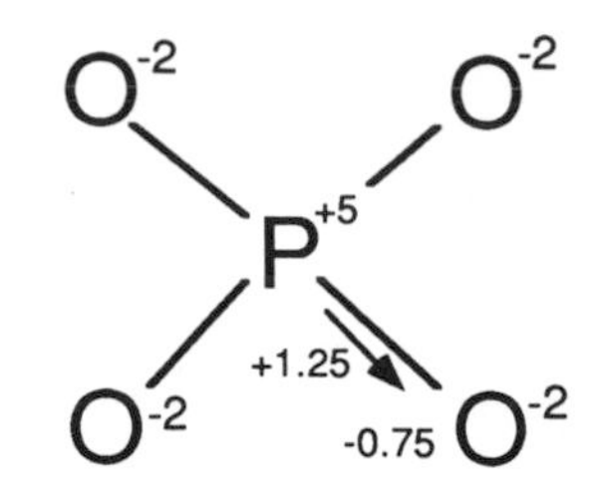

Figure 4.6. Schematic diagram of charge sharing by the P^{5+} atom with oxygen in the phosphate anion.

considered because there are covalency contributions to the strength of the oxyanion-surface bond. For example, chromate bonds more strongly than selenate on Fe hydroxide at any particular pH even though these two oxyanions have the same "shared charge."

With the exception of fluoride, F^-, the halide anions listed in Table 4.2 bond by outer-sphere electrostatic attraction, and are therefore adsorbed only if variable-charge mineral surfaces are positively charged. This means that all halides but F^- rank low on the scale of selectivity of minerals for anions. The strong F^--surface bond is explained by the high electronegativity and the small charge to radius ratio of F^-, ensuring an energetic anion-surface association of the inner-sphere type.

In summary, those anions that rank high on Table 4.2 are chemisorbed by ligand exchange on soil minerals, while those that rank lower tend to be adsorbed by non-specific anion exchange. Anions of high rank readily displace those of lower rank by a competitive process of ligand exchange, unless adsorption sites are available in excess. For example, phosphate displaces arsenate from bonding sites in soils, and sulfate suppresses chromate adsorption by soil minerals.

Several of the oxyanions, including phosphate, arsenate, and selenite, are believed to chemisorb on oxides by a binuclear bridging mechanism:

$$\begin{matrix} >S{-}OH \\ >S{-}OH \end{matrix} + MO_y^{n-} = \begin{matrix} >S{-}O \\ >S{-}O \end{matrix}\!\!>MO_{y-2}^{(n-2)-} + 2OH^- \tag{4.32}$$

Because of the energy of such an association, nonreversibility in adsorption-desorption behavior is both expected and observed. The association is further stabilized by entropy, since desorption requires the improbable event that two bonds be simultaneously broken. These binuclear anion-surface complexes may form slowly and decompose at an even slower rate.

Since anions associate with protons more or less easily depending on their acidity, and proton dissociation from the anion is a necessary step in the adsorption process, anion acidity and pH fundamentally influence the chemisorption of anions. For example, silicate in solution takes the form of uncharged monosilicic acid, $Si(OH)_4$, unless the pH is quite high. Consequently, Si adsorption on oxides increases as the pH is raised from 3 to about 9, a result of the dissociation of the weakly acidic $Si(OH)_4$ to silicate anions at high pH. Similarly, boron in solution takes the form of the neutral $B(OH)_3^0$ molecule, a very weak acid. Boron adsorbs to a greater extent at pH 9 than at lower pH, as the higher pH favors conversion to the borate anion, $B(OH)_4^-$. In contrast, most of the oxyanions listed in Table 4.2, being anions of strong acids (H_3PO_4, H_2SO_4, HNO_3, etc.), form negatively charged species in solution even at low pH, and their dissociation is sufficiently easy that it does not impede chemisorption. These anions therefore adsorb most effectively at *low* pH, as the general anion adsorption equations (4.30 and 4.31) predict should be the case. Given that the effect of pH on adsorption can be different for different anions, generalizations about the preference of mineral surfaces for one anion over another must be qualified. The result is the following set of rules for adsorption on silicates and oxides:

> *Oxyanions of weak acids chemisorb optimally at moderate to high pH.*
> *Oxyanions of strong acids chemisorb optimally at low pH.*
> *Oxyanion adsorption has a tendency to maximize at a pH near the* pK_a *value of the protonated form of the adsorbing anion.*
> *Anion adsorption at high pH is disfavored by competition with* OH^- *and carbonate and by surface charge.*

4.2b. A Mathematical Model of Anion Adsorption

Consider an adsorbing variable-charge mineral with a concentration of reactive surface hydroxyl groups, $\{S_T\}$, defined earlier in this chapter (section 4.1c). As before, at any particular pH these groups can be protonated, uncharged, or deprotonated, so that equation 4.15 applies. The association and dissociation reactions of the groups are defined as before (reactions 4.7 and 4.8) using the equilibrium constants, K_1 and K_2. The assumptions and weaknesses inherent to this approach are described in section 4.1c.

Now an anion, A^{n-}, is introduced into the system. For simplicity, we will assume that only the monodentate surface bonding reaction (4.30) is important.[1] The equilibrium constant for chemisorption of this anion is then

$$K_A^1 = \frac{\{S{-}A\}[OH^-]}{\{S{-}OH\}[A^{n-}]} \tag{4.33}$$

where the brackets are defined as in section 4.1c. Equation 4.15 must now be modified to account for surface sites involved in anion adsorption:

$$\{S_T\} = \{S{-}OH_2^+\} + \{S{-}OH\} + \{S{-}O^-\} + \{S{-}A\} \tag{4.34}$$

Entering expressions for $\{S{-}OH_2^+\}$, $\{S{-}OH\}$, and $\{S{-}O^-\}$ from equations 4.9, 4.10, and 4.33 into equation 4.34 gives

$$\{S_T\} = \frac{\{S{-}A\}[OH^-][H^+]}{K_A^1K_1[A^{n-}]} + \frac{\{S{-}A\}[OH^-]}{K_A^1[A^{n-}]} + \frac{\{S{-}A\}[OH^-]\,K_2}{K_A^1[A^{n-}][H^+]} + \{S{-}A\} \tag{4.35}$$

which becomes upon rearrangement:

$$\frac{\{S{-}A\}}{\{S_T\}} = \frac{K_A^1K_1[A^{n-}]}{K_W + (K_1K_W/[H^+]) + (K_1K_2K_W/[H^+]^2) + K_A^1K_1[A^{n-}]} \tag{4.36}$$

1. This is for the sake of mathematical simplification rather than the belief that the binuclear reaction is not important.

where $K_W = [H^+][OH^-]$ is the dissociation constant of water. For adsorption at a fixed pH, equation 4.36 can be greatly simplified to

$$\frac{\{S-A\}}{\{S_T\}} = \frac{[A^{n-}]}{C + [A^{n-}]} \tag{4.37}$$

where C is a constant defined by the equation

$$C = \frac{K_W}{K_A^1}\left(\frac{1}{K_1} + \frac{1}{[H^+]} + \frac{K_2}{[H^+]^2}\right) \tag{4.38}$$

Equation 4.37 is the mathematical description of the adsorption isotherm of an anion, A^{n-}, at a controlled pH. It has the form of the well-known Langmuir equation used to describe chemisorption.

To apply equation 4.37 to a real situation, we can use sulfate adsorption on Al hydroxide as the example. For this oxide, K_1 and K_2 are fairly well known ($K_1 = 10^{-5.3}$ moles/liter, $K_2 = 10^{-8.1}$ moles/liter), and assuming that sulfate adsorption can be adequately described by the monodentate reaction:

$$>Al-OH + SO_4^{2-} = >Al-O-SO_3^- + OH^- \qquad K_A^1 = 10^{-5.8} \tag{4.39}$$

then the constant, C, in equation 4.37 is determined by the values of K_1, K_2 and K_A^1. The resulting isotherms for sulfate adsorption on $Al(OH)_3$ at three chosen pH values then take the form revealed in Figure 4.7. These isotherms are the Langmuir type,

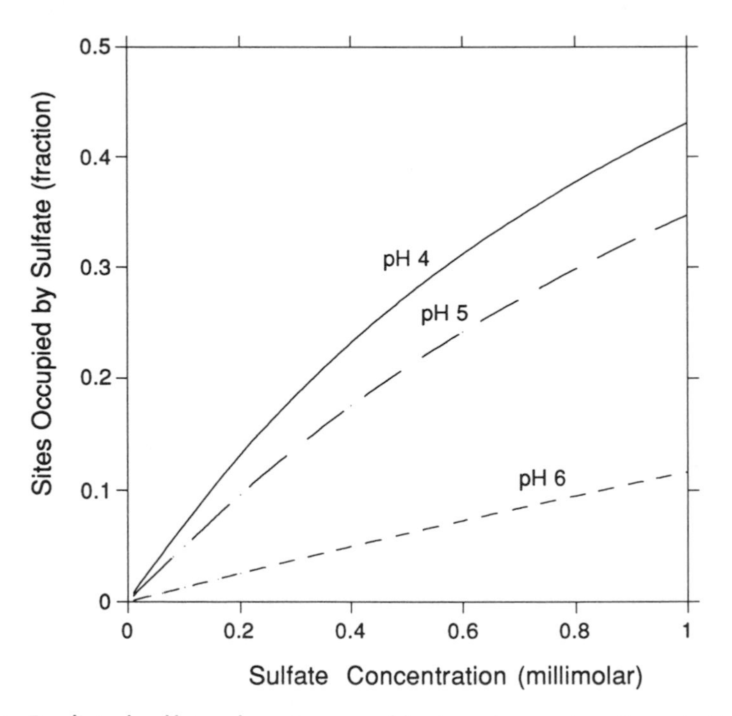

Figure 4.7. Predicted sulfate adsorption on Al hydroxide as a function of pH and sulfate concentration in solution. (See text for dissociation and complexation constants used in the calculation.)

which can be shown by defining Θ_A as the fraction of reactive sites occupied by adsorbed anions:

$$\Theta_A = \frac{\{S-A\}}{\{S_T\}} \tag{4.40}$$

The rearrangement of equation 4.37 then gives the familiar form of the Langmuir function:

$$[A^{n-}] = C\left[\frac{\Theta_A}{1-\Theta_A}\right] \tag{4.41}$$

Strongly retained anions such as phosphate tend to adsorb on soil minerals according to this function, where the saturation of sorption sites is interpreted to mean that $\Theta_A = 1$. The success of equation 4.41 for many soils means that phosphate solubility is often controlled by the degree of saturation of sorption sites. Furthermore, adherence to the Langmuir function by anions that chemisorb strongly means that, as the adsorption capacity of a soil is approached, solubility increases abruptly. Consequently, as soils approach their saturation limit for adsorption, they are much more likely to allow anions to leach.

In most soil systems, anion adsorption shows more complex behavior than that implied by this model because the adsorption capacity seems to increase over time. Such a time-dependent phenomenon may be the consequence of slow precipitation reactions superimposed on the more rapid chemisorption. Figure 4.8 illustrates that,

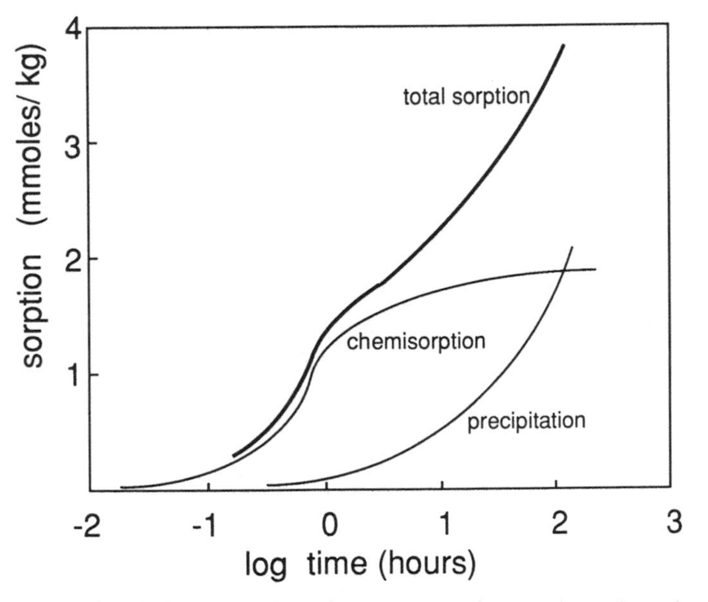

Figure 4.8. Time-dependence of phosphate sorption by a soil, attributed to the difference in chemisorption and precipitation kinetics. Total sorption is taken to be the sum of chemisorption and precipitation. (Adapted from S.E.A.T.M. Van der Zee and W. H. Van Riemsdijk. 1991. Model for the reaction kinetics of phosphate with oxides and soil. In G. H. Bolt et al. (eds.), *Interactions at the Soil Colloid-Soil Solution Interface*. Dordrecht: Kluwer.)

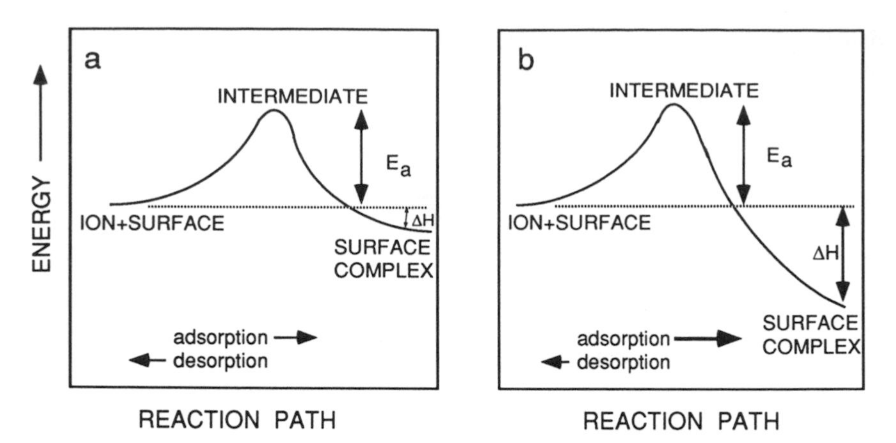

Figure 4.9. Energy profile of more reversible (a) and less reversible (b) adsorption reactions with comparable activation energies(E_a), showing the effect of overall energy of reaction (ΔH) on reversibility.

initially, the total sorption of an anion like phosphate may be largely due to chemisorption, but as time progresses, sorption results increasingly from precipitation. A discussion of the relationship between precipitation and sorption will follow later in this chapter.

4.2c. Desorption of Chemisorbed Anions

Compared with cation adsorption on minerals, anion adsorption seems more reversible. For example, selenite and borate reversibly adsorb on reactive soil minerals. A sizable fraction of phosphate adsorbed by soils is rapidly converted to a nonlabile[1] form, however, and phosphate desorption is slow. Even so, a labile fraction remains that is capable of rapid exchange with dissolved phosphate. In the case of phosphate, both chemisorption and precipitation may be occurring, with the latter reaction perhaps accounting for much of the nonlabile fraction.

In summary, it appears that chemisorbed anions often retain a higher lability than chemisorbed cations, perhaps because anion adsorption, as expressed by reactions 4.30 and 4.31, may involve little reaction energy. The exchange of the OH^- ligand on a mineral by an oxyanion is, after all, merely the replacement of one oxygen ligand by another at the metal coordination site. As suggested by the ranking of anions in Table 4.2, OH^- should be a strong competitor with oxyanions for metal coordination sites. The resulting small reaction energy produces the situation diagrammed in Figure 4.9a, where the reverse (desorption) reaction has a low activation energy and is therefore kinetically favorable. This is contrasted to reactions in which a weak bond is broken and a strong bond formed, creating the less reversible situation of Figure 4.9b.

1. Nonlabile is defined here as unable to exchange with phosphate in solution.

4.3. METAL COMPLEXATION ON SOIL ORGANIC MATTER

4.3a. Nature and Stability of Metal-Organic Complexes

In one way, metal cation bonding on organic matter can be viewed as an ion exchange process between H^+ and the metal at acidic functional groups, L:

$$M^{x+} + >LH_y = >L-M^{(x-y)+} + yH^+ \qquad (4.42)$$

Chapter 3, Section 3.4 deals with metal adsorption on humus as simple ion exchange. However, the high degree of selectivity shown by organic matter for certain metals signifies that some metals coordinate directly (i.e., form inner-sphere complexes) with the functional groups. Strong ionic and covalent bonds are formed in these complexes. Metal adsorption on organic matter must therefore be viewed as a form of chemisorption, having the properties of chemisorption outlined in section 4.1b. A typical affinity sequence of metals for soil organic matter at pH 5 is given in Table 4.3. The metals listed first tend to form inner-sphere complexes with organic matter, while those toward the end of the list are inclined to retain a hydration shell and remain freely exchangeable. Unfortunately, no consistent rule of metal selectivity for organic matter can be given, as selectivity depends on a number of factors beyond the properties of the metals themselves, including:

1. The chemical nature of the organic ligands (type of functional group)
2. The level of adsorption on the organic matter
3. The pH at which adsorption is measured (some metals compete more effectively with H^+ for bonding on functional groups than others)
4. The ionic strength of the solution in which adsorption is measured (this determines intensity of competition by other cations for the complexing sites)

The last two factors are dealt with in the next section, but the first two merit some comment at this point. Soil organic matter contains many different types of functional groups that act as Lewis bases in metal bonding reactions. They include carboxylic, phenolic, amine, carbonyl, and possibly sulfhydryl groups, with the first two being the most abundant. These ligands can be classified as Lewis bases according to their hardness (see Chapter 1). In general, the order of hardness in Lewis bases is O > N > S, where S is in the reduced state (valence of −2). Ca^{2+}, for example, is a "hard" acid, preferentially complexing with O-containing groups such as carboxylate. Conversely, Cd^{2+}, an ion with the same charge and radius as Ca^{2+}, is a "soft" acid and preferentially complexes with S^{2-}-containing groups such as sulfhydryl. If

Table 4.3. Observed Order of Affinity of Divalent Metal Ions for Soil Organic Matter Related to Electronegativity

Affinity sequence	Cu >	Ni >	Pb >	Co >	Ca[a] >	Zn >	Mn >	Mg
Electronegativity (Pauling)	2.0	1.91	1.87	1.88	1.00	1.65	1.55	1.31

[a]The affinity of humus for the essential macronutrients, Ca^{2+} and K^+, is (fortunately for plants and animals) higher than electronegativity would predict, suggesting that certain complexing or chelating groups in humus select for these metals on the basis of ionic size.

Cd^{2+} adsorption behavior in soils is now considered, it is found that low levels of the metal are adsorbed with very high preference over Ca^{2+}, but at high levels, preference over Ca^{2+} diminishes to nearly equal affinity. It seems that the soil contains a limited number of "soft" bases (ligands), and once these complex with Cd^{2+}, the remaining bases (presumably the harder carboxylic and phenolic groups) fail to show selectivity for Cd^{2+} over Ca^{2+}. A general rule of chemisorption of metal cations on organic matter seems to apply:

> *The degree of preference for one metal over another on soil organic matter reduces with increasing adsorption of the preferred metal.*

This rule seems to echo the one stated in section 4.1 regarding metal adsorption on soil mineral surfaces.

For any particular type of Lewis base, say an amine or carboxylate group, metals of smaller radius generally form the stronger complex. This produces the well-known *Irving–Williams series* of complexing strength for the divalent metal ions:

$$\mathrm{Ba} < \mathrm{Sr} < \mathrm{Ca} < \mathrm{Mg} < \mathrm{Mn} < \mathrm{Fe} < \mathrm{Co} < \mathrm{Ni} < \mathrm{Cu} > \mathrm{Zn}$$

The Irving–Williams series places Cu^{2+} at the apex of complex stability, consistent with the observed order of affinity for organic matter displayed in Table 4.3. At the same time, the metal "softness" factor cannot be ignored, as ionic radius fails to explain tendencies such as the ability of Pb^{2+} to complex strongly with soil organics. The metal "softness" factor measures ability to form covalent bonds with organic bases, and is related to electronegativity. In fact, it appears that electronegativity gives a fair indication of tendency of metals to complex with organic matter (see Table 4.3). Metals with high electronegativities draw electrons in Lewis bases toward them with the greatest energy, favoring "softer" bases. Because metals such as Cu^{2+} and Ni^{2+} are more electronegative (are "softer" acids) than Mg^{2+}, Ca^{2+}, or Mn^{2+}, they have a greater tendency to complex with amine (or other less "hard") ligands in humus.

A further contribution to stability in metal-organic matter complexes is the "chelate effect," that is, the entropy created by the complexation reaction. This is only a consideration if the metal bonds to two or more functional groups, for example:

$$M^{x+} + 2\,{>}LH = \begin{matrix} {>}L \diagdown \\ \quad M^{(x-2)+} \\ {>}L \diagup \end{matrix} + 2H^{+} \tag{4.43}$$

There is a net release of 1 mole of ions into solution by this reaction, because 2 moles of protons are released for each mole of metal ions complexed. The greater degrees of rotational and translational freedom associated with this release contributes a positive entropy term and hence a more negative free energy for reaction 4.43.[1] Consequently, the reaction is likely to be spontaneous in the direction written.

1. For a polydentate ligand with n complexing groups, this "chelate effect" adds (n − 1) log 55.5 to the logarithm of the stability constant, relative to the stability constant for a complex with n unidentate ligands of chemically similar type.

4.3b. Complexation Model of Metal Adsorption on Organics

The stability of any soluble metal-ligand complex, ML_n, is defined by the equation

$$K_{ML} = \frac{(ML_n)}{(M^{x+})(L^{y-})^n} \tag{4.44}$$

where the round brackets denote activities. K_{ML} is a true thermodynamic stability constant, remaining invariable for different pH and ionic strength conditions. This form of constant is not, and cannot, be employed in the study of soil organic matter-metal complexation for several reasons. First, for solid soil organics, ML_n is a surface complex rather than a dissolved molecule, and "activity" is not straightforwardly defined for such a species. Second, the effect of pH on metal complexation is often of primary interest in soil studies, yet the stability constant defined by equation 4.44 reveals no apparent involvement of pH. The reason for this is that pH affects M^{x+} and L^{y-} activity through various "side-reactions" that are not explicitly accounted for in equation 4.44. By controlling the activities of these two ions, the free metal and the uncomplexed but fully dissociated ligand, pH alters the degree of metal complexation while maintaining accordance with equation 4.44.

For many metals of concern in soils, pH has to be viewed as a "master variable," controlling the degree of metal hydrolysis and metal hydroxide precipitation, for example, by the reaction

$$M^{x+} + xOH^- = M(OH)_x \tag{4.45}$$

The solubility product of the metal hydroxide, K_{SO}, then determines the free metal activity:

$$K_{SO} = (M^{x+})(OH^-)^x \tag{4.46}$$

Furthermore, most complexing ligands on soil organic matter are weak or very weak acids and must first dissociate to form the free ligand:

$$LH_y = L^{y-} + yH^+ \tag{4.47}$$

before metal complexation is possible. The dissociation constant for this reaction is given by

$$K_{DISS} = \frac{(L^{y-})\,(H^+)^y}{(LH_y)} \tag{4.48}$$

Now equations 4.46 and 4.48 can be used to reexpress the equation for the metal-ligand stability constant, K_{ML}, in a form that directly considers pH effects. It then becomes evident that, if the pH is raised to an alkaline value, the rise in OH^- activity and fall in H^+ activity have opposing effects on metal complexation. The lowered metal ion activity, (M^{x+}), resulting from precipitation as the hydroxide, disfavors complexation. Conversely, the raised free ligand activity, (L^{y-}), resulting from dissociation at higher pH favors complexation. At extreme pH, complexation diminishes because of metal hydroxide precipitation (high pH) and association of the ligands with H^+ (low pH). Optimal complexation of metals with organic matter must then occur at some intermediate pH.[1]

Consider the more specific case of a 1:1 ligand/metal complex, ML, on soil organic matter. Now $n = 1$ in equation 4.44, and (L^{y-}) can be replaced by an expression from equation 4.48. This gives the equation

$$K_{ML} = \frac{(ML)(H)^y}{K_{DISS}(M)(LH_y)} \tag{4.49}$$

where the charge on the individual ions is not shown for convenience. Because LH_y and ML are the undissociated free ligand and complexed ligand, respectively, and both exist in solid form, their activities can only be estimated. If it is assumed that the activities of such species are proportional to their mole quantities, then equation 4.49 becomes

$$K_{DISS}K_{ML} = \frac{\{ML\}(H)^y}{\{LH_y\}(M)} \tag{4.50}$$

where { } brackets denote molar concentrations in the solid phase.

Now we are ready to consider the practical application of metal-ligand stability constants to metal adsorption on organic matter. First we will take the case of a metal-humus system that is undersaturated with respect to the metal hydroxide.

Case A: *Metal Complexation Without Precipitation*

The Cu^{2+} ion is used as an example of a metal that complexes with a high degree of selectivity on polyphenolic groups of soil organic matter. One of these groups is the biphenol (catechol) type, which dissociates to form the ligand L^{2-}:

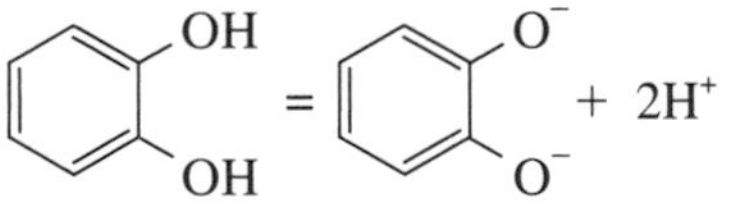

$$C_6H_4(OH)_2 = C_6H_4(O^-)_2 + 2H^+ \tag{4.51}$$

phenol (LH_2) phenolate (L^{2-})

Consider a sample of organic matter with 100 mmoles/kg of these complexing groups. Suppose 10 g of the organic matter is suspended in a liter of 10^{-5} *M* $CuCl_2$. The problem is to calculate the fraction of Cu^{2+} complexed in the pH range of 4 to 6.

First, the two-step dissociation of catechol must be considered:

$$\begin{aligned} LH_2 &= LH^- + H^+ \qquad K_1 = 10^{-9.23} \\ LH^- &= L^{2-} + H^+ \qquad K_2 = 10^{-13.0} \end{aligned} \tag{4.52}$$

Since the sum of the two steps produces the complete dissociation (given in general form by reaction 4.47), then the product of the K values, K_1K_2, equals K_{DISS}, the overall dissociation constant. The value of K_{DISS} is then $10^{-9.23} \times 10^{-13.0} = 10^{-22.23}$. The

1. In actual experiments, because "adsorption" is operationally defined as the removal of metal ions from solution, complexation and precipitation are not separated, and adsorption seems to continue to increase at high pH.

stability constant, K_{CuL}, defined by equation 4.44, has the value of $10^{13.9}$ for the reaction[1]

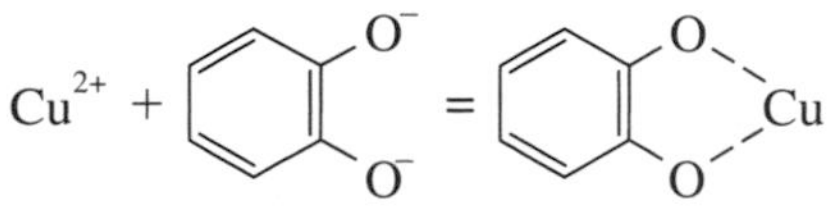

$$Cu^{2+} + \text{(catecholate: benzene ring with two } O^- \text{)} = \text{(benzene ring with two O bonded to Cu)} \tag{4.53}$$

Equation 4.50, written for Cu^{2+} complexation by catechol groups, becomes

$$K_{DISS}K_{CuL} = \frac{\{CuL\}(H)^2}{\{LH_2\}(Cu)} = 10^{-22.23}10^{13.9} \tag{4.54}$$

where round brackets signify activities of the free Cu^{2+} and H^+ ions and the { } symbols denote concentrations in the solid phase with units of millimoles per kilogram. From a mass balance analysis, the total ligand concentration, $\{L_T\}$, is seen to be

$$\{L_T\} = \{CuL\} + \{LH_2\} + \{LH^-\} + \{L^{2-}\} \tag{4.55}$$

but because catechol is a very weak acid, $\{LH^-\}$ and $\{L^{2-}\}$ are negligible compared with $\{LH_2\}$ in the pH 4 to 6 range. Consequently, $\{LH_2\}$ can be equated to $\{L_T\} - \{CuL\}$, and equation 4.54 becomes

$$\frac{\{CuL\}}{\{L_T\} - \{CuL\}} \frac{(H)^2}{[Cu]} = 10^{-8.33} \tag{4.56}$$

where the Cu^{2+} activity has been equated to Cu^{2+} concentration (moles/liter, denoted by square brackets), assuming that the ionic strength of the solution is low.

Mass balance can also be applied to the Cu, since total Cu in the system, Cu_T, is fixed by the amount initially in solution. That is, Cu_T must equal (10^{-5} moles/liter) $\times$ 1.0 liter = 10^{-2} millimoles. After adsorption, this quantity is partitioned between the solution and adsorbed phases:

$$Cu_T = 10^{-2} \text{ millimoles} = \{CuL\} \cdot W + [Cu] \cdot V \tag{4.57}$$

where W is the weight of the adsorbent (kilograms) and V is the solution volume (ml). Manipulation of equation 4.57 gives

$$[Cu] = \frac{Cu_T - \{CuL\} \cdot W}{V} = 10^{-5}(1 - \{CuL\}) \tag{4.58}$$

Substituting this expression and the value of $\{L_T\}$ = 100 mmoles/kg into equation 4.56 gives

$$\frac{\{CuL\}}{100 - \{CuL\}} \frac{1}{10^{-5}(1 - \{CuL\})} = 10^{-8.33}/(H)^2 \tag{4.59}$$

1. The constants, K_{DISS} and K_{ML}, are strictly valid equilibrium constants only when the metals and ligands are in solution. They are used in this calculation because corresponding constants have not been measured for catechol-type groups at solid surfaces. Consequently, the solution to this problem, while valid for catechol in solution, may be less than accurate for catechol-like ligands attached to humus.

Rearrangement of this equation produces a quadratic equation in {CuL}, where (H) is treated as a constant in the expression

$$0.04677\{CuL\}^2 - \{CuL\}[4.724 + 10^{12}(H)^2] + 4.677 = 0 \tag{4.60}$$

The solutions to this equation at pH 4, 5, and 6 are

$$\begin{aligned} \text{pH } 4, \{CuL\} &= \text{negligibly low} \\ \text{pH } 5, \{CuL\} &= 0.0447 \text{ mmoles/kg} \\ \text{pH } 6, \{CuL\} &= 0.823 \text{ mmoles/kg} \end{aligned}$$

Since $Cu_T = 10^{-2}$ mmoles, complete adsorption of Cu^{2+} from solution would correspond to $\{CuL\} = 10^{-2}$ mmoles/10 g = 1.0 mmole/kg.

It is concluded that nearly complete adsorption of trace concentrations of Cu^{2+} on catechol-type bonding sites of humus should be expected at pH 6.[1] At low pH, H^+ association with phenolic groups suppresses Cu^{2+} adsorption.

Case B: *Metal Complexation with Metal Hydroxide Precipitation*

In this case, Al^{3+} will be used to exemplify metals strongly predisposed toward precipitation as well as complexation with humus. Like Cu^{2+}, Al^{3+} complexes energetically with catechol-type and other polyphenolic ligands. The stability constant, K_{AIL}, for the solution reaction:

$$Al^{3+} + C_6H_4(O^-)_2 = C_6H_4O_2Al^+ \tag{4.61}$$

is known to be $10^{16.3}$, and will be used here as a measure of the ability of Al^{3+} to complex with catechol-type biphenolic groups on the surface of soil organic matter.

The system in this case is the same 10 g of organic matter suspended in a liter of solution as described in case A. In this example, however, the solution is assumed to be saturated with respect to $Al(OH)_3$. By analogy with Cu^{2+} in case A, equation 4.54 becomes

$$K_{DISS}K_{AiL} = \frac{\{AlL\}(H^+)^2}{\{LH_2\}(Al^{3+})} = 10^{-22.23}10^{16.3} \tag{4.62}$$

where the solid concentrations and solution activities are defined as before. Because the solution is saturated with respect to the precipitation of $Al(OH)_3$, equation 4.46 applies, which for noncrystalline aluminum hydroxide is

$$K_{SO} = 10^{-31.2} = (Al^{3+})(OH^-)^3 \tag{4.63}$$

Since the dissociation constant of water is $(H^+)(OH^-) = 10^{-14}$, equation 4.63 is equivalent to

$$(Al^{3+}) = 10^{10.8}(H^+)^3 \tag{4.64}$$

1. The most numerous complexing sites on humus, carboxylate groups, bond Cu^{2+} much less selectively, and are unlikely to compete with polyphenolic groups for Cu^{2+} retention except at low pH or at high Cu^{2+} levels that saturate the polyphenolic complexing sites.

Now this expression can be used to replace (Al^{3+}), the activity of free uncomplexed Al^{3+}, in equation 4.62. The result is a simpler equation:

$$\frac{\{AlL\}}{\{LH_2\}} = 10^{4.87}(H^+) \tag{4.65}$$

By the same reasoning used for the Cu problem, $\{LH_2\}$ is approximated by $\{L_T\}$ − $\{AlL\}$, and since $\{L_T\}$ = 100 mmoles/kg, equation 4.65 becomes

$$\frac{\{AlL\}}{(100 - \{AlL\})} = 10^{4.87}(H^+) \tag{4.66}$$

Collecting the terms in $\{AlL\}$ gives the desired equation that quantifies the dependence of Al^{3+} complexation on pH:

$$\{AlL\} = \frac{10^{6.87}(H^+)}{1 + 10^{4.87}(H^+)} \tag{4.67}$$

This relationship, plotted in Figure 4.10, shows that significant amounts of Al^{3+} can remain in surface-complexed form well above pH 5, but as the pH is raised, precipitation forces the removal of Al^{3+} from complexation sites.

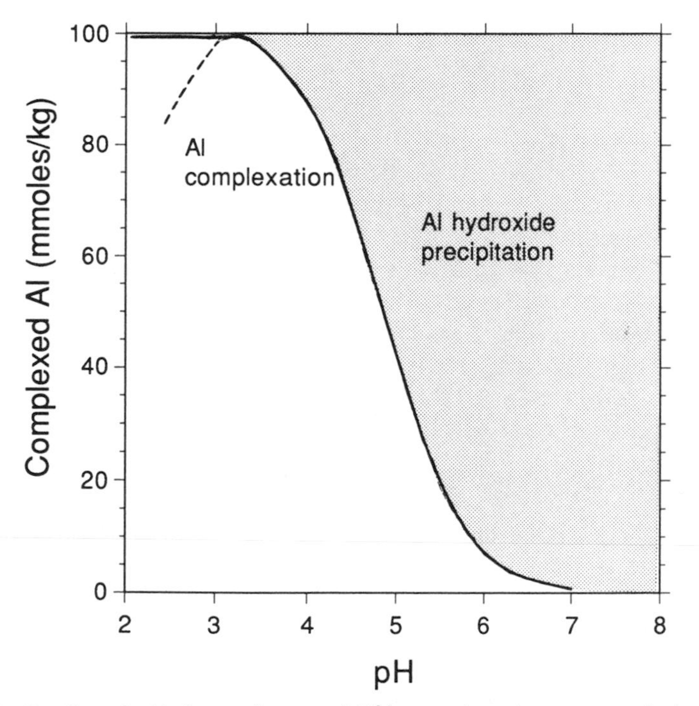

Figure 4.10. Predicted pH-dependence of Al^{3+} complexation on catechol-type groups of organic matter, assuming that $Al(OH)_3$ precipitation-dissolution controls Al^{3+} solubility. Below pH 3, complexation based on the assumption of a large excess of $Al(OH)_3$ (solid line) and the assumption of a limited quantity of $Al(OH)_3$ (broken line) is depicted. The 100 mmoles/kg level is the maximum complexing capacity of this model humic material.

At lower pH, the surface biphenolic groups are predicted by equation 4.67 to be almost completely occupied by Al^{3+} between pH 3 and 4, as dissolution of the mineral phase, $Al(OH)_3$, supplies Al^{3+} until complexation by organic matter is satisfied. Because no upper limit has been placed on the quantity of $Al(OH)_3$ in the system, dissolution would be predicted to elevate the solution concentration of Al^{3+} to levels limited only by the solubility product of this mineral (equation 4.63). Consequently, equation 4.67 predicts 100 percent complexation of surface groups even below pH 2. Realistically, however, as sources of readily dissolved Al^{3+} become exhausted during acidification, protonation of the phenolic complexing groups probably forces desorption of Al^{3+} from organic matter at very low pH.

This pattern of complexation and precipitation behavior is qualitatively similar for all metal ions that complex readily with humus and hydrolyze to form insoluble hydroxides or oxides. For example, Fe^{3+} complexed by humus at very low pH is probably bonded by polyphenols but as the pH is raised, precipitation of $Fe(OH)_3$ reduces iron solubility and strips Fe^{3+} from organic matter complexes.

4.3c. Desorption of Metals from Organic Complexes

Because metal ion release from bonding sites controls metal mobility (and therefore plant availability) in soils, the rate of desorption is at least as important to the understanding of metal behavior as the rate of adsorption. Generally, those metal cations that bond most strongly to soil organic matter (e.g., Pb^{2+}, Cu^{2+}) tend also to be the most rapidly adsorbed. This is a likely result of the tendency for chemical reactions that are furthest from equilibrium to be the fastest. Conversely, these strongly bonding metals are the most slowly desorbed, with dissociation from surfaces being several orders of magnitude slower than adsorption. Sluggish desorption is characteristic of the inner-sphere multidentate complexes formed by strongly adsorbed cations. A large activation energy may be needed to break the ligand-metal bond of these nonlabile complexes. The degree of lability of metal-surface complexes is a key predictor of how mobile and available the metal will be in soils.

Lability of metal-surface complexes can be described by the rate at which bound metal ions exchange with metal ions of the same element in solution, according to the self-exchange reaction:

$$M^*(H_2O)_n^{x+} + M{-}L = M(H_2O)_n^{x+} + M^*{-}L \tag{4.68}$$

The asterisk indicates that the metal ions initially in solution have somehow been labeled (perhaps by a radioactive or stable isotope) to experimentally distinguish them from the metal ions complexed with the surface groups, L. How quickly reaction 4.68 mixes the labeled with the unlabeled ions depends on the rate at which the hydrated metals undergo ligand exchange with the surface group. The actual exchange process involves the displacement of a water ligand by the ligand L at a metal coordination site. An estimate of the time required for this exchange is made from the known water-water exchange rate for the metal cation of interest, that is, for the reaction

$$M(H_2O)_n^{x+} + H_2O^* = M(H_2O^*)(H_2O)_{n-1}^{x+} + H_2O \tag{4.69}$$

The principle used here is that, for any particular metal, the ligand exchange rates of reaction 4.68 and 4.69 are likely to be correlated, because it is the radius, charge, and electronic properties of the metal cation more than the identity of the ligand that determines these rates. As is clear from the water exchange rates presented in Figure 4.11, metal cations of low charge and large size generally have the fastest ligand exchange. This should not be surprising, because the M—L and M—OH_2 bonds are weakest for such cations. Consequently, as a group, the alkali metals and most of the alkaline earth metals have rapid exchange. Transition metals, however, do not follow this simple trend because of *d*-orbital effects on bonding. For example, $Cr(H_2O)_6^{3+}$ has such an extremely slow water exchange rate (half-time greater than 10^5 seconds!) that it cannot be plotted on the scale of Figure 4.11. The $Cu(H_2O)_6^{2+}$ ion, in contrast, has an unusually rapid exchange rate for a small divalent cation, attributed to distortion in its octahedral geometry. The very low bioavailability of Cr^{3+} in soils is perhaps attributable in part to the kinetic stability of Cr^{3+}-humus complexes.

The exchange rates of Figure 4.11 should be applied with caution to the issue of metal-surface complex lability. Taking the example of Cu^{2+} again, the water exchange rate predicts a high degree of lability in Cu^{2+}-humus complexes. The fact that Cu^{2+}-humus complexes are found to be even *less* labile than Co^{2+} and Ni^{2+} complexes belies the rapid exchange rates of water ligands coordinated to Cu^{2+}, and demands a more careful analysis. It is easy to understand this apparent contradictory behavior when it is realized that there are really two kinds of coordination sites on the distorted octahedral (tetragonal) $Cu(H_2O)_6^{2+}$ ion, axial and equatorial. The axial water ligands are bonded loosely to Cu^{2+}

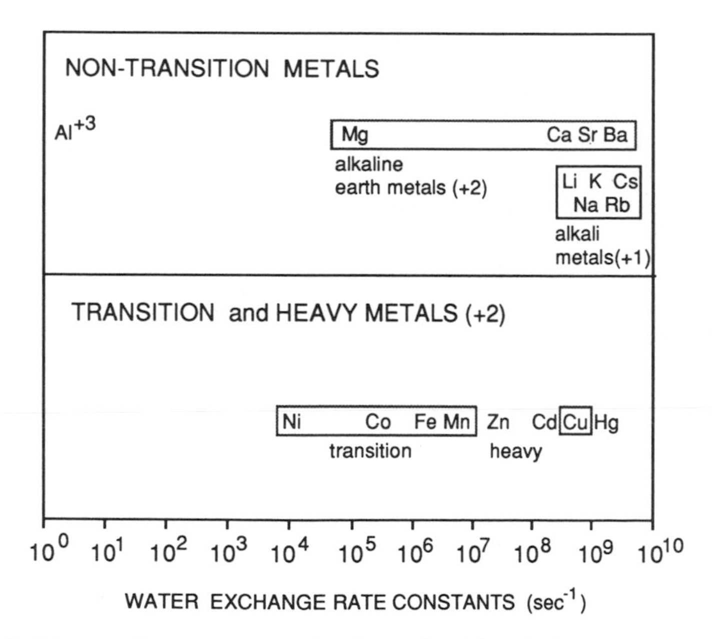

Figure 4.11. Measured rate constants for the H_2O-H_2O substitution reaction in the inner sphere of hydrated cations. (Adapted from F. A. Cotton and G. Wilkinson. 1980. *Advanced Inorganic Chemistry*. 4th ed. New York: Wiley.)

via a rather long bond, and therefore possess the rapid exchange rates reported in Figure 4.11. The equatorial water, in contrast, is tightly bonded and would exchange quite slowly with other ligands. It is at these equatorial positions that organic groups of humus bond, so that the lability of Cu^{2+}-humus complexes is more accurately gauged by the nonlabile nature of equatorial water rather than the labile axial water. Generally, any transition metal complexes that are distorted by *d*-orbital effects can display complicated behavior of this kind.

Few observations have been made on the rates of metal-organic matter dissociation in soils. Those made on Cu^{2+} and Fe^{3+} indicate that a portion of these metals, when complexed to humus, do not self-exchange (according to reaction 4.68) within one day, an indication of low lability associated with sluggish ligand exchange. Humus-bound Ni^{2+} and Co^{2+}, on the other hand, retain a high degree of lability after adsorption as indicated by fairly fast self-exchange. Interestingly, the partially hydrolyzed Fe^{3+} species, $Fe(H_2O)_5OH^{2+}$, forms complexes with organic ligands much more rapidly than $Fe(H_2O)_6^{3+}$, possibly because of charge reduction "felt" by the metal upon hydrolysis.

Measurements of soluble organic ligand (fulvic acid) dissociation from Al^{3+} in soil solution reveal a slow process that takes from minutes to hours. This is consistent with the slow water ligand exchange rate reported for Al^{3+} (see Figure 4.11). Slow ligand exchange rates for this phytotoxic metal may or may not be advantageous in practice. For example, humus-bound Al^{3+} is notoriously slow-reacting with liming materials added to soils, but organic matter added to strongly acid mineral soils protects plant roots from the toxic effects of Al^{3+}.

4.4 TERNARY COMPLEXES

The previous sections describe specific adsorption (chemisorption) of cations and anions as completely separate processes. In reality, soil solutions contain numerous metal cations and anions simultaneously, each of which may be able to chemisorb on variable-charge surfaces. Because the same type of hydroxyl group may be responsible for both cation and anion adsorption on these surfaces, there can be competition for these sites both among and between groups of cations and anions. What is sometimes found, however, is metal ion adsorption enhanced by the presence of certain anions, and vice versa. This appears to be a synergetic process, in which adsorption of anions and adsorption of cations are not competitive, and are sometimes greater in combination than separate. The explanation of this synergism is based on the concept of *ternary complex* formation at variable-charge mineral surfaces, schematically described as follows:

$$>S-OH + M + A = >S-O-M-A + H^+ \quad (4.70)$$

$$>S-OH + M + A = >S-A-M + OH^- \quad (4.71)$$

Reaction 4.70 forms a type A ternary complex, in which the metal ion, M^{x+}, links the surface to the anion, A^{n-}. Reaction 4.71 forms a type B ternary complex, in which the anion forms a bridge between the metal ion and the surface. Type A complexes seem to be the more common of the two.

Examples of observed ternary complexes include Cu^{2+}, Pb^{2+}, Cd^{2+}, and Zn^{2+} with PO_4^{3-} on Fe and Al hydroxide surfaces, in which the presence of phosphate in solution promotes trace metal adsorption and lowers metal solubility. In soils, the presence of exchangeable Ca^{2+} lowers the solubility of phosphate, perhaps because of Ca-phosphate ternary complexes formed on minerals. Many type A complexes of trace metals with chelating organic ligands are known. One example of this kind, glycine-Cu—$Al(OH)_3$, is pictured in Figure 4.12. Ternary complexes are also believed to form on soil organic matter, as multivalent cations such as Al^{3+} and Fe^{3+} bond simultaneously to the functional groups of humus and to anions such as phosphate. This affords a way for soil organic matter to adsorb certain anions, albeit to a limited degree.

Ternary complexes seem to form only between multivalent cations (particularly transition and heavy metals) and those anions (or uncharged molecules) possessing at least two metal-coordinating ligand positions. Consequently, organic ligands with chelating tendencies such as oxalate, bipyridine, glycine, and ethylenediamine form particularly stable type A complexes. Organic ligands with three or more coordinating positions may not form ternary complexes, at least when present in excess of the metal, since the metal cation is forced to dissociate from the surface to maximize

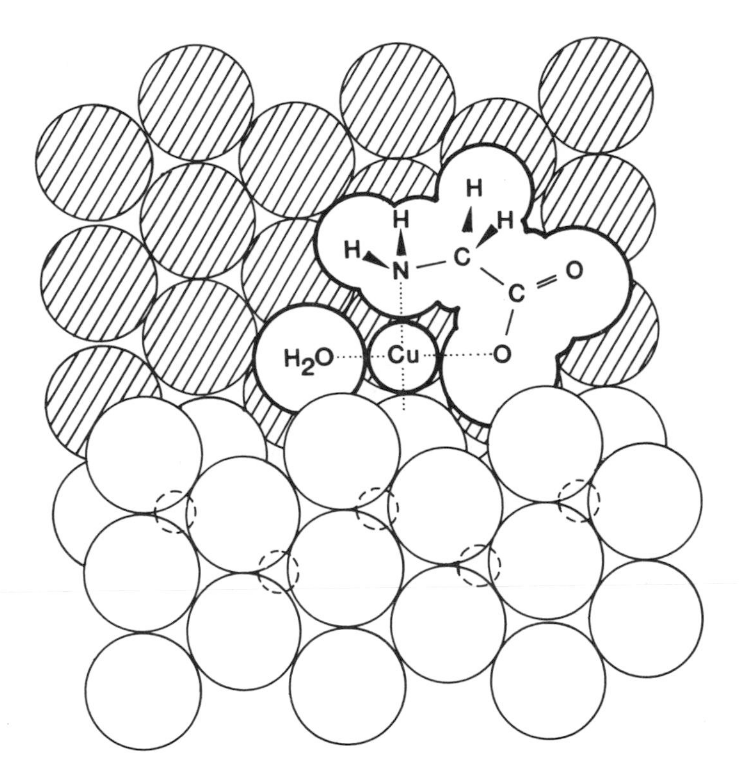

Figure 4.12. Depiction of a likely structure of the glycine-Cu^{2+}-$Al(OH)_3$ ternary complex. This is a type A ternary complex, as the metal ion bonds simultaneously with the organic ligand and surface oxygens located at steps and edges of gibbsite crystals. Large open and shaded circles symbolize structural OH^- groups, while smaller broken circles represent structural Al^{3+} ions.

bonding with the ligand. The metal-chelating anion, EDTA, tends to solubilize chemisorbed metals, bringing them into solution as metal-EDTA complexes. In general

> *If any anion is able to form a soluble complex with a metal cation, that anion competes with adsorbing surfaces to diminish metal ion adsorption.*

Whether an anion or ligand enhances metal cation adsorption by ternary complex formation or suppresses it by competition with the surface may depend on the anion/metal ratio. A large mole excess of the anion generally suppresses metal cation adsorption, while molar parity with the metal favors adsorption by ternary complex formation.

Given the immense number of metal-ligand pairs possible in systems as complex as soils, the question arises: How do we decide in which cases a ternary complex might influence metal solubility? It is reasonable to expect that those particular cations and anions that are predisposed to form ion pairs in solution will display this same tendency to pair on adsorptive surfaces, forming ternary complexes. This is a statement of the general principle that:

> *Metal-ligand complex stability at surfaces is correlated to metal-ligand complex stability in solution.*

The direct consequence of ternary complex formation in soils is likely to be that solubilities of numerous anions and trace metal cations are lowered below those expected from either chemisorption or precipitation.

4.5. PRECIPITATION AND CO-PRECIPITATION OF INORGANIC IONS

Chemisorption reactions in soils, which are two-dimensional surface processes, can rarely be separated experimentally from the three-dimensional nucleation and precipitation reactions. It is perhaps best to view the removal of adsorbate ions from solution, broadly termed "sorption," as a continuous process that ranges from chemisorption (at the low end of solubility) to precipitation (at the high end of solubility). Unless a new solid phase can be detected, the onset of precipitation and termination of chemisorption during sorption is usually not recognized by experimentalists. For this reason, an understanding of sorption necessitates some knowledge of precipitation reactions, which will be outlined here.

4.5a. Precipitation Theory

As the concentration of an ion in a solution is increased, precipitation of a new solid phase will not occur until the solubility product of that phase has been exceeded. That is, some degree of supersaturation is required because crystal nuclei can only be formed after an energy barrier has been overcome. Put in another way, the solubility of crystal nuclei initially formed in homogeneous solutions is higher than the solubility of larger crystals that grow from these nuclei. This difference in solubility of the same solid phase arises from the fact that the tiny nuclei have higher surface energy than the larger crystallites.

The status of the solution phase with respect to precipitation can be quantified by comparing the *ion activity product,* IAP, to the solubility product of the crystalline solid, K_{SO} (see Chapter 1). The ratio, IAP/K_{SO}, defines *undersaturation* (IAP/$K_{SO} < 1$), *saturation* (IAP/$K_{SO} = 1$), and *supersaturation* (IAP/$K_{SO} > 1$) of the solution with respect to a particular solid phase. Because of the higher solubility product of smaller crystallites and nuclei, precipitation can only begin in homogeneous solutions if the solution is supersaturated by a large margin (IAP/$K_{SO} > 100$). Extreme supersaturation leads to a high rate of formation of crystal nuclei and produces many very small crystallites or even noncrystalline solids. Minimal supersaturation, on the other hand, can give exceedingly low rates of nuclei formation, so that crystal growth is at a few nuclei only and a highly crystalline product with large crystals results (if any crystals form at all!).

In soil solutions, *heterogeneous nucleation*[1] is much more likely that *homogeneous nucleation* because mineral and organic surfaces are present that can catalyze the nucleation step of crystallization. The energy barrier to nucleation is reduced or removed by these surfaces, especially in cases where there is crystallographic similarity between the surface and the precipitating phase. This reduces the extent of supersaturation necessary for precipitation to be initiated. For example, solutions supersaturated with respect to gibbsite do not always form a precipitate, but the presence of smectite promotes gibbsite precipitation in these same solutions. Similarly, $CaCO_3$ in soils seems to promote the heterogeneous nucleation of $CdCO_3$, thereby preventing solutions in Cd^{2+}-contaminated calcareous soils from becoming supersaturated. Nevertheless, precipitation reactions are often much slower than chemisorption reactions so that time-dependent "sorption" of metals and other ions is often characteristic of precipitation in soils.

4.5b. Precipitation in the Soil Environment

For many of the more abundant elements that are found in soils, such as Al, Fe, Si, Mn, Ca, and Mg (and perhaps P and S), precipitation of mineral forms is common and can control the solubility of these elements. For most of the trace metals, however, precipitation is less likely than chemisorption because of their low concentrations in the soil. Only when soils become heavily loaded with these metals is the solubility raised to a level that supports precipitation of pure mineral phases. The ensuing discussion is relevant if a high level of a trace metal has accumulated in the soil, either from pollution or from natural geochemical processes.

Table 4.4 lists the solubility products of some of the least soluble minerals, minerals that are most likely to precipitate in the chemical environment of a soil, thereby setting an upper limit on trace and heavy metal solubility. This table shows, for example, that Cd^{2+} or Pb^{2+} solubility is most likely to be limited by carbonate or sulfide precipitation, depending on the redox potential of the soil. The solubility of

1. Heterogenous nucleation is the formation of crystal nuclei at surfaces of a different solid that is present before the initiation of precipitation.

Table 4.4. Solubility Products of Metal Carbonates, Oxides, and Sulfides[a]

Carbonates: $K_{SO} = (M^{2+})(CO_3^{2-})$										
	Pb	Cd	Fe	Mn	Zn	Ca				
$-\log K_{SO}$	13.1	11.7	10.7	10.4	10.2	8.42				
Oxides and Hydroxides: $K_{SO} = (M^{n+})(OH^-)^n$										
	Fe^{3+}	Al^{3+}	Hg^{2+}	Cu^{2+}	Zn^{2+}	Pb^{2+}	Fe^{2+}	Cd^{2+}	Mn^{2+}	Mg^{2+}
$-\log K_{SO}$	39	31.2	25.4	20.3	16.9	15.3	15.2	14.4	12.8	11.2
Sulfides: $K_{SO} = (M^{2+})(S^{2-})$										
	Hg	Cu	Pb	Cd	Zn	Fe	Mn			
$-\log K_{SO}$	52.1	36.1	27.5	27.0	24.7	18.1	13.5			

[a]Data from C. F. Baes and R. E. Mesmer. 1976. *The Hydrolysis of Cations.* New York: Wiley; and from W. Stumm and J. J. Morgan. 1981. *Aquatic Chemistry.* 2nd ed. New York: Wiley.

Hg^{2+} may also be controlled by sulfide in reducing soils, while the relatively insoluble oxide can form in nonacidic aerobic soils. Unfortunately, the hydrolysis of Hg^{2+} produces $Hg(OH)_2^0$, a moderately soluble molecule that could maintain a hazardous concentration of mercury in solution even though the concentration of the free Hg^{2+} cation is extremely low in the neutral pH range. This example illustrates the pitfalls of using solubility products alone to assess potential mobility and toxicity without also considering speciation of the dissolved metal. Chapter 9 details the important role of speciation in controlling bioavailability and mobility of elements in soils.

Other precipitates that can sometimes form in soils include silicates, phosphates, and sulfates. In isolated instances, these solids may limit the solubility of certain trace metals. In the case of Pb^{2+} and Zn^{2+}, phosphate concentrations in well-fertilized soils may be high enough to favor metal phosphate precipitates over oxides or carbonates. Precipitates of more than one anion are sometimes less soluble than simple pure solids, at least as measured by the metal activity. This is certainly true for Cu^{2+} and Zn^{2+}, both of which form hydroxycarbonates that can be less soluble than the simple hydroxides or carbonates, depending on the CO_2 gas pressure prevailing in the soil. Hydroxysulfates and hydroxyphosphates of Al are also known to be stable precipitates in soils.

4.5c. Precipitation of Phosphate—A Case Study

Consider now the case of $AlPO_4 \cdot 2H_2O$ (variscite) precipitation in soil as a model system. This example is of interest because of the potential control of Al^{3+} solubility in acid mineral soils by phosphate fertilizer additions, and because of the observation that phosphate solubility in these soils tends to be limited by the known solubility of variscite.

The dissolution reaction of variscite is

$$AlPO_4 \cdot 2H_2O = Al^{3+} + H_2PO_4^- + 2OH^- \qquad K_{SO} = 10^{-30.5} \tag{4.72}$$

so that the solubility expression:

$$K_{SO} = (Al^{3+})(H_2PO_4^-)(OH^-)^2 = 10^{-30.5} \tag{4.73}$$

is valid at any pH. If we consider a mineral soil that is well buffered at pH 5, with a low ionic strength in soil solution, then solution activities can be replaced by con-

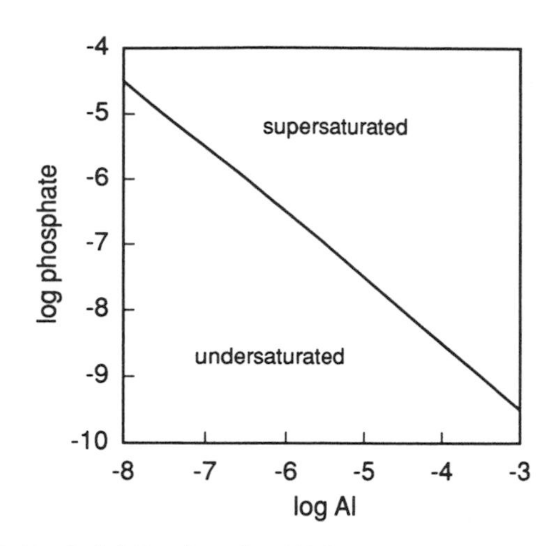

Figure 4.13. Solubility line for $AlPO_4 \cdot 2H_2O$ (variscite) at pH 5.

centrations in equation 4.73 without introducing much error, and the $H_2PO_4^-$ concentration in solution can be equated to $[P_T]$, the total soluble phosphate concentration.[1] These approximations give the equation

$$[Al^{3+}][P_T] = \frac{10^{-30.50}}{10^{-18}} = 10^{-12.50} \tag{4.74}$$

which is plotted in Figure 4.13, showing the "fields" of undersaturation and supersaturation defined by this precipitation reaction. Generally, then, if soluble phosphate is added to the soil at a constant pH of 5, Al^{3+} solubility should diminish, and vice versa.

Equation 4.73 suggests that, as soils are made very acid (pH < 5), the increase in Al^{3+} activity from clay mineral dissolution and the decrease in OH^- activity from the pH change could result in either a lower or higher solubility of phosphate. Because it is commonly assumed that Al^{3+} activity is controlled in acid soils by $Al(OH)_3$ precipitation, that is, by the solubility product

$$K_{SO} = (Al^{3+})(OH^-)^3 = 10^{-33.5} \tag{4.75}$$

then equation 4.73 takes the form

$$(H_2PO_4^-) = \frac{10^{-10.5}}{(H^+)} \tag{4.76}$$

and phosphate solubility is predicted to *decrease* as the pH is lowered. In part, the conventional "wisdom" that the liming of very acid soils and lowering of Al^{3+} activity should improve phosphate availability to plants is based on this solubility argument. Actually, soils containing significant quantities of high-charge clays (e.g., smectite, vermiculite) or humus may not conform to this expectation because of the inadequacy of the $Al(OH)_3$ model of aluminum solubility for these soils. In fact,

1. H_3PO_4, $H_2PO_4^-$, and HPO_4^{2-} have acid dissociation constants of $10^{-2.12}$, $10^{-7.20}$, and $10^{-12.33}$, respectively. Thus, $H_2PO_4^-$ is the only significant solution species at pH 5.

phosphate solubility in these soils may *increase* on acidification. This behavior is attributable to the rapid adsorption of Al^{3+}, released by mineral dissolution, by organic and clay colloids.

It is often the case that in nonacid soils, Ca phosphate precipitation controls the solubility of phosphate; Ca phosphates are less soluble than Al phosphates if the soil pH is much higher than 6. Ca^{2+} becomes a more dominant exchangeable and soluble ion relative to Al^{3+} as the pH rises in soils. However, this dependence on pH does not extend as far as alkaline soils containing high exchangeable Na^+ and elevated concentrations of soluble bicarbonate and carbonate. In these high pH soils, phosphorus solubility is higher than in calcium-rich soils for at least two reasons. Not only does the high Na^+ suppress Ca phosphate precipitation (Na^+ phosphate is a soluble salt), but also the alkaline anions (HCO_3^-, CO_3^{2-}, OH^-) displace phosphate from chemisorption sites on variable-charge minerals.

The raised solubility of phosphate in some types of soils at both low and high pH has led to the observation that phosphorus solubility is *lowest* near neutral pH, contrary to the model of solubility based on simultaneous Al hydroxide, Al phosphate, and Ca phosphate precipitation. The latter model predicts phosphorus to be most soluble between pH 6 and 7, although there is little evidence from soils to confirm this prediction. While precipitation in very acid and in Ca-rich mineral soils is undoubtedly a key process in determining the solubility of phosphate, oversimplified precipitation models for this and other ions can lead to misleading generalizations about solubility in a material as complex as soil.

4.5d. Co-Precipitation of Trace Elements in Solid Solutions

The "solubility product" approach to ion solubility, outlined in the previous sections, is most successful for elements with moderate to high total concentration in soils (Fe, Al, Ca, P, etc.). Such an approach is rarely successful for the trace elements (e.g., Cu, Zn) unless the soil is grossly contaminated with the element in question. At low (or even moderate) concentrations, trace metal solubility is usually much lower than that expected from the solubility product of likely precipitates. A case in point is the zinc ion, Zn^{2+}, added to soils adjusted to different pH values. The Zn^{2+} solubility levels in the soil solutions, plotted in Figure 4.14, reveal undersaturation with respect to all known pure solid precipitates of Zn^{2+} that could reasonably be expected to form in the soil. Possible explanations for the low solubility include:

1. The chance that solid phases not previously considered might control Zn^{2+} solubility at an even lower level
2. The possibility that chemisorption sites are not saturated, or that cation exchange is involved, so that adsorption limits Zn^{2+} solubility at a lower level than precipitation
3. The existence of mixed solid phases in which the trace ion lowers its solubility by co-precipitating into another component.

The first explanation can probably be ruled out despite thermodynamic data indicating that certain spinel-type structures such as $ZnFe_2O_4$ and Zn_2SiO_4 are less soluble (more stable) than any of the pure solid phases represented in Figure 4.14. There is no evidence that these minerals can crystallize in the pressure-temperature regime typical for soils (Tiller and Pickering, 1974). They are sometimes found in soils as

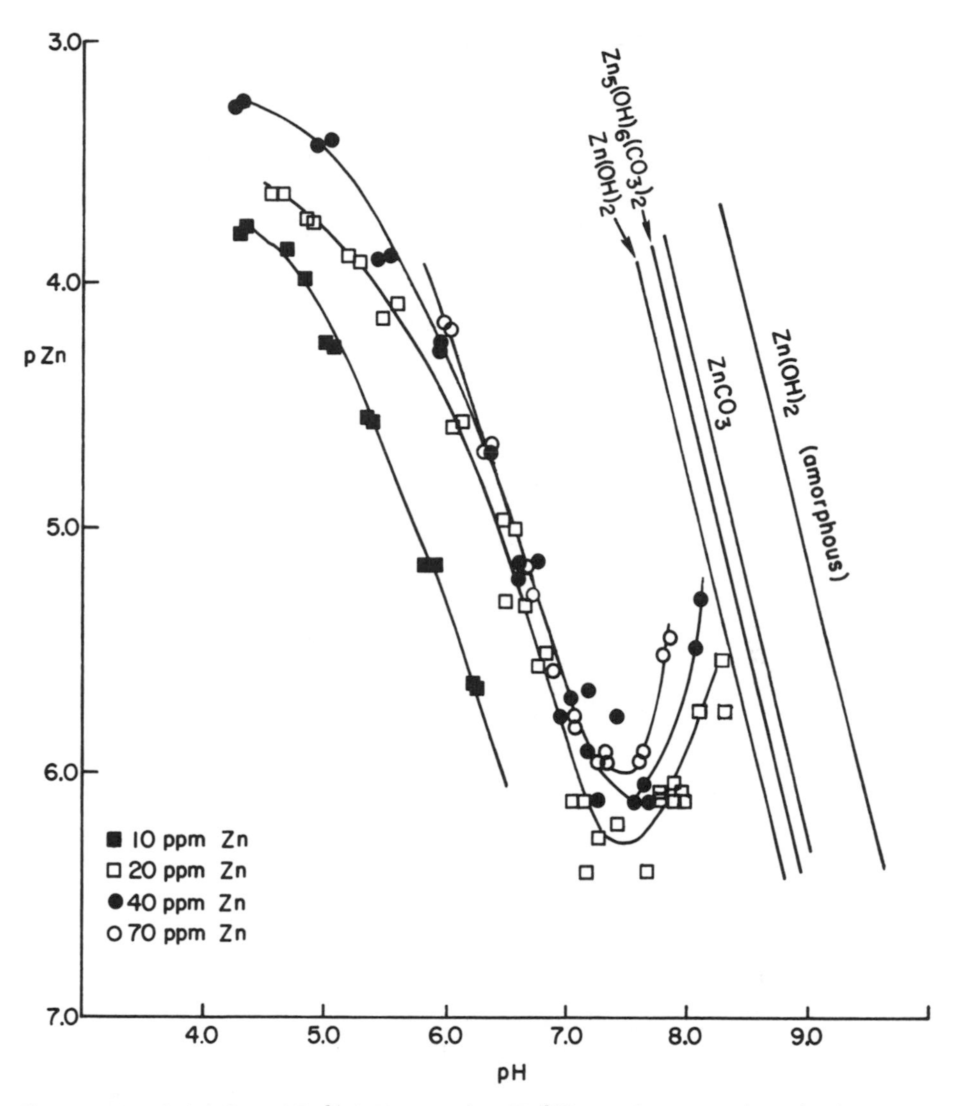

Figure 4.14. Solubility of Zn^{2+} (pZn = $-\log(Zn^{2+})$) as a function of pH for four different concentrations of Zn^{2+} added to a soil. Solubility lines of some precipitates likely to form in soils are positioned assuming that the soil is in equilibrium with atmospheric CO_2.

residual minerals, evidently having formed under metamorphic conditions. These structures often violate the radius ratio rules (see Chapter 2), suggesting that extreme conditions are necessary to their crystallization.

The second explanation is probably operative in most situations. Cation exchange could lower the solubility of metals that have a high selectivity over the main exchange cations, but its involvement is commonly suppressed experimentally by the introduction of a "bathing electrolyte," such as 10^{-2} *M* $CaCl_2$. This still allows specific adsorption reactions to lower metal solubility.

There are numerous examples where the third explanation is at least possible.

How the formation of co-precipitates can lower the solubility of trace elements is the subject of this section.

The theory of solid solutions predicts that

> *The solubility of an ion can be lowered in a mixed ionic compound relative to the solubility of the pure compound.*

Consider, for example, a trace metal cation, B, isomorphously substituted into a solid composed of metal cations, A, and anions, Y. The chemical formula, $A_{1-x}B_xY$ is variable because x can range from 0.0 to 1.0 if AY and BY, the solid-solution end members, form a continuous series. Unlike ionic compounds of fixed composition, solid solutions do not have constant solubility products. Instead, equations for both components, AY and BY, must be specified:

$$(A)(Y) = K_{SO}^{AY} a_{AY} \tag{4.77}$$

$$(B)(Y) = K_{SO}^{BY} a_{BY} \tag{4.78}$$

Here, parentheses symbolize the activities of ions in aqueous solution, while a_{AY} and a_{BY} are the activities of the components AY and BY in the solid solution. K_{SO}^{AY} and K_{SO}^{BY} are the solubility products of *pure* AY and BY. Since the trace metal B, when incorporated at low levels into the solid, results in a low value of a_{BY}, then equation 4.78 predicts an effective solubility product, (B)(Y), that is lower than that of pure BY. Consequently, solid solution formation is an effective means of lowering the solubility of a trace element.

At equilibrium, the distribution of A and B ions between the aqueous (aq) and solid (s) phase is represented by the reaction

$$AY\ (s) + B\ (aq) = A\ (aq) + BY\ (s) \tag{4.79}$$

If the solid phase is homogeneous, the reaction can be described by

$$\frac{a_{BY}}{a_{AY}} = \boldsymbol{D}\frac{(B)}{(A)} \tag{4.80}$$

where $\boldsymbol{D}$ is the distribution coefficient that specifies the relative degree to which cations A and B are incorporated into the solid. It is easily shown, from equations 4.77, 4.78 and 4.80, that $\boldsymbol{D}$ equals the ratio of solubility products of the pure phases, K_{SO}^{AY}/K_{SO}^{BY}.

To form an ideal solid solution of formula $A_{1-x}B_xY$, the two cations A and B must be closely matched such that one cation can substitute for the other with no change in energy of the solid structure. The solid solution is then said to have *zero heat of mixing*. On the other hand, the theoretical *entropy of mixing* in this ideal solution is not generally zero, but is dependent on the mole fractions of A and B in the solid according to the equation

$$S = 2.303R[x \log x + (1 - x) \log (1 - x)] \tag{4.81}$$

It can be shown from this equation that the activities of the components in ideal solid solutions are given simply by

$$a_{AY} = 1 - x \tag{4.82}$$

$$a_{BY} = x \tag{4.83}$$

That is, activities of AY and BY can be equated to mole fractions of AY and BY in the solid. Symbolizing these mole fractions as X_{AY} and X_{BY}, equation 4.80 becomes

$$\frac{X_{BY}}{X_{AY}} = \boldsymbol{D}\frac{(B)}{(A)} \tag{4.84}$$

For dilute aqueous solutions, (B) and (A) in this equation can be replaced by [B] and [A], the mole concentrations in solution, giving

$$\frac{X_{BY}}{X_{AY}} = \boldsymbol{D}\frac{[B]}{[A]} \tag{4.85}$$

A very large or small $\boldsymbol{D}$ means that a solid solution precipitated from solution cannot be homogeneous because one of the two metals, A or B, is selectively scavenged from solution as the solid precipitates. This more preferred ion ends up occluded in the center of the growing crystallites, while the less preferred ion concentrates in the outer layers.

The significance of this segregating mechanism in controlling metal solubility and bioavailability in soils can be easily imagined, as it provides a means of literally "burying" toxic or essential metals in a form inaccessible to desorption. Furthermore, the solubility of a toxic or essential metal could be lowered well below that predicted from the solubility of the pure solid phase. That is, if the trace metal of interest is symbolized by B, the solubility of B in equilibrium with the solid solution would be lower than that in equilibrium with pure BY. This fact is demonstrated by combining equations 4.78 and 4.83 to give

$$(B)(Y) = xK_{SO}^{BY} \tag{4.86}$$

Consequently, a trace metal substituted at low levels ($x \ll 1$) into the structure produces an "effective" solubility product, (B)(Y), that can be orders of magnitude less than the K_{SO} of the pure solid, BY. Whether such a lowered solubility is realized depends on the identity of the metal and the chemical properties of the soil.

There is evidence that solid solution formation plays a role in Cd^{2+} and Mn^{2+} sorption in suspensions of calcite ($CaCO_3$). Calcite adsorbs these cations initially by a fast reaction that seems to involve exchange of Ca^{2+} by the trace metal at the surface, that is,

$$Cd^{2+} + CaCO_3\,(s) = Ca^{2+} + CdCO_3\,(s) \tag{4.87}$$

Generally, in reactions between two metal carbonates as exemplified by reaction 4.87, the metal cation of the least soluble carbonate (in this case, Cd^{2+}) is preferentially adsorbed at the carbonate surface. The reaction can be viewed as a chemisorption process rather than precipitation because

1. It proceeds to a degree determined by the calcite surface area.
2. It occurs even when the suspension is undersaturated with respect to solid $CdCO_3$.

Subsequent slow removal of Cd^{2+} from solution is observed, and may be the result of recrystallization in a thin calcite surface layer to form a Cd^{2+}/Ca^{2+} solid solution at the surface. This would explain the observed reduction in extractability and exchangeability of the Cd^{2+} over time. Because calcite is in a dynamic state of dissolution and recrystallization in suspension, even when at equilibrium, a solid solution can form at the surface fairly quickly. If it is assumed that the $Ca_{1-x}Cd_xCO_3$ solid solution is ideal, then the distribution coefficient equals the quotient of solubility products of the pure carbonate solids:

$$\boldsymbol{D} = \frac{K_{SO}^{CaCO_3}}{K_{SO}^{CdCO_3}} = \frac{10^{-8.47}}{10^{-11.3}} = 680 \tag{4.88}$$

This large $\boldsymbol{D}$ value signifies a strong tendency for the recrystallizing calcite to incorporate Cd^{2+}.

In soils, the prevalence and importance of solid solution formation in controlling cation and anion solubility has not yet been determined. Certain minerals readily incorporate only those metal ions with radii similar to the radius of the structural metal ion. For calcite, this means that Mn^{2+}, Cd^{2+}, and Fe^{2+} readily enter the calcite structure on precipitation while smaller ions such as Cu^{2+} and Zn^{2+} do not. Even so, it is not clear that metal selectivity demonstrated by pure solid solutions has much control in soils and sediments. A case in point is the lack of evidence for a strong association of Cd^{2+} and Pb^{2+} with calcite in natural sediments, despite the favorable radii of these metals compared with Ca^{2+}. On the other hand, both Cd^{2+} and Pb^{2+} are associated with hydroxyapatite in nature because they fit well into the Ca^{2+} site of this mineral. This association poses a problem for the long-term use of phosphate fertilizers in soils.

For oxides, as for other minerals, a key factor controlling which cations are able to form solid solutions is the ionic radius, which must be small enough for the cations to enter octahedral sites. In this regard, it is interesting to note that Pb^{2+}, which chemisorbs strongly on aluminum hydroxide, does not readily substitute into the hydroxide during co-precipitation. This means that the rules that determine selectivity in chemisorption at a solid surface do not apply in general to substitution within the same solid.

The trace metals Cu^{2+}, Zn^{2+}, Ni^{2+}, and Co^{2+} are excluded from hematite during its crystallization, but substitute more readily into magnetite ($Fe_2^{3+}Fe^{2+}O_4$). Small trivalent metal ions like Cr^{3+} and Mn^{3+} (and perhaps V^{3+}) substitute for Fe^{3+} and Al^{3+} in precipitating oxides and hydroxides. Certain trace metal cations, notably Co^{2+}, are often found in association with Mn oxides in soils, a fact attributed to the ready substitution of these metals into structural octahedral sites. Small divalent cations, including Zn^{2+}, Cu^{2+}, and Mg^{2+}, form solid solutions with aluminum hydroxide, while larger cations like Pb^{2+}, Mn^{2+}, and Ca^{2+} do not. In fact, solid solutions with the general formula $[M_{1-x}^{2+}Al_x(OH)_2]^{x+}A_{x/n}^{n-}$, where A^{n-} is an anion, are well known and referred to as *hydrotalcites,* layered structures possessing the unusual property of permanent positive charge that is manifested as a high anion exchange capacity.

The short-term potential for solid solution formation is low for minerals with small solubility products, such as aluminum oxides and aluminosilicates, because spontaneous dissolution and recrystallization is very slow in these minerals. Without recrystallization, trace metals cannot be incorporated into the mineral structures. Movement of metal ions into these mineral crystals by solid diffusion is not possible on the time scale of adsorption experiments; ionic diffusion into most crystalline solids is negligibly slow at all but extremely high temperatures. Nevertheless, metals could diffuse into imperfect solids along interstices, pores, or other structural defects.

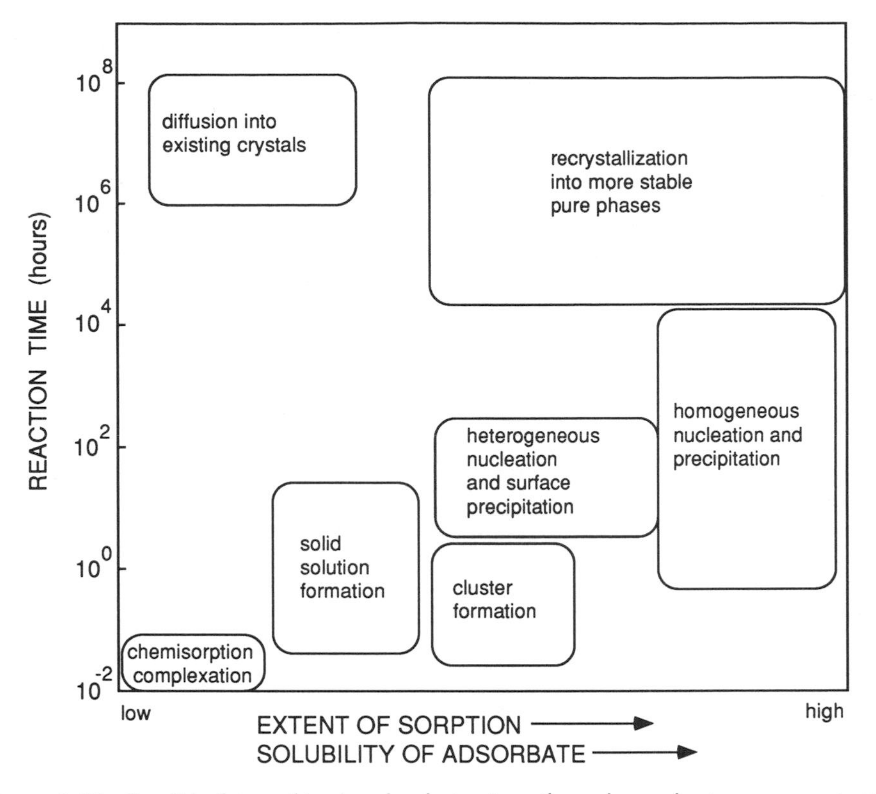

Figure 4.15. Possible fates of ionic adsorbates in soils as dependent on concentration and time.

The opportunity for co-precipitation and solid solution formation is higher with Fe and Mn oxides than Al oxides or aluminosilicates. The reason for this is the higher solubility of the former two minerals under anaerobic conditions (see Chapter 7). Soil reduction generates the soluble ions, Fe^{2+} and Mn^{2+}, which then reoxidize to again form the insoluble oxides once the soil is aerated. The possibility exists for co-precipitation of trace metals during these cycles of alternating reduction and oxidation. Furthermore, the fresh precipitates may also be effective adsorbates for trace metals.

Whether solid solutions actually do form in soils is probably limited more by slow rates of soil mineral dissolution—a necessary preliminary step for co-precipitation—than by lack of thermodynamic favorability.[1] Although the overall impact of solid solution formation on metal solubility in soils remains to be determined, certain features of metal sorption are consistent with (but do not prove) solid solution formation. These include a sorption capacity that is ill-defined and increasing with time, decreasing reversibility of sorption with time that follows a decreasing lability of the

1. The entropy of mixing always favors the dispersion of trace metals throughout the solid phases of the soil, even though the heat of mixing may be unfavorable if the structures do not easily accommodate the metal. The fact that Al-substituted FeOOH is common in soils suggests the thermodynamic favorability of this particular solid solution.

sorbed ion, selectivity for metals based largely on their ionic radii, and solubilities below those predicted from the solubility products of the pure metal-bearing solid phases.

The diagram in Figure 4.15 emphasizes the continuity of precipitation and co-precipitation processes with chemisorption, both in time and space. Low levels of adsorbate (whether metal cations or anions) are usually bound by chemisorption, higher levels by the formation of solid solutions or by the nucleation of small adsorbate "clusters" at surfaces. The highest levels of adsorbate lead to precipitation of separate mineral phases, a process that can be viewed as an extension of cluster growth that allows a new solid phase to become detectable.

References

Baes, C. F. and R. E. Mesmer. 1976. *The Hydrolysis of Cations.* New York: Wiley.

Cotton, F. A. and G. Wilkinson. 1980. *Advanced Inorganic Chemistry.* 4th ed. New York: Wiley.

Manceau, A., L. Charlet, M. C. Boisset, B. Didier and L. Spadini. 1992. Sorption and speciation of heavy metals on hydrous Fe and Mn oxides. From microscopic to macroscopic. *Applied Clay Science* 7:201–223.

McBride, M. B. 1989. Reactions controlling heavy metal solubility in soils. In B. A. Stewart (ed.), *Advances in Soil Science* 10:1–56. Springer-Verlag, New York.

McKenzie, R. M. 1980. The adsorption of lead and other heavy metals on oxides of manganese and iron. *Aust. J. Soil Res.* 18:61–73.

Perrott, K. W., B.F.L. Smith, and R.H.E. Inkson. 1976. The reaction of fluoride with soils and soil minerals. *J. Soil Sci.* 27:58–67.

Schindler, P. W. and W. Stumm. 1987. The surface chemistry of oxides, hydroxides and oxide minerals. In W. Stumm (ed.), *Aquatic Surface Chemistry.* New York: Wiley, pp. 83–110.

Stumm, W. and J. J. Morgan. 1981. *Aquatic Chemistry.* 2nd ed. New York: Wiley.

Tiller, K. G., J. Gerth, and G. Brummer. 1984. The relative affinities of Cd, Ni and Zn for different soil clay fractions and goethite. *Geoderma* 34:17–35.

Tiller, K. G. and J. G. Pickering. 1974. The synthesis of zinc silicates at 20°C and atmospheric pressure. *Clays and Clay Minerals* 22:409–416.

Van der Zee, S.E.A.T.M. and W. H. Van Riemsdijk. 1991. Model for the reaction kinetics of phosphate with oxides and soil. In G. H. Bolt, M. F. DeBoodt, M.H.B. Hayes, and M. B. McBride (eds.), *Interactions at the Soil Colloid-Soil Solution Interface.* Dordrecht, The Netherlands: Kluwer.

Suggested Additional Reading

Schindler, P. W. and G. Sposito. 1991. Surface complexation at (hydr)oxide surfaces. In G. H. Bolt, M. F. DeBoodt, M.H.B. Hayes, and M. B. McBride (eds.), *Interactions at the Soil Colloid-Soil Solution Interface.* Dordrecht, The Netherlands: Kluwer, pp. 115–145.

McBride, M. B. 1991. Processes of heavy and transition metal sorption by soil minerals. In Bolt et al. (eds.), *Interactions at the Soil Colloid-Soil Solution Interface,* pp. 149–175.

Van Riemsdijk, W. H. and S.E.A.T.M. Van der Zee. 1991. Comparison of models for adsorption, solid solution and surface precipitation. In Bolt et al., (eds.), *Interactions at the Soil Colloid-Soil Solution Interface,* pp. 241–256.

Swift, R. S. and R. G. McLaren. 1991. Micronutrient adsorption by soils and soil colloids. In Bolt et al., (eds.), *Interactions at the Soil Colloid-Soil Solution Interface,* pp. 257–292.

MacCarthy, P. and E. M. Perdue. 1991. Complexation of metal ions by humic substances: fundamental considerations. In Bolt et al., (eds.), *Interactions at the Soil Colloid-Soil Solution Interface,* pp. 469–489.

Questions

1. Use the definition of the proton dissociation constants for oxide surfaces, K_1 and K_2, to show that PZC, defined in Chapter 3, is given by the equation

$$PZC = -0.5 (\log K_1 + \log K_2)$$

Since the PZC of silica, SiO_2, is believed to be about 2, use this equation to estimate K_1 for silica (obtain a value of K_2 from Table 4.1). What does this result tell you about the nature of surface $-Si-OH_2^+$ groups?

2. How is it possible to explain the observation that those minerals that are most reactive in chemisorption of metal cations are also most reactive in chemisorption of anions?

3. Calculate the fraction of Cu^{2+} removed by precipitation of $Cu(OH)_2$ from 10^{-3}, 10^{-4}, and 10^{-5} *M* solutions of $CuCl_2$ as the pH is raised in steps of 0.5 units from 5 to 8. Plot the data in the same format as shown in Figure 4.5 and compare the shape of these curves with those in the figure.

4. Explain in chemical terms why oxyanions that are weak acids tend to adsorb on Al and Fe oxides at a pH that is near their pK_a values. In what pH range would monosilicic acid, H_4SiO_4, be expected to adsorb optimally?

5. The borate anion, $B(OH)_4^-$, forms from boric acid by the reaction

$$B(OH)_3 + OH^- = B(OH)_4^- \qquad K = 10^{-9}$$

(a) According to the "shared charge" concept, which of the two species, $B(OH)_3$ or $B(OH)_4^-$, might be expected to adsorb to a greater extent on minerals, assuming that both adsorb without loss of protons? (Remember that each associated proton shares a charge of +1 with the oxygen.)
(b) How would the prediction in (a) be changed if surface bonding required one of the protons to be displaced?
(c) Given the fact that boron adsorption on most clay minerals is greatest at pH 8 to 10, which of the above two alternate descriptions of boron bonding seems to be most likely?

6. In high-selenium soils of some arid regions, selenate is a problem pollutant in the drainage water of irrigated fields, while selenite is not. Develop a plausible argument based on chemical principles to explain this observation.

7. Studies of organic acid adsorption on Fe oxides indicate that the maximum surface density of chemisorbed anions on these minerals is about 2 molecules/100 Å^2.
(a) Since CO_2 can chemisorb on oxides as HCO_3^-, calculate the greatest conceivable quantity of inorganic carbon that could be tied up in an oxidic soil that has a surface area of 50 m^2/g. Express your answer in units of kilograms per hectare. (Assume a 10 m depth of soil material with a bulk density of 1.5 g/cm^3.)
(b) How does this inorganic carbon "pool" compare in size with the organic carbon "pool," if the soil contains an average of 2 percent (by weight) organic carbon in the top 30 cm and negligible organic carbon below 30 cm.

(c) Write the sequence of reactions that will occur in the oxidic soil as CO_2 in the soil's atmosphere increases.
(d) What competing processes could limit CO_2 adsorption by soil oxides?

8. The following OH^- release data were obtained for a number of different minerals, calculated from the rise in pH that results when the mineral is immersed in a 0.85 M NaF solution (data from Perrott et al., 1976).

Mineral	OH^- Release (moles/kg)
montmorillonite	0.22
illite	0.17
kaolinite	0.12
imogolite	12.0
allophane	19.0
quartz	0.01
silica (amorphous)	17.6
gibbsite	0.51
alumina (amorphous)	21.0
goethite	0.09
hematite	0.14
ferric oxide (amorphous)	5.5

(a) Write the general chemical reaction that explains the pH rise.
(b) Estimate the final pH if 1.0 g of allophane is immersed in 25 ml of 0.85 *M* NaF solution.
(c) Explain the relative magnitudes of OH^- release based on your knowledge of these mineral structures (see Chapter 2).
(d) What do these data imply about the relative capacity of the listed minerals to chemisorb phosphate?
(e) Would a concentrated NaCl solution have the same effect on pH as NaF? Explain.

9. Copper, Cu^{2+}; cadmium, Cd^{2+}; and lead, Pb^{2+}, all show enhanced adsorption on Al hydroxide after the surface has adsorbed phosphate. Describe a bonding mechanism that would explain this effect.

10. The logarithm of the association constants, log K_1^M, for selected trace metals with acetate and ethylenediamine at 25°C are given below:

Metal	Acetate	Ethylenediamine
Ni^{2+}	1.43	7.32
Cu^{2+}	2.22	10.48
Zn^{2+}	1.57	5.66
Cd^{2+}	1.93	5.41

(a) Why do all of these metals, regardless of electronic properties and size, complex much more strongly with ethylenediamine than with acetate?
(b) Why does Cu^{2+} complex most strongly with both ligands, particularly in the case of ethylenediamine?
(c) Which metal shows the largest relative preference for ethylenediamine over acetate? Why?

(d) Why does Ni^{2+}, a transition metal, show relatively more affinity for ethylenediamine than either Zn^{2+} or Cd^{2+}, which are not transition metals?
(e) Which ligand type, acetate or ethylenediamine, is more representative of the bulk of functional groups in soil organic matter?

11. The Mn^{2+} ion bonds with humus by interaction with dissociated carboxylic groups. These groups could be modeled by a monoprotic acid such as acetic acid, or alternatively by a diprotic acid such as phthalic acid. Bonding to acetate would involve a single metal-carboxylate association, while bonding to phthalate could form a chelate with two metal-carboxylate associations. The relevant first and second dissociation (K_1 and K_2) and association (K_1^{Mn}) constants at 25°C are known for these ligands and their 1:1 Mn^{2+} complexes and are given below:

	$\log K_1$	$\log K_2$	$\log K_1^{Mn}$
Acetic acid	−4.75	—	1.40
Phthalic acid	−2.95	−5.41	2.74

Calculate, for a 10-g sample of soil organic matter suspended in a liter of 10^{-4} *M* $MnCl_2$, the fraction of Mn^{2+} complexed in the pH range of 4 to 9, using first acetic and then phthalic acid to model the actual functional groups of the organic matter. (The carboxylic acid content of the organic matter can be taken to be 800 mmoles/kg.)

12. Consider a waterlogged soil with a pH of 6.5 that is known to be polluted with high levels of heavy metals, including cadmium, from industrial sources. The gas pressures of CO_2 (P_{CO_2}) and hydrogen sulfide (P_{H_2S}) are measured in this soil and found to be 10^{-2} and 10^{-4} atmospheres, respectively.
(a) Derive an equation analogous to the one for the carbonate equilibrium:

$$(CO_3^{2-}) = \frac{10^{-18.15}\, P_{CO_2}}{(H^+)^2}$$

relating the sulfide ion activity, (S^{2-}), to the H_2S gas pressure. Use the reaction equilibrium constants listed in Chapter 7 (reactions 7.60 to 7.62).
(b) Determine what the Cd^{2+} activity in the soil solution would be if $CdCO_3$ were in equilibrium with the solution (see Table 4.4).
(c) Determine what the Cd^{2+} activity in the soil solution would be if CdS were in equilibrium with the solution (see Table 4.4).
(d) Which solid phase is more stable in this soil, limiting Cd^{2+} solubility?

13. A 2-g sample of $CaCO_3$ is suspended in 25 ml of water containing 5×10^{-6} *M* $CdCl_2$. The calcite has a specific surface area of 0.5 m^2/g. The solubilities of Ca^{2+} and Cd^{2+} are measured at 5×10^{-4} and 10^{-7} *M*, respectively, after one day of reaction.
(a) Use equation 4.85 to calculate the final mole fraction of Cd^{2+} expected in the calcite.
(b) Does this mole fraction apply to the surface or to the entire bulk of the solid? Why?
(c) Calculate the density of Cd^{2+} ions sorbed at the surface (Cd^{2+} ions per square Ångstrom) based on the measured amount of sorption. Compare this with an estimated density of Ca^{2+} at the calcite crystal surface. (Use the density of calcite, 2.71 g/cm^3, to estimate an average area occupied by each Ca^{2+} ion assuming a simple crystal geometry.)

(d) What fraction of surface Ca^{2+} ions were actually replaced by Cd^{2+}, according to the analysis of part (c)? How does this result compare with the predicted mole fraction in part (a)?

14. Arsenic in the oxidized form (H_3AsO_4) adsorbs strongly on variable-charge minerals, with maximum adsorption near pH 4. Arsenic in the reduced form ($As(OH)_3$) adsorbs less strongly, with maximum adsorption near pH 7. Given the fact that H_3AsO_4 is a much stronger acid than $As(OH)_3$, explain this behavior.

5

Soil Acidity

Acidity is a major limitation to soil productivity in much of the world. Although acidification is a natural process in many soil environments, agricultural practices and pollution from industrial, mining, and other human activities have accelerated the process. It is important that acidity be understood in terms of its fundamental chemistry so that soil management and remediation schemes are based on sound principles rather than empirical knowledge that may only be locally relevant. This chapter attempts to provide this understanding, and uses some example field data to illustrate important principles.

5.1. SOIL pH

Arguably, pH is the single most diagnostic chemical measurement made on soil. Although soil pH is rarely low or high enough to indicate direct harmful effects of H^+ or OH^- on microorganisms and roots, it signals potential harm from the indirect effects of extreme H^+ activities. For example, soil pH values below 5.0 to 5.5 warn that soluble levels of certain metals, particularly Al^{3+} and Mn^{2+}, may be high enough to be biologically toxic. Conversely, pH values above 7 are often associated with very low solubility of micronutrient metal cations such as Zn^{2+}. Even more extreme pH values signify the presence of particular minerals or ions in the soil—a pH above 8.5 is generally associated with high soluble or exchangeable Na^+, whereas a pH below 3 usually indicates metal sulfides.

To soil chemists, pH is the "master variable," controlling ion exchange, dissolution/precipitation, reduction/oxidation, adsorption, and complexation reactions. Typically, pH is plotted as the independent variable against which some soil chemical process or reaction is measured.

5.1a. Theory and Practice of Measurement

In homogeneous aqueous solutions, pH is ideally defined by

$$pH = -\log (H^+) \tag{5.1}$$

where (H^+) is the proton activity. In reality, pH is determined relative to a standard buffer with a defined proton activity. In dilute solutions of electrolytes, such as may be expected in soil solution, the measured pH closely approximates the pH defined by equation 5.1 (within ± 0.02 pH units).

The measurement of solution pH is typically done with a glass membrane electrode, called the "glass electrode," inserted in the solution along with a reference electrode and its associated "salt bridge." The glass membrane surface generates an electrical potential in response to the H^+ activity in the external solution, whereas the reference electrode generates a known and constant potential that is (ideally) not sensitive to the composition of the solution. The two electrodes are connected to a potentiometer (pH meter) that measures the potential, ΔE, between them. The reference electrode is connected to the glass electrode by the *salt bridge,* which is usually some type of junction that allows ions to pass between the internal electrode solution and the measured (external) solution, but does not allow the solutions to mix. The complete electrochemical cell is then:

glass electrode (reversibly responsive to H^+) | unknown or standard buffer solution || salt bridge | reference electrode

From the Nernst equation (see Chapter 1), the potential between the two electrodes, ΔE, is related to pH at room temperature by the equation

$$pH = \frac{\Delta E - \text{constant}}{0.059} \tag{5.2}$$

where ΔE is expressed in units of volts. This means that the ideal "Nernstian" response of the glass electrode is 0.059 volt (59 millivolts) for each unit change in the solution pH.

The salt bridge by necessity allows some ions to diffuse out of the reference electrode's internal solution. The particular salt chosen for the internal solution is critical because an unbalanced outward diffusion of anions and cations generates a *junction potential.* KCl is usually chosen as the salt because of the well-matched current-carrying abilities of K^+ and Cl^-.[1] When measuring the pH of solutions, this junction potential is of little consequence either because it is negligible or because it is a constant factor that is canceled out by the fact that solution pH is measured *relative* to that of a standard buffer. However, when measuring the pH of colloidal suspensions such as soil clays, the attraction of K^+ for the negatively charged particles creates a concentration gradient and unbalances the rates of K^+ and Cl^- diffusion from the salt bridge, producing a significant junction potential. This potential, not present during calibration of the electrode in the standard buffer, creates an *apparent* pH decrease. Conversely, in positively charged colloidal suspensions, the apparent pH would be shifted higher. The error in pH reading caused by charged, suspended particles is called the *junction potential effect* or *suspension effect.* It was thought at one time to be a real pH effect and contributed to the idea of clay surfaces being more acidic than the surrounding solution.

1. The equivalent (molar) conductances of K^+ and Cl^- at infinite dilution are 74 and 76, respectively.

The problem of the suspension effect can be avoided in the glass electrode–reference electrode cell by inserting the reference electrode with its salt bridge junction into the supernatant, above the settled soil colloidal particles. In any event, the effect can be anticipated to produce an important error only in soil suspensions with a large content of colloidal material and low salt concentration. If the pH is measured in a salt such as 0.01 *M* $CaCl_2$, the suspension effect is suppressed, apparently by the ability of the excess cations in soil solution to lower the tendency for K^+ from the salt bridge to diffuse to exchange sites. This is one reason that soil pH is sometimes measured in 0.01 *M* $CaCl_2$ rather than in distilled water.

5.1b. pH and Reserve Acidity

Although pH is viewed by soil chemists as the "master variable" in soils, it is sensitive to the acidity retained on solids in both explicit (proton) and latent (proton-generating) forms. The important forms of surface acidity are:

1. Organic acids that release acidity by dissociation:

$$R{-}H = R^- + H^+ \tag{5.3}$$

2. Al^{3+}-organic complexes that release acidity by hydrolysis, e.g.:

$$\begin{matrix} R \\ & \diagdown \\ & & Al^{3+} \\ & \diagup \\ R \end{matrix} = \begin{matrix} R \\ & \diagdown \\ & & AlOH^{2} \\ & \diagup \\ R \end{matrix} + H^+ \tag{5.4}$$

3. Exchangeable H^+ and Al^{3+}, released as acidity by cation exchange and hydrolysis, e.g.:

$$K^+ + H^+\text{-smectite} = H^+ + K^+\text{-smectite} \tag{5.5}$$

$$K^+ + Al^{3+}\text{-smectite} = AlOH^{2+} + K^+\text{-smectite} + H^+ \tag{5.6}$$

4. Nonexchangeable forms of acidity on minerals:

$$>Fe{-}OH_2]^{+1/2} = >Fe{-}OH]^{-1/2} + H^+ \tag{5.7}$$

$$>Al{-}OH_2]^{+1/2} = >Al{-}OH]^{-1/2} + H^+ \tag{5.8}$$

This last form of acidity can build up at low pH on the surfaces of variable-charge minerals (Al and Fe oxides, allophane, and Al-hydroxy and Fe-hydroxy polymers adsorbed on layer silicate clays).

All of these forms of acidity comprise what is referred to as the *reserve acidity.* They can be slow to respond chemically to a change in the concentration of H^+ and Al^{3+} in soil solution, termed the *active acidity,* because of slow ionic diffusion through micropores of the soil particles and slow dissociation of Al^{3+} complexes. The relationship of reserve acidity, a quantity that represents the *buffer capacity* of the soil, to active acidity or pH, a measure of acid *intensity* in the soil, is diagrammed in Figure 5.1. Reactions of bases (e.g., lime) added to the soil occur first with the active acidity in soil solution. Subsequently, the "pool" of reserve acidity gradually releases acidity into the active form.

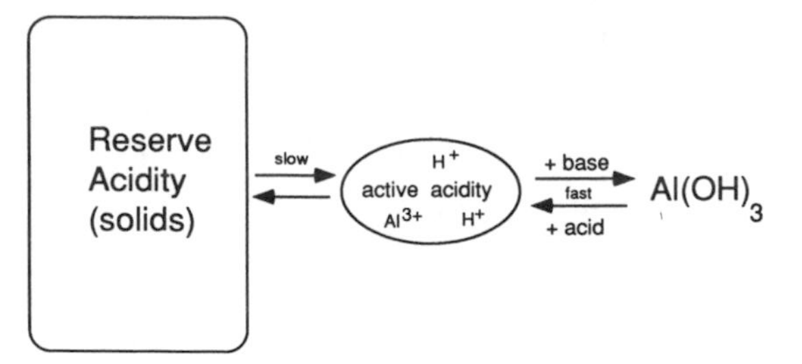

Figure 5.1. Connection between soil reserve acidity, active acidity, and acid or base inputs.

5.1c. Relation of pH to Exchangeable Al^{3+}

From a simple perspective of cation exchange (see Chapter 3), it is easy to see how pH could be related to the extent of H^+ occupation of exchange sites. If the Ca^{2+} ion is allowed to represent all of the so-called "base cations" (Ca^{2+}, Mg^{2+}, K^+, Na^+), then the soil clay's initial response to an input of H^+ is the exchange reaction:

$$2H^+ + Ca^{2+}\text{-clay} = Ca^{2+} + 2H^+\text{-clay} \tag{5.9}$$

The exchange coefficient, K_s, has the form

$$K_s = \frac{(Ca^{2+})}{(H^+)^2} \cdot \frac{N_{H+}^2}{N_{Ca2+}} \tag{5.10}$$

where the round brackets denote solution activities of ions and N_M symbolizes the fraction of exchange sites occupied by a particular ion M. By taking the logarithm of both sides of equation 5.10, and realizing that N_{H+}^2/N_{Ca2+} in this simplified two-cation exchange system is really a single variable that is a function of the fraction of exchange sites occupied by the base cation Ca^{2+}, the relationship is found that

$$\text{pH} - \tfrac{1}{2}\text{pCa} = f\,(\%\ \text{base saturation}) + \text{constant} \tag{5.11}$$

Here, *percent base saturation* is simply the percentage of total cation exchange capacity (CEC) that is balanced by the base cations, and $f(x)$ symbolizes an unspecified function of some variable x. It is this equation that is in part the basis for the importance given historically to percent base saturation in soil. In actual fact, the relationship between pH and percent base saturation in soils, while statistically significant, is weak.[1] As a result, percent base saturation and pH are unreliable predictors of one another, although there is clearly a tendency for soils with low pH to have low percent base saturation.

The pH − ½pCa value of equation 5.11 is termed the *lime potential* of a soil because it relates directly to the activity of $Ca(OH)_2$, a form of lime, in soil solution. Lime potential has been shown to be less dependent than pH on the electrolyte concentration of soils, and is therefore argued to be a more diagnostic measure of a soil's

1. This is true even when $\text{pCa} = -\log(Ca^{2+})$ is forced to a constant value by measuring soil pH in 0.01 M $CaCl_2$.

acid-base status than pH itself. Theoretically, lime potential increases as the percent base saturation increases.

The reason for the poor correlation between percent base saturation and pH is apparent in the assumptions needed to obtain equation 5.11; that is, the H^+-Ca^{2+} exchange model oversimplifies the buffering response of soil solids to acid inputs. Specifically, the complicating involvement of Al^{3+} in soil acidity must be considered. Reaction 5.9 cannot proceed very far before adsorbed H^+ begins to attack and dissolve the mineral on which it is adsorbed. This serves to consume H^+, releasing structural Al into soluble form:

$$H^+ + \begin{matrix} O \quad O \quad OH \\ \diagdown \; | \; \diagup \\ Al \\ \diagup \; | \; \diagdown \\ O \quad O \quad OH \end{matrix} \rightarrow Al^{3+} + H_2O \tag{5.12}$$

The soluble Al^{3+} is then adsorbed on remaining exchange sites. As a result, the H^+ exchange form of clay is converted into the Al^{3+} form. Figure 5.2 reveals that the H^+ exchange form of layer silicate clays is a strong acid that rapidly "ages" to the weak acid (Al^{3+}) form, consistent with reaction 5.12.

Hydrolysis of Al^{3+}, as discussed in Chapter 1, generates 3 moles of acidity per mole of Al^{3+}:

$$Al^{3+} + 3H_2O = Al(OH)_3\,(s) + 3H^+ \tag{5.13}$$

These Al^{3+} hydrolysis reactions buffer the soil pH in the 4.5 to 5.0 range. Above this range, Al tends to form the precipitated solid, $Al(OH)_3$. Below this range, Al tends to convert to the soluble free cation form, Al^{3+}. As soils are acidified by inputs of H^+ from natural processes or pollution, reaction 5.13 proceeds in the reverse direction, and the pH does not necessarily decrease noticeably while Al is being converted from

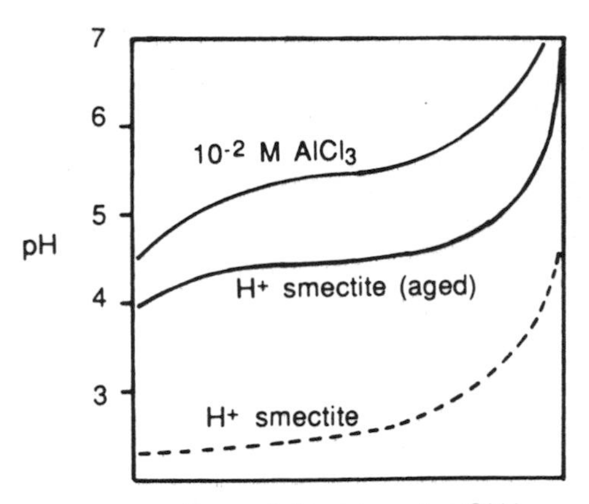

Figure 5.2. The titration curves of freshly prepared H^+-smectite, aged H^+-smectite, and 10^{-2} *M* $AlCl_3$. (From G. H. Bolt, M.G.M. Bruggenwert, and A. Kamphorst. 1976. Adsorption of cations by soil. In G. H. Bolt and M.G.M. Bruggenwert (eds.), *Soil Chemistry*. Amsterdam: Elsevier.)

insoluble to soluble (and phytotoxic) forms. This is because reaction 5.13 occurs within a fairly narrow pH range (near 5.0).

In mineral soils at least, Al^{3+} and Al-hydroxy cations are the dominant acidic exchange cations. Their importance, and the relative scarcity of exchangeable H^+, is explained by reaction 5.12 combined with the high selectivity of permanent-charge exchange sites for Al^{3+} over H^+. In fact, Al cations can be difficult to exchange by salts such as KCl, particularly if they are bonded in interlayers of vermiculites or complexed with organic matter, so that *exchangeable Al* and *exchangeable acidity* are not easily defined terms.

If the ill-defined nature of exchangeable acidity were not enough to discourage attempts to universally apply the concept of percent base saturation, it is found that many soils composed largely of variable-charge minerals (particularly highly weathered soils) have ill-defined and quite low CEC values. The estimates of exchangeable Al in these soils depend on the properties of the cation used to displace Al. Much of the Al release and adsorption that occurs in soils of this type may be controlled by processes *other than* cation exchange, such as precipitation and dissolution, making the base saturation concept misleading or even irrelevant. Given these facts, percent base saturation seems not to be a particularly useful concept, and might be better replaced by separate measures of exchangeable base cations and exchangeable Al. This point of view is strengthened by observations that the phytotoxicity of acid soils is correlated better with soluble free Al^{3+} than with percent base saturation.

5.2. ALUMINUM SOLUBILITY IN SOILS

5.2a. The Chemistry of Aluminum Hydrolysis

The solubility and chemical form of Al in pure water is determined by a sequence of hydrolysis steps that ultimately results in the precipitation of $Al(OH)_3$ above pH 5, and the dissolution of $Al(OH)_3$ as the aluminate anion above pH 8. The most important of these hydrolysis reactions in dilute solutions of Al^{3+} are

$$Al^{3+} + H_2O = AlOH^{2+} + H^+ \qquad K_1 = 10^{-4.97} \tag{5.14}$$

$$AlOH^{2+} + H_2O = Al(OH)_2^+ + H^+ \qquad K_2 = 10^{-4.93} \tag{5.15}$$

$$Al(OH)_2^+ + H_2O = Al(OH)_3^0\,(aq) + H^+ \qquad K_3 = 10^{-5.7} \tag{5.16}$$

$$Al(OH)_3^0\,(aq) + H_2O = Al(OH)_4^- + H^+ \qquad K_4 = 10^{-7.4} \tag{5.17}$$

The concentrations of the dissolved species of Al, calculated from the reaction constants above, are plotted as a function of pH in Figure 5.3, assuming that crystalline $Al(OH)_3$ (gibbsite) is present and controls solubility according to the dissolution reaction

$$Al(OH)_3 + 3H^+ = Al^{3+} + 3H_2O \qquad \log K_{SO} = 8.5 \tag{5.18}$$

From Figure 5.3 it is apparent that Al solubility in solutions should be minimized in the pH 6 to 7 range. The reaction scheme for this system is diagrammed in Figure 5.4.

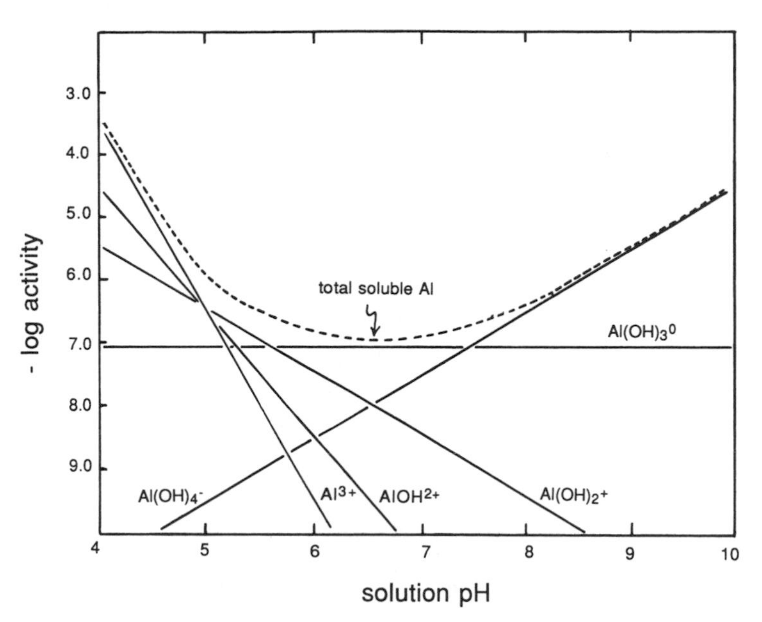

Figure 5.3. Concentrations of the most significant species of soluble Al in an aqueous solution of $AlCl_3$, assuming that gibbsite is present as the solid phase at all pH values. The broken line depicts total soluble Al (sum of all species concentrations). Polymeric Al-hydroxy cations are not significant species under the conditions imposed on this system.

5.2b. Exchangeable and Soluble Al in Soil

On the basis of the behavior of Al revealed in Figure 5.3, it is not surprising that, once soil pH is lowered much below 5.5, aluminosilicate clays and Al hydroxide minerals begin to dissolve, releasing Al-hydroxy cations and Al^{3+} that then exchange other cations from soil colloids. The result, as illustrated by Figure 5.5, is that the fraction of exchange sites occupied by Al^{3+} and its hydrolysis products can become large once the soil pH falls below 5.0. Furthermore, as the pH is lowered, the concentration of soluble Al builds up. This is seen for two acid mineral soils in Figure

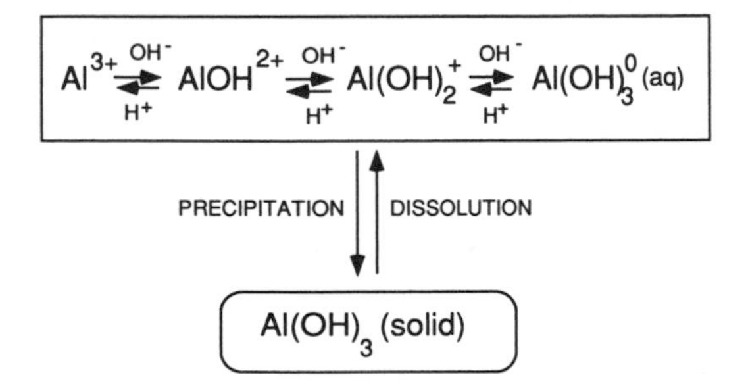

Figure 5.4. Reaction scheme for Al species formed in response to pH change in water.

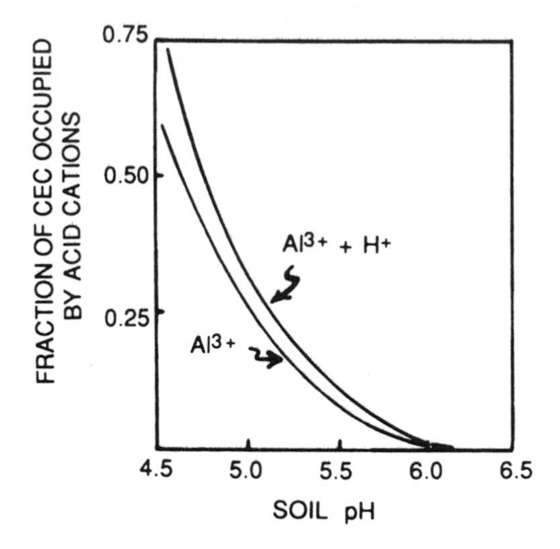

Figure 5.5. Prevalence of the acidic cations, Al^{3+} and H^+, on soil exchange sites as a function of pH. CEC, cation exchange capacity.

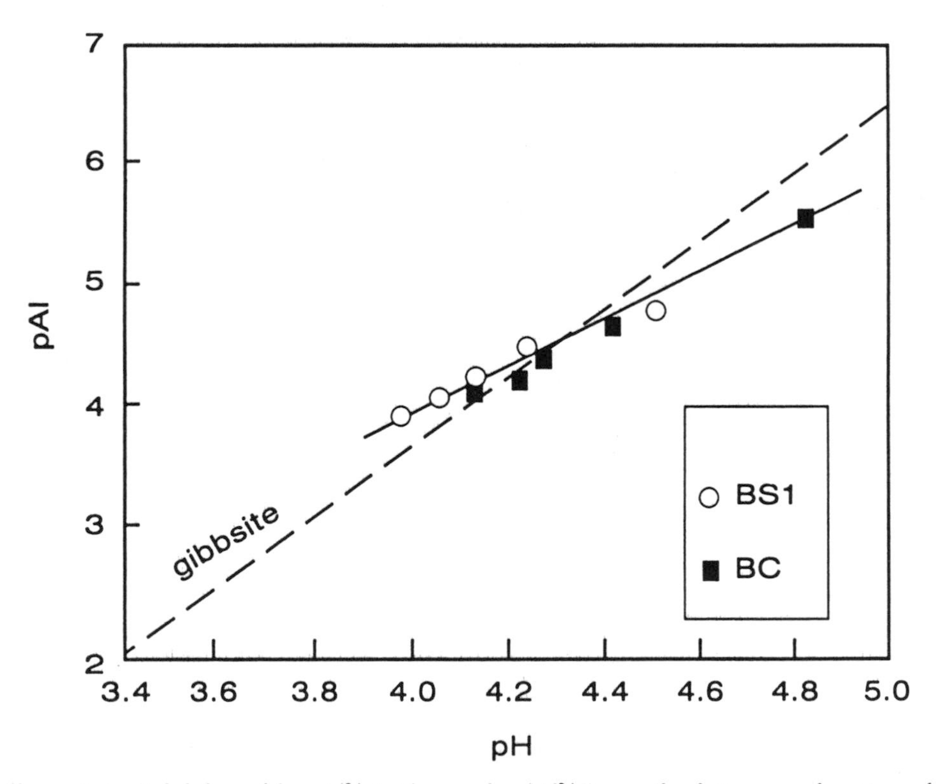

Figure 5.6. Solubility of free Al^{3+} (pAl = $-\log[Al^{3+}]$) in soil solution as a function of pH in two spodosol B horizons. (Data from M. G. Johnson and M. B. McBride. 1991. Solubility of aluminum and silicon in acidified spodosols. In R. J. Wright et al. (eds.), *Plant-Soil Interactions at Low pH*. Dordrecht: Kluwer.)

5.6, where dissolved Al^{3+} as measured by $-\log [Al^{3+}]$[1] follows (approximately) the solubility equation of gibbsite, which is derived from equation 5.18 as follows:

$$K_{SO} = \frac{[Al^{3+}]}{[H^+]^3} = 10^{8.5}$$
$$\log K_{SO} = \log [Al^{3+}] - 3 \log [H^+] = 8.5 \quad (5.19)$$
$$pAl = 3\ pH - 8.5$$

Data points above the gibbsite solubility line in Figure 5.6 represent soil solutions that are *undersaturated* with respect to gibbsite; that is, the solutions have an *ion activity product,*[2] $[Al^{3+}]/[H^+]^3$, that is less than $10^{8.5}$. Data points below the line represent soil solutions that are *supersaturated* with respect to gibbsite; that is, the solutions have an ion activity product that is greater than $10^{8.5}$.

The solid phase of soils can consist of many Al-bearing materials, including organic matter; oxides and hydroxides of Al; noncrystalline aluminosilicates; layer silicate clays; and various primary minerals. Any model of Al solubility that is based on the dissolution reaction of only one of these materials is likely to be too simple. Nevertheless, some of these models will now be considered in turn.

Gibbsite Solubility Model. This model employs equation 5.19 to predict Al^{3+} solubility, but even the limited data in Figure 5.6 show significant deviation from the expected solubility. The soil solutions tend to be undersaturated (relative to gibbsite precipitation) at low pH and supersaturated at high pH. There is no compelling evidence that soils approach well-defined solubility products of common Al-bearing minerals after long periods of time. However, studies of some acid forest soils of humid temperate climates have produced circumstantial evidence for the control of Al^{3+} solubility by noncrystalline aluminosilicates (imogolite and allophane).

Cation Exchange Model. The simplest exchange models assume that Al^{3+} exchange is stoichiometric, as for example, in Ca^{2+} displacement:

$$2Al^{3+} + 3Ca^{2+}\text{-clay} = 2Al^{3+}\text{-clay} + 3Ca^{2+} \quad (5.20)$$

and that this reaction obeys an exchange equation based on mass-action behavior:

$$K_s = \frac{N_{Al}^2}{N_{Ca}^3} \cdot \frac{(Ca^{2+})^3}{(Al^{3+})^2} \quad (5.21)$$

where N_M symbolizes the fraction of exchange sites occupied by the metal, M, and parentheses denote solution activities. This equation fails in practice, overestimating the ability of Ca^{2+} (and other base cations) to displace Al^{3+} from exchange sites when the solution concentration of Ca^{2+} (base cation) is increased. In other words, K_s is not constant but is a function of the Ca^{2+} concentration in solution (see Chapter 3 for a more detailed analysis). Possibly related to this observation is the fact that layer silicate clays, especially kaolinite, tend to adsorb polymeric Al-hydroxy cations that

1. No correction is made here for activity coefficient, so that $[Al^{3+}]$ denotes Al^{3+} concentration, and pAl, the negative logarithm of the Al^{3+} activity, is taken to equal $-\log[Al^{3+}]$.

2. This activity expression is actually a quotient, but could be written as $[Al^{3+}][OH^-]^3$, the true activity product of $Al(OH)_3$, without changing the interpretation.

"block" exchange sites and inhibit Ca^{2+} adsorption. In effect, this reduces the measurable CEC of clays in acid soils. Exchange models are therefore unlikely to be successful in quantitatively predicting the extent of base cation displacement from soil exchange sites by aluminum.

Organic Matter Complexation Model. Soil humus contains a variety of acidic functional groups, some of which have high specificity for Al complexation or chelation (see section 4.3b). Al-organic complex formation requires that the acidic group, RH_n, dissociate:

$$RH_n = R^{n-} + nH^+ \tag{5.22}$$

The R^{n-} ligand then forms one or more coordination bonds with Al^{3+}:

$$Al^{3+} + R^{n-} = Al{-}R^{(3-n)+} \tag{5.23}$$

Because two reaction steps are involved, there are two separate factors that control the tendency of Al to complex with acidic groups of humus:

1. The ease of dissociation of the functional groups. This is determined by how acidic the group is. Carboxylic groups, having relatively low pK_a values, dissociate readily compared to phenolic groups.
2. The strength of the functional group$-Al^{3+}$ bond. This is determined by the electron-donating power[1] of the R^{n-} ligand. Generally, the bond with phenolic groups is stronger than that with carboxylate groups.

The first factor determines that Al^{3+} complexation with, or dissociation from, organic matter is strongly dependent on pH. At any chosen pH there is an order of preference of functional groups for Al^{3+}. For example, at pH 4.5, the expected preference of Al^{3+} for some aromatic organic ligands is

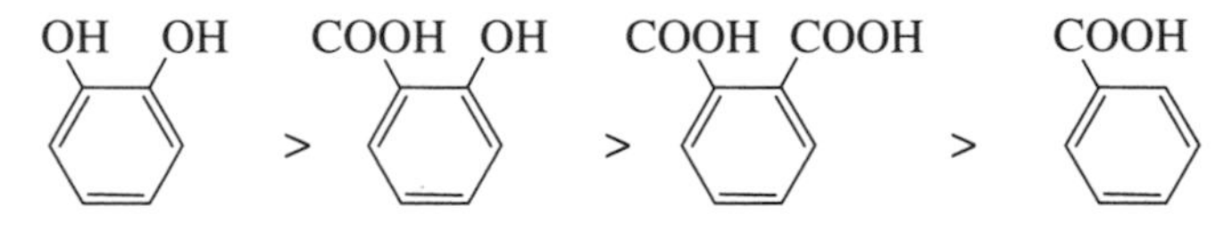

This order results from the fact that phenolic groups have higher affinity for Al^{3+}; that is, the second factor overrides the first at pH 4.5. In contrast, Ca^{2+} (and other base cation) affinity for these same groups depends largely on the extent of dissociation of the groups, which at pH 4.5 follows the order

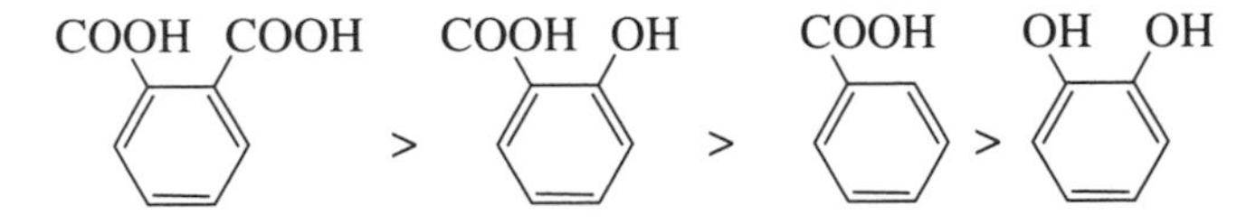

This means that for Ca^{2+}, the first factor overrides the second, presumably because the Ca^{2+}-organic bond is a fairly weak (long-range) electrostatic one.

These chemical factors combine to provide Al^{3+} with the ability to force a much larger degree of dissociation of humus (reaction 5.22) than might be expected otherwise. The process driving this dissociation is reaction 5.23, which sequesters func-

1. Measured by the Lewis basicity of the ligand.

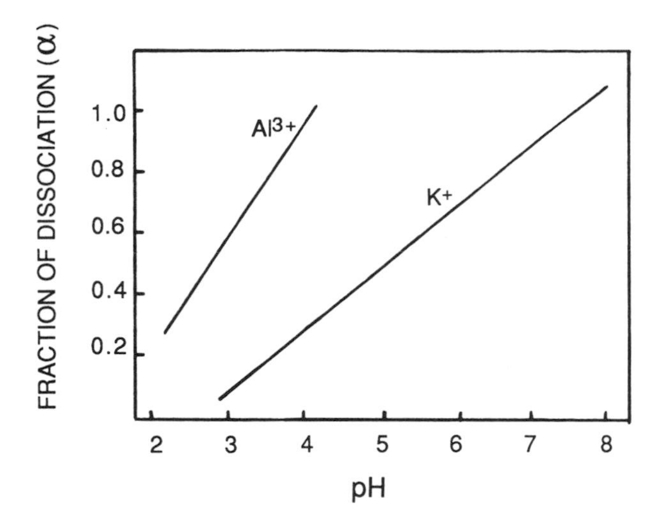

Figure 5.7. Extent of dissociation of the protonated functional groups (expressed as the fraction, α) of an acid peat soil, in the presence of Al^{3+} (as $AlCl_3$) or of K^+ (as KCl). (Adapted from P. Bloom and M. B. McBride. 1979. Metal ion binding and exchange with hydrogen ions in acid-washed peat. *Soil Sci. Soc. Am. J.* 43:687–692.)

tional groups and counters the negative charge generated by dissociation. As shown in Figure 5.7, the extent of dissociation of acidic groups in humus is much higher when Al^{3+} is in solution as opposed to situations when only base cations such as K^+ are in solution. This does not translate into a greater CEC of the humus; the "dissociated" groups are strongly bonded to Al^{3+} and reluctant to enter into exchange reactions with base cations.

By combining equations 5.22 and 5.23, the overall Al bonding reaction with organic matter is obtained:

$$Al^{3+} + RH_n = Al{-}R^{(3-n)+} + nH^+ \tag{5.24}$$

which has a complexation "constant" given by

$$K_{Al-R} = \frac{\{Al{-}R^{(3-n)+}\}(H^+)^n}{\{RH_n\}(Al^{3+})} \tag{5.25}$$

where { } brackets denote "activities" of the bracketed species in the solid phase. Equation 5.25 in logarithmic form becomes

$$\log K_{Al-R} - \log \frac{\{Al{-}R^{(3-n)+}\}}{\{RH_n\}} = pAl - n\,pH \tag{5.26}$$

The activity ratio, $\{Al{-}R^{(3-n)+}\}/\{RH_n\}$, may be considered to be a fairly constant property of any particular soil organic material, reflecting the extent of Al occupation of organic functional groups.[1] A particular activity ratio would correspond to one specific position on the dissociation curve of an Al-complexed humus (Figure 5.7).

1. As this ratio increases, pAl must decrease; that is, Al solubility is raised. But the quantity of Al released to solution is not sufficient to alter the ratio significantly.

As an approximation, for short-term studies of Al^{3+} solubility, equation 5.26 can be simplified to

$$pAl = n\,pH + K' \qquad (5.27)$$

where K' is a constant defined by the parameters on the left side of equation 5.26. Studies of Al^{3+} solubility in acid organic soils have shown the value of n to be about 2, suggesting that each Al^{3+} ion complexes with a functional group (or groups) possessing two dissociable protons (on average). Even in many surface mineral soils, equation 5.27 with $n = 2$ seems to be more successful in describing Al^{3+} solubility than the gibbsite solubility product (equation 5.19).

It is possible then, on the basis of this indirect evidence, that organic matter *does* control Al^{3+} solubility in many surface soils. Other less subtle experimental observations seem to support this hypothesis. For example, additions of organic matter to acid mineral soils lower the solubility of Al^{3+} and reduce exchangeable Al^{3+}. Acidification of soils low in organic matter increases exchangeable Al^{3+} more than the same degree of acidification of soils high in organic matter. Soils low in organic matter commonly have $(Al^{3+})(OH^-)^3$ ion activity products more in agreement with gibbsite solubility than soils high in organic matter.

The organic complexation model represented by equation 5.27 is not expected to succeed over a wide range of soil conditions or over a long period of time because

1. Many kinds of functional groups exist in humus, each one presumably having a different value of K_{Al-R} and different n.
2. Large changes in Al solubility would follow from a significant change in the fraction of organic groups bonded with Al. The assumption of constant K' in equation 5.27 would then be invalid.
3. The complexation reaction is heterogeneous, involving a solution and a solid phase. It is not known how to quantify thermodynamic "activities" of the solid-phase species, RH_n and $Al{-}R^{(3-n)+}$ in equation 5.25.

More complex mathematical models have been developed to describe complexation of soluble and insoluble organic matter with Al^{3+} and other metals. These attempt to predict the strong pH-dependence of Al^{3+} complexation by organic matter resulting from two important H^+-generating reactions: the displacement of H^+ from a range of weak acid groups and the hydrolysis of organically complexed Al to form an organic-Al hydroxy complex. Given the complexity of soil organic matter, further advances in understanding Al-humus chemistry may be slow.

5.2c. Speciation of Soluble Al

The chemistry of Al^{3+} in soil solutions is complicated by the fact that soluble inorganic and organic ligands form complexes with Al^{3+}. Since many of these complexes are soluble, the effect of the ligands is to increase the concentration of total soluble Al. A case in point is the fluoride anion, which strongly complexes with Al^{3+}. The first step of complexation between Al^{3+} and F^- is written

$$Al^{3+} + F^- = AlF^{2+} \qquad K = 1.05 \times 10^7 \qquad (5.28)$$

showing that stable Al-fluoride ion pairs are able to elevate the concentration of dissolved Al. The effect of 10^{-3} M oxalate (a naturally occurring organic acid in soils) on Al solubility is shown in Figure 5.8. Comparison with Figure 5.3 reveals the dramatic effect of this chelating organic in boosting Al solubility, especially in the pH range of 5 to 7. On the other hand, the equilibrium activities of Al^{3+} and the Al-hydroxy species in solution are *not* affected (in principle) by the presence of this organic acid because gibbsite is assumed to be present and controlling the solubility of these species. If, however, an Al-bearing mineral phase such as $Al(OH)_3$ were not in equilibrium with the solution, either because of the absence of the solid phase or because of a slow dissolution reaction, then the addition of complexing ligands such as oxalate to soil solution would immediately *decrease* the concentration of free Al^{3+} and Al-hydroxy ions in solution.

Whether a ligand increases or decreases Al solubility depends on the particular Al-ligand complex and its tendency to remain in solution or precipitate. Ligands that *increase* the overall solubility of Al include F^-, oxalate^{2-}, citrate^{3-}, fulvic acid, and silicate(monomeric). Ligands that *decrease* the overall solubility of Al include phosphate, sulfate, silicate(polymeric), and hydroxyl. Figure 5.9 summarizes the influence on Al of common soluble ligands that may be encountered in soil solution.

It is usually the case that, when the speciation of Al in soil solutions is investigated, a large fraction of the soluble Al is found to be in the form of organic and fluoride complexes. Some of the Al may also be complexed with soluble silicate, although stability constants for such aluminosilicate ions are not known. There is

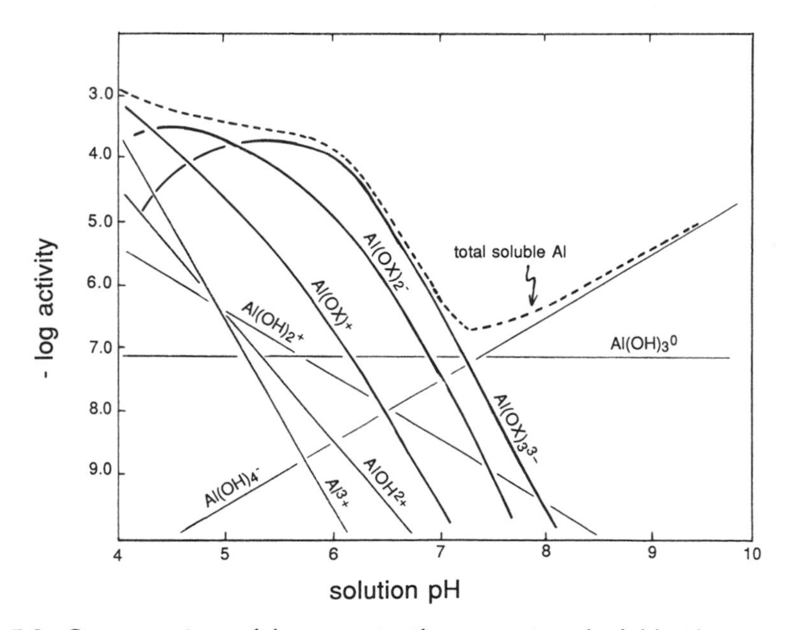

Figure 5.8. Concentrations of the most significant species of soluble Al in an aqueous solution of $AlCl_3$ and 10^{-3} M oxalic acid, assuming that gibbsite is present as the solid phase at all pH values. The broken line depicts total soluble Al (sum of all species concentrations). The oxalate anion (charge = −2) is symbolized as OX.

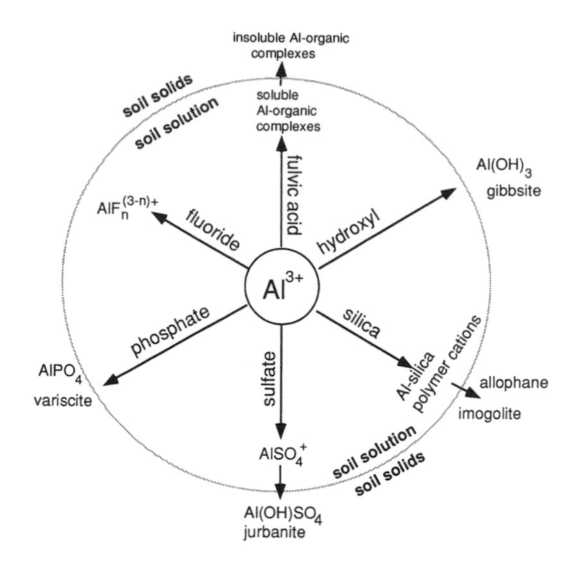

Figure 5.9. Possible complexation reactions of Al^{3+} with common ligands in soil solution, and the tendency of these complexes to form insoluble precipitates.

evidence that these various complexed forms of Al are much less phytotoxic than soluble Al^{3+} or Al-hydroxy cations. In fact, the Al^{3+} activity in soil solution is better correlated to diminished root growth in acid soils than is total soluble Al or exchangeable Al as a fraction of cation exchange sites.[1]

5.3. SOIL BUFFERING

5.3a. Lime Requirement

As described in section 5.1b, the reserve acidity of the soil functions as a buffer when alkaline materials such as lime ($CaCO_3$) are added. For example, lime reacts with carboxylic acid groups in organic matter:

$$2R{-}COOH + CaCO_3 \rightarrow \begin{matrix} R{-}COO^-] \\ \quad\searrow \\ \quad\quad Ca^{2+} \\ \quad\nearrow \\ R{-}COO^-] \end{matrix} + H_2O + CO_2 \qquad (5.29)$$

or with exchangeable Al^{3+} on clays:

$$2Al^{3+}\text{-clay} + 3CaCO_3 \rightarrow 2Al(OH)_3 + 3Ca^{2+}\text{-clay} + 3H_2O + 3CO_2 \qquad (5.30)$$

These reactions, which apply for acid soils, result in 1 mole of $CaCO_3$ neutralizing 2 moles of acidity (that is, 2 moles of H^+ or ⅔ mole of Al^{3+}). The Ca^{2+} is immobilized

1. There is some evidence to suggest that the $Al_{13}O_4(OH)_{24}(H_2O)_{12}^{7+}$ polymeric cation is highly toxic to plants. However, this is a metastable species that may not exist in soil solutions, being formed by localized and transitory high pH conditions created during the titration of Al salt solutions with strong base. It has a high affinity for layer silicate surfaces.

at exchange sites made available by the neutralization reaction, so that Ca^{2+} introduced into acid soils by liming is unlikely to be very mobile.

If the soil is less acid, then bicarbonate may be formed by reaction with lime, for example:

$$2R{-}COOH + 2CaCO_3 \rightarrow \left.\begin{matrix} R{-}COO^- \\ R{-}COO^- \end{matrix}\right] Ca^{2+} + 2HCO_3^- + Ca^{2+} \qquad (5.31)$$

and half the Ca^{2+} is in the soluble bicarbonate salt form. This salt may leach through the soil or react elsewhere with soil acidity.

The *lime requirement* of soils was at one time defined as the amount of $CaCO_3$ (or its equivalent in any other alkaline material) needed to neutralize all exchangeable acidity and bring the soil to 100 percent base saturation. This would raise the soil pH to 7 or higher, and is generally neither economical nor agronomically advisable.[1] Lime requirement is now more rationally based on the quantity of lime required to reduce exchangeable and soluble Al^{3+} to a nontoxic level for the particular crop to be grown.

The desirable pH increment, ΔpH, can be decided on the basis of a function similar to that shown in Figure 5.5, which establishes for each soil how high the pH must be to eliminate Al toxicity for a crop. Of course, the optimum pH varies for different crops because of plant-specific Al tolerances and Ca requirements. The lime requirement of any soil can be calculated from the slope of the titration curve of that soil, as diagrammed in Figure 5.10. Conversely, the *buffer capacity* of the soil is the reciprocal of this slope; that is, buffer capacity is the quantity of lime (alkali) added to the soil that achieves a unit change in pH.

The soil pH rises nearly linearly with the quantity of lime (or alkali) added. In this sense, soil pH is *not* a logarithmic parameter, unlike the pH of solutions. The relatively featureless titration curve (Figure 5.10) illustrates the fact that numerous forms of weak acids exist in the soil, each with a particular effective pH range for buffering. Unlike simple acids, then, soils reveal no titration endpoint. Furthermore, they react slowly with the added alkaline material. The liming reaction has been found to be much faster if neutral salts are present in the soil, suggesting that the rate-limiting reaction step may be release of Al^{3+} from interlayer regions of layer silicate clays or from complexes with humus. Salts such as KCl encourage Al^{3+} exchange into solution:

$$3KCl + Al^{3+}\text{-humus} \overset{\text{slow}}{\rightleftharpoons} 3K^{+}\text{-humus} + Al^{3+} + 3Cl^{-} \qquad (5.32)$$

where reaction with lime (alkali) takes place rapidly:

$$2Al^{3+} + 3CaCO_3 + 3H_2O \overset{\text{fast}}{\rightarrow} 2Al(OH)_3 + 3Ca^{2+} + 3CO_2 \qquad (5.33)$$

1. Micronutrient deficiencies can be induced by the lower solubility of many trace elements at higher pH. If the soil is Ca^{2+} deficient, however, liming may be advisable for reasons other than pH correction.

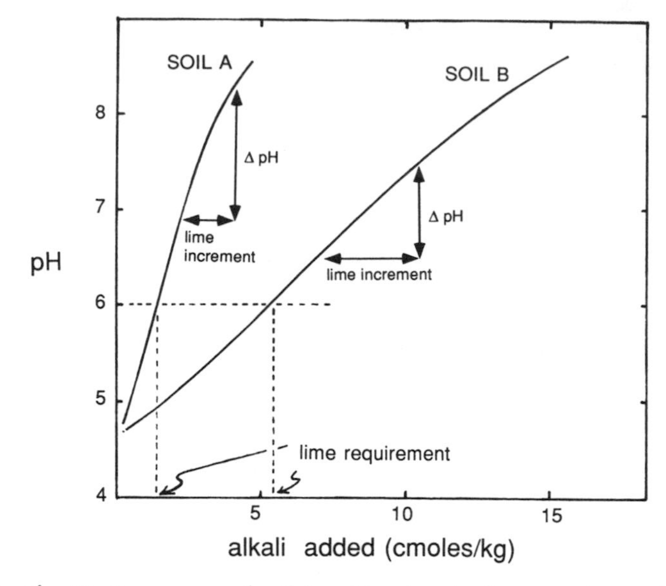

Figure 5.10. The titration curve of soil A, with a low buffer capacity, and soil B, with a high buffer capacity. The lime requirements of these soils are indicated (in units of centimoles of base per kilogram of soil) assuming that they are to be limed to pH 6.0.

Evidence for the slow overall reaction of lime with soil solids can be seen in Figure 5.11. The pH of the soil solution increases dramatically once lime is mixed into acid soils, an indication of the fast reaction between $CaCO_3$ and the active acidity (see Figure 5.1). Subsequent slow release of H^+ and Al^{3+} from the reserve acidity in soil particles gradually lowers the solution pH to an equilibrium value. In the field, where mixing of the lime into the soil is likely to be incomplete, the pH may drift downward

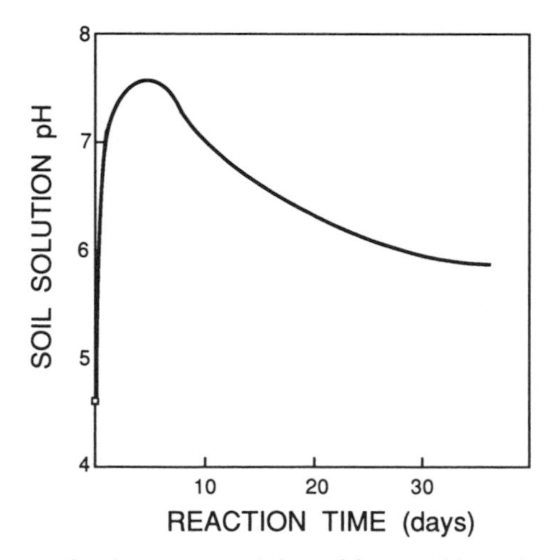

Figure 5.11. Effect on soil solution pH of the addition of lime ($CaCO_3$) to an acid mineral soil at the rate of 6 metric tons per hectare.

from an initially high value for months. Part of the lime may even be sequestered by reaction 5.33, as rapidly precipitating $Al(OH)_3$ coats and protects $CaCO_3$ particles from further dissolution by soil acidity.

5.3b. Buffer Reactions of Soils with Acids

Soils resist pH change whether acidity or alkalinity has been introduced in one form or another. There are numerous mechanisms potentially involved in this ability to buffer pH. Some of the most important ones, listed in order of the soil pH range at which they function, are:

1. **Carbonate mineral buffering.** Free carbonate minerals in the soil, most commonly Ca and Mg carbonates, are a reserve of alkalinity that can neutralize natural soil acidity (carbonic acid, organic acids) or acidity introduced as pollutants (e.g., acid rain). Soils containing free Ca or Mg carbonates are termed *calcareous* and usually have a pH above 7. Acid neutralization reactions generate bicarbonate salts, for example,

$$H^+ + CaCO_3 \rightarrow HCO_3^- + Ca^{2+} \tag{5.34}$$

This reaction, combined with subsequent leaching of the bicarbonate salts from the soil surface, slowly depletes the carbonate buffering capacity of the soil, and ultimately the soil pH may be lowered. However, the carbonate buffer capacity of many calcareous soils is so large that it would take centuries of acid inputs (natural and anthropogenic) to have any effect on pH. From the principle of solubility products, these carbonates must be completely dissolved out of the soil before the pH can drop.

2. **Exchangeable base cation buffering.** Soils in the intermediate pH range (5.5–7.0) do not contain free carbonates but can still buffer pH because added acid cations (H^+, Al^{3+}) exchange base cations from clay or humus exchange sites, for example,

$$2H^+ + Ca^{2+}\text{-clay} = 2H^+\text{-clay} + Ca^{2+} \tag{5.35}$$

The capacity of this mechanism to buffer depends on the quantity of exchangeable base cations in the soil. In comparing soils at the same pH, the soil with the highest organic matter and layer silicate clay content should possess the greatest base cation buffer capacity. The CEC of the soil is directly proportional to base cation buffering, assuming that CEC is measured as the sum of base cations.

It should be stressed that the exchangeable base cation buffering mechanism does not actually neutralize the acidity but merely stores it in the soil's reserve acidity "pool." If the soil is then subjected to leaching, the cost of buffering against acid inputs by this mechanism is depletion of the exchangeable base cations in the soil.

3. **Buffering by aluminosilicate mineral decomposition.** In moderately to strongly acid soils (pH < 5.5), variable-charge mineral surfaces as well as layer silicate edges accept protons to generate anion exchange sites, for example at Al sites:

$$H^+ + {>}Al{-}OH]^{-1/2} = {>}Al{-}OH_2]^{+1/2} \tag{5.36}$$

This mechanism is only able to accommodate a finite surface concentration of adsorbed protons before mineral dissolution begins:

$$nH^+ + {>}Al{-}OH_2]^{+1/2} \rightarrow Al^{3+} + H_2O \tag{5.37}$$

In general, high-surface-area minerals such as layer silicate clays and allophane dissolve at significant rates in strongly acid soils:

$$nH^+ + \text{alumino-silicate clays} \rightarrow Al^{3+} + Si(OH)_4^0 \tag{5.38}$$

In these reactions, part of the acidity is consumed in breaking Si—O—Si, Al—OH—Al, and other bonds in the minerals. This is demonstrated by the acid attack on two corner-shared silica tetrahedra in a layer silicate as follows:

$$(O)_3Si{-}O{-}Si(O)_3 \xrightarrow{H^+} (O)_3Si{-}OH^+{-}Si(O)_3 \xrightarrow{H_2O} (O)_3Si{-}OH + HO{-}Si(O)_3 \tag{5.39}$$

The product of this reaction, soluble monosilicic acid, $Si(OH)_4^0$, is readily leached out of the soil surface, representing export of some of the strong acidity in a weak acid form that is environmentally innocuous. At the same time, part of the acidity accumulates in the soil in more toxic forms—soluble and exchangeable Al^{3+}.

Primary mineral dissolution is also a potential buffer system because primary silicate minerals are, in effect, insoluble alkaline salts of silicate anions. The tendency for primary minerals to have large particle size (low surface area) limits their reaction rate with acids. Reactions such as feldspar decomposition:

$$KAlSi_3O_8 + H^+ + 7H_2O \rightarrow Al(OH)_3 + 3Si(OH)_4^0 + K^+ \tag{5.40}$$

could theoretically neutralize acidity without release of soluble Al^{3+}, but reaction 5.38 is much faster. Consequently, in acid soils with any appreciable clay content, Al^{3+} is generally released into solution by acid additions, and clay decomposition has a greater role than primary mineral dissolution. To a limited degree, the clay can then readsorb and control the solubility of Al^{3+}.

Figure 5.12 depicts a hypothetical soil pH-buffer capacity curve. It is clear from this relationship that the soils most likely to reach very low pH and toxic Al solubilities even with modest inputs of acidity are those with low inherent buffer capacities (arising from the lack of carbonates, clay, or humus content) or those whose buffers have already been expended in neutralizing past acid inputs. These are termed "acid-sensitive" soils, and are typically acid soils with little clay and humus. They are the most likely soils to reach extremely low pH (<4) on exposure to acid rain or other sources of acidity. Since primary and secondary aluminosilicate mineral dissolution is very pH sensitive, phytotoxic concentrations of soluble Al^{3+} are increasingly likely as the pH lowers.

5.3c. Interrelating the Various Forms of Soil Acidity

A possibly confusing array of different forms of soil acidity has been described in this section. These forms are all interrelated, but in a complex way that depends on soil properties. The *total acidity* (titratable acidity) of the soil, as measured by titration with a base such as $Ca(OH)_2$, encompasses all the various other forms of acidity, including, in decreasing order of magnitude:

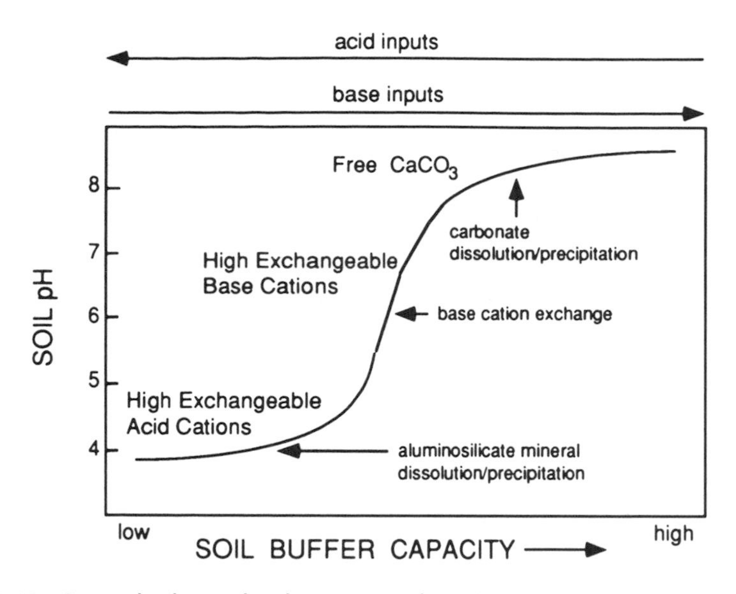

Figure 5.12. General relationship between soil pH (a measure of acid intensity) and quantity of acid or base added. The inverse of the slope of this curve provides a measure of soil buffer capacity.

Reserve acidity. All titratable acidity in the soil that is associated with the solid phase. It equals in magnitude the sum of the nonexchangeable and exchangeable acidity.

Non-exchangeable acidity. Bound H^+ and Al^{3+} that is not displaced, or is extremely slowly displaced, into solution by a concentrated neutral salt, usually 1.0 *M* KCl. This form is associated with weak acid groups on humus, organically complexed Al, and Al-hydroxy cations strongly retained at mineral surfaces.

Exchangeable acidity. Bound H^+ and Al^{3+} that is displaced into solution by a concentrated neutral salt, usually 1.0 *M* KCl. This form is associated with organically complexed Al, easily dissociating acid groups on humus, and Al^{3+} cations retained at clay exchange sites. It is predominantly composed of *exchangeable* Al^{3+} in mineral soils, and *exchangeable* H^+ in organic soils. In some soils, exchangeable acidity can exceed the nonexchangeable acidity.

Active acidity. All titratable acidity in the soil that is associated with the solution phase (dominantly free Al^{3+} and H^+ ions). It is the difference between the total and reserve acidity. Because this difference is a very small number, active acidity is best calculated directly from a measure of pH and Al^{3+} concentration in soil solution.[1]

The lime requirement is most closely related to total acidity, because lime added to soil neutralizes all forms of acidity. However, lime requirement is adjusted in prac-

1. Soluble Al complexes may or may not contribute to titratable acidity in solution. For example, Al-citrate is composed of an acid (Al^{3+}) and a base (citrate), and would neutralize alkalinity at a pH high enough to decompose the complex and form $Al(OH)_3$.

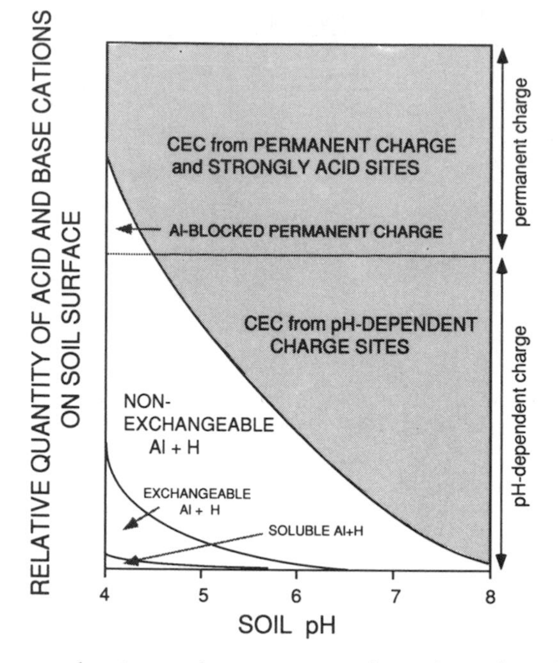

Figure 5.13. Sources of cation exchange sites in a hypothetical acid soil that is limed from pH 4 to 8. The shaded region depicts the magnitude of the cation exchange capacity (CEC) at each pH.

tice to prevent complete neutralization of total acidity, so that practical lime requirements are less than the quantity of lime needed to neutralize total acidity.

Neutralization of these various forms of acidity by liming proceeds selectively; the most available and labile forms are reacted early in the process, while the nonexchangeable forms react sluggishly and at somewhat higher pH. This is illustrated in Figure 5.13, where the elimination of exchangeable and nonexchangeable acidity is seen to create cation exchange sites occupied by base cations. Liming, then, increases the effective CEC of most soils.

5.4. AGENTS OF SOIL ACIDIFICATION

Soils that form in humid climates tend to become more acidic over time.[1] The natural and anthropogenic (agricultural and industrial) agents of soil acidification are identified in this section.

5.4a. Natural Acidification

Plant roots remove nutrients from the soil in the form of cations (K^+, NH_4^+, Ca^{2+}, Mg^{2+}, etc.) and anions (PO_4^{3-}, NO_3^-, SO_4^{2-}, etc.). The form of nitrogen utilized by the

1. On the other hand, soils of arid and semiarid climates often become alkaline, as is explained in Chapter 8.

plant generally tips the balance toward either excess cation or excess anion uptake. This apparent imbalance is not real, of course, because electroneutrality must be maintained. Either protons or OH^- (and HCO_3^-) are exuded from the root into soil solution, balancing the moles of anion and cation charge that are taken up. Thus, for NH_4^+-feeding plants:

$$\text{(Nutrient cation uptake)} - \text{(nutrient anion uptake)} = H^+ \text{ ions exuded} \quad (5.41)$$

and for NO_3^--feeding plants:

$$\text{(Nutrient anion uptake)} - \text{(nutrient cation uptake)} = OH^- \text{ or } HCO_3^- \text{ ions exuded} \quad (5.42)$$

These equations are valid only when the uptake is measured as moles of ionic *charge* rather than moles of ions.

In natural soil-plant systems, equation 5.41 is often the relevant one, so that nutrient uptake acidifies the soil in the vicinity of the roots. However, to the extent that these nutrients are returned to the soil (as plant residues), there may be little long-term acidification *on balance.* In agricultural systems, harvesting can remove much of the plant material, leading to permanent acidification.

Acidification is also attributable to the metabolic activity of roots, microorganisms, and other living organisms in the soil. Metabolic processes generate CO_2, soluble organic acids, and acidic organic residues, all of which behave as weak acids in the soil, displacing base cations from exchange sites. For example, the reaction sequence in the case of CO_2 might be:

$$\begin{aligned} CO_2 + H_2O &= H_2CO_3 \text{ (carbonic acid)} \\ H_2CO_3 + \tfrac{1}{2}Ca^{2+}\text{-clay} &= \tfrac{1}{2}Ca(HCO_3)_2 + H^+\text{-clay} \end{aligned} \quad (5.43)$$

The base cations may then be leached from the soil in the form of salts of bicarbonate and organic anions. Reaction 5.43 is favored by high levels of CO_2 in soil "air," which can contain as much as 1 to 5 percent CO_2 by volume (compared with only 0.03% in the atmosphere). Reaction 5.43 is less important in very acid soil solutions (pH < 5) because carbonic acid (and weak organic acids as well) does not readily dissociate at low pH. Therefore, acidity accumulation within the soil by this type of mechanism is believed to become important only when the soil pH exceeds 5.

Oxidation of reduced forms of S and N can also acidify the soil. Insoluble sulfides in the soil matrix react when exposed to air:

$$FeS_2 \text{ (pyrite)} \xrightarrow{O_2} 4H^+ + 2SO_4^{2-} + \tfrac{1}{2}Fe_2O_3 \quad (5.44)$$

The oxidation of small sulfide particles is rapid once soils are aerated, so that acidification is immediate and extreme when anaerobic sulfide-bearing soils are drained. Long-term low-level acidification may occur in soils containing large sulfide particles (perhaps pyrites inherited from the parent material). Measurement of the oxidative dissolution of pyrite and other sulfides has produced the following rate law:

$$\begin{aligned} \frac{d[SO_4^{2-}]}{dt} &= \frac{K_1 A [O_2][H^+]^{0.2}}{V} \\ &\approx 10^{-11}\text{–}10^{-12} \text{ moles/g-sec} \end{aligned} \quad (5.45)$$

where A and V are particle surface area and volume, respectively.

Pyrites occur in many soil parent materials, including some soils of the Adirondack Mountain region of New York, a fact that may explain the large increases in sulfate concentration of waters as they leach through these soils. Acid "mine spoils" of coal and metal mining regions contain sufficient sulfides to generate extreme acidity in water leaching through these materials, posing an environmental hazard from dissolved heavy metals.

Soils with fluctuating water tables can acidify when the soil drains and is exposed to O_2, according to reaction 5.44, then rise in pH again when flooding creates a reducing environment. However, because sulfate may leach from this system, it is likely that the overall effect would be acidifying. In wetlands along the ocean, the high SO_4^{2-} concentration in seawater provides an ample source of reducible sulfur in the soil, so that secondary sulfide minerals, including pyrite, are formed from S^{2-} and reduced iron. If such soils are subsequently drained, the sulfides reoxidize and extremely low pH values (<3) can develop, at least temporarily. These are the *acid sulfate* soils, which will be discussed further in Chapter 7 as products of redox reactions.

Naturally occurring ammonium, or ammonium-forming organic materials, are potential acid formers in soils, since biological *nitrification* generates acidity:

$$NH_4^+ + 2O_2 \rightarrow NO_3^- + 2H^+ + H_2O \qquad (5.46)$$

This reaction may not actually acidify unless the NO_3^- is lost from the soil by leaching along with exchangeable base cations, leaving H^+ to occupy exchange sites. Other possible fates of the NO_3^-, such as plant uptake (see equation 5.42), generate alkalinity in the soil and may result in no net acidification. Similarly, biological *denitrification* of the NO_3^-:

$$NO_3^- + 6H^+ + 5e^- = \tfrac{1}{2}N_2 + 3H_2O \qquad (5.47)$$

could consume the acidity. It is evident, then, that: *statements about the acidifying effect of isolated chemical reactions in soils are meaningless without considering the entire system.*

Other natural soil acidifying agents are found in the atmosphere. Carbon dioxide has a concentration of only 350 parts per million (by volume) in the open atmosphere, so that rainwater in equilibrium with this air is a very dilute solution of carbonic acid (10^{-5} *M*). This source of H_2CO_3 is of much less consequence than the carbonic acid generated from biological activity within soils. Lightning can oxidize atmospheric nitrogen, N_2, to produce NO which further oxidizes to nitric acid:

$$N_2 \xrightarrow{O_2} NO \xrightarrow{O_2} NO_2 \xrightarrow{H_2O} HNO_3 \qquad (5.48)$$

The relatively small quantities of strong acid produced by this reaction are then washed into the soil by rainfall.

5.4b. Anthropogenic Sources of Acidity

Since it was shown in the previous section that ammonium is potentially acid forming, then all NH_4^+ salts applied to soils as fertilizers (NH_4NO_3, $(NH_4)_2SO_4$, etc.) may acidify the soil. Even anhydrous ammonia, which initially reacts with water in soil to produce alkalinity:

$$NH_3 + H_2O = NH_4^+ + OH^- \qquad (5.49)$$

is ultimately nitrified according to reaction 5.46, so that the net potential effect is generation of one mole of H^+ per mole of NH_3 applied to the soil. Again, whether this potential acidification is realized or not depends on the subsequent fate of the nitrate formed by nitrification. Based on the principles presented here, nitrate fertilizers such as $Ca(NO_3)_2$ should not acidify soils.

Phosphate salts, applied to soils as fertilizers, may acidify over the long term. For example, superphosphate fertilizers contain $Ca(H_2PO_4)_2$, which dissolves readily in water:

$$Ca(H_2PO_4)_2 = Ca^{2+} + 2H_2PO_4^- \qquad (5.50)$$

Since the dissociation of $H_2PO_4^-$ occurs near pH 7:

$$H_2PO_4^- = H^+ + HPO_4^{2-} \qquad \log K = -7.2 \qquad (5.51)$$

this phosphate salt does not immediately acidify the soil. However, a much less soluble phosphate mineral eventually precipitates, producing acidity. In nonacid soils the reaction is likely to be of the type

$$2H_2PO_4^- + 3Ca^{2+} \rightarrow Ca_3(PO_4)_2\,(s) + 4H^+ \qquad (5.52)$$

whereas in acid soils the reaction might be similar to

$$H_2PO_4^- + Al^{3+} \rightarrow AlPO_4\,(s) + 2H^+ \qquad (5.53)$$

Thus, highly soluble phosphate fertilizers can actually precipitate and inactivate phytotoxic Al^{3+} in very acid soils, thereby counteracting aluminum toxicity without raising the soil pH.[1]

Sometimes materials are added to soils in a deliberate effort to acidify them. The materials used include elemental sulfur, which gradually oxidizes to form sulfuric acid, and sulfate salts of Al^{3+} or Fe^{3+}, which generate H^+ ions once $Al(OH)_3$ or $Fe(OH)_3$ precipitates in the soil:

$$Al_2(SO_4)_3 \rightarrow 2Al(OH)_3 + 6H^+ + 3SO_4^{2-} \qquad (5.54)$$

Finally, an unintended addition of acidity to soils occurs in the form of acid rain, the acidity of which is largely due to sulfuric and nitric acids. These strong acids in dilute form have a very complex interaction with the plant canopy and the soil, which is discussed in the next section.

It is informative at this stage to compare estimates of soil acidification attributable to the several natural and anthropogenic sources discussed in this section. Estimates are given in Table 5.1 of the most important sources of acidity in a forested soil of the southeastern United States. It is clear that acid rain is an important, but not necessarily dominant, contributor to the soil's gradual acidification. The analysis is different in areas where rainfall is even more acidic, and is affected by change in land use from forestry to agriculture. The annual contribution to soil acidification is seen in Table 5.1 to be a very small fraction of the total soil acidity at this particular

1. Phosphate fertilizer has been called an expensive alternative to limestone for suppressing Al toxicity.

Table 5.1. Annual Accumulation of H^+ (kmoles/hectare) Compared with Exchangeable and Total Acidity in a Forested Soil of Tennessee

Biological acid production (bicarbonate leaching) = 0.6
Excess base uptake by roots = 1.1
Acid rainfall (rainwater pH = 4.3–4.4) = 0.67
KCl-exchangeable acidity = 100–275
Total titratable acidity = 500–800

Source: D. D. Richter. 1986. Sources of acidity in some forested udults. *Soil Sci. Soc. Am. J.* 50:1584–1589.

site, and it would take centuries for the reserve acidity of this very acid soil to increase noticeably.

5.4c. Examples of Acidified Soil Systems

Soils acidified by strong mineral acids are likely to have unacceptably high concentrations of soluble Al^{3+} as well as very low pH. This has been seen in a forested soil of the Netherlands, for example, where deposits of $(NH_4)_2SO_4$ arise from the reaction between sulfuric acid and ammonia pollutants in the atmosphere. The ammonium is nitrified in the soil (see reaction 5.46), leading to strong acid inputs of about 3 to 7.5 kmoles/hectare/year. Analysis of soil solutions for the various forms of soluble Al produced the results shown in Figure 5.14. Most of the Al in these soil solutions was in the phytotoxic Al^{3+} form; relatively little was complexed with soluble organic matter, sulfate, or fluoride ligands that could keep it in less toxic forms.

The acid inputs at this Netherlands forested site are much larger than those at the Tennessee location (Table 5.1), and in this more extreme case of acid pollution, the soil's buffering capacity was severely tested (soil pH was between 3.8 and 4.5). In fact, drainage water from this soil contained dissolved Al, an indication that dissolution

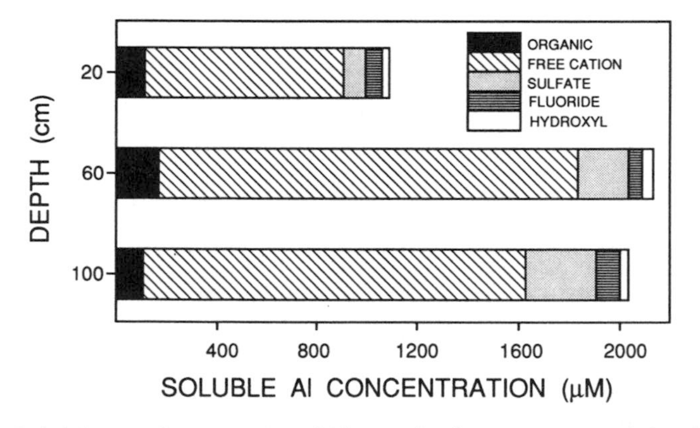

Figure 5.14. Solubility and speciation of Al in soil solution at several depths of an acid (pH = 3.8–4.5) forest soil of the Netherlands. Air pollution provides acid inputs of about 3 to 7.5 kmol/ha/yr. (Adapted from Mulder et al. 1987. Impacts of acid atmospheric deposition on woodland soils in the Netherlands. *Soil Sci. Soc. Am. J.* 51:1640–1646.)

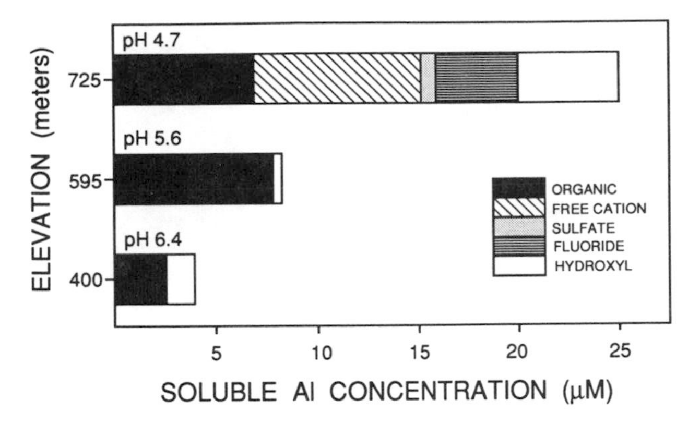

Figure 5.15. Solubility and speciation of Al in soil solution at several elevations of the Hubbard Brook Forest. (Adapted from C. T. Driscoll et al. 1985. Aluminum chemistry in a forested spodosol. *Soil Sci. Soc. Am. J.* 49:437–444.)

of Al-bearing minerals was the buffer mechanism and the soil had reached an advanced stage of acidification (see Figure 5.12).

In contrast, as shown by Figure 5.15, acid soils of the Hubbard Brook Forest in New Hampshire had much lower soluble Al levels, and a large fraction of this Al was complexed with organic and inorganic ligands. Since the highest concentrations of inorganic Al were detected in soil solutions at high elevation sites of Hubbard Brook, the suggestion has been made that inputs of atmospheric acid are greater at higher elevations. Other factors could also be responsible for the trend. For example, water seepage could be leaching base cations, bicarbonate, and other products of mineral weathering downslope, accounting for the higher pH of the low-elevation soils. Furthermore, spatial differences in Al solubility can result from a variation in type of vegetation (deciduous vs. coniferous forest, for example) and mineralogical composition of the parent material.

5.4d. Dynamics of Cation Leaching by Acid Precipitation

A discussion of aluminum leaching can be restricted to the most acid-sensitive soils—soils that are already acid throughout the profile. The reason for this is that Al^{3+} mobilized by acid inputs into the surface is effectively adsorbed or precipitated in the lower part of the profile of soils with nonacid lower horizons. Commonly, then, a very acidic surface soil ($pH < 5$) depleted of base cations lies immediately above a much less acid soil ($pH > 6$) with higher exchangeable base cations and virtually no exchangeable Al^{3+}. That is, soil acidification tends to creep downward along a "front." Unlike Al^{3+} leaching, base cation leaching is not hindered by the less acidic layer below. Loss of base cations into groundwater, streams, and lakes is a phenomenon that occurs in all soils subjected to leaching by rainfall.

Field studies of cation movement through plant canopies and acid soil profiles have been conducted in which pH and ionic composition of the rainfall, throughfall (rain that has fallen through the plant canopy), soil solution, and stream water have been monitored. Such studies try to assess the degree to which net cation loss from

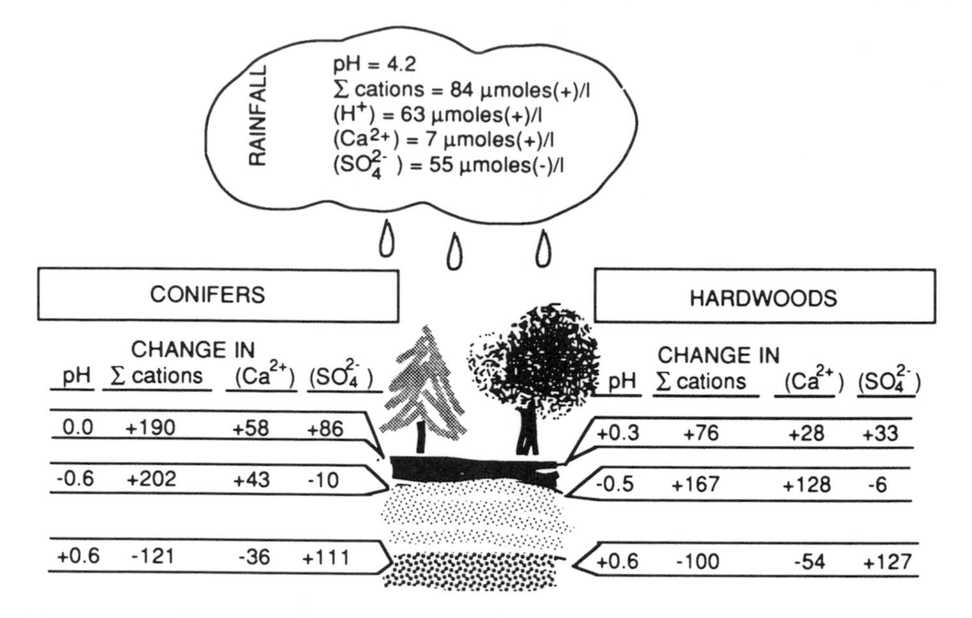

Figure 5.16. Changes (relative to rainfall composition) in the ionic concentrations of leachates at several positions in the Adirondack forest canopy/soil profile. These positions are: below the forest canopy; below the soil surface organic (litter) layer; and below the soil profile. All of the changes are reported in units of micromoles per liter. (Adapted from A. V. Mollitor and D. J. Raynal. 1982. Acid precipitation and ionic movements in Adirondack forest soils. *Soil Sci. Soc. Am. J.* 46:137–141.)

the soil can be attributed to natural dissolution processes as opposed to the effects of strong acid pollutants in the rainfall.

If paired soil-plant ecosystems, similar in all respects except for the quantity of strong acid pollutants in the rain, were available for direct comparison, this assessment would not be too difficult. Such ideal comparisons are rarely, if ever, possible because soil ecosystems can differ by vegetation, rainfall, mineralogy, profile morphology, and numerous other properties. Nevertheless, even without the benefit of a "control" treatment for comparison, it is informative to follow in detail the composition of rainwater as it falls through a forest canopy and then leaches through an acid soil below. As is summarized in Figure 5.16 for two forested sites (hardwood and conifer) in the Adirondack Mountains, dramatic changes in the water chemistry occur. Notable chemical changes *within* the forest canopy are:

1. The rainfall picks up relatively large quantities of cations and anions as it passes through the canopy. In particular, Ca^{2+} and SO_4^{2-} increase. Evidently the ions are dissolved from dust on the foliage and from the plant tissue itself.
2. The pH of the acid rainfall is raised by contact with the hardwood canopy, but not by contact with the conifers.

Once the water leaches through the organic (surface) layer of the forest floor, the most notable changes are:

1. The pH decreases substantially in both conifer and hardwood forests. This is attributable to the extremely acidic nature of the surface organic layer of these soils.
2. Large increases occur in the total cation concentration of the leachate. In the hardwood forest, much of this increase is due to Ca^{2+}. Since little relative change is seen in the SO_4^{2-} concentration (it actually decreases slightly), some other anion must be introduced from the organic layer (presumeably anionic forms of fulvic acid) to accompany the leaching cations.

Finally, as the leachate passes through the layer of mineral soil, the most apparent changes are:

1. A large increase in pH, as much of the acidity is neutralized by buffering reactions in the soil
2. A decrease in total soluble cations, including Ca^{2+}, probably indicating some kind of precipitation reaction
3. A large increase in soluble SO_4^{2-}, possibly arising from SO_4^{2-} desorption or sulfide oxidation in the mineral soil.

These patterns of behavior are fairly typical for acid soils of the northeastern United States and eastern Canada, although the increase in soluble SO_4^{2-} in leachate seems to be a regional phenomenon. In fact, in many acid soils, the SO_4^{2-} concentration actually decreases as the leachate passes through the soil profile, evidence that SO_4^{2-} is adsorbed on clays, particularly oxides and allophanes, in the B-horizon.

Analysis of leachate for soluble Al at the same Adirondack hardwood and conifer sites described by Figure 5.16 gave the results summarized in Table 5.2. It is clear that the acid rainwater dissolves Al in the canopy and in the organic and upper mineral horizons. Much of this dissolved Al is in the organically complexed form, which tends to be more mobile than free Al^{3+} in soils. Yet some of this Al is readsorbed or precipitated out of solution in the B-horizon. The vertical change in Al solubility is consistent with the changes in pH as leachate passes through the soil (see Figure 5.16). Furthermore, the conifer leachate dissolved more Al from the soil than did the hardwood leachate. This is probably related to the acidifying and Al-complexing properties of organic acids derived from conifer leaf litter. Those organic acids that

Table 5.2. Change in Leachate Concentration of Total Soluble Al at the Adirondack Forest Site

	Change in Soluble Al (μmoles(+)/liter)	
	Conifer	Hardwood
Canopy throughfall	+12	+12
Organic litter leachate	+68	+40
Upper mineral horizon (E) leachate	+43	+10
Lower mineral horizon (B) leachate	−33	−19

Source: M. B. David and C. T. Driscoll. 1984. Aluminum speciation and equilibria in soil solutions of a haplorthod in the Adirondack Mountains (New York, U.S.A.). *Geoderma* 33:297–318.

strongly chelate Al^{3+} can aggressively dissolve and mobilize Al from soil minerals.[1] It is often noted that soil pH is lower, and soluble Al is higher, for soils under conifer forest compared with nearby soils under deciduous forest. This effect of vegetation complicates attempts to experimentally isolate and quantify the various natural and anthropogenic sources of acidity.

Overall, the soils referred to in Table 5.2 are net exporters of Al, mostly in the form of soluble Al-organic complexes. If the lower mineral horizons are considered separately from the surface organic litter layer, these B horizons are a net accumulator (sink) of Al-organic complexes and a net generator (source) of inorganic Al.

In summary, the changes in the chemistry of water as it moves through a plant-soil ecosystem into groundwater are complex. The field observations reported in Figure 5.16 and Table 5.2 suggest that several kinds of chemical reactions are involved, including dissolution and precipitation of minerals, ion exchange, and dissociation of organic acids. Simple models that attempt to quantify soil acidification are likely to fail, yet such models can be useful in pointing out properties of the soil-water system that are most critical to predicting how the soil will respond to long-term inputs of acids.

5.4e. A Simplified Model of Acidification in Mineral Soil

As the discussion to this point has revealed, hydrogen ions originate from various sources, some external and some internal to the soil. The soil solids, including mineral and organic ion exchangers, Fe and Al oxides, noncrystalline aluminosilicates, and carbonates all accept these H^+ ions from solution and neutralize them or add them to the pool of reserve acidity. Eventually, as the soil profile becomes acidified throughout, H^+ ions entering the soil are merely converted stoichiometrically into acidity in a weaker form (particularly Al^{3+}), part of which is exported from the soil in drainage water.

In modeling this acid buildup, we might begin with the chemical reactions expected to produce soil acidity internally. This was shown earlier in this chapter to be due in part to the dissolution of biologically generated CO_2 (or organic acids) in water. The relevant reactions of CO_2 with water are discussed in Chapter 3 (see equations 3.55 and 3.56). The equilibrium expressions from these reactions are:

$$K_H = \frac{[H_2CO_3^*]}{P_{CO_2}} = 10^{-1.41} \tag{5.55}$$

and

$$K_1 = \frac{[H^+][HCO_3^-]}{[H_2CO_3^*]} = 10^{-6.3} \tag{5.56}$$

These two equations are combined to give the equation

$$[H^+][HCO_3^-] = K_1 K_H P_{CO_2} \tag{5.57}$$

1. In fact, some organic acids are more efficient in dissolving Al from soil minerals and humus than are strong inorganic acids such as HNO_3.

Assuming that Ca^{2+} is the predominant base cation in soil solution and NO_3^- is the only significant anion besides HCO_3^- (soluble Al species, fulvic acid, CO_3^{2-}, and OH^- are ignored for the sake of simplicity), then charge balance in soil solution requires the equality:

$$[H^+] + 2[Ca^{2+}] = [HCO_3^-] + [NO_3^-] \tag{5.58}$$

where square brackets denote molar concentrations.

Consider now the simplest case: pure water in equilibrium with CO_2 in the soil air. In this instance, $[H^+] = [HCO_3^-]$, because the only significant source of acidity in the soil water is carbonic acid dissociation. Equation 5.57 then simplifies to

$$[H^+]^2 = K_1 K_H P_{CO_2} \tag{5.59}$$

or, in logarithmic form:

$$pH = 3.855 - \tfrac{1}{2} \log P_{CO_2} \tag{5.60}$$

This equation is plotted in Figure 5.17 as a straight line (labeled "0"). It shows that:

pH is an inverse function of the CO_2 gas pressure in the atmosphere.

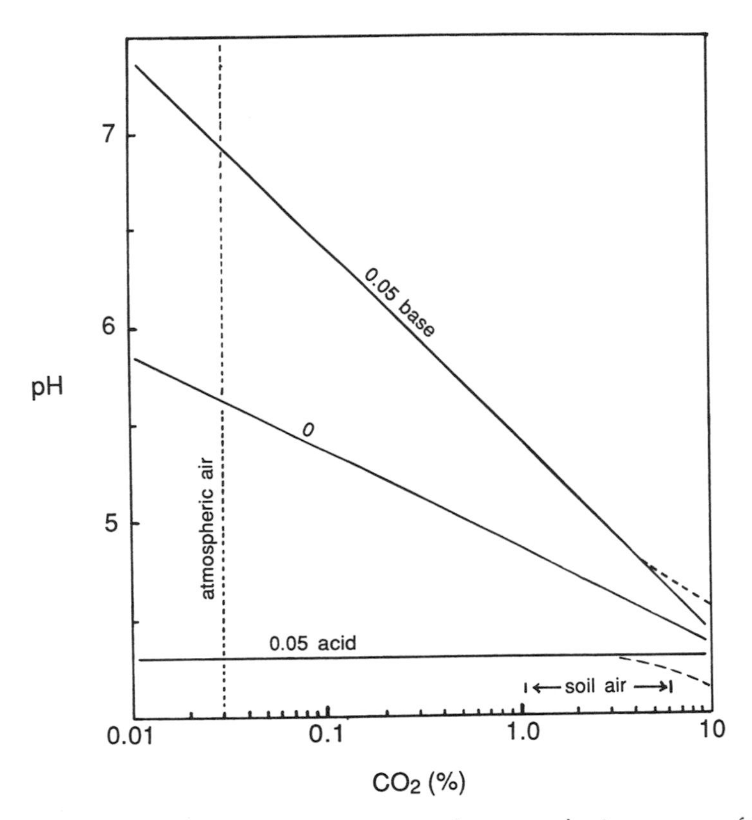

Figure 5.17. Dependence of pH in solution on the atmospheric pressure of CO_2 (expressed as % by volume), and the addition of strong acid or base. (Adapted from J. O. Reuss and D. W. Johnson. 1985. Effect of soil processes on the acidification of water by acid deposition. *J. Environ. Qual.* 14:26–31.)

This general principle, while derived here for a system much simpler than soil solution in equilibrium with soil air, is qualitatively valid for soils not exposed to strong acid inputs.

Consider now the case where strong acids have been added to the water—for example, HNO_3 from acid rain. Since equation 5.57 reveals that an increase in $[H^+]$ must cause $[HCO_3^-]$ to decrease for any fixed CO_2 pressure in the soil, it is necessary to conclude that in this situation, $[H^+] > [HCO_3^-]$. In fact, if $[H^+] \gg [HCO_3^-]$, charge balance (equation 5.58) requires that $[H^+] \approx [NO_3^-]$; that is, $[H^+]$ is no longer a function of P_{CO_2}. The result is the nearly horizontal line in Figure 5.17 (labeled "0.05 acid" to indicate the effect of 0.05 mM HNO_3 on pH). This line only begins to become nonlinear as P_{CO_2} reaches high levels and $[HCO_3^-]$ and $[H^+]$ increase relative to the other cations and anions in solution.

If a base (e.g., $Ca(HCO_3)_2$) is added to the solution, then equation 5.57 requires that $[HCO_3^-] > [H^+]$. Furthermore, charge balance requirements imply that $[Ca^{2+}] > [H^+]$.[1] Since no acids such as HNO_3 have been added, it is assumed that $[HCO_3^-] \gg [NO_3^-]$, and therefore (from equation 5.58) that $2[Ca^{2+}] \approx [HCO_3^-]$. The Ca^{2+} concentration in this simple aqueous system is determined only by the quantity of base (alkali) added to solution, so now $[HCO_3^-]$ is seen to become more or less independent of P_{CO_2}. Consequently, equation 5.57 simplifies to:

$$[H^+] = \frac{K_1 K_H P_{CO_2}}{[HCO_3^-]} \tag{5.61}$$

or in logarithmic form:

$$pH = -\log\{K_1 K_H\} - \log P_{CO_2} + \log [HCO_3^-] \tag{5.62}$$

Specifically for 0.05 mM HCO_3^- alkalinity added to water:

$$pH = 7.71 - \log P_{CO_2} - 4.30 = 3.41 - \log P_{CO_2} \tag{5.63}$$

Equation 5.63 is plotted in Figure 5.17 and labeled "0.05 base."

What becomes evident from this simple chemical model is that pH is not a very sensitive indicator of the *quantity* of strong acid or base in water *when the CO_2 pressure is high.* This revelation may help to explain the following apparently contradictory observations:

> Observation 1: *The pH of solutions in forest soils of northern temperate climates is commonly well below 5, with or without inputs of strong acids from pollution. In fact, the pH of acid rain is commonly* higher *than that of the soil solution receiving it.*
>
> Observation 2: *The pH of water draining from these same forest soils is notably lower if the soils have been exposed to inorganic acids from pollution.*

These observations are reconciled when it is realized that, once the water leaves the CO_2-charged soil and enters streams and lakes (which are closer to equilibrium

1. For bases in general, this is equivalent to saying that the soluble base cations (Ca^{2+}, Mg^{2+}, K^+, Na^+) exceed H^+ in concentration.

with the much lower CO_2 pressure of the open atmosphere), much of the total acidity in the water is lost in the form of degassed CO_2:

$$H_2CO_3 \text{ (dissolved)} \rightarrow CO_2 \text{ (gas)} + H_2O \tag{5.64}$$

The effect of this loss of CO_2 on solution pH is deduced from Figure 5.17; if no strong acids are present, the pH rises above 5.5 on degassing. On the other hand, if strong acids *are* present, the effect on pH of degassing CO_2 is much less dramatic. As a result, a wide range in pH can be found in the drainage water from acid soils depending on the relative contribution of carbonic acid and strong acids to total solution acidity.[1]

The lesson from this analysis is that:

Dilute solutions of strong acids added to naturally acid soils can further acidify the drainage water, even when the pH of the soil solution is as low as (or lower than) the pH of the added solutions.

The pH of a solution, per se, is not the best measure of its potential to acidify, nor is the pH of a soil the most reliable measure of the degree to which it is acidified.

Once a solid mineral phase is introduced into this simple model and cation exchange or dissolution/precipitation reactions are allowed for, the relationships among P_{CO_2}, pH, and added acidity or basicity are fundamentally changed. For example, the H^+ ions can react with the acid mineral soil to produce soluble Al^{3+}. Figure 5.6 showed that the solubility of Al^{3+} in the B and C horizons of some acid mineral soils is described fairly well by the solubility product of crystalline gibbsite ($K_{SO} = [Al^{3+}][OH^-]^3 = 10^{-34}$),[2] so that the approximate relationship between $[Al^{3+}]$ and $[H^+]$ is

$$[Al^{3+}] = 10^8[H^+]^3 \tag{5.65}$$

The Al^{3+} thereby generated from mineral dissolution becomes available to displace Ca^{2+} from soil exchange sites. Unfortunately, the mass-action cation exchange equation (5.21) does not adequately describe Ca^{2+}-Al^{3+} exchange on layer silicate clays (see Chapter 3). Instead, a Langmuirlike function is found to be useful:

$$\log [Al^{3+}] = \log K_{Al} + \log \left\{\frac{N_{Al}}{(1 - N_{Al})}\right\} \tag{5.66}$$

where N_{Al} is the fraction of clay exchange sites occupied by Al^{3+}. The Al bonding coefficient, K_{Al}, remains fairly constant as long as the Ca^{2+} concentration in solution does not vary too widely. For low Ca^{2+} concentrations, such as might be found in acid soils, $\log K_{Al} \approx -5.3$.

Now, combining the conditions set by equations 5.57, 5.58, 5.65, and 5.66, where 5.58 is modified to account for soluble Al^{3+}, several "scenarios" of soil acidification can be visualized:

1. The contribution from organic acids has been ignored here to keep the system simple, but if these weaker acids constitute a large fraction of the soil solution acidity, degassing CO_2 will not raise the pH of the drainage water as much as suggested by Figure 5.17.

2. By contrast, the organic surface layer of the same soil was highly undersaturated with respect to gibbsite, so that this approximation of solubility is not valid for organic-rich soils.

1. **Unpolluted rainfall.** In this case, $[H^+] \approx [HCO_3^-]$, $[Al^{3+}]$ is considered to be negligible in the charge balance equation, and, assuming that the rainfall is essentially free of salts (i.e., $[NO_3^-] \approx 0$), the equations are combined to give

$$\log \left\{ \frac{N_{Al}}{(1 - N_{Al})} \right\} = 1.73 + 1.5 \log P_{CO_2} \tag{5.67}$$

This produces the result shown in Figure 5.18, in which increasing CO_2 pressure in the soil air lowers the pH of the soil solution, increases the solubility of Al^{3+}, and raises the fraction of exchange sites occupied by Al^{3+}. Implicit in this model is the assumption that CO_2 from the soil air is not depleted by reactions with the soil minerals; that is, a constant CO_2 pressure is maintained as equilibrium is reached. This probably causes an overestimation of the acidification that is likely to occur within a specific period of time because CO_2 dissolution in water is slow. Even so, the highest concentration of free Al^{3+} that could be achieved by this model within the range of realistic CO_2 pressures is $<10^{-5}$ *M*. On the other hand, if soluble organic acids were present, total soluble Al could greatly exceed the free Al^{3+} concentrations plotted in Figure 5.18. As was demonstrated in Figure 5.8, oxalic acid, one of the most common organic acids in soil solutions, raises the solubility of Al.

An interesting consequence of the presence of soil solids is that, as higher levels of CO_2 in the soil generate more H^+ and HCO_3^- from carbonic acid dissociation, H^+

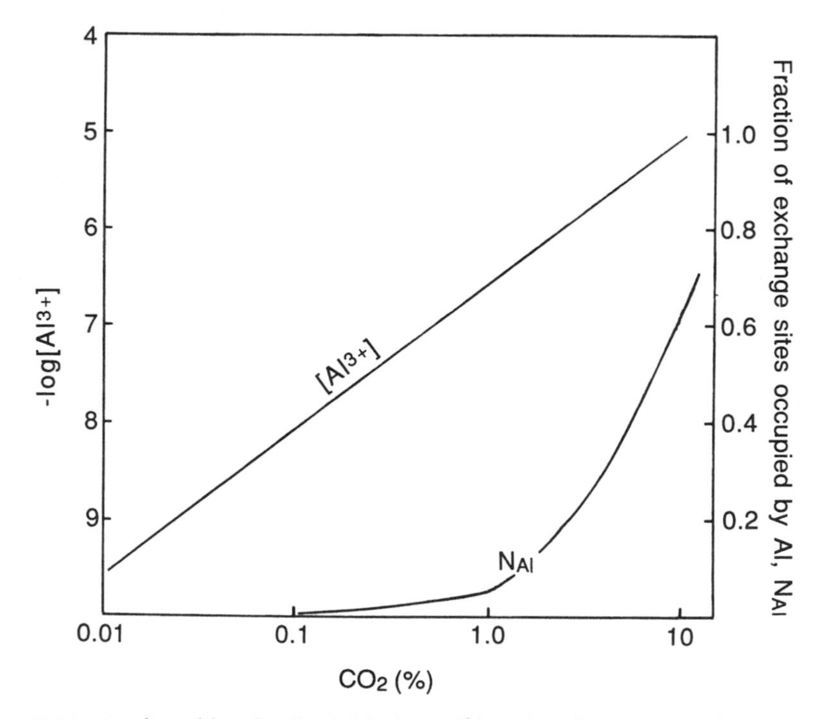

Figure 5.18. Predicted levels of soluble free Al^{3+} and exchangeable Al^{3+} as a function of the CO_2 concentration (volume %) in the soil air, based on a simple carbonic acid model of soil acidification.

is either adsorbed by ion exchange or consumed by dissolution reactions, and the *total alkalinity* in soil solution, defined in this system as

$$\text{Total dissolved alkalinity} = [HCO_3^-] - [H^+] - 3[Al^{3+}] \tag{5.68}$$

is *increased*. However, leaching processes then cause a loss of alkalinity from the soil and net acidification of the whole soil system.

2. **Acid-polluted rainfall.** In this case, suppose that 0.05 m*M* HNO_3 enters the soil in rainfall. This acidity could dissolve $Al(OH)_3$:

$$Al(OH)_3 + 3H^+ = Al^{3+} + 3H_2O \tag{5.69}$$

according to the solubility product (equation 5.65). As shown in Figure 5.17, this level of acidity makes $[H^+]$ almost independent of P_{CO_2}. But reaction 5.69 shows that 3 moles of H^+ are consumed for every 1 mole of Al^{3+} generated. At equilibrium, if x moles/liter of Al^{3+} are solubilized, then $(5 \times 10^{-5} - 3x)$ moles/liter of H^+ remain in solution. Entering these unknown quantities into equation 5.65 gives the relation

$$x = 10^8(5 \times 10^{-5} - 3x)^3 \tag{5.70}$$

By trial and error, the solution to this equation is found to be $x \approx 0.45 \times 10^{-5}$ *M*. That is, there should be 0.45×10^{-5} *M* Al^{3+} in solution, and the pH should be 4.4. These are higher Al^{3+} concentrations and lower pH than would result from natural acidification by carbonic acid, except at very high CO_2 pressures (see Figures 5.17 and 5.18). This high dissolved Al^{3+} could persist only if acids continued to enter the soil and the source of dissoluble Al was not exhausted. In any case, the dissolved Al^{3+} would displace base cations from exchange sites, so that soil acidification would be accompanied by the leaching of the soluble nitrate salts of Ca^{2+} and Mg^{2+} from the soil.[1]

The extent and severity of this soil acidification process would depend on the total quantity of strong acids, including HCl and H_2SO_4 as well as HNO_3, to enter the soil.[2]

In Table 5.3, the result of adding increasing quantities of the strong acid, HCl, to an acid mineral soil is shown. The added H^+ increases both total soluble Al and soluble free Al^{3+}, due to release of Al^{3+} from soil particles. Displacement of base cations by Al^{3+} from exchange sites follows. Much of the difference between total soluble Al and free Al^{3+} is attributed to organically complexed forms of Al. These organic forms of Al represent a complication that has not been considered in the simple model of soil acidification developed here. This is in large part because Al-organic complexation in humic and fulvic substances cannot be simply quantified by a single stability constant (see subsection 5.2b). Numerous types of complexing ligands and ligand groups are probably involved in binding Al^{3+}. As a result, the control of Al solubility by humus is not subject to simple analysis, but two key vari-

1. In natural soil acidification, base cations are lost largely in the form of salts of bicarbonate and organic anions.

2. If salts of base cations (such as Ca nitrate or Ca sulfate) are present in acid rain, they can to some degree mitigate the effects of strong acids (HNO_3 and H_2SO_4) on soil, suppressing the tendency of H^+ to exchange base cations from clays and organic matter. Their possible influence has not been built into this model because of the ease with which Al^{3+} displaces base cations from exchange sites. A more complete model would, however, take them into account.

Table 5.3. Analysis of Soil Solution after Adding H^+ in the Form of HCl to an Adams B_S Horizon

H^+ Added (mmoles/kg soil)	pH	Al Concentration ($M \times 10^5$) Total	Free Al^{3+a}
0.0	4.85	0.48	0.33
1.0	4.50	1.8	1.5
2.0	4.36	4.2	3.6
3.0	4.27	7.9	6.3
4.0	4.25	11.1	8.9

[a]Estimated from the rapidity of complexation with a chelating ligand.

Source: Data from M. G. Johnson and M. B. McBride. 1991. Solubility of aluminum and silicon in acidified spodosols: Evidence for soluble aluminosilicate. In R. J. Wright et al. (eds.), *Plant-Soil Interactions at Low pH*. Dordrecht: Kluwer, pp. 15–24.

ables are pH and the degree of saturation of the complexation sites by Al. At lower pH, H^+ ions associate with ligands such as carboxylate and displace Al^{3+} into solution. At higher Al saturation of complexation sites, Al release into solution becomes easier, not only from the standpoint of probability (the entropy factor), but also because the types of functional groups having a high bonding energy for Al^{3+} are already bonded to Al^{3+}.

5.4f. Base Cation Leaching and the Mobile Anion Concept

One clear prediction from the simple acidification model developed in the last section is the loss of alkalinity (as defined by equation 5.68) and base cations from soil solution by leaching. High P_{CO_2} in the soil should naturally result in base cation loss from the upper soil layer in the form of bicarbonate salts. Strong acid additions to the soil should produce additional base cation loss. The result in both cases is the accumulation of acidic cations in the soil, and eventually, escape of soluble Al in drainage from the soil.

Anions that are transported through soils without being retained to any great degree are termed *mobile anions.* They are anions of strong acids, and include chloride, nitrate, and, to a lesser extent, sulfate. Since charge balance requires that negative charge of soluble anions be balanced by positive charge of the associated soluble cations, anion mobility always implies cation mobility. Cations, however, are prone to adsorption and various other immobilizing reactions so that they are not generally transported through soils without at least some retention. Of course, charge balance dictates that any particular cation (e.g., Ca^{2+}) that is removed from solution and retained is simply replaced by another (e.g., H^+, Mg^{2+}, Al^{3+}) in solution. Ultimately, as drainage water passes through soils, the cation composition of the water shifts in favor of the least strongly retained or most abundant cations on the soil particles, while anion composition may be less variable. The total moles (of charge) of cations and anions do not change as the soil solution migrates unless dissolution or precipitation reactions occur.

It is sometimes stated that the addition of "mobile anions" to soils is the cause of base cation loss from the soil because these anions act as "carriers" for the cations. This way of thinking is, in effect, invoking charge balance requirements as a driving force or mechanism to explain reactions. Since "mobile anions" cannot be added to soils without the simultaneous addition of an equivalent number (in terms of charge) of cations, it is more logical from a mechanistic standpoint to argue that it is the added cations that displace cations into solution and mobilize them.

Consider the simple example of the same mole quantity of sulfate added to an acid soil, in one case in the form of K_2SO_4 and in the other case as H_2SO_4. The experiment is depicted diagrammatically in Figure 5.19 and assumes few specific adsorption sites for sulfate exist in the soil. Now in the K_2SO_4 case, K^+ displaces some base cations as well as H^+ and Al^{3+} from exchange sites, and the released Al^{3+} may generate more H^+ in solution by hydrolysis. As a result, the leachate from the soil has a quite low pH even though K_2SO_4 is a neutral salt. In the H_2SO_4 case, the added H^+ ions exchange with base cations, Al^{3+} and soil H^+, so that again, base cations as well as H^+ and Al^{3+} appear in solution and the leachate has a quite low pH. At first glance there may appear to be little difference in the effect of H_2SO_4 and K_2SO_4 on the soil and leachate, leading to the implication that it is the concentration of the "mobile anion," sulfate, that is the controlling factor. Closer inspection shows that K_2SO_4 actually *decreases* the reserve acidity in the soil (increases exchangeable base content), whereas H_2SO_4 *increases* the soil acidity reserve (decreases exchangeable base content). The degree to which the composition and pH of the two leachates depicted in Figure 5.19 are similar depends on the buffering capacity of the soil. In a weakly buffered soil, H_2SO_4 will produce a leachate that is markedly more acidic and lower in base cations than will K_2SO_4.

The point here is that cation exchange and other surface processes control ion mobility and soil acidification. The direct cause of soil acidification is the import or internal generation of acids, not the presence of "mobile anions." The fact that these anions are often (but not always) present in solutions and leachates of strongly acidified soils is a *symptom* and indicator of the possible input of strong acids into the soil, and should not be seen as a cause of acidification.

The carbonate (CO_3^{2-}), bicarbonate (HCO_3^-), phosphate (PO_4^{3-}), and hydroxyl (OH^-) anions are considered to be immobile anions in soil, and are not viewed as potential cation "carriers." This is because they react with adsorbed H^+, Al^{3+}, and

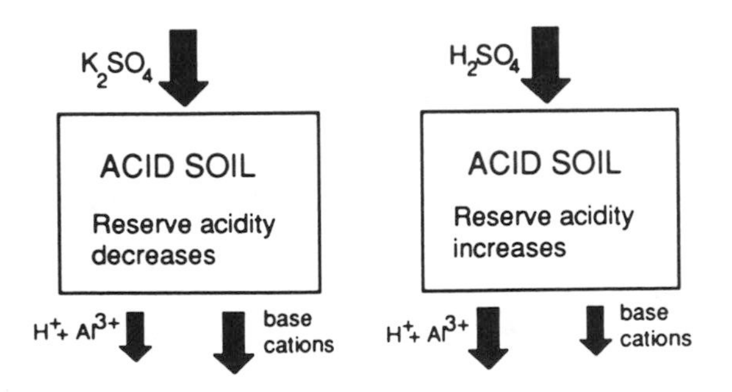

Figure 5.19. Effect of equimolar additions of K_2SO_4 and H_2SO_4 to an acid mineral soil on the soil's reserve acidity and composition of the leachate.

other soil components to change surface charge or form insoluble precipitates. For example, $CaCO_3$ added to acid soils can be consumed by a reaction with acid organic matter:

$$CaCO_3 + 2H^+\text{-organic} = Ca^{2+}\text{-organic} + H_2O + CO_2 \quad (5.71)$$

In the process, Ca^{2+} is immobilized on the newly formed exchange sites. As a result, when liming materials such as $CaCO_3$ and $Ca(OH)_2$ are added to acid soils, Ca mobility is very limited. In less acid soils, Ca added as lime can be mobilized in the form of soluble $Ca(HCO_3)_2$.

In contrast, if gypsum ($CaSO_4$) is added to acid soils possessing few cation exchange sites, Ca^{2+} is mobile (unless sulfate is adsorbed to a large degree). This fact is demonstrated in Figure 5.20, where gypsum applied to the surface of an acid oxisol allowed deeper penetration of Ca^{2+} than $CaCO_3$. The greatest Ca^{2+} mobility was found for $CaCl_2$, as expected for this salt of a "mobile anion." The Cl^- adsorbs to a lesser extent than sulfate, even though these very acid soils do have some surface positive charge before they are limed.

These observations on anion mobility suggest again that "mobile anions" are indicators but not agents of cation mobility. Those anions that are strong bases (Lewis or Brønsted) tend to be immobile because they react with acids (such as Al^{3+}) on surfaces to create sites for cation adsorption. It is useful to classify anions as "mobile" and "immobile" as long as the surface reactions controlling mobility are understood.

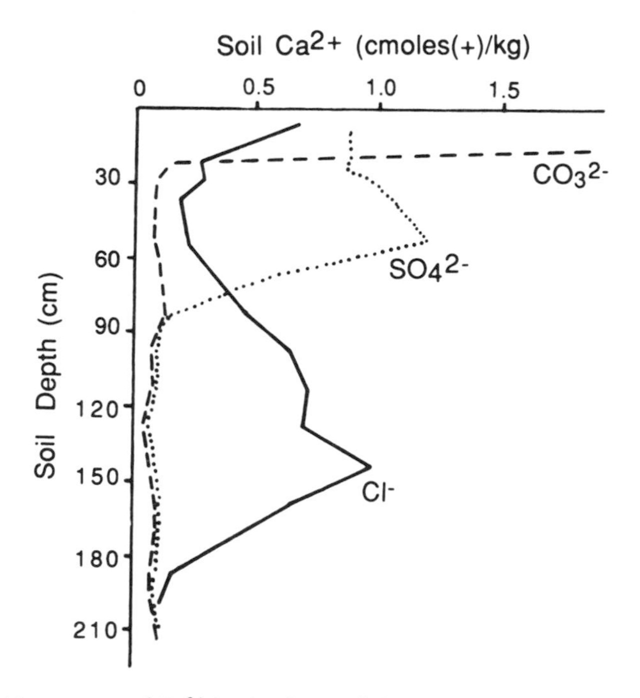

Figure 5.20. Movement of Ca^{2+} in the form of $CaCO_3$, $CaCl_2$, and $CaSO_4$ by water leaching through an oxisol. The Ca was initially added to the soil surface. (Adapted from K. D. Ritchey et al. 1980. Calcium leaching to increase rooting depth in a Brazilian savannah oxisol. *Agron. J.* 72:40–44.)

References

Bloom, P. and M. B. McBride. 1979. Metal ion binding and exchange with hydrogen ions in acid-washed peat. *Soil Sci. Soc. Am. J.* 43:687–692.

Bolt, G. H., M.G.M. Bruggenwert, and A. Kamphorst. 1976. Adsorption of cations by soil. In G. H. Bolt and M.G.M. Bruggenwert (eds.), *Soil Chemistry, A. Basic Elements.* Amsterdam: Elsevier, pp. 54–90.

David, M. B. and C. T. Driscoll. 1984. Aluminum speciation and equilibria in soil solutions of a haplorthod in the Adirondack Mountains (New York, USA). *Geoderma* 33:297–318.

Driscoll, C. T., N. van Breemen, and J. Mulder, 1985. Aluminum chemistry in a forested spodosol. *Soil Sci. Soc. Am. J.* 49:437–444.

Johnson, M. G. and M. B. McBride. 1991. Solubility of aluminum and silicon in acidified spodosols: Evidence for soluble aluminosilicate. In R. J. Wright, V. C. Baligar, and R. P. Murrmann (eds.), *Plant-Soil Interactions at Low pH.* Dordrecht, Netherlands: Kluwer, pp. 15–24.

Mollitor, A. V. and D. J. Raynal. 1982. Acid precipitation and ionic movements in Adirondack forest soils. *Soil Sci. Soc. Am. J.* 46:137–141.

Mulder, J., J.J.M. van Grinsven, and N. van Breemen. 1987. Impacts of acid atmospheric deposition on woodland soils in the Netherlands: III. Aluminum chemistry. *Soil Sci. Soc. Am. J.* 51:1640–1646.

Reuss, J. O. and D. W. Johnson. 1985. Effect of soil processes on the acidification of water by acid deposition. *J. Environ. Qual.* 14:26–31.

Richter, D. D. 1986. Sources of acidity in some forested udults. *Soil Sci. Soc. Am. J.* 50:1584–1589.

Ritchey, K. D., D.M.G. Souza, E. Lobato, and O. Correa. 1980. Calcium leaching to increase rooting depth in a Brazilian savannah oxisol. *Agron. J.* 72:40–44.

Suggested Additional Reading

Adams, F. 1984. *Soil Acidity and Liming.* Agronomy Monograph Series 12, 2nd ed. Madison, Wis.: American Society of Agronomy.

Gillman, G. P. 1991. The chemical properties of acid soils with emphasis on soils of the humid tropics. In Wright et al. (eds.), *Plant-Soil Interactions at Low pH,* pp. 3–14.

Sposito, G. 1989. *The Chemistry of Soils.* New York: Oxford University Press, chap. 11.

Questions

1. (a) Calculate the pH of a 10^{-3} *M* solution of $AlCl_3$, assuming that the reaction forming $AlOH^{2+}$ is the only important hydrolysis reaction in this system.
   (b) Determine the fraction of total soluble Al that is hydrolyzed in (a).
   (c) Is the solution in (a) undersaturated, saturated, or supersaturated with respect to gibbsite?
   (d) Suppose the solution in (a) is diluted by a factor of 10 with pure water. Again determine the pH, fraction of Al hydrolyzed, and state of saturation relative to gibbsite. What do these results suggest about the tendency of Al^{3+} to precipitate on dilution?

2. A mineral soil with a pH of 5.2 retains 70 mmoles(+)/kg of exchangeable bases and 10 mmoles(+)/kg of exchangeable Al^{3+}. At pH 7, the CEC is 100 mmoles(+)/kg.
   (a) What is the percent base saturation at pH 5.2?
   (b) What fraction of the total acidity at pH 5.2 is nonexchangeable?
   (c) What is the approximate lime requirement (metric tons/hectare) of the soil?
   (*Note:* assume the lime reacts with soil to a 15 cm.-depth, and that the bulk density of the soil is 1.2 g/cm^3)

3. A 100-g sample of acid soil (pH = 4.5) is mixed with 0.5 g of pure $CaCO_3$ and allowed to react for several months in a moist condition. The final pH is found to be 5.8.
   (a) Calculate the quantity of acidity (mmoles/kg) neutralized by the $CaCO_3$.
   (b) Calculate the buffer capacity of the soil.
   (c) Estimate the quantity of $CaCO_3$ that would be needed to raise the pH to 7.0. What assumption must you make to do this estimate?

4. Calculate the relative acidifying tendencies of 100 kg/ha of nitrogen applied to soil in the form of $(NH_4)_2SO_4$, NH_4NO_3, and NH_3 fertilizers. Express your answer in units of kilomoles of H^+ per hectare. How do these values compare in magnitude with natural acidification by biological activity in the soil?

5. Calculate the pH of acid rainwater that contains 90 μmolar HNO_3 and is in equilibrium with atmospheric CO_2.

6. (a)Show how equation 5.67 was derived. What assumptions were made to get this relationship?
   (b) Adjust equation 5.67 for the case when 10^{-2} M $Ca(NO_3)_2$ is present in soil solution. (The sorption coefficient, log K_{Al}, shifts to -4.1 at this Ca^{2+} concentration.)
   (c) Plot $[Al^{3+}]$ and N_{Al} as a function of log P_{CO_2} for this higher Ca^{2+} level.
   (d) How does the plot in (c) compare with that of Figure 5.17, and what does this suggest about the effect of Ca^{2+} on the soil acidification process?

7. (a) Estimate the maximum quantity (kg/ha) of Ca^{2+} that could be exchanged and leached out of an acid soil in one year by the natural H_2CO_3 acidity in unpolluted rainfall, assuming 90 cm of annual rainfall. ($[H_2CO_3] = K_H P_{CO_2}$, where log $K_H = -1.41$ and P_{CO_2} is in units of atmospheres.)
   (b) Repeat this calculation of base cation leaching using the assumption that actual levels of CO_2 pressures in the soil can be as high as 0.03 atmospheres or more because of biological activity.
   (c) Estimate base cation leaching by rainwater for the case of acid rainfall (assume rainwater with pH = 4).
   (d) What do these answers suggest about the relative soil acidifying potentials of natural acidity in rain and biological activity in soil?

8. (a) Derive the mathematical relationship between the activity of $Ca(OH)_2$ in solution and the lime potential, pH − ½pCa.
   (b) If the solubility product, K_{SO}, of $Ca(OH)_2$ is $10^{-5.19}$, what is the lime potential of a soil saturated with $Ca(OH)_2$?
   (c) Can the lime potential in (b) be maintained in natural soils? Explain.
   (d) Show that the lime potential of a calcareous soil has a fixed value as long as the CO_2 gas pressure in the soil remains constant.

9. From equation 5.21 for $Al^{3+}-Ca^{2+}$ exchange and equation 5.19 for gibbsite solubility, derive an equation that relates percent Ca^{2+} saturation of exchange sites to lime potential. (Assume an ionic medium of 0.01 M $CaCl_2$ and $K_s = 30$ for the exchange reaction in this medium.)

10. Two grams of an acid mineral soil are mixed into 50 ml of 1.0 M KCl solution. The resulting suspension has a pH of 4.0 and a dissolved Al concentration of 3.0 μg/ml. It takes 6.5 ml of 0.01 M $Ca(OH)_2$ to adjust the suspension pH to 7.
   (a) Calculate the exchangeable acidity in units of millimoles per kilogram.
   (b) Calculate the "total" acidity (based on titration to pH 7) in units of millimoles per kilogram.
   (c) Compare the magnitudes of total and exchangeable acidity. Why are they different?

6

Mineral Weathering and Formation

The structures of the important primary and secondary minerals are described in Chapter 2. In the present chapter, the dynamic relationship between these two groups of minerals will be discussed. In particular, the physical-chemical factors that determine:

1. which primary minerals weather most readily in soils
2. which secondary minerals tend to form, and
3. the rates of these processes

are all of interest in soil chemistry, since they ultimately control the elemental composition, mineralogy, chemical characteristics, and morphology of soils.

Mineral weathering generally occurs whenever water comes into contact with primary mineral particles. The overall processes of weathering are not reversible even though individual reaction steps are often treated as reversible equilibria. The main chemical mechanisms of weathering are:

1. exchange—displacement of base cations in structures by solution cations, most notably H^+ generated from water or carbonic acid dissociation
2. hydration—addition of water to the mineral structure by processes that hydrolyze metal-oxygen bonds
3. oxidation—electron removal from the mineral, usually by molecular oxygen (see Chapter 7)

The actual secondary mineral products formed are determined by the relative rates of these weathering reactions at the surface of the particular primary mineral, by the relative rates of crystallization of possible metastable and stable secondary minerals, and by the extent to which the soil environment allows mobile chemical components to be lost by leaching.[1]

1. That is, the degree to which the soil can be treated as an open thermodynamic system.

6.1. PRIMARY MINERAL STABILITY

The primary minerals crystallize from high-temperature melts (magma), often under conditions of high pressure beneath the earth's crust. The sequence of crystallization of the common minerals in forming igneous rocks is called the Bowen reaction series. As diagrammed in Figure 6.1, it is seen that there are actually two reaction series:

1. A discontinuous series of ferromagnesian minerals. The first-formed phases (e.g., olivine) react with the melt to form new phases of different structure and composition (e.g., pyroxenes, amphiboles).
2. A continuous series of feldspars. The feldspar crystals change continuously in composition by reaction with the remaining melt.

As the vertical axis of Figure 6.1 indicates, those minerals that crystallize at lower temperature tend to be more stable and resistant to weathering at earth surface (low temperature, 1 atmosphere pressure) conditions. The geochemical properties associated with low-temperature stability are:

1. Increasing silicate (SiO_4) linkage, indicative of greater Si—O—Si bonding and more open structures

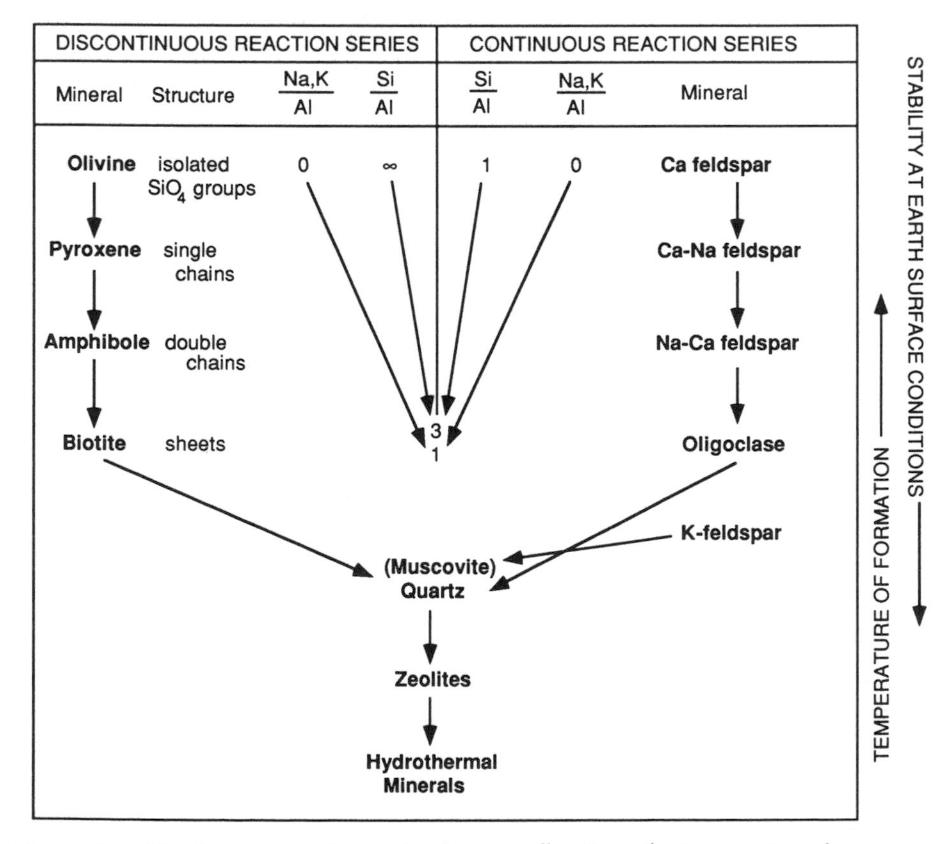

Figure 6.1. The Bowen reaction series for crystallization of primary minerals. (Adapted from B. Mason. 1966. *Principles of Geochemistry*. 3rd ed. New York: Wiley.)

2. Increasing replacement of Si by Al in the discontinuous series; decreasing replacement of Si by Al in the feldspars (continuous series)
3. Increasing Na and K content relative to Al and relative to Ca and Mg
4. Presence of OH^- and F^- anions in the structure (e.g., amphiboles, biotite)

Substitution of Al for Si *within* any given structure of either the continuous or discontinuous series raises the mineral to a higher position in the series. Because Al for Si substitution leads to greater negative charge in the silicate lattice, and that charge must be balanced by alkali or alkaline earth metal cations, it is found that high Ca and Mg contents in particular are associated with minerals of low weathering stability.

From the Bowen reaction series, it is seen that the more silica-rich minerals crystallize at the lowest temperatures; quartz (SiO_2) is consequently one of the most stable and weathering-resistant minerals known.

6.2. MECHANISMS OF PRIMARY MINERAL DISSOLUTION

6.2a. Feldspar Weathering

Feldspars are the most abundant rock-forming minerals in the earth's crust, warranting a great deal of study in order to understand the process by which they weather. It is believed that feldspar weathering proceeds by the following steps:

1. H^+ ions cause the hydrolysis of surface Si—O—Al or Si—O—Si bonds, breaking open the "silicate cage" and increasing the accessibility of the alkali metal ion:

$$\begin{array}{c}\text{Si}\\ \backslash\\ K^+\text{--}O\\ /\\ \text{Al}^-\end{array} + H^+ \rightarrow \begin{array}{c}\text{Si}\\ \backslash\\ K^+\text{--}OH\\ /\\ \text{Al}^-\end{array} + \rightarrow \begin{array}{c}\text{Si}\\ \backslash\\ K^+\text{--}OH\\ \\ \text{Al}-OH_2\end{array}$$

2. The metal ion is released to solution once the structural charge is balanced by the proton.
3. A "metal-leached" residual layer of no more than 15 to 25 Å thick is created at the feldspar surface.
4. Weathering reaches a steady state in which the leached layer dissolves at a rate equal to the rate of replacement of alkali metals by H^+ at the interface between the leached layer and the intact structure.

Steps 1, 2, and 3 account for the initially high rate of dissolution of alkali and alkaline earth metals (Na^+, K^+, Ca^{2+}) from feldspars, and the relative lack of silica and aluminum dissolution. This is termed *incongruent (nonstoichiometric) dissolution,* meaning that a portion of the mineral structure is dissolving selectively, leaving a residue enriched in silica and alumina. Step 4 accounts for the later stage of *congruent (stoichiometric) dissolution,* in which the elements are released into solution in proportion to their mole fractions in the structure.

The rate of release of the chemical constituents of feldspar into solution is essentially constant, so that concentrations of the dissolved species increase linearly with time. This means that the process follows a zero-order rate where the dissolution constant is given by $dm/dt = K_{diss}$. Here, dm/dt has units of moles dissolved per square meter of mineral surface area per second. Observed laboratory rates of feldspar (and other silicate mineral) weathering fall in the range of 10^{-11} to 10^{-12} moles of silicate oxygen/m^2/sec at 25°C. Rates of weathering in natural soil systems appear to be much slower than this, perhaps because laboratory weathering experiments are of such short duration that the mineral dissolution is far from equilibrium. As dissolution products such as K^+ build up in solution, back reactions begin to lower the overall rate and a zero-order rate law is no longer obeyed.

Dissolution reactions of silicate minerals are more rapid at higher temperatures. The temperature dependence could indicate that diffusion processes limit the weathering rate, or that an activation energy must be surmounted for dissolution to occur. The latter explanation is most likely for mica and feldspar weathering because these reactions show a high sensitivity to temperature. Feldspar dissolution is more rapid in very acid solutions, is more or less constant in the pH 5 to 8 range, and increases again above pH 8. The nearly constant dissolution rate in the mid-pH range may mean that H_2O or H_2CO_3 rather than H^+ itself is the agent of hydrolysis at the surface under less acid conditions.

Feldspars weather directly to ionic components by hydrolysis and H^+ attack; however, secondary minerals may form as products of the weathering reaction, depending on the chemical environment in the soil. In a closed system where feldspar is not exposed to acidity generated within or outside the soil, hydrolysis is the dominant reaction, producing an alkaline solution:

$$KAlSi_3O_8 + 8H_2O \rightarrow Al(OH)_4^- + 3Si(OH)_4^0 + K^+ \qquad (6.1)$$

The pH from this reaction can exceed 9 for K feldspar and approaches 10 for Na feldspar.

In acidic soil environments that are open to CO_2 and other acidifying agents of weathering, K feldspar could, in principle, form only soluble species, as indicated by the reaction

$$KAlSi_3O_8 + 4H^+ + 4H_2O = Al^{3+} + 3Si(OH)_4^0 + K^+ \qquad (6.2)$$

This dissolution reaction converts strong acids (H^+) into less strongly acidic forms (Al^{3+} and monosilicic acid). Because monosilicic acid is a very weak acid, it can leach through the soil in the undissociated ($Si(OH)_4$) form. In reality, however, the Al component is usually conserved as a solid phase in the soil; that is, soluble Al is typically so low that equation 6.2 is replaced by a more realistic one in which Al^{3+} precipitates as gibbsite, kaolinite, or allophane, for example,

$$KAlSi_3O_8 + H^+ + 7H_2O \rightarrow \underset{\text{gibbsite}}{Al(OH)_3\ (s)} + 3Si(OH)_4^0 + K^+$$

$$KAlSi_3O_8 + H^+ + 4.5H_2O \rightarrow \underset{\text{kaolinite}}{\tfrac{1}{2}Al_2Si_2O_5(OH)_4\ (s)} + 2Si(OH)_4^0 + K^+ \qquad (6.3)$$

It is not unusual to find these secondary minerals in close association with the surfaces of weathered feldspar particles. Only in cases of extreme soil acidification can significant loss of Al from the soil to drainage and groundwater be expected.

6.2b. Mica Weathering

Micas lose their structural K^+ relatively rapidly by a process of cation exchange with H^+ or metal ions in the weathering solution. This initial reaction is an alteration that is not necessarily very sensitive to pH. The pH, however, tends to rise during alteration as H^+ is adsorbed. The secondary mineral formed directly is a vermiculite, which inherits more or less intact the 2:1 layer structure of the mica parent. An ideal biotite mica might have the formula:

$$K_2(Fe^{2+}, Mg)_6(Si_6Al_2)O_{20}(OH)_4$$

possessing the ideal mica layer charge of 2.0 per unit cell (2 moles of negative charge per formula weight). On the other hand, a vermiculite formed by weathering of this biotite might have the formula:

$$Mg_{0.84}(Mg_{5.05}, Fe^{3+}_{0.9})(Si_{5.48}Al_{2.52})O_{20}(OH)_4$$

in which the layer charge is only 1.7 per unit cell. Evidently, weathering has effected structural changes in addition to the replacement of K^+ by Mg^{2+} in the interlayer region of the structure. Specifically, structural Fe^{2+} has undergone oxidation to Fe^{3+}, a process that reduces the overall negative charge of the layers. Furthermore, Al^{3+} has apparently entered some of the tetrahedral sites, increasing the negative charge in this part of the structure. The net effect of weathering, however, is layer charge *reduction,* and the degree of this reduction determines the swelling properties of the 2:1 clay mineral formed.

In some weathering environments, Fe and Mg are ejected from the mica structure to counter the excess positive charge built up in the trioctahedral sheet as Fe^{2+} is oxidized. In the process, Mg may form a $Mg(OH)_2$ sheet between the 2:1 layers, forming *chlorite* as an intermediate weathering product.

In general, weathering conditions tend to favor the gradual conversion of trioctahedral layer silicate clays (with dominantly Mg^{2+} and Fe^{2+} in octahedral sites) to dioctahedral layer silicates (with dominantly Al^{3+} and Fe^{3+} in octahedral sites).[1] Layer charge reduction accompanies this process. Expected weathering sequences of the two common micas, biotite and muscovite, in leaching environments are:

$$\text{biotite} \underset{-K}{\rightarrow} \text{chlorite} \underset{-Mg}{\rightarrow} \text{vermiculite (trioctahedral)} \rightarrow \text{smectite}$$

$$\text{muscovite} \underset{-K}{\rightarrow} \text{illite} \underset{-K}{\rightarrow} \text{vermiculite (dioctahedral)} \rightarrow \text{smectite}$$

The tendency, then, is toward decreasing particle size, decreasing layer charge, increasing expandability in water, reduced structural Fe^{2+} and Mg, and displacement of tetrahedral Al by Si. The formation of smectites from vermiculites by this process of structural *alteration* is very slow, requiring perhaps centuries or millenia. This alteration process is inferred from long-term observations of soil weathering, or more commonly from patterns of mineral abundance from the soil surface to the subsoil, which often represents a gradient of weathering intensity.

1. It is a general rule that octahedrally coordinated Al is favored in low-temperature, low-pressure (secondary) minerals, whereas tetrahedrally coordinated Al tends to be found in high-temperature, high-pressure minerals.

Smectites also form by precipitation of dissolved ions from alkaline solution, a process referred to as *neoformation* because none of the structure is inherited from weathering parent minerals. Neoformation of minerals in soils will be discussed in more detail later in this chapter.

Because micas are an important source of K^+ for plant growth in soils, much effort has gone into the study of K^+ release into solution from micas of different composition. Certain rules govern the ease of K^+ release from micas by reaction with acidic or saline solutions, determining which micas are likely to be the most persistent in soils. The rules are listed and explained below.

1. **Trioctahedral micas release K^+ more rapidly than dioctahedral micas.** As a result, most biotites are much easier to deplete of interlayer K^+ than muscovite. The explanation is based on structure; the OH^- groups that lie directly beneath the K^+ ions embedded in the "hexagonal hole" of the silicate layers are oriented differently

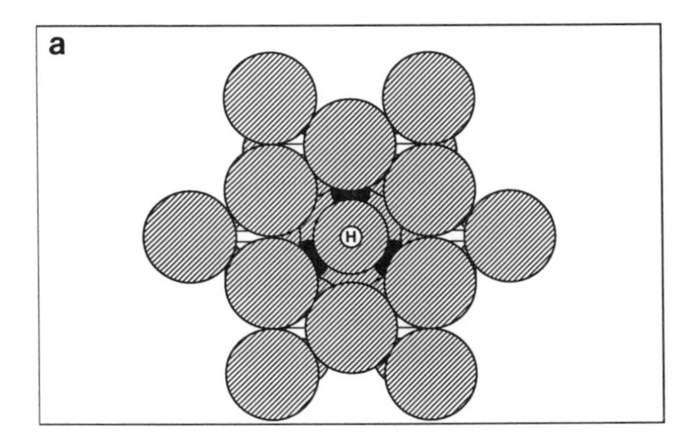

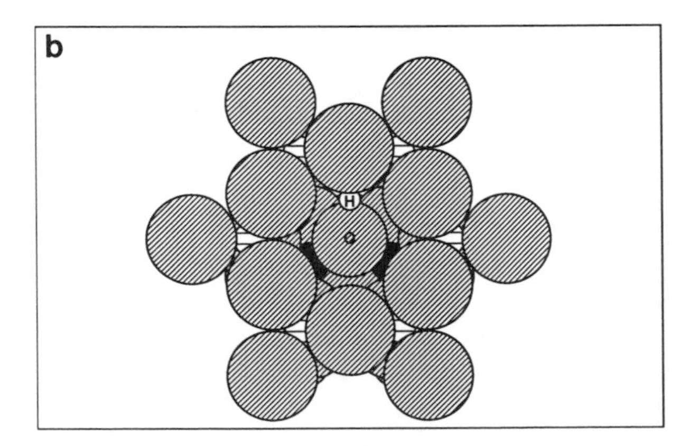

Figure 6.2. View from directly above the hexagonal hole of (a) trioctahedral mica, and (b) dioctahedral mica. The proton of the structural OH is labeled with an "H", while the O and octahedral metal ions are symbolized by large shaded and smaller black circles, respectively.

in these two forms of mica. In trioctahedral micas, the O—H bond orients perpendicular to the plane of the silicate layer, as shown in Figure 6.2a, minimizing the repulsion between the proton of the OH^- group and the three octahedrally coordinated M^{2+} ions ($M^{2+} = Fe^{2+}$ or Mg^{2+}) that are symmetrically distributed around the hydroxyl. However, this orientation maximizes the $K^+ \cdots H{-}O$ electrostatic repulsion and destabilizes the mineral somewhat, lowering the energy barrier to layer separation and the entry of other cations into the interlayer.

In dioctahedral micas, one of three octahedral sites is vacant, so that the O—H bond tilts toward this vacant site, as pictured in Figure 6.2b, to reduce repulsion between the proton and the two octahedrally coordinated M^{3+} ions ($M^{3+} = Fe^{3+}$ or Al^{3+}). As a result, the K^+ ion is further from the structural proton and is stabilized in the hexagonal hole. The energy barrier to K^+ removal is consequently high in muscovite.

Studies of potassium release from micas in solutions of 1 *M* NaCl have revealed that the concentration ratio of K^+ to Na^+ in solution, $[K^+]/[Na^+]$, ranges from 1×10^{-5} for resistant dioctahedral micas to 1.2×10^{-3} for easily weathered biotites (Newman, 1969).

2. **High fluoride content in trioctahedral micas impedes K^+ release.** This is actually a special case and an exception to the first rule. Since F^-, which can proxy for OH^- in mica structures, is not a dipole, it attracts K^+ electrostatically regardless of whether the mica is trioctahedral or dioctahedral. As a result, trioctahedral micas in which F^- isomorphously substitutes for much of the structural OH^- release K^+ with difficulty. That is, this particular type of trioctahedral mica behaves much like a dioctahedral mica with respect to K^+ removal.

3. **Large mica particles release K^+ very slowly by a particle diffusion-limited process. Small particles release K^+ much more quickly, but less completely, by a layer diffusion process.** The effect of particle size on the rate of K^+ replacement from muscovite by Na^+ is shown in Figure 6.3. Large particles release K^+ along a weathering front around the edge of the particle. This front migrates inward toward the center as diagrammed in Figure 6.4a. If the concentration of K^+ in solution is controlled at a very low and essentially constant value, either by continuous leaching with fresh salt solution or by precipitation of K^+,[1] a steady-state weathering process may be attained. The release of K^+ from large mica particles is described under these conditions by the equation

$$\frac{Q}{Q_0}\left(1 - \ln\left[\frac{Q}{Q_0}\right]\right) = 1 - \frac{kt}{r} \tag{6.4}$$

where Q is the quantity of K^+ remaining in the mica after time, t, of weathering; Q_0 is the quantity of K^+ in the unweathered mica; k is a constant; and r is the particle radius. A plot of $Q/Q_0\,(1 - \ln[Q/Q_0])$ against t should produce a straight line, which Figure 6.5 shows to be the case, at least for the larger mica particles.

Clay-size mica particles, as Figure 6.3 illustrates, release as much as half their K^+ almost instantaneously, but resist the subsequent release of the remaining K^+. It

1. In the accelerated laboratory weathering of mica, Na tetraphenylboron is an extremely effective reagent because it reacts with released K^+ to form the insoluble K tetraphenylboron precipitate, thereby maintaining a very low concentration of K^+ ions in solution.

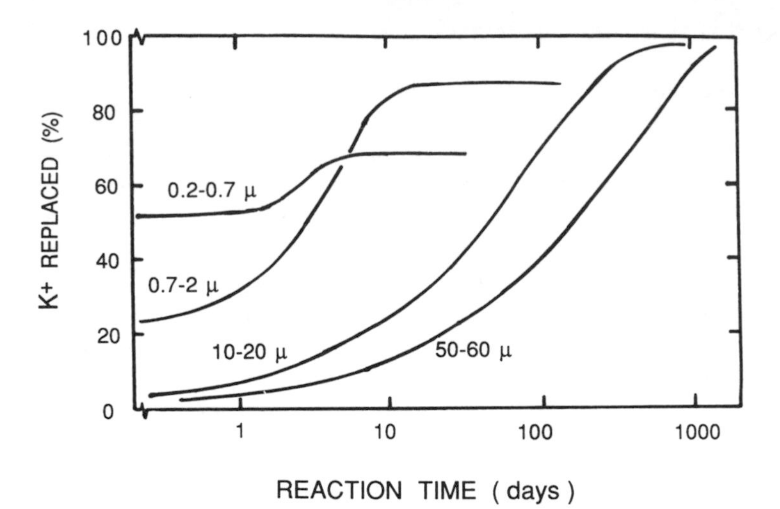

Figure 6.3. Potassium ion extraction from muscovite samples of different particle size by Na tetraphenylboron solutions. (Data from A. D. Scott. 1968. Effect of particle size on interlayer potassium exchange in micas. *Transactions of Ninth International Congress of Soil Science*. Vol. 2, pp. 649–660.)

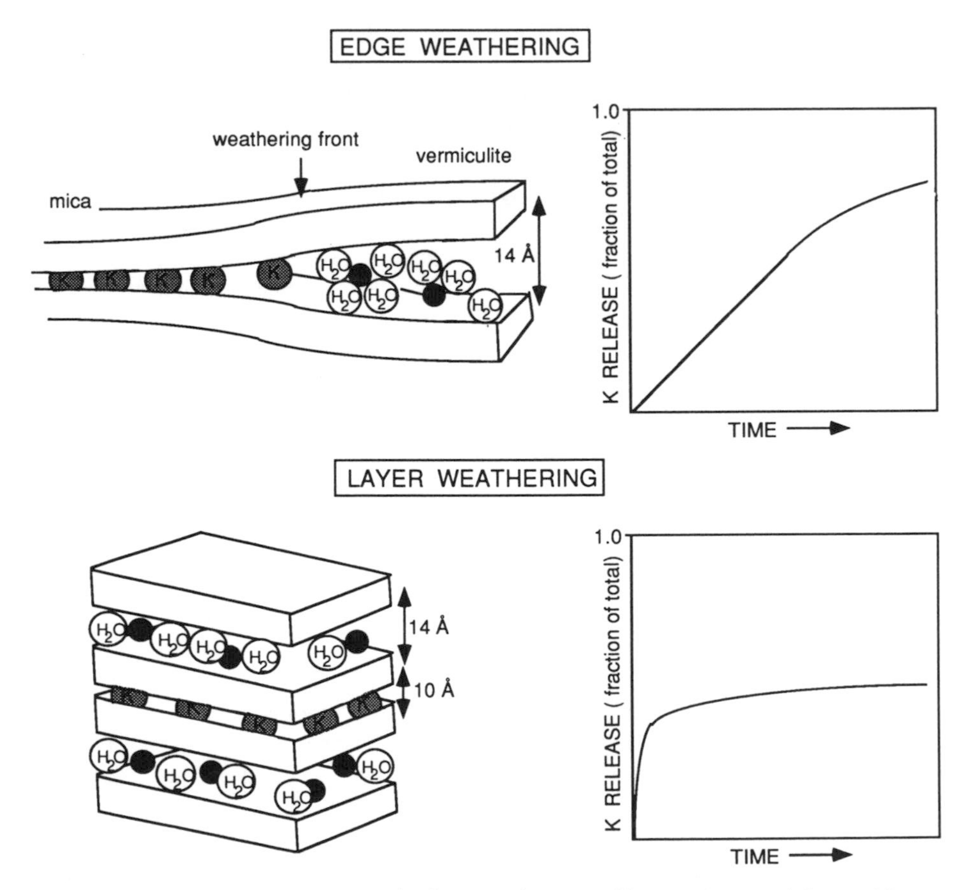

Figure 6.4. Schematic picture of edge weathering of large mica particles and layer weathering of small mica particles.

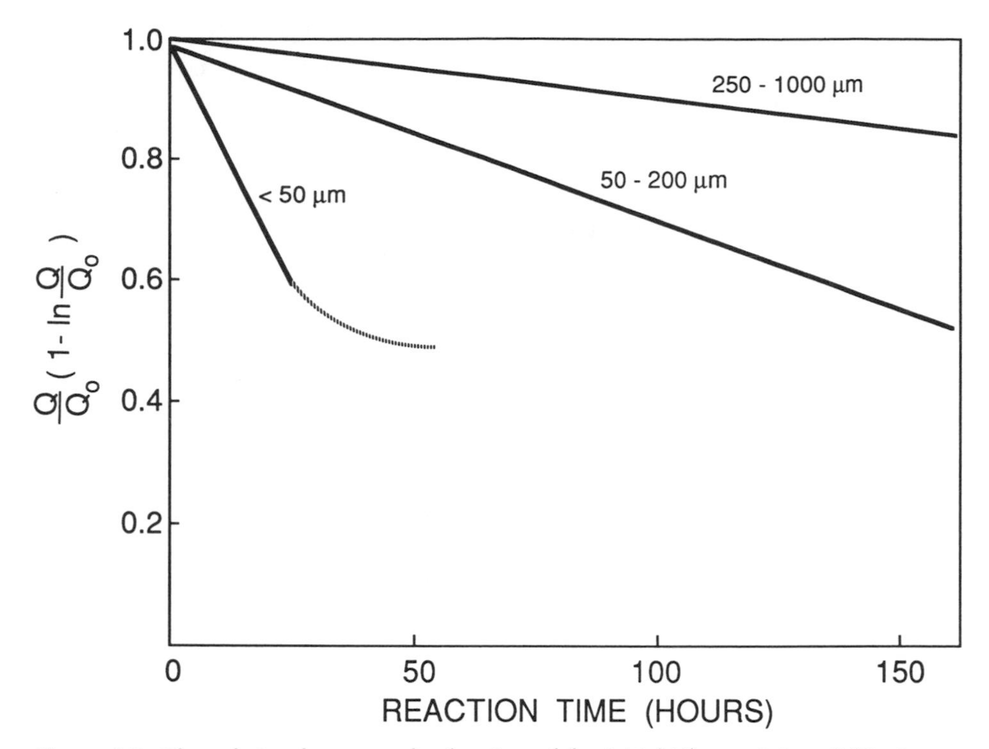

Figure 6.5. The relation between the fraction of the initial K^+ remaining, Q/Q_o, in biotite mica samples of different particle size and the time of reaction with a NaCl-Na tetraphenylboron solution. (Data from M. G. Reed and A. D. Scott. 1962. Kinetics of potassium release from biotite and muscovite in sodium tetraphenylboron solutions. *Soil Sci. Soc. Am. Proc.* 26:437–440.)

seems that K^+ release from clay-size particles does not occur on a front but by complete layers, as illustrated in Figure 6.4b.[1] The tendency for incomplete release of K^+ from clay-sized micas might explain the longevity of illites, with their strongly held interlayer K^+, in soils.

4. **Iron oxidation in biotites impedes K^+ release.** Since Fe^{2+} oxidation decreases overall layer charge, one might think that this would facilitate layer expansion and continued mica weathering. In fact, the $Fe^{2+} \rightarrow Fe^{3+}$ transformation in the octahedral sheet may ultimately confer dioctahedral character to the mica (see rule 1) and render K^+ release more difficult.

5. **K^+ release continues only if there is a mechanism to remove the K^+ from solution.** Even low concentrations of K^+ in solution allow the back reaction, the read-

1. This "all or nothing" process causes some interlayers to completely lose their K^+ ions while others (often adjacent) do not. The reason for this behavior is not known for certain, but structural strain created by the interlayer expansion necessary for K^+ exchange by the replacing cation may promote continued rapid K^+ removal from that same interlayer. Meanwhile, the two adjacent unexpanded interlayers on either side may be stabilized by modifications induced in the pair of facing silicate layers exposed to chemical weathering. In biotite, Fe^{2+} oxidation in these two layers causes reorientation of the structural OH groups, which in turn could cause the K^+ in the two adjacent interlayers to become more difficult to remove. The common observation of regularly interstratified weathered biotites, with alternating 10-Å and 14-Å layer thicknesses and a 24-Å c-axis repeat distance, could thereby be explained.

sorption of K^+, to proceed at a rate comparable to that of K^+ release, blocking any further weathering. The K^+ concentration at which this occurs depends on the particular type of mica or layer silicate, as shown in Figure 6.6. The solution concentrations of K^+ that block K^+ replacement by Na^+ (1 *M* NaCl) beyond the 5 percent level are:

Phlogopite	30 μg K^+/ml
Biotite	15 μg K^+/ml
Illite	7 μg K^+/ml
Muscovite	0.3 μg K^+/ml

These "threshold concentrations" are indicators of the relative tenacity with which each of these minerals retains K^+.

Plant uptake, leaching, and exchange or fixation processes tend to maintain a very low K^+ concentration in soil solutions. Even so, the K^+ release curves of Figure 6.6 reveal that trioctahedral micas are much better K^+ suppliers than dioctahedral micas. Vermiculite, which is formed by removal of K^+ from mica, does not retain subsequently added K^+ nearly as tenaceously as its mica "parent." In other words,

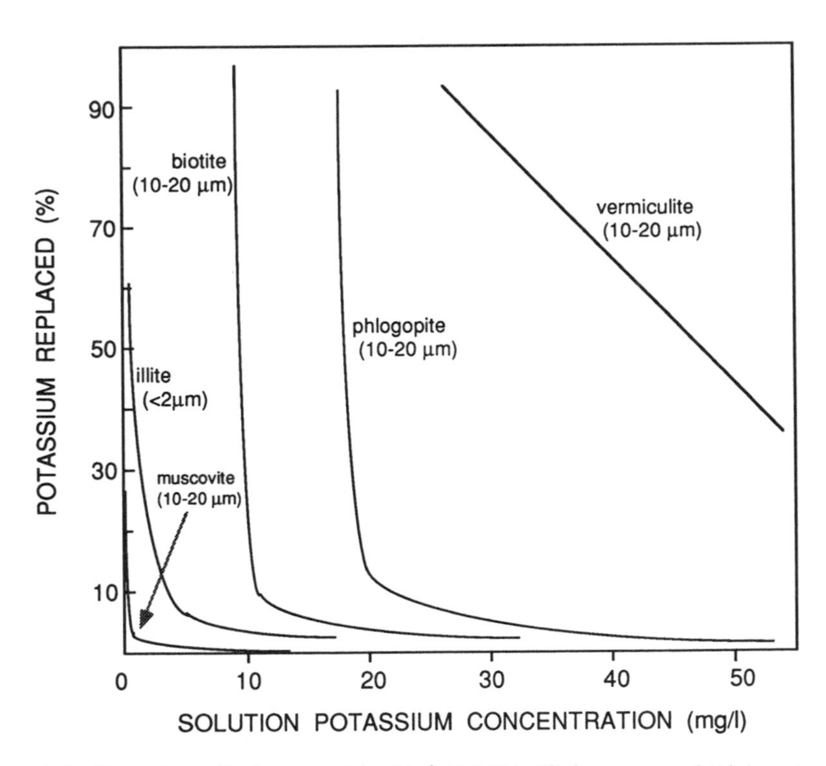

Figure 6.6. Potassium displacement by Na^+ (1 *M* NaCl) from several K^+-bearing silicates as a function of the solution concentration of K^+. (Adapted from A. D. Scott and S. J. Smith. 1966. Susceptibility of interlayer potassium in micas to exchange with sodium. In *Fourteenth National Conference on Clays and Clay Minerals.* New York: Pergamon Press.)

the mica-vermiculite weathering transition is not reversible in soils, but rather should be written:

$$\text{mica} \underset{+\,Mg^{2+}}{\overset{-K^{+}}{\rightarrow}} \text{vermiculite} \underset{-K^{+},\ +Mg^{2+}}{\overset{+\,K^{+},\ -Mg^{2+}}{\rightleftarrows}} K^{+}-\text{vermiculite}$$

even though mica and $K^{+}-$vermiculite have similar c-axis spacings (~ 10 Å).

Vermiculites in soil are said to "fix" K^{+}, trapping it between the collapsed 2:1 layers in nonexchangeable form. This K^{+} is, however, available for plant uptake over the long term.

6.2c. Ferromagnesian Mineral Weathering

Olivines, pyroxenes, and amphiboles are typically rich in reduced iron (Fe^{2+}), weathering rapidly in aerated soil environments as the Fe^{2+} is oxidized to Fe^{3+}:

$$\begin{matrix}\text{Olivines}\\ \text{Pyroxenes}\\ \text{Amphiboles}\end{matrix} \xrightarrow[+H^{+}]{+O_2} \begin{matrix}Fe(OH)_3 \text{ (precipitated)}\\ +\ Mg^{2+},\ Ca^{2+} \text{ (leached)}\\ +\ \text{silica residue}\end{matrix}$$

There is not much resistance to weathering in these minerals because of the relative lack of Si—O—Si bonding, especially in island silicates such as olivine. Layer silicate minerals rich in Mg^{2+} (e.g., trioctahedral smectites, chlorite, serpentine) may form from the siliceous residue if leaching does not deplete Mg^{2+} in the weathering zone.

Island and chain silicate mineral dissolution rates are limited by the inherent ligand-exchange rates of the octahedrally coordinated metals (see Chapter 4 for a discussion of water ligand exchange on metal cations). The reason for this is that the H^{+}-promoted dissolution of a structural metal ion from a mineral surface requires the breaking and reformation of a metal-oxygen bond. Because large metal cations generally have faster ligand exchange rates than small cations of the same charge, Ca^{2+} silicates dissolve faster than Mg^{2+} silicates. The island silicate Ca_2SiO_4 has a dissolution rate of about 10^{-5} moles/m^2/sec (Casey and Westrich, 1992), at least six orders of magnitude faster than the rate of dissolution of more resistant silicates such as feldspar.

For silicates composed of transition metal cations, crystal field stabilization energy (CFSE) affects dissolution rates. Thus, the rate of dissolution in acidic water of structurally similar orthosilicates follows the order:

$$Ca_2SiO_4 > Zn_2SiO_4 \approx Mn_2SiO_4 > Fe_2SiO_4 > Co_2SiO_4 > Ni_2SiO_4$$

and spans about five orders of magnitude (Casey and Westrich, 1992). Because Zn^{2+} and Mn^{2+} have d^{10} and d^{5} electronic states, respectively, they possess no CFSE. In contrast, Ni^{2+} has the greatest CFSE of the metals listed.[1]

The overall oxidation reaction of ferromagnesian minerals can be expressed in general terms as

$$\{Fe^{2+}, \tfrac{1}{2}Mg^{2+}\}_{\text{silicate}} + H^{+} + \tfrac{1}{4}O_2 \rightarrow \{Fe^{3+}\}_{\text{silicate}} + \tfrac{1}{2}Mg^{2+} + \tfrac{1}{2}H_2O \quad (6.5)$$

1. This principle relating mineral dissolution rates also extends to secondary minerals such as oxides. The Cr^{3+} ion, if substituted for Fe^{3+} in FeOOH, inhibits dissolution of the oxide, presumably because Cr^{3+} is a d^{3} ion with high CFSE, while Fe^{3+} is a d^{5} ion with zero CFSE.

The weathering process, as suggested by the reaction written, consumes dissolved oxygen most readily at low pH. However, the mechanism of oxidation at the silicate surface may involve the initial transfer of an electron to aqueous (aq) Fe^{3+}:

$$Fe^{3+}(aq) + \{Fe^{2+}\}_{silicate} = Fe^{2+}(aq) + \{Fe^{3+}\}_{silicate} \quad (6.6)$$

The reduced iron in solution then reacts with O_2[1]:

$$Fe^{2+}(aq) + \tfrac{1}{4}O_2 + \tfrac{5}{2}H_2O \rightarrow Fe(OH)_3 + 2H^+ \quad (6.7)$$

The weathering of other Fe^{2+}-bearing primary minerals such as pyrite (iron sulfide) also requires the presence of O_2:

$$FeS_2 + \tfrac{7}{2}O_2 + H_2O \rightarrow Fe^{2+} + 2SO_4^{2-} + 2H^+ \quad (6.8)$$

but again it seems that the direct oxidant is not O_2 but Fe^{3+} formed from the oxidation of Fe^{2+}. In fact, once the pH becomes very low from reaction 6.8, Fe^{3+} is sufficiently soluble that an Fe^{3+}-mediated surface oxidation begins:

$$FeS_2 + Fe^{3+}(aq) \rightarrow \text{oxidized pyrite} + Fe^{2+}(aq) \quad (6.9)$$

This reaction is faster than oxidation by O_2, and is *autocatalytic;* that is, protons generated by reaction 6.9 dissolve more Fe^{3+} from the oxidized surface and accelerate the reaction.

6.3. SECONDARY MINERAL FORMATION

Silica, alumina, iron, and the various base cations that are dissolved by primary mineral weathering can precipitate as new low-temperature minerals—a process known as *neoformation.* This should be distinguished from weathering processes, termed *alteration,* in which part of the parent mineral structure is inherited by the weathering product. Examples of alteration are given in the previous section.

The particular secondary minerals formed by neoformation are controlled by the leaching intensity of the local soil environment, and to what extent the environment is confining, allowing the more soluble weathering products (silica, base cations) to accumulate. The temperature and moisture conditions also have an effect on the specific secondary minerals that form. To understand in general terms the complex balance of factors that dictates which secondary minerals tend to form, it is first necessary to appreciate the chemical factors that limit solubility of two weathering products—alumina and silica—the major constituents of clay minerals.

6.3a. Solubility of Silica and Alumina

As long as the total concentration of silica in water is less than 2×10^{-3} *M* at 25°C (and this is the case in most soils), at equilibrium, silica is in the form of monomers

1. This reaction pathway may be favored because O_2, although a potent oxidant thermodynamically, is kinetically rather inert because of its electronic state.

of $Si(OH)_4$, *monosilicic acid.* This is because amorphous silica has a fairly high solubility, expressed by the reaction

$$SiO_2\ (\text{amorphous}) + 2H_2O = Si(OH)_4^0 \qquad \log K = -2.7 \tag{6.10}$$

The solubility line for amorphous silica is then

$$K = 10^{-2.7} = [Si(OH)_4^0] \tag{6.11}$$

which on a plot of $Si(OH)_4^0$ concentration against pH is a horizontal line as shown in Figure 6.7. However, because $Si(OH)_4^0$ is a weak acid, it begins to dissociate to silicate anions at high pH:

$$Si(OH)_4^0 = SiO(OH)_3^- + H^+ \qquad \log K = -9.46 \tag{6.12}$$

The silicate ions are much more soluble in water than the uncharged $Si(OH)_4^0$ molecule, explaining the higher solubility of silica in water above pH 9.

Quartz, a crystalline form of silica, is much less soluble than amorphous silica, as shown by the K value of its dissolution reaction:

$$SiO_2\ (\text{crystalline}) + 2H_2O = Si(OH)_4^0 \qquad \log K = -3.7 \tag{6.13}$$

This means that soil solutions in equilibrium with amorphous silica are oversaturated with respect to quartz. Nevertheless, because quartz crystallizes extremely slowly at ambient temperatures, the solubility of amorphous silica effectively sets the upper limit of dissolved silica in natural waters.

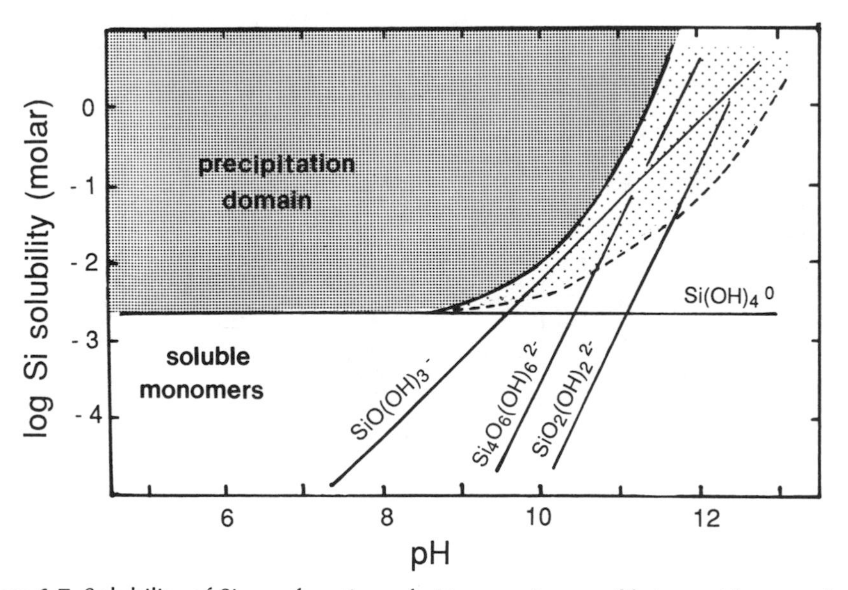

Figure 6.7 Solubility of Si as a function of pH, assuming equilibrium with amorphous silica. The dotted region indicates conditions promoting soluble silica polymer formation, while the shaded region denotes supersaturated conditions favorable to polymerization and precipitation. (Adapted from W. Stumm and J. J. Morgan. 1981. *Aquatic Chemistry*. 2nd ed. New York: Wiley.)

If the total concentration of monosilicic acid in water (at 25°C) is raised above 2×10^{-3} moles of Si per liter, the solution becomes supersaturated with respect to amorphous silica, and the excess $Si(OH)_4^0$ begins to form polymers:

$$Si(OH)_4 + Si(OH)_4 \rightarrow (OH)_3Si{-}O{-}Si(OH)_3 + H_2O \qquad (6.14)$$

Consequently, supersaturation eventually results in polymerization to form colloidal suspensions of amorphous silica, but equilibrium is reached slowly because reaction 6.14 has a high activation energy.

In soils, aluminosilicate clays precipitate from siliceous solutions if alumina is available: the solubility of these clays in terms of dissolved $Si(OH)_4^0$ is less than the solubility of amorphous silica.

In contrast to silica, alumina solubility is influenced profoundly by the fact that aluminum hydroxide, a common product of mineral weathering, is amphoteric. It can react in acid solutions to yield Al^{3+} ions, and in alkaline solutions to form $Al(OH)_4^-$:

$$\underset{\text{soluble}}{Al^{3+}} \underset{+H^+}{\overset{+OH^-}{\rightleftarrows}} \text{Al-hydroxy cations} \underset{+H^+}{\overset{+OH^-}{\rightleftarrows}} \underset{\text{insoluble}}{Al(OH)_3} \underset{+H^+}{\overset{+OH^-}{\rightleftarrows}} \underset{\text{soluble}}{Al(OH)_4^-} \qquad (6.15)$$

This means that noncrystalline alumina is least soluble in the mid-pH range of about 5 to 7, as depicted in Figure 6.8. Crystalline $Al(OH)_3$ in the form of gibbsite extends insolubility over a wider range of pH. Therefore, as freshly precipitated alumina is allowed to age, its structure becomes more ordered and its insolubility field is extended.

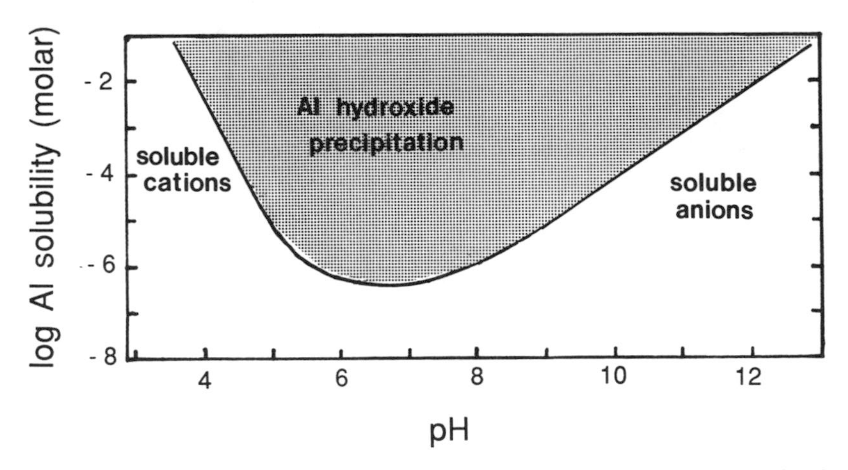

Figure 6.8. Solubility of Al (in units of total molar concentration of all dissolved species) as a function of pH, assuming equilibrium with gibbsite. The shaded region denotes supersaturated conditions favorable to precipitation. (Adapted from C. F. Baes and R. E. Mesmer. 1976. *The Hydrolysis of Cations.* New York: Wiley.)

In summary, the generalizations that can be made about the solubility of silica and alumina in soil solutions are:

1. In acid solutions, Al is more soluble than Si.
2. In neutral solutions, Al is insoluble but Si remains soluble.
3. In alkaline solutions, Al and Si solubilities increase with pH.
4. If Al and Si are in solution together, they co-precipitate as aluminosilicates in the pH range of 4 to 11, lowering both of their solubilities relative to either one alone.

6.3b. Aluminosilicate Mineral Precipitation

The kinds of minerals that form in soil solution depend on the chemical environment, which can be broadly classified under the following headings.

Strongly Alkaline, Confined Environments. Because Si is most soluble in the form of silicate anions, it is not surprising that smectites form most rapidly in dilute alkaline solutions of sodium silicate and magnesium chloride. The layer silicate structure evolves by the precipitation of a planar $Mg(OH)_2$ (brucite) sheet, on which monomeric silicate ions condense. Silica polymers are unable to reorganize into layered structures. It seems that the metal hydroxide sheet must be layered to begin with, a fact that necessitates 6-coordination of the metal ion that is to form the octahedral sheet of the 2:1 layer. Smectites can be formed from hydroxide sheets of the following cations (listed in order of increasing radius):

$$(Al^{3+}\ Fe^{3+})\ Mg^{2+}\ Ni^{2+}\ Co^{2+}\ Fe^{2+}\ Zn^{2+}\ (Mn^{2+})$$

Those cations whose size does not ideally suit them for octahedral sites are denoted by parentheses. Although all of these cations are capable of 6-coordination with OH^- and O^{2-}, and can therefore form layered hydroxides, Al^{3+} and Fe^{3+} are tetrahedrally coordinated (as $Al(OH)_4^-$ and $Fe(OH)_4^-$) in alkaline solutions. Consequently, the synthesis of smectites with Al^{3+} or Fe^{3+} in the octahedral sheet is difficult and seems to require some Fe^{2+} or Mg^{2+} to aid sheet formation (Harder, 1978). At the other size extreme, Mn^{2+} is too large for a comfortable fit in octahedral sites, so that smectite with octahedral Mn^{2+} is also difficult to synthesize. The Cu^{2+} ion is notably absent from the list of ions that are known to form the octahedral sheet of layer silicates, perhaps because of its electronegativity or because it tends to prefer a highly distorted (tetragonal) arrangement of coordinating oxygen atoms.[1]

Artificial synthesis of smectites, especially those with Mg^{2+} in the octahedral sheet, is relatively easy. A solution saturated with monomeric silica, to which an equal number of moles of $MgCl_2$ has been added, is adjusted to pH 11 with NaOH. This yields a crystalline precipitate of trioctahedral smectite[2] at room temperature. Magnesium silicate with the 1:1 layer structure (serpentine) results if the Si/Mg mole

1. Cu^{2+} has a high electronegativity, which tends to mitigate against substitution of this metal into silicates, which have largely ionic bonds.

2. This smectite, termed stevensite, has the formula $Na_{2y}\,(Mg_{3-y}\square_y)Si_4O_{10}(OH)_2$, where □ represents octahedral vacancies that create the layer charge.

ratio in solution is lowered. Crystalline smectites have been made recently from much less alkaline solutions (pH > 8) that contain Ca^{2+} and Fe^{2+} and are kept anoxic (Wilson, 1992). The Ca^{2+} is necessary to balance layer charge as crystals grow.

The conditions necessary for artificial synthesis of smectites suggest the kind of natural environment in which these minerals will be found. This environment is alkaline as a result of restricted drainage and/or evaporative salt accumulation (see Chapter 8), so that the normally mobile alkaline and alkaline earth ions (Na, K, Ca, Mg) accumulate, along with silica. Neoformation of smectites from Na-rich saline water, and possibly illites from more K-rich water, is favored under these conditions. Other very silica-rich silicate minerals, such as attapulgites, sepiolites, and zeolites, are also known to form under these conditions. Less alkaline conditions may be necessary for Fe and Al-rich (dioctahedral) smectite formation, and a reducing environment may assist crystallization.

Moderately Alkaline, Weak-Leaching Environment. Only a portion of the mobile weathering products (silica, base cations) is lost by leaching in this situation. Aluminum and iron hydroxide are the least soluble weathering products, so these react with the soluble silica and base cations to produce 2:1 layer silicates, including dioctahedral smectites and illites. Chlorites can be formed in this situation as well.

Acidic, Strong Leaching Environment. The only weathering products of primary minerals to be conserved in this environment are alumina and Fe hydroxides or oxides. As a result, the trend is toward "desilication" of the initial weathering products and complete loss of base cations. If the leaching is not extreme or prolonged, some dissolved silica can react with the alumina to form allophane and imogolite (noncrystalline aluminosilicates), kaolinite or halloysite. If leaching is very intense or prolonged, desilication results in oxides of Fe and Al being the only important secondary minerals remaining in the soil. For example, the weathering of feldspar may follow the sequence

$$\text{feldspar} \xrightarrow{-\text{Si}} \text{kaolinite} \xrightarrow{-\text{Si}} \text{gibbsite} \tag{6.16}$$

Kaolinite can be neoformed from solutions of Al and silica only at acid pH; in alkaline solutions, the tetrahedrally coordinated Al ion in $Al(OH)_4^-$ is unable to form the gibbsitelike sheet that is the precursor of the 1:1 aluminosilicate layer. Even in acid solutions, however, kaolinite crystallization is very slow at room temperature, probably because $Al(OH)_3$ is insoluble except at very low pH. Natural organic chelating agents, such as oxalate, increase the solubility of Al at acidic pH, thereby increasing the rate of kaolinite neoformation.

Fe oxide accumulation is perhaps the most visually obvious process in soils that have been subjected to leaching and desilication over many thousands of years. Hematite (Fe_2O_3), with its deep red color, is commonly seen in intensely weathered soils of tropical and subtropical climates, as well as in some soils of very dry climates. Goethite (FeOOH), having a yellow-brown color, is geographically widespread in many different soil types, but often is the only soil-formed Fe oxide in temperate and cold regions. Its lighter color can be masked by the more intense color of hematite. Climatic change to wetter conditions can transform hematite to goethite in surface soils, probably by reductive dissolution of the hematite, followed by reoxidation of Fe^{2+} and precipitation of goethite (or perhaps lepidocrocite). This would explain the tendency of soils with high organic matter to contain goethite rather than hematite. There is no evidence that direct interconversion between the hematite and goethite structures occurs in nature without dissolution. In the laboratory, hematite synthesis is known to be favored in the pH 6 to 9 range, while goethite forms preferentially at higher and lower pH. Higher temperatures always favor hematite formation.

The noncrystalline hydrated iron oxide, ferrihydrite, is indicative of "young" deposits of Fe oxides in soils, forming by the fast hydrolysis of Fe^{3+} salts or rapid oxidation of Fe^{2+} solubilized by reduction reactions. It is more reddish than goethite, but less reddish than hematite. It is stabilized relative to the more crystalline oxides by adsorbed cations and anions, notably silicate. There is some uncertainty about its structure, but it is composed of tiny (30–70 Å) spherical particles that may have short-range hematitelike or goethitelike order. Ferrihydrite is probably a precursor of more crystalline oxides such as hematite.

A flow diagram of the weathering sequences that can be expected for the most important rock-forming minerals is depicted in Figure 6.9. This diagram incorporates the processes of alteration, neoformation by addition (silica added to Fe and Al oxides in confined environments), and neoformation by subtraction (silica removed

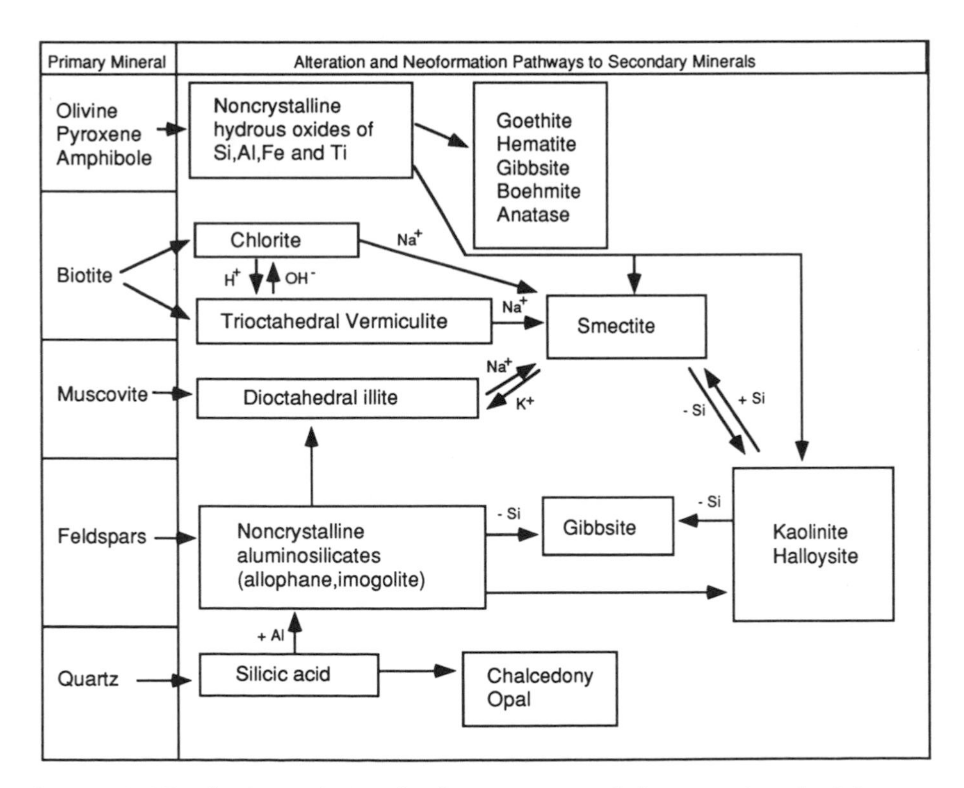

Figure 6.9. Weathering pathways for the common rock-forming minerals of the earth's crust. (Adapted from B. Mason. 1966. *Principles of Geochemistry*. 3rd ed. New York: Wiley.)

from layer silicates in leaching environments). The diagram is not exhaustive, however, leaving out some alteration and neoformation pathways that could be significant.

6.4. CLAY MINERAL ASSOCIATIONS AND WEATHERING STAGES IN SOILS

From the principles of primary mineral stability, mineral alteration, and neoformation developed in this chapter, it is evident that as soils evolve from the parent rock material their elemental composition begins to depend more on the weathering environment than on the composition of the parent minerals. For this reason, clay mineral associations in soils are seen as indicators of the degree of soil development (which may or may not be a function of the chronological age of the soil). The Jackson and Sherman sequence of soil development is given in Table 6.1, a sequence based on the premise that soils develop in nonconfined environments where leaching ultimately leads to depletion of silica and base cations. This sequence is not meant to imply that individual soils progress through all thirteen stages. Instead, it indicates the mineral associations that tend to be found in the clay fractions of soils with histories of weathering that vary in time and intensity. A weathering sequence such as muscovite → illite → kaolinite → gibbsite is feasible, according to this scheme, while the reverse is not. This assumes that climatic or drainage factors do not change drastically over time to create confining environments; such changes could cause "regressive" transformations of secondary minerals at the more advanced stages of soil development. For example, 1:1 minerals could react with silica and base cations introduced at a later time, neoforming 2:1 minerals. This would mean regression

Table 6.1. Stages of Soil Development as Indicated by the Presence of Clay Mineral Groups

Soil Development Stage	Prominent Clay Minerals
1	Gypsum, sulfides, soluble salts
2	Calcite, dolomite, apatite
3	Olivine, amphiboles, pyroxenes
4	Micas, chlorite
5	Feldspars
6	Quartz
7	Muscovite
8	Vermiculite, illite
9	Smectites, allophane
10	Kaolinite, halloysite
11	Gibbsite
12	Goethite, hematite
13	Titanium oxides, zircon corundum

Source: M. L. Jackson and G. D. Sherman. 1953. Chemical weathering of minerals in soils. In A. G. Norman (ed.), *Advances in Agronomy,* Vol. 5. New York: Academic Press, pp. 221–317.

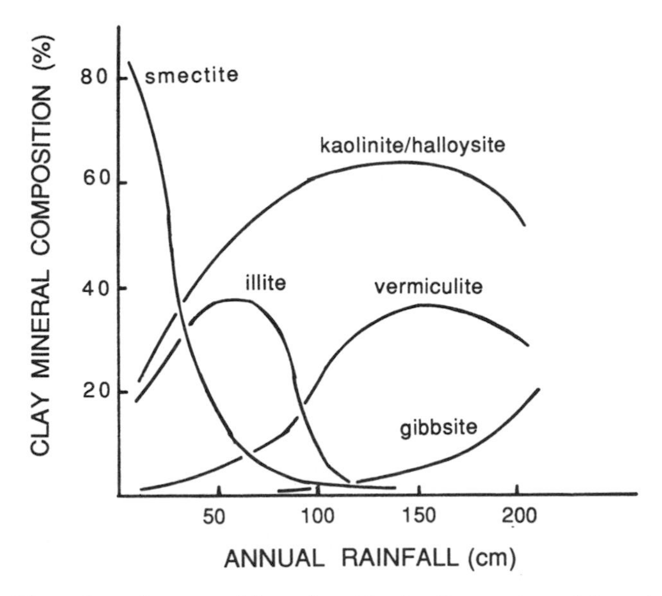

Figure 6.10. Clay mineral composition of residual soils weathered from igneous rock under different moisture regimes. (Adapted from I. Barshad. 1966. The effect of variation in precipitation on the nature of clay mineral formation in soils from acid and basic igneous rocks. *Proc. Int. Clay Conf. (Jerus.)* 1:167–173.)

from stage 10 to stage 9. Generally, the Si/(Al+Fe) mole ratio of minerals is an index of the progress of weathering, with low ratios indicating an advanced stage.

Climates with high rainfall represent more intensely leaching environments than those with low rainfall. With comparable parent materials subjected to weathering over the same time period, wet climates cause more advanced soil development than dry climates. This principle is confirmed by the dependence of clay mineral composition of soils on rainfall, as has been shown by numerous examples such as the one illustrated in Figure 6.10.

6.5. WEATHERING RATES

On the basis of the scheme of mineral weathering that has been developed to this point, the factors determining mineral weathering rates might be expected to be:

1. Intrinsic structural stability (resistance to hydrolysis by H^+ attack)
2. pH
3. Presence of complexing ligands (organic acids, inorganic anions)
4. Temperature
5. Specific surface area (m^2/g) of mineral
6. Efficiency of removal of soluble weathering products (by precipitation, leaching, etc.)

Dissolution of silicate minerals is now believed to follow a zero-order rate law (Wood and Walther, 1983):

$$\frac{dm}{dt} = K_{diss} \qquad (6.17)$$

where dm/dt is the moles of silicate dissolved per square meter of surface area per second, and K_{diss} is the dissolution constant for a particular temperature, pH, and mineral surface area. This means that a clay-sized mineral weathers at a much faster rate than silt or sand-sized particles of the same mineral. Since primary minerals in soils are typically coarse grained, while secondary minerals tend to be clay sized, weathering of the secondary minerals is potentially much faster under strongly acidic or alkaline conditions.

Weathering rates of the framework and layer silicate minerals (feldspars, quartz, micas) fall in the range of $K_{diss} = 10^{-11}$ to 10^{-12} moles of silicate oxygen/m^2/sec at 25°C, while rates for island and chain silicates are even greater. K_{diss} is a function of both temperature and pH, being larger at extreme pH and higher temperature. The general dependence of silicate mineral weathering on pH is sometimes expressed as $dm/dt = k \cdot (H^+)^n$, where n varies from 0 to 1 under acid conditions, and n is less than 0 under neutral to alkaline conditions. This means that silicates dissolve fastest in acid and alkaline solutions. The exact value of n is pH and mineral dependent, but it is usually at low pH (<5) that mineral weathering rates become very sensitive to pH (n approaches unity).

The temperature dependence of K_{diss} can be approximated by the Arrhenius expression:

$$\log K_{diss}\,(\text{moles/m}^2\text{/sec}) = \frac{-2900}{T} - 2.85 \qquad (6.18)$$

This temperature dependence reveals that the activation energy (E_a) for layer and framework silicate weathering reactions is high (>12 kilocalories or 50 kilojoules/mole), evidence that the dissolution process is not diffusion controlled,[1] but is limited by a slow surface reaction such as the protonation of a surface group or the rupture of a metal-oxygen bond.

Not much is known for certain about the rate of weathering of secondary minerals. Gibbsite dissolves in acid solutions (pH 2–3) at a rate *per unit of surface area* that is not greatly different from that of feldspars ($K_{diss} \approx 10^{-11}$ moles/m^2/sec). Naturally, because gibbsite crystals tend to be much smaller (micron-sized) than particles of feldspars and other primary minerals, *actual* dissolution rates of feldspars and gibbsite on a mass basis are very different. Consequently, when H^+ ions enter acid soils, they are more likely to react with clays and thereby dissolve Al^{3+}, rather than be consumed by the process of primary mineral (e.g., feldspar) weathering.

Layer silicate clay minerals typically dissolve incongruently, with the octahedral sheet being more susceptible to hydrolysis and decomposition by acid attack than the tetrahedral (silica) sheet. Consequently, base cations from exchange sites and from the octahedral sheet initially dissolve out of these minerals at much faster rates than silica. This means that the residue of weathering is typically a siliceous material depleted in Ca, Mg, K, and Na.

Complexing ligands, both organic and inorganic, enhance weathering rates of pri-

1. Diffusion-controlled reactions have activation energies less than 20 kilojoules/mole.

mary and secondary minerals. They may complex with metals (e.g., Fe^{3+}, Al^{3+}) at the mineral surface, facilitating detachment of the metal. In this way, many organic acids found in soil solution (citric, oxalic, etc.) increase aluminosilicate mineral weathering, particularly if they are strong Al^{3+} chelators. Phosphate and sulfate enhance gibbsite dissolution, probably by a similar mechanism involving the complexation of Al.

The last-mentioned factor influencing weathering rates of minerals in soils (see list above), efficiency of removal of soluble weathering products, warrants further comment. In confined soil environments, dissolved silica and base cations accumulate from weathering reactions. So, for example, the neoformation of kaolinite from smectite:

$$\text{Smectite} + H^+ = \text{kaolinite} + \text{base cations} + \text{silica} \tag{6.19}$$

should continue as long as there is a source of H^+ and the base cations and silica are removed by leaching or precipitation. If, on the other hand, the soluble products accumulate, then the back-reaction can begin to reform the smectite and the overall reaction rate approaches zero. This means that the dissolution of smectite continues to be significant only if the soil solution remains well undersaturated with respect to this particular mineral. Clearly, a leaching environment (wet, well-drained conditions) favors this undersaturated state.

Consider the case of gibbsite dissolution:

$$Al(OH)_3\,(s) = Al^{3+} + 3OH^- \tag{6.20}$$

where the solubility product, K_{SO}, is defined by

$$K_{SO} = 10^{-34} = (Al^{3+})(OH^-)^3 \tag{6.21}$$

The *actual* activities of Al^{3+} and OH^- in soil solution, $(Al^{3+})_s$ and $(OH^-)_s$, define the *ion activity product,* or IAP:

$$IAP = (Al^{3+})_s(OH^-)_s^3 \tag{6.22}$$

The relative degree of saturation of the soil solution with respect to gibbsite precipitation is then given by the ratio, IAP/K_{SO}. This could be called the *saturation index,* indicating the degree of solution saturation as follows:

$IAP/K_{SO} > 1$	Supersaturated solution
$IAP/K_{SO} = 1$	Saturated solution
$IAP/K_{SO} < 1$	Undersaturated solution

The solubility line of Figure 6.11 depicts these three conditions for gibbsite.

A highly undersaturated condition in soil solutions speeds up mineral dissolution because the back-reaction (reverse of reaction 6.20) is negligible. The dissolution rate might be described by an equation such as:

$$\frac{d[Al^{3+}]}{dt} = k_f \boldsymbol{S}\,[H^+] - k_b[Al^{3+}][OH^-]^3 \tag{6.23}$$

where $\boldsymbol{S}$ is the surface area of the gibbsite and k_f and k_b are the forward and backward reaction rate constants. If gibbsite dissolution is monitored only in the early stages,

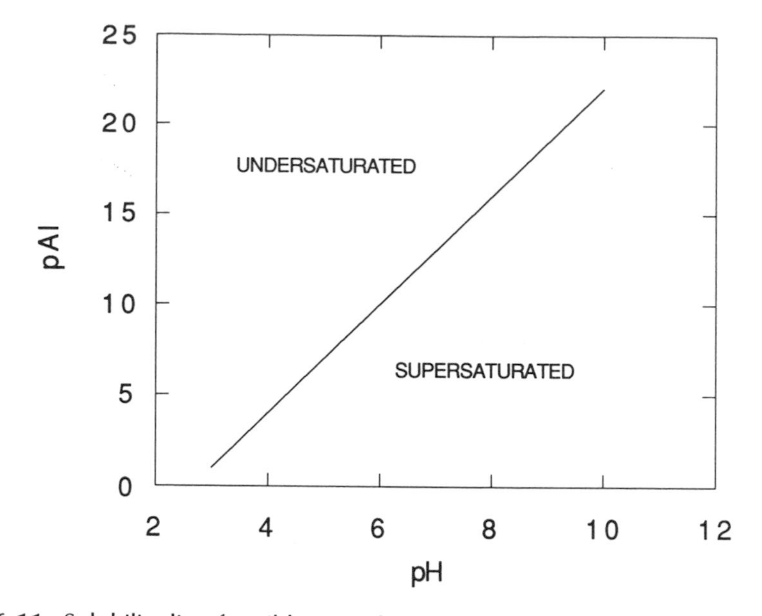

Figure 6.11. Solubility line for gibbsite with $K_{SO} = 10^{-34}$.

soluble Al^{3+} is sufficiently low that the last term in equation 6.23 can be neglected, and dissolution follows zero-order kinetics at a particular pH. In general, the proton-dependent term of equation 6.23 for various mineral dissolution reactions has to be written as $[H^+]^n$, where a fractional value of n is usually needed to fit the observed data.

Similar principles regarding the effect of weathering product removal apply to primary mineral dissolution rates as well. In this case, however, the weathering reaction is unlikely to be reversible under the temperature and pressure conditions prevailing in soil environments, as will be demonstrated in the next section.

6.6. THERMODYNAMIC STABILITY OF MINERALS

6.6a. Limitations of the Thermodynamic Approach

Thermodynamic principles are used by soil scientists and geochemists to define solution conditions favorable for the formation of particular minerals. Consider dissolution of the feldspar, microcline, for example:

$$2KAlSi_3O_8 + 8H^+ + 8H_2O \rightarrow 2K^+ + 2Al^{3+} + 6Si(OH)_4^0 \qquad (6.24)$$

In soils, if acid leaching is not too intense, kaolinite may form as a weathering product from the dissolved Al^{3+} and silica:

$$2Al^{3+} + 2Si(OH)_4^0 + H_2O = \text{kaolinite} + 6H^+ \qquad (6.25)$$

The overall reaction is then:

$$2KAlSi_3O_8 + 2H^+ + 9H_2O \rightarrow \text{kaolinite} + 2K^+ + 4Si(OH)_4^0 \qquad (6.26)$$

Using the known free energies of formation (ΔG_f^0) of the reactants and products in equation 6.26, it is possible to calculate the standard free energy of the reaction, ΔG_r^0:

$$\Delta G_r^0 = \Sigma\{\Delta G_f^0 \text{(products)}\} - \Sigma\{\Delta G_f^0 \text{(reactants)}\} \tag{6.27}$$

where Σ symbolizes the summation for all products and reactants. The equilibrium constant for the reaction, K_r, is then determined from the thermodynamic relation

$$\Delta G_r^0 = -RT \ln K_r \tag{6.28}$$

and found to be $10^{-5.05}$. But since the equilibrium condition for reaction 6.26 is described by the expression that defines the K_r value:

$$K_r = \frac{(K^+)^2(\text{Si(OH)}_4^0)^4}{(\text{H}^+)^2} = 10^{-5.05} \tag{6.29}$$

it is evident that the value of K_r, calculated from equation 6.28, can be entered into equation 6.29 to specify the solution conditions under which kaolinite and feldspar coexist at equilibrium. Thermodynamic principles appear, then, to define the K^+, H^+, and dissolved silica concentrations that cause feldspar to dissolve, precipitate, or exist at equilibrium with kaolinite.

If equation 6.29 is converted to logarithmic form, the "boundary condition" separating the stability field of feldspar from kaolinite is defined by the condition for equilibrium between feldspar and kaolinite:

$$\log K_r = -5.05 = 2 \log (K^+) + 2 \text{ pH} + 4 \log (\text{Si(OH)}_4^0) \tag{6.30}$$

This condition is represented as a plane on the mineral stability field diagram of Figure 6.12. As is obvious from the nature of reaction 6.26, the diagram predicts that the conditions favoring kaolinite formation and feldspar dissolution are low pH and low soluble K^+ and silica.

The problem with this analysis is that, even in a confined environment where equilibrium has some chance to be approached, there is no evidence that dissolution reactions of most primary minerals, including reaction 6.26, are reversible. Feldspars (and most other primary minerals), once dissolved in water under earth-surface conditions, do not re-precipitate in the same environment. This should not be surprising given the fact that feldspars form from high-temperature melts. Therefore, the question arises: Is the boundary condition shown in Figure 6.12 physically meaningful? Thermodynamics does not consider the effect of activation energy barriers on reactions (see Chapter 1), and accordingly indicates that the formation of feldspars should proceed under conditions of high pH and elevated concentrations of dissolved K^+ and silica. But the organization of the three-dimensional framework silicate structure of feldspar requires Si—O—Si and Si—O—Al bonds to form, thereby linking together monomeric Si(OH)_4 and Al ions. This is not favored at earth-surface conditions.[1] As Millot (1970) stated *"The guiding role in the arrangement of silica in the hydrosphere is ensured by cations, among which aluminum is the first to intervene."*

1. Microcrystalline feldspathoids and zeolites, both frame-work silicate structures, can precipitate from very alkaline solutions near 20°C at the expense of layer silicates, but true crystalline feldspars do not form at measurable rates in this way.

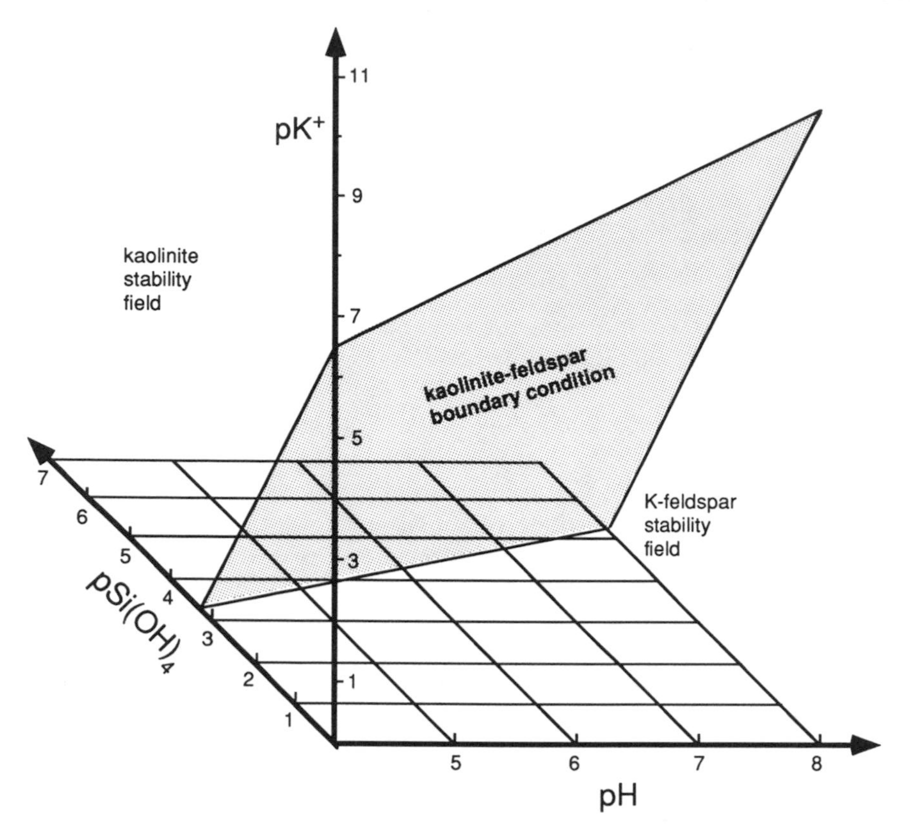

Figure 6.12. The stability fields of kaolinite and feldspar in pH-p$Si(OH)_4$-pK space.

By this hypothesis, Millot argued that aluminum tends to direct silica toward a two-dimensional (sheet) rather than three-dimensional (framework) structure under conditions usually prevalent in soil solutions (although this assumes that conditions are not strongly alkaline). Thus, feldspars are not known to neoform (by, say, the reverse of reaction 6.26) except perhaps under geological conditions favorable to metamorphosis. Yet thermodynamically less stable minerals such as zeolites *do* form under the alkaline conditions that are presumably favorable to the formation of feldspar. Zeolites are framework silicates with an arrangement of tetrahedra unlike that of feldspar, and are considered to be metastable with respect to feldspar. We find, then, that in weathering processes, the final mineral products may not be, in the thermodynamic sense, the most stable products.

It is necessary to conclude that many of the mineral stability diagrams that are commonly constructed to explain primary mineral weathering (such as the one for feldspar in Figure 6.12) have no quantitative value—they are useful only to the extent that they gauge the tendency of the weathering reaction to proceed in a forward direction. If, for example, feldspar and kaolinite coexist in a soil, overall equilibrium between the two minerals is not possible when a realistic temperature and time frame is considered. The reaction is irreversible; if it proceeds, feldspar must decompose and kaolinite must precipitate. It is true that back reactions such as the readsorption of K^+ from solution may slow feldspar dissolution, but such reactions do not reform the original primary mineral structure. For example, mica weathering is sensitive to K^+ readsorption, but the original mica is not reconstituted by the back reaction.

The proof of reversibility in primary mineral weathering would be instances where primary mineral structures have formed under earth-surface conditions. There are reports that "secondary" quartz can slowly precipitate at room temperature from solutions supersaturated with monosilicic acid. More typically, however, precipitated silica in soils is structurally disordered, in the form of chalcedony or opal. In fact, as long as alumina is present, silica does not precipitate as a separate phase, reacting instead to form aluminosilicates (layer silicates, imogolite, or allophane).

Based on the above arguments, one might logically question whether thermodynamics and mineral stability diagrams have any validity or use in understanding mineral formation and decomposition. Secondary mineral associations actually found in soils are often metastable—that is, they are persistent but are not the most stable associations according to thermodynamics. Furthermore, true equilibrium seems to exist in soils only on a local scale. It is common for one mineral phase to precipitate on another, coating it completely, with the result that the coated mineral is no longer in equilibrium with the solution. The equilibrium is only partial in this case.

6.6b. Stability Fields of Secondary Minerals

Despite concerns regarding the rigorous application of equilibrium concepts to soils, stability diagrams for secondary minerals have some worth as indicators of the general sequences of weathering that can be expected. To construct stability diagrams for the relevant secondary minerals, those minerals that could potentially control the solubility of silica, aluminum, and other ions in soil solution need to be considered. At a minimum, the minerals that must be considered as potentially controlling Al and Si solubility are gibbsite (or some less crystalline form of Al hydroxide), noncrystalline silica,[1] kaolinite (or less crystalline but metastable forms of aluminosilicate such as imogolite), and a 2:1 layer aluminosilicate (such as beidellite).

The stability line for gibbsite is calculated from the known solubility product of gibbsite, defined by equation 6.21. The dissociation constant of water:

$$K_w = (H^+)(OH^-) = 10^{-14} \tag{6.31}$$

is then used to convert the solubility expression to one in which (H^+) rather than (OH^-) appears explicitly:

$$\frac{(Al^{3+})}{(H^+)^3} = \frac{K_{SO}}{K_w^3} = 10^{8.0} \tag{6.32}$$

The logarithmic form of equation 6.32 is then an equation for a straight line:

$$pAl^{3+} - 3\ pH = -8.0 \tag{6.33}$$

1. Quartz is less soluble than noncrystalline forms of silica and is therefore more stable, but forms too slowly in soil environments to control Si solubility.

when pAl^{3+} (the negative logarithm of the Al^{3+} activity) is plotted as a function of pH. This stability line for crystalline gibbsite is shown in Figure 6.11.

For kaolinite, the dissolution reaction is the reverse of reaction 6.25 and has an equilibrium constant estimated to be $10^{7.63}$. The equilibrium expression for this reaction is then

$$K_K = \frac{(Al^{3+})^2(Si(OH)_4^0)^2}{(H^+)^6} = 10^{7.63} \tag{6.34}$$

The logarithmic form of this expression is a three-variable equation:

$$pAl^{3+} + pSi(OH)_4^0 - 3\ pH = -3.81 \tag{6.35}$$

representing the stability plane of kaolinite when plotted three-dimensionally as shown in Figure 6.13. This plane defines the activities of Al^{3+}, H^+, and monomeric

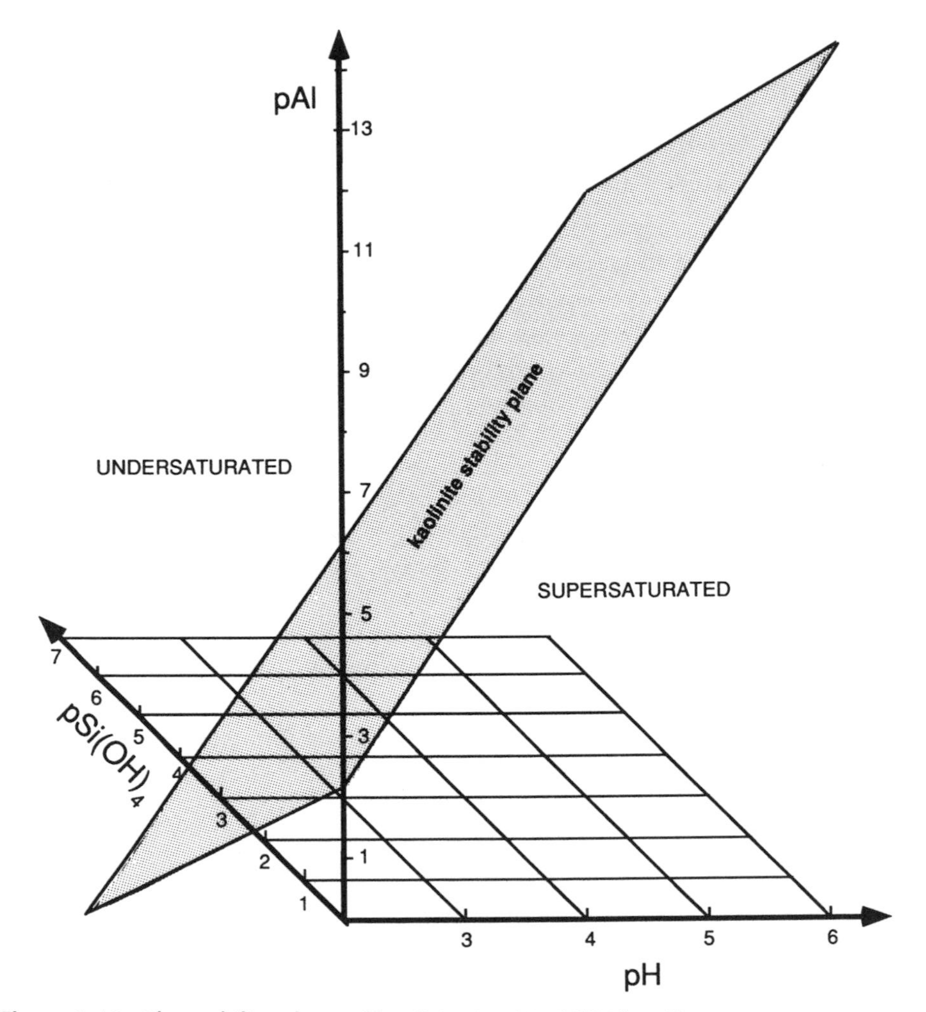

Figure 6.13. The stability plane of kaolinite in pH-pSi(OH)$_4$-pAl space.

silica in soil solution that are consistent with the presence of kaolinite in the soil at equilibrium.

Finally, the dissolution of a 2:1 layer silicate such as beidellite can also be involved in controlling the solubility of Al and Si. If Mg^{2+} is assumed to occupy the exchange sites of this clay, the dissolution reaction is

$$6Mg_{0.167}Al_{2.33}Si_{3.67}O_{10}(OH)_2 + 44H^+ + 16H_2O = Mg^{2+} + 14Al^{3+} + 22Si(OH)_4^0 \quad (6.36)$$

The equilibrium expression for this reaction is

$$K_B = \frac{(Mg^{2+})(Al^{3+})^{14}(Si(OH)_4^0)^{22}}{(H^+)^{44}} = 10^{36.6} \quad (6.37)$$

which in logarithmic form becomes

$$pMg^{2+} + 14pAl^{3+} + 22pSi(OH)_4^0 - 44\,pH = -36.6 \quad (6.38)$$

Because this equation has four variables, one of them must be eliminated or fixed in order to plot a stability relation on the same three-variable axis system used for kaolinite (Figure 6.13). Since in this exercise the stabilities of gibbsite, kaolinite, and beidellite are being compared, only one of which contains Mg^{2+}, it is logical to eliminate pMg^{2+} as a variable by fixing it at some reasonable value. A reasonable activity for Mg^{2+} in temperate-region soils is 10^{-4}, so that $pMg^{2+} = 4.0$. Equation 6.38 then becomes:

$$7pAl^{3+} + 11pSi(OH)_4^0 - 22\,pH = -20.3 \quad (6.39)$$

This equation defines the stability plane for beidellite, intersecting that for kaolinite, as shown in Figure 6.14. The points on the line of intersection between these planes represent solution compositions (pH, pAl^{3+}, $pSi(OH)_4^0$) that are consistent with beidellite and kaolinite coexisting at equilibrium in soils. This assumes, of course, that both minerals dissolve congruently and reversibly and that some process such as ion exchange "buffers" the Mg^{2+} activity at 10^{-4}.

When the stability planes for three minerals are considered simultaneously (say, gibbsite, kaolinite, and beidellite), the three planes must intersect at a point (unless they happen to be parallel to one another). This intersection point represents the *single* solution composition consistent with all three minerals coexisting in equilibrium. In other words, equilibrium among these three mineral phases would fix the pH and the activities of Al^{3+} and $Si(OH)_4^0$ in solution at one unique *invariant point.* Since the gibbsite solubility expression (equation 6.33) can be written as

$$pH - \tfrac{1}{3}pAl^{3+} = 2.67 \quad (6.40)$$

it is apparent that the stability plane of gibbsite is normal to the plane of the pAl^{3+}-pH axis system, running parallel to the $pSi(OH)_4^0$ axis. Finding the intersection of the three planes at the invariant point is equivalent to solving equations 6.40, 6.39, and 6.35 simultaneously.

An analysis of the gibbsite-kaolinite-beidellite-solution system can be made using the Gibbs phase rule. The system has four distinct phases (three different minerals plus solu-

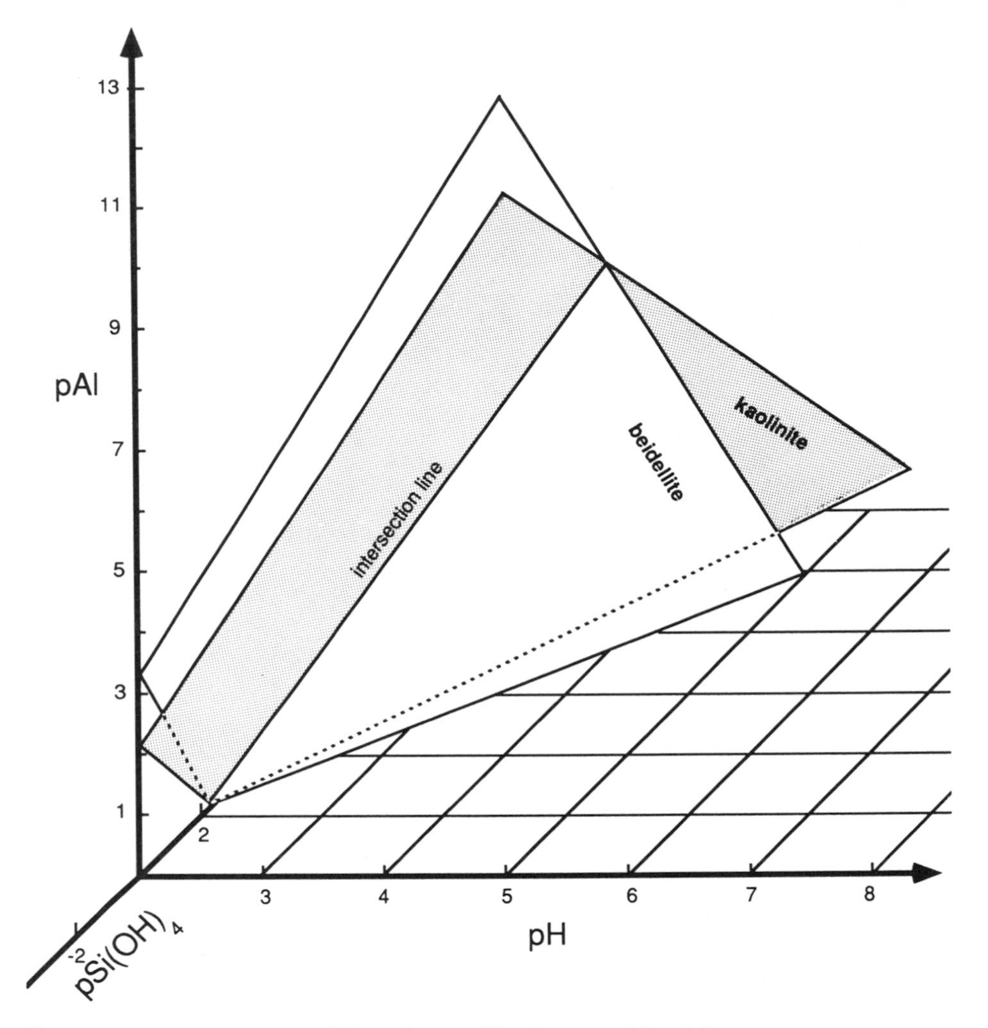

Figure 6.14. Intersecting stability planes of kaolinite and beidellite in pH-pSi(OH)$_4$-pAl space. The Mg^{2+} activity in solution is arbitrarily fixed at 10^{-4}.

tion) so that the number of phases, P, equals 4. The chemical components, C, of these aluminosilicate clays in water can be represented as Al_2O_3, SiO_2, MgO and H_2O because all chemical species can be formed (in principle) by reactions of these components. Gibbs' phase rule states that the number of degrees of freedom, F, in a system is given by

$$F = C - P + 2 \tag{6.41}$$

Since the weathering environment of the soil determines the temperature (about 20°C) and pressure (1 atmosphere), the number is effectively reduced by two, so that

$$F = C - P \tag{6.42}$$

This means that, in this four-component system, the maximum number of phases that can coexist at equilibrium is $P = C - F = 4 - 0 = 4$. In other words, the three minerals in equilibrium with solution produce an invariant point with F = 0 (the intersection of three stability planes). If a fourth mineral were present in the system, this would be an indication that the mineral phases were not all at equilibrium. It is probably common for such a nonequilibrium situation to exist in soils, since more than three clay mineral phases are often found within a small soil volume. One or more mineral phases may be kinetically "inert," undergoing dissolution and re-precipitation at such a slow rate that it has no influence on the solution composition.

In order to plot the three-mineral (gibbsite-kaolinite-beidellite) system on the three-axis diagram of Figure 6.14, it is necessary to reduce the number of components to three by fixing pMg. This is offset by the assumption in soil systems that acid or base components are added into the system, so that the pH is not fixed internally by mineral equilibria. Consequently, the number of degrees of freedom remains at zero.

Other 3-mineral assemblages, such as montmorillonite-kaolinite-silica, produce additional invariant points in the system.

Because three-dimensional diagrams of mineral stability planes are difficult to show and interpret visually, the number of axes is usually reduced from three to two either by fixing one of the variables or, more commonly, by representing two variables on one axis. For aluminosilicates, the combined variable, pH $-$ $\frac{1}{3}$pAl, is often plotted against $pSi(OH)_4^0$. In that case, each mineral's stability field is represented by a straight line. A typical diagram of this sort is shown in Figure 6.15, where the sol-

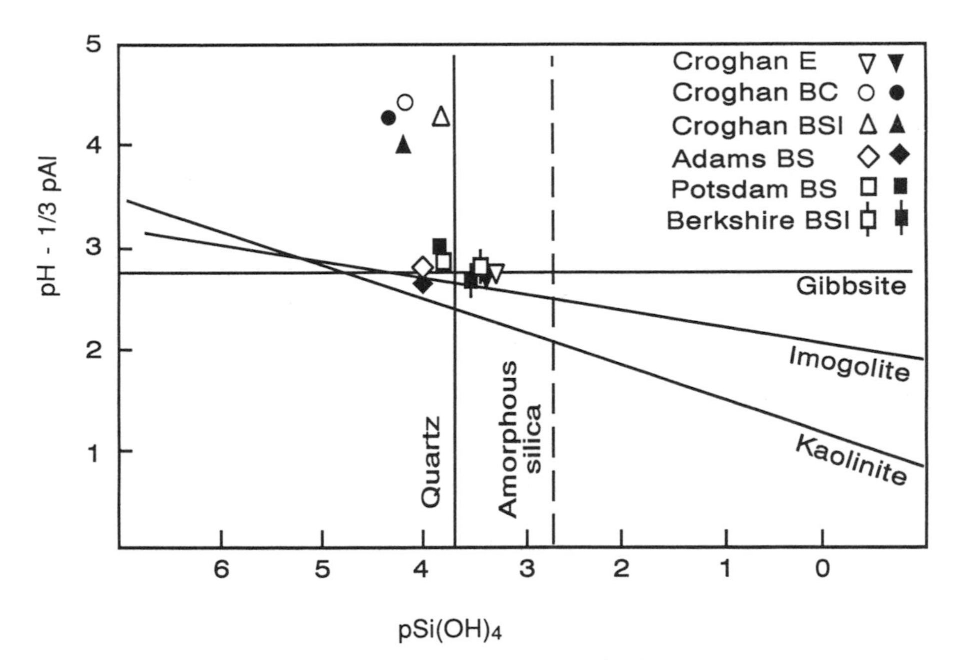

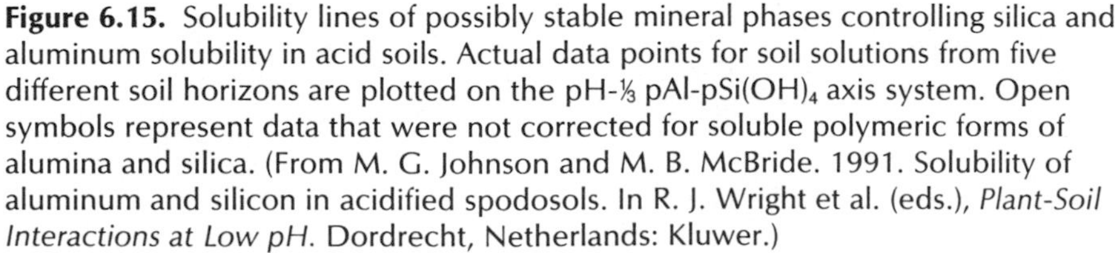

Figure 6.15. Solubility lines of possibly stable mineral phases controlling silica and aluminum solubility in acid soils. Actual data points for soil solutions from five different soil horizons are plotted on the pH-$\frac{1}{3}$ pAl-$pSi(OH)_4$ axis system. Open symbols represent data that were not corrected for soluble polymeric forms of alumina and silica. (From M. G. Johnson and M. B. McBride. 1991. Solubility of aluminum and silicon in acidified spodosols. In R. J. Wright et al. (eds.), *Plant-Soil Interactions at Low pH.* Dordrecht, Netherlands: Kluwer.)

ubility of amorphous silica, quartz, gibbsite, kaolinite, and imogolite are all plotted on common axes. The points plotted on this diagram denote measured soil solution activities of Al^{3+}, $Si(OH)_4^0$, and H^+. Many of these points, particularly those for solutions taken from the acid soil's B horizon, are positioned close to the intersection point of the gibbsite and imogolite solubility lines. Although this observation is consistent with the gibbsite-imogolite equilibrium controlling aluminum solubility, imogolite was detected in these soils while gibbsite was not. At the silica concentrations measured in these soil solutions, Figure 6.15 shows kaolinite to have a lower solubility than imogolite. Kaolinite is therefore the more stable of the two aluminosilicates. The presence of imogolite is another example where metastable minerals prevail in soils, perhaps because they have faster kinetics of crystallization.

It is quite possible that other mineral assemblages (including, for example, Al-hydroxy interlayered vermiculites, which have been detected in these same soils) not considered in Figure 6.15 could explain the observed Al and Si solubility as well. Coincidence of solubility data with the solubility line of a particular mineral is not proof of the existence of that mineral.

In Figure 6.16, the effect of adding a 2:1 clay mineral (montmorillonite, in this case) to the solubility diagram is shown. Because this clay has Mg^{2+} in its octahedral sheet, it becomes more stable (that is, less soluble), as the activity of Mg^{2+} is increased. It is necessary, then, to fix Mg^{2+} activity at a reasonable value (pMg^{2+} = 3.7). Furthermore, montmorillonite becomes more stable as the solution pH is raised. To incorporate this fact into the solubility diagram (since the *y*-axis reflects

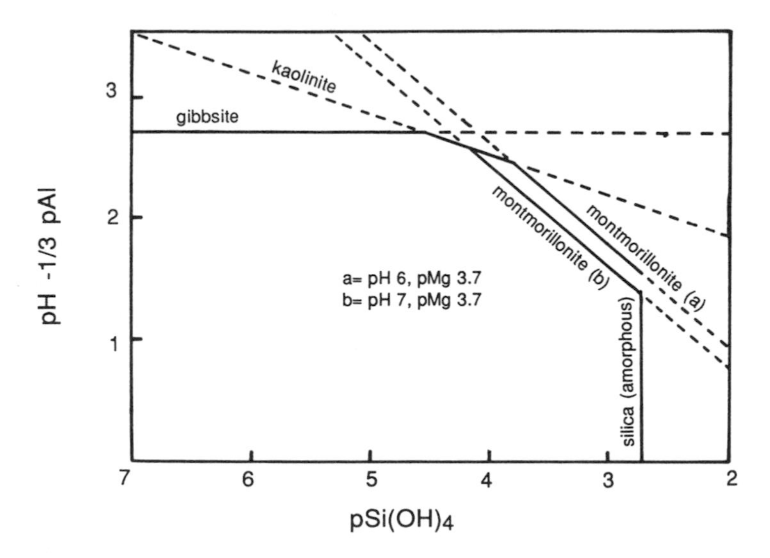

Figure 6.16. Solubility lines of potentially stable mineral phases controlling silica and aluminum solubility in acid soils. The montmorillonite lines represent the Belle Fourche clay at two different pH values, assuming constant Mg^{2+} activity, Fe^{3+} activity controlled by hematite, and Na^+ as the exchange cation. (Adapted from D. Rai and J. A. Kittrick. 1989. Mineral equilibria and the soil system. In J. B. Dixon and S. B. Weed (eds.), *Minerals in Soil Environments*. Madison, Wis.: Soil Science Society of America.)

pH only within the mixed variable, pH − ⅓pAl), the line representing montmorillonite solubility is shifted downward as the pH is raised. The cause of this shift is lowered aluminum and/or silica solubility resulting from the precipitation of montmorillonite.

The solid portions of the stability lines for each mineral signify the range of $pSi(OH)_4^0$ for which that mineral is the most stable solid phase of those being considered in Figure 6.16. This means that, as silica in solution increases, first gibbsite, then kaolinite, and finally montmorillonite becomes the most stable mineral phase in the soil. Furthermore, smectites in general are stabilized by high pH and higher concentrations of base cations (Mg^{2+} in the specific case of montmorillonite). Those conditions that favor montmorillonite stability reduce the range of kaolinite stability at the same time. Figure 6.16 also reveals that amorphous silica cannot exist in equilibrium with gibbsite or kaolinite, nor can montmorillonite attain equilibrium with gibbsite. Minerals that can coexist in equilibrium are termed *compatible pairs,* namely, montmorillonite-kaolinite, montmorillonite-silica, and kaolinite-gibbsite. The condition of their equilibrium is represented by the intersection point of their respective stability lines.

The validity of using equilibrium concepts to describe mineral weathering and soil genesis has been questioned by some. Soils are generally in a state of overall disequilibrium, with steady-state conditions only locally or partially established. Reasons for this include slow rates of mineral neoformation and alteration, incongruent and slow mineral dissolution reactions, and the fact that soils are typically open systems that experience intermittent leaching, which disrupts solution-solid equilibria. The solid phases that do form and appear to approach equilibrium with solution are often compositionally complex (such as the beidellite described above), and may only be metastable. That is, minerals of even lower solubility should ultimately form. However, from a practical standpoint, if a mineral can precipitate and congruently dissolve rapidly enough to control solubility, and is stable in soil for the time period of interest, that mineral can be treated as a true solid phase that attains equilibrium with solution.

Although true multiphase equilibrium may rarely be achieved in the complex mineral systems that are typical of soils, the approach to equilibrium signifies the direction of change in soil composition and mineralogy. For example, Figure 6.16 helps provide a qualitative understanding of long-term weathering and clay genesis in soils. Because silica is a relatively mobile component of soils, the $pSi(OH)_4^0$ axis can be visualized as the "arrow" of weathering intensity (time). Larger values of $pSi(OH)_4^0$ denote lower soluble silica and more extreme desilication (weathering) of the soil. The weathering sequence predicted from Figure 6.16 is

$$\text{amorphous silica} \rightarrow \text{smectite} \rightarrow \text{kaolinite} \rightarrow \text{gibbsite}$$

as long as silica is exported from the soil by leaching. This means that the elemental composition of the soil, as measured by the ratio Si/(Al + Fe), shifts with time toward smaller values. Ultimately, weathering enriches the soil in iron and aluminum oxides and hydroxides even as it depletes it of silicates. Only the rejuvenation of ancient soils by processes of surface deposition (vulcanism, air and water erosion), or dramatic changes in climate, can reverse this trend that eventually strips the soil of its inherent fertility.

References

Baes, C. F. and R. E. Mesmer. 1976. *The Hydrolysis of Cations.* New York: Wiley.

Barshad, I. 1966. The effect of variation in precipitation on the nature of clay mineral formation in soils from acid and basic igneous rocks. *Proc. Int. Clay Conf. (Jerusa.* 1:167–173.

Casey, W. H. and H. R. Westrich. 1992. Control of dissolution rates of orthosilicate minerals by divalent metal-oxygen bonds. *Nature* 355:157–159.

Harder, H. 1978. Synthesis of iron layer silicate minerals under natural conditions. *Clays and Clay Minerals* 26:65–72.

Jackson, M. L. and G. D. Sherman. 1953. Chemical weathering of minerals in soils. In A. G. Norman (ed.), *Advances in Agronomy,* Vol. 5. New York: Academic, pp. 221–317.

Johnson, M. G. and M. B. McBride. 1991. Solubility of aluminum and silicon in acidified spodosols: Evidence for soluble aluminosilicate. p. 15–24. In R. J. Wright, V. C. Baligar, and R. P. Murrmann (eds.), *Plant-Soil Interactions at Low pH.* Kluwer, Dordrecht: Kluwer, pp. 15–24.

Lasaga, A. C. 1984. Chemical kinetics of water-rock interactions. *J. Geophys. Res.* 89:4009–4025.

Mason, B. 1966. *Principles of Geochemistry.* 3rd ed. New York: Wiley.

Millot, G. 1970. *Geology of Clays.* New York: Springer-Verlag.

Newman, A.C.D. 1969. Cation exchange properties of micas. I. The relation between mica composition and potassium exchange in solutions of different pH. *J. Soil Sci.* 20:357–373.

Rai, D. and J. A. Kittrick. 1989. Mineral equilibria and the soil system. In J. B. Dixon and S. B. Weed (eds.), *Minerals in Soil Environments.* Madison, Wis.: Soil Science Society of America, pp. 161–198.

Reed, M. G. and A. D. Scott. 1962. Kinetics of potassium release from biotite and muscovite in sodium tetraphenylboron solutions. *Soil Sci. Soc. Am. Proc.* 26:437–440.

Scott, A. D. 1968. Effect of particle size on interlayer potassium exchange in micas. *Trans. 9th Int. Congress Soil Sci.* 2:649–660.

Scott, A. D. and S. J. Smith. 1966. Susceptibility of interlayer potassium in micas to exchange with sodium. *Fourteenth National Conference on Clays and Clay Minerals.* New York: Pergamon Press, pp. 69–81.

Stumm, W. and J. J. Morgan. 1981. *Aquatic Chemistry,* 2nd ed. New York: Wiley.

Wilson, M. J. 1992. Pedological factors influencing the distribution and properties of soil smectites. Agronomy Abstracts, p. 383. American Society of Agronomy, Madison, WI.

Wood, B. J. and J. V. Walther. 1983. Rates of hydrothermal reactions. Science 222:413–415.

Suggested Additional Reading

Allen, B. L. and B. F. Hajek. 1989. Mineral occurrence in soil environments. In J. B. Dixon and S. B. Weed (eds.), *Minerals in Soil Environments.* Madison Wis.: Soil Science Society of America, pp. 199–278.

Schott, J. and J.-C. Petit. 1987. New evidence for the mechanisms of dissolution of silicate minerals. In W. Stumm (ed.), *Aquatic Surface Chemistry.* New York: Wiley, pp. 293–315.

Stumm, W. and J. J. Morgan. 1981. *Aquatic Chemistry.* 2nd. ed. New York: Wiley, ch. 9.

Questions

1. Why do island and chain silicates have faster dissolution rates in acidic solutions than framework and layer silicates?

2. Suggest mechanisms to explain the more rapid weathering of aluminosilicate minerals at low and at high pH.

3. (a)The activation energy for dissolution of K-feldspar is 38 kilojoules/mole in the pH range relevant to soils. Calculate how much faster this mineral will dissolve at 25°C compared with 5°C. (Use the Arrhenius rate equation, $r = Ae^{-E_a/RT}$, described in Chapter 1.) Is this temperature effect important for soil mineral weathering?
   (b) The activation energy for dissolution of quartz is 77 kilojoules/mole. Is quartz dissolution more or less dependent on temperature than K-feldspar dissolution? What reason might you give to explain the difference in activation energy for these two framework silicates?

4. (a)Suggest explanations based on structural differences for the following dissolution rates (moles/m^2/sec) measured at 25°C and pH 5 for pure minerals (data from Lasaga, 1984):

Quartz	4.1×10^{-14}
K-feldspar	1.7×10^{-12}
Na-feldspar (albite)	1.2×10^{-11}
Ca-feldspar (anorthite)	6.6×10^{-9}

   (b) Calculate the initial rate of dissolution at 25°C and pH 5 of a gram of sand-sized (1 mm diameter) and clay-sized (1 μm diameter) quartz. (Assume particles to have a cubic shape to simplify the calculation. The density of quartz is 2.65 g/cm^3.)
   (c) Why are actual dissolution rates of minerals in soil environments much lower than those listed in (a)?

5. Micas weather by the loss of K^+ from interlayer sites, but this reaction, when conducted in concentrated salts such as 1 *M* NaCl, is not usually very sensitive to pH.
   (a) What is the mechanism of K^+ replacement in salt solutions?
   (b) Is mica weathering best described as congruent or incongruent dissolution? Why?
   (c) Can thermodynamics be used to predict K^+ solubility in a soil where mica is weathering to illite? Explain.
   (d) If a dioctahedral mica can maintain a solution ratio of $[K^+]/[Na^+]$ of about 10^{-5} when immersed in NaCl solution, explain the fact that this ratio is higher for trioctahedral micas.

6. Find the solution condition at which gibbsite, kaolinite, and beidellite can theoretically coexist in equilibrium. Is this condition likely to occur in soil?

7. Reaction 6.2, describing microcline dissolution, is found to have an apparent equilibrium constant of $K = 10^{1.29}$.
   (a) In a closed system in the absence of CO_2, what equilibrium pH would this weathering reaction produce? What concentration of dissolved silica? Would this concentration of silica persist in solution?
   (b) What mineral products could eventually result from the dissolution of microcline if the soil environment were a closed (confined) system? How would this answer differ for an open soil system?

8. Smectite and gibbsite have occasionally been found together in soils. Is this possible under any conditions of equilibrium? Explain.

9. Growing plants can accelerate the weathering of K^+-bearing minerals such as mica. How does this occur? Which types of mica would be most and least susceptible to this form of weathering?

10. The laboratory synthesis at 20°C of trioctahedral (Mg-bearing) layer silicate clays from solution has been found to be much easier than that of dioctahedral clays (e.g., kaolinite, montmorillonite). Provide a mechanism-based explanation for this, given that $Al(OH)_3$ is least soluble in the pH range of 5–9, while $Mg(OH)_2$ is least soluble above pH 10.

7

Oxidation-Reduction Reactions

The importance of pH as a "master variable" controlling chemical reactions in soils has been stressed in previous chapters. However, soils subjected to fluctuations in water content come under the influence of another master variable: the *reduction-oxidation* (or redox) *potential.* Under conditions of water saturation, the lack of molecular oxygen can result in a sequence of redox reactions that changes the soil pH. In this sense the redox state of the soil exerts control over the pH. The nature of redox reactions will be discussed in this chapter, as these reactions profoundly influence metal ion solubility and the chemical form of ions and molecules dissolved in soil solution. The reader is referred to section 1.2f in Chapter 1 for a review of the basic chemical principles necessary for the understanding of redox reactions.

7.1. REDOX POTENTIALS

7.1a. Theory

For any electron-transfer half-reaction of general form

$$\text{Oxidized molecule} + m\text{H}^+ + n \text{ electrons} = \text{reduced molecule} \qquad (7.1)$$

the Nernst equation can be written:

$$E_h = E_h^0 - \frac{0.059}{n} \log \frac{(\text{reduced molecule})}{(\text{oxidized molecule})\,(\text{H}^+)^m} \qquad (7.2)$$

where:

E_h = potential for the half-reaction (volts, V)
E_h^0 = standard state potential for the half-reaction (volts, V)
n = moles of electrons involved in the reaction as written
m = moles of protons involved in the reaction as written

and the parentheses denote chemical activities of the reduced and oxidized molecules.

Table 7.1 lists a number of half-reactions important in soils and their corresponding standard-state potentials. The more positive the E_h^0, the stronger the tendency for

Table 7.1. Standard-State Reduction Potentials of Half-Reactions Involving Important Elements in Soils

Reaction	E_h^0 (volts)[a]
$Mn^{3+} + e^- = Mn^{2+}$	1.51
$MnOOH(s) + 3H^+ + e^- = Mn^{2+} + 2H_2O$	1.45
$\frac{1}{5}NO_3^- + \frac{6}{5}H^+ + e^- = \frac{1}{10}N_2(g) + \frac{3}{5}H_2O$	1.245
$\frac{1}{2}MnO_2(s) + 2H^+ + e^- = \frac{1}{2}Mn^{2+} + H_2O$	1.23
$\frac{1}{4}O_2(g) + H^+ + e^- = \frac{1}{2}H_2O$	1.229
$Fe(OH)_3(s) + 3H^+ + e^- = Fe^{2+} + 3H_2O$	1.057
$\frac{1}{2}NO_3^- + H^+ + e^- = \frac{1}{2}NO_2^- + \frac{1}{2}H_2O$	0.834
$Fe^{3+} + e^- = Fe^{2+}$	0.711
$\frac{1}{2}O_2(g) + H^+ + e^- = \frac{1}{2}H_2O_2$	0.682
$\frac{1}{8}SO_4^{2-} + \frac{5}{4}H^+ + e^- = \frac{1}{8}H_2S + \frac{1}{2}H_2O$	0.303
$\frac{1}{6}N_2(g) + \frac{4}{3}H^+ + e^- = \frac{1}{3}NH_4^+$	0.274
$\frac{1}{8}CO_2(g) + H^+ + e^- = \frac{1}{8}CH_4(g) + \frac{1}{4}H_2O$	0.169
$H^+ + e^- = \frac{1}{2}H_2(g)$	0.000

[a]The E_h^0 can be converted to the equilibrium constant for the half-reaction, K, using the equation $E_h^0 = (0.059/n) \log K$.

these half-reactions to proceed as written. This principle is evident from equation 7.2. Since the stipulation of standard-state conditions fixes the activities of the reduced molecule, oxidized molecule, and proton at unity (by definition), then the last term in equation 7.2 is zero, and $E_h = E_h^0$. That is, the reaction's potential, E_h, is equal to its standard-state potential, E_h^0, only under conditions where the activities of the molecules and the proton in reaction 7.1 are unity. Such conditions are obviously not realistic in soils, where activities of electron-donating and -accepting ions are typically below 10^{-4} and the pH ranges from 4 to 9. It is more useful, then, to present the potentials of Table 7.1 in graphic form, where the H^+ activity is seen as an independent variable with an important effect on redox potential.[1] This is done in Figure 7.1, imposing the additional constraint that most dissolved ions and molecules be assigned an activity of 10^{-5}. (The exception here is the dissolved gases, O_2, N_2, CH_4, CO_2, and H_2. The first two are given partial pressures of 0.2 and 0.8 atmosphere, respectively, while the last three are assigned a partial pressure of 0.001 atmosphere). While these activity values are arbitrary, they are more realistic than the unit activities of standard-state conditions.

Consider the reduction of Mn oxide to soluble Mn^{2+} as an example half-reaction:

$$MnO_2\,(s) + 4H^+ + 2e^- = Mn^{2+} + 2H_2O \qquad E_h^0 = 1.23 \text{ V} \tag{7.3}$$

Then, according to the Nernst equation:

$$E_h = E_h^0 - \frac{0.059}{2} \log \frac{(Mn^{2+})}{(H^+)^4} \tag{7.4}$$

1. Since E_h can affect pH directly through reaction 7.1, it may be arbitrary whether E_h or pH is chosen as the dependent (y-axis) variable. But soil pH is sometimes determined by processes unrelated to the redox reaction itself, in which case pH is justifiably selected to be the independent (x-axis) variable.

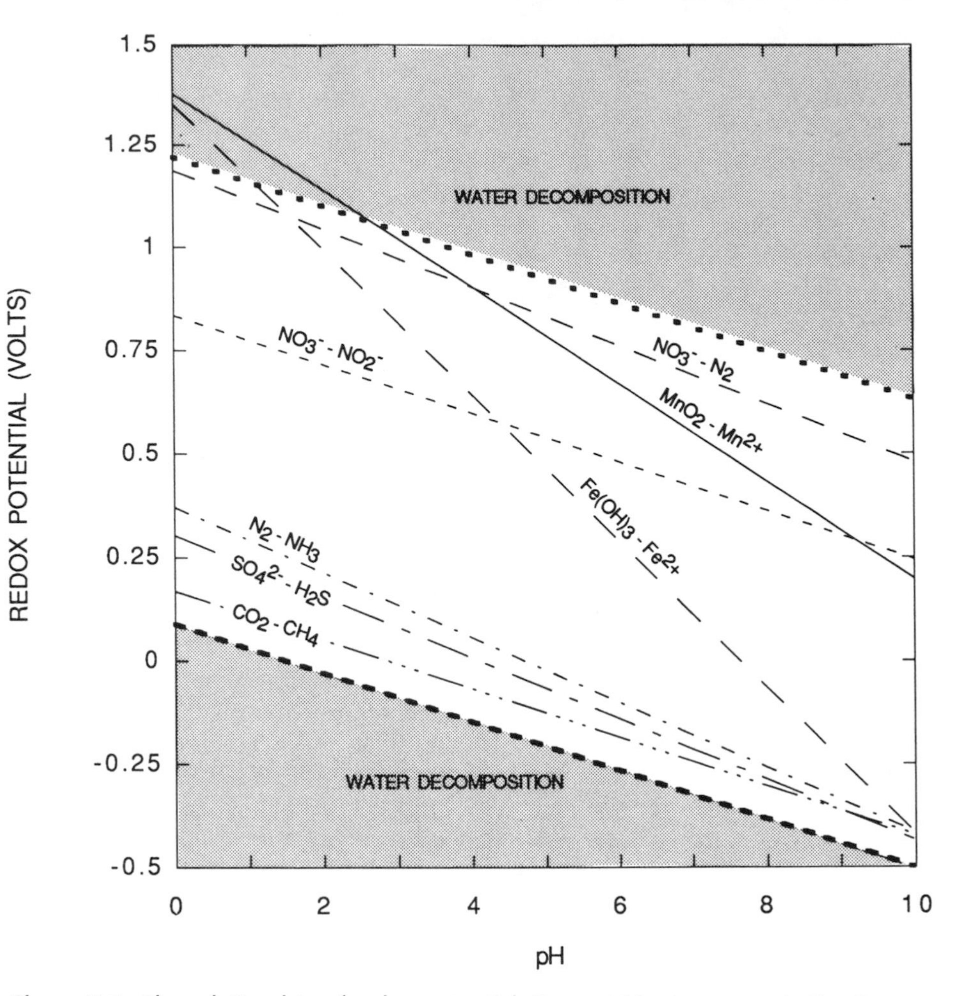

Figure 7.1. The relationship of redox potential, E_h, to pH for important half-cell reactions in water. The bold broken lines denote the E_h at which water is oxidized to O_2 (upper line) or reduced to H_2 (lower line).

That is:

$$E_h = 1.23 - 0.0295 \log (\mathrm{Mn}^{2+}) - 0.118 \text{ pH} \tag{7.5}$$

And, assuming that $(\mathrm{Mn}^{2+}) = 10^{-5}$, then:

$$E_h = 1.378 - 0.118 \text{ pH} \tag{7.6}$$

This equation describes a straight line on the E_h-pH axis system, and is plotted in Figure 7.1 along with the analogous equations for several other half-reactions of interest. It should be noted that the slope of each line is determined by the stoichiometric ratio of protons to electrons in the particular half-reaction, specified by the value of m/n in reaction 7.1 For $m/n = 1$, the slope is -59 mV/pH unit. In general, for any value of m/n, the slope is $(m/n) \times (-59 \text{ mV/pH unit})$.

7.1b. Reduction Reaction Sequence and E_h in Waterlogged Soils

Figure 7.1 is useful in understanding the sequence of reactions likely to occur as a soil becomes waterlogged and molecular oxygen, O_2, is depleted in the soil due to its low solubility in water (8 μg/ml) and its consumption by continued biological activity. Before the soil is fully depleted of oxygen, residual O_2, even under much less than fully aerated conditions, should maintain the redox potential of the soil solution at a high (strongly oxidizing) value. This means that all half-reactions listed below the O_2-H_2O reaction in Table 7.1, *if* they are at equilibrium with ("coupled to") the O_2-H_2O reaction, are driven from right to left. For example, essentially all of the iron in the soil would be maintained in oxidized forms (Fe oxides), and the activity of Fe^{2+} would be extremely low. This situation implies equilibrium and a spatially constant redox potential, unlikely in practice because of slow diffusion of O_2 into water-filled micropores of soil aggregates, and because redox reactions in general tend to be sluggish. The O_2 molecule, despite being a potentially powerful oxidant, often reacts very slowly with substrates because of kinetic limitations. Thus, certain reduced species such as Mn^{2+} and Fe^{2+} are often present in soil solution along with dissolved oxygen, an indication that conditions of nonequilibrium are common.

Microbial activity changes from aerobic to anaerobic when the dissolved oxygen diminishes to trace levels (about 10^{-6} *M*). The observable change that follows waterlogging, once O_2 is consumed in the water-saturated soil by biological activity, is a decrease in redox potential accompanied by a rise in pH toward 7. Most reduction reactions consume H^+, accounting for this pH change. In fact, it is CO_2 in the reduced soil that buffers pH in the 6 to 7 range via the $H_2CO_3-HCO_3^-$ reaction. Without CO_2, the soil pH would go well above 7 as reduction progressed and continued to consume H^+. If, however, the pH of the aerated soil is initially higher than 7, as is the case in calcareous and sodic soils, then reduction reactions actually *lower* the pH toward 7 because metal ions made soluble by reduction can precipitate as carbonates, hydroxides, or sulfides. For example Mn^{2+} and Fe^{2+} precipitate as $MnCO_3$ and $FeCO_3$ (see reactions 7.7a and b), generating protons to counter those consumed by reduction.

Anaerobic organisms are able to use oxidized chemical species such as nitrate (NO_3^-) as electron acceptors in place of molecular oxygen. Consequently, microbial reduction of nitrate to N_2 (a process termed "denitrification") occurs in the early stages of soil reduction, as does Mn oxide reduction to Mn^{2+}. Reduced species such as nitrite (NO_2^-) and Mn^{2+} then appear in solution. As more extreme reducing conditions develop, ammonium (NH_4^+) accumulates from nitrogen reduction reactions, and iron solubility increases in the form of Fe^{2+}. The elevated iron and manganese solubility is ultimately limited by precipitation of the rather insoluble carbonates of Fe (siderite) and Mn (rhodochrosite) if the soil pH is not too low:

$$Fe^{2+} + H_2CO_3 = FeCO_3 + 2H^+ \tag{7.7a}$$

$$Mn^{2+} + H_2CO_3 = MnCO_3 + 2H^+ \tag{7.7b}$$

These reactions prevent the pH from rising much above 7 even as reduction reactions continue in the soil. The process of Fe reduction in soils dissolves silica and phosphate adsorbed on ferric oxides and precipitated in Fe^{3+}-rich minerals. The fainter colors of reduced mineral forms of Fe and Mn compared with their oxidized counterparts account for the pallid appearance of waterlogged soils.

Ultimately, prolonged anoxic conditions cause sulfate to be reduced to sulfide, which then precipitates with Fe^{2+} and other metals as insoluble sulfide minerals such as pyrite (FeS_2). Manganese sulfides form if the reducing conditions are extreme; otherwise $MnCO_3$ is the more likely precipitated form of reduced Mn in anaerobic soils. Consequently, *prolonged* waterlogging of soils tends to bring Fe and Mn solubilities back down to low levels. If the redox potential decreases to low positive or somewhat negative values (< 200 mV), organic carbon can undergo reduction, forming methane. At these extreme conditions of soil reduction, hydrogen gas may be generated in the soil.

When considering the actual processes by which electrons are transferred in these reduction reactions, it should not be assumed that all reductions are caused directly by microorganisms. For example, bacteria may exude organic chemicals such as polyphenols, which in turn chemically reduce Mn oxides and other easily reducible compounds. Therefore, Mn oxide may in this sense be an indirect electron acceptor, whereas NO_3^- accepts electrons by a direct enzymatically catalyzed reaction within the cell.

Whether the process is direct or indirect, all of the reduction reactions require soil flooding to saturate pores combined with organic matter to support microbial activity and create anaerobic conditions. However, once water is drained out of the soil, many of the reduction reactions reverse. As O_2 reenters pores, the redox potential increases and iron and manganese oxidize to darkly colored oxides, forming red, brown, and black stains and deposits in the soil. In contrast to reduction, oxidation reactions as a group are H^+-generating, so that soil aeration generally causes the pH to shift lower. Only if the soil is alkaline prior to flooding can subsequent aeration be expected to raise the pH.

Reduction-oxidation episodes in soils cause appreciable redistribution of elements as some minerals dissolve and new minerals precipitate. It is unlikely that the surface chemical properties of the soil are preserved through the reduction and oxidation processes. In particular, freshly precipitated Fe and Mn oxides seem to provide recently reoxidized soils with new reactive surfaces for heavy metal and organic sorption.

E_h, or redox potential, is a measure of the *intensity* of the reducing or oxidizing conditions; that is, the tendency of the soil solution to donate electrons to or accept electrons from a chemical species or electrode introduced into the solution. Conversely, the *capacity* of the soil to be reduced or oxidized is measured by the quantities of potentially reactive electron donors (e.g., S^{2-}, Fe^{2+}, Mn^{2+}) and acceptors (e.g., $Fe(OH)_3$, MnO_2, NO_3^-, SO_4^{2-}) in the soil. Once a soil is flooded and O_2 is consumed, the redox potential (E_h) begins to decrease. How much the E_h decreases depends on the quantity of reactive electron acceptors in the soil compared with the quantity of electrons generated by biological reactions, particularly organic matter oxidation. A large reserve of electron acceptors such as Mn^{4+} or Fe^{3+} in the form of oxides means that the redox potential tends to maintain a more nearly constant value as reduction of acceptors continues; the system is then said to be "well poised" with regard to E_h in much the same way that pH is well buffered in soils having a large capacity to donate or accept protons. Even so, a high organic matter level in soils usually means that the electron acceptors are overwhelmed by microbial activity, and the E_h decreases rapidly.

7.1c. Redox Potential Measurement in Soil by Electrode

The potentials listed in Table 7.1 and plotted in Figure 7.1 are thermodynamic values and do not necessarily correspond to actual values measured in solutions, for reasons that will be discussed in this section. For example, aerobic soil solutions typically have measured E_h values of about 400 to 500 mV, and O_2 disappears from the soil at about 350 mV. These are much lower values than expected theoretically, since dissolved O_2 should in principle maintain potentials closer to 1000 mV (see Figure 7.1). To understand discrepancies between theory and measurement, some knowledge of the principles involved in redox potential measurement is needed.

The commonest method of measuring redox potential is to immerse a platinum (Pt) electrode along with a reference electrode into a solution; the electrodes are connected to a potentiometer that measures the potential difference between the two electrodes. Reduced species in solution tend to donate electrons to the conducting Pt electrode, while oxidized species tend to accept electrons from the electrode. This creates electron flow in the electrode. However, to cause any *detectable* movement of electrons in the electrode, the reaction between the oxidized (ox) and reduced (red) form of an element in solution:

$$\text{ox} \underset{-e^-}{\overset{+e^-}{\rightleftarrows}} \text{red} \tag{7.8}$$

must be shifted away from equilibrium. At the steady state of equilibrium, the forward reaction rate of equation 7.8 equals the backward rate, and the *net* electron flow (net current) is zero. Even at equilibrium, however, electron flow *in either direction,* termed the *exchange current,* i_0, is not zero.

A very small shift of the electrode potential away from its equilibrium value causes the half-reaction (equation 7.8) to proceed either to the left or right, as illustrated in Figure 7.2, thereby creating a net current that can be measured. How precisely this measurement can determine the equilibrium potential of a particular half-reaction depends on how steeply the net current deviates from zero near the equilibrium potential. The greater the magnitude of the exchange current, i_0, the more steep is the net current function. This, in turn, is a function of the concentration of reduced and oxidized species near the electrode surface, as is illustrated in Figure 7.2 for the example of the $Fe^{3+}-Fe^{2+}$ couple at two different concentrations. For this particular half-reaction, if either Fe species is at a concentration less than about 10^{-5} *M*, the exchange current is sufficiently low to prevent accurate measurement of the redox potential using the Pt electrode.

In the particular case of the O_2-H_2O half reaction:

$$\tfrac{1}{2}O_2 + 2H^+ \underset{-2e^-}{\overset{+2e^-}{\rightleftarrows}} H_2O \tag{7.9}$$

it is found that, even when the Pt electrode is shifted well away from the equilibrium potential, relatively little exchange current is generated, as shown in Figure 7.3. This

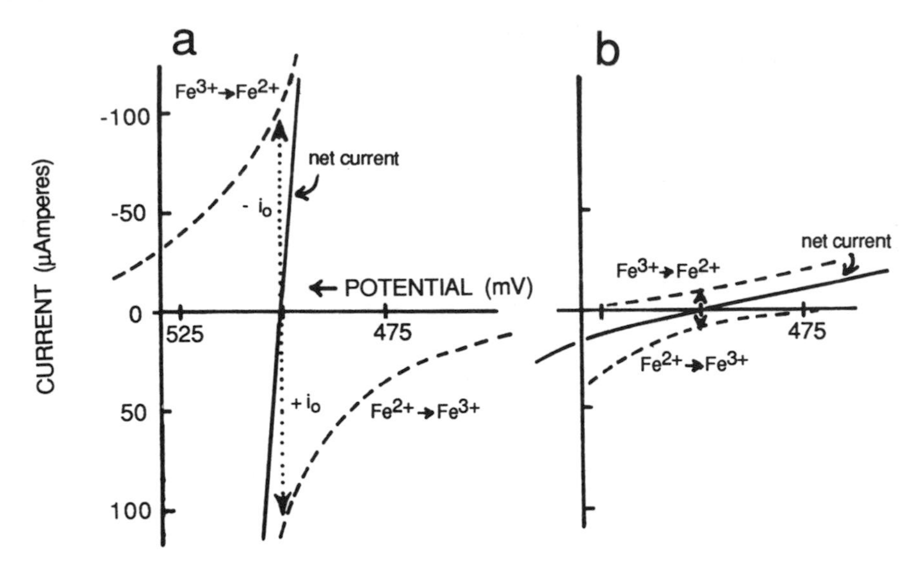

Figure 7.2. Measurement of the platinum electrode potential in relation to the electrode current for the Fe^{3+}-Fe^{2+} redox pair at pH 2 in water, under conditions of (a) $[Fe^{3+}] = [Fe^{2+}] = 10^{-3}$ *M*, and (b) $[Fe^{3+}] = [Fe^{2+}] = 10^{-4}$ *M*. (Adapted from W. Stumm and J. J. Morgan. 1981. *Aquatic Chemistry*. 2nd ed. New York: Wiley. Used with permission.)

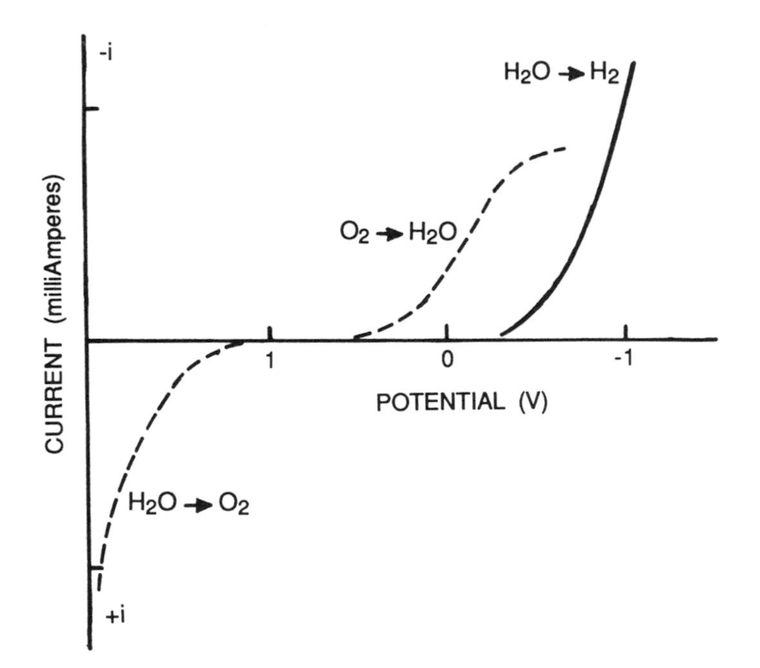

Figure 7.3. Electrode current in pure water (aerated) as a function of potential. Net current is close to zero over a wide range of electrode potential, making it difficult to locate the equilibrium potential, E_h. (Adapted from W. Stumm and J. J. Morgan. 1981. *Aquatic Chemistry*. 2nd ed. New York: Wiley. Used with permission.)

makes it difficult, if not impossible, to determine the value of the equilibrium potential (E_h) for the O_2-H_2O reaction.

A further problem in complex systems such as soil solutions, in which several redox couples coexist, is that it is quite possible that the exchange current in either direction is the sum of the exchange currents for two or more half-reactions. Consider the $Fe^{3+}-Fe^{2+}$ and O_2-H_2O redox couples, which can occur together in soil solutions. As Figure 7.4 illustrates, the measured potential, E_m, at which the net current is zero might then be the potential at which the rate of O_2 reduction at the electrode surface equals the rate of Fe^{2+} oxidation. This situation would be likely if the concentration of Fe^{2+} in solution exceeded that of Fe^{3+} (which is usually the case in reduced soils because of the generally low solubility of Fe^{3+}). In instances such as this, the two redox couples are not in equilibrium with one another, and the measured potential is termed a *mixed potential*, E_m, representing neither the E_h of the O_2-H_2O nor the $Fe^{3+}-Fe^{2+}$ couple. Thus, this measured potential is not informative about the extent of oxidation or reduction for *individual* redox couples at equilibrium. Because of this problem of interpreting mixed potentials, and because many redox couples (such as CO_2-CH_4 and $NO_3^- - NO_2^-$) do not react in a reversible manner at the electrode surface, redox potentials measured in soil solution cannot be relied on to reflect the thermodynamic reaction potentials such as those listed in

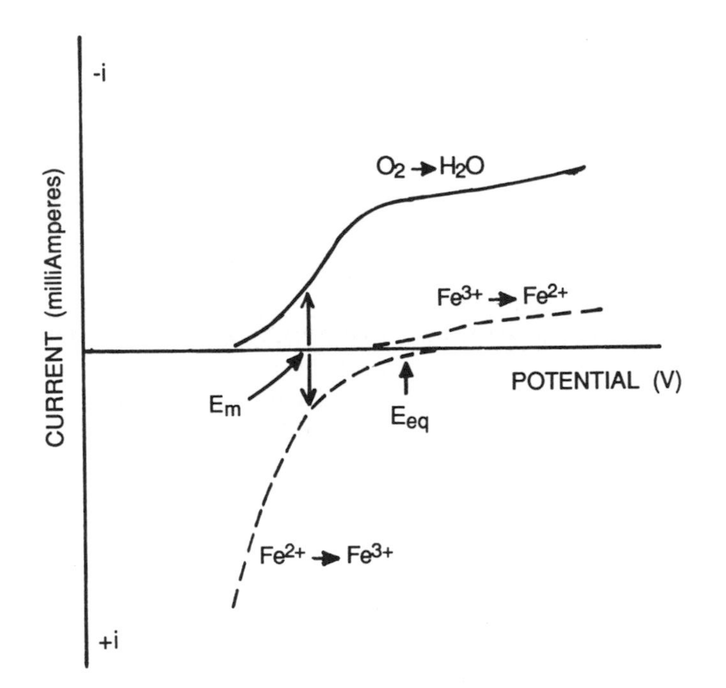

Figure 7.4. Electrode current in aerated water containing Fe^{2+}. The measured "equilibrium" potential is a mixed potential, E_m, obscuring the true equilibrium potential, E_{eq}, of the Fe reaction. The mixed potential is not subject to the usual interpretation because the Fe^{2+}-Fe^{3+} and O_2-H_2O redox pairs are not in equilibrium. (Adapted from W. Stumm and J. J. Morgan. 1981. *Aquatic Chemistry*. 2nd ed. New York: Wiley. Used with permission.)

Table 7.1. This may be why, in aerated soil solutions, measured electrode potentials tend to be around 500 mV, while theoretical O_2 reduction potentials are considerably higher.

We find, then, that the Pt electrode responds well to some redox couples (such as $Fe^{3+}-Fe^{2+}$) and not to others. If all couples were at equilibrium with one another, each would "report" the same potential to the electrode. That is, all redox couples would be leveled to a single E_h value by electron exchange among them. In reality, certain couples fail to enter into this equilibrium because of inert behavior. A good example is the $NO_3^- - N_2$ couple; nitrate is expected to be a strong oxidant by virtue of the high reduction potential of this half reaction (see Table 7.1). However, because this reaction is extremely slow, the theoretical E_h (calculated from equation 7.2) of the $NO_3^- - N_2$ couple has no influence on the redox potentials of soil solutions. Otherwise, redox potentials of soil solutions containing measurable nitrate would always be high. Actual reduction of NO_3^- to N_2 in soils tends to begin at an electrode-measured E_h below 400 mV, much lower than theoretical. Figure 7.5 compares the theoretical E_h of important redox pairs (solid lines) with the electrode-measured E_h range over which these pairs are found to be active in accepting or donating electrons in soils (open and shaded boxes). It is notable that NO_3^- disappears from anaerobic soils at a considerably higher E_h than that at which NH_4^+ appears. The reason for this is the favorability of the denitrification reaction ($NO_3^- \rightarrow N_2$) over NO_3^- reduction to NH_4^+. As a result, only a small fraction of the NO_3^- in a soil is converted to NH_4^+ on waterlogging; most of the nitrogen is lost as gaseous products (N_2O, N_2).

The O_2-H_2O couple is the redox pair controlling reactions in aerated solutions, so that reaeration of anoxic soils drives reduced species (e.g., Fe^{2+}) toward the oxidized state. The range of redox potentials over which Fe^{2+}, Mn^{2+}, and NH_4^+ have been found to oxidize and disappear on aeration of a reduced soil are denoted by the open boxes in Figure 7.5. Nitrate reappearance on aeration is also depicted by an open box. The measured redox potentials that follow re-aeration do not directly reflect the O_2-H_2O equilibrium state but rather the status of redox couples having faster electron exchange rates. Furthermore, while each redox couple would be expected (in theory) to undergo complete conversion to the reduced form (in flooded soils) or to the oxidized form (in re-aerated soils) before the adjacent redox couple on the E_h scale became active, actual behavior in soils is much less ideal. Several redox reactions are typically active simultaneously. This may reflect spatial variability in the aeration (and redox potential) of soil aggregates, caused by slow diffusion processes in micropores.

Large discrepancies between theoretical and measured E_h are obvious in Figure 7.5 for the O_2-H_2O, $NO_3^- - N_2$, and Mn oxide-Mn^{2+} couple. Although the first two are explained by inert behavior of O_2, NO_3^-, and N_2 at the Pt electrode, the discrepancy for the Mn couple may have more to do with an inappropriate choice of Mn oxide (MnO_2) to represent soil Mn in the oxidized state. The exact nature of this Mn is not known, but if it is in the form of Mn oxides, they are likely to be impure, noncrystalline, and of mixed oxidation state (between +3 and +4), with ill-defined reduction potentials.

It is clear, after considering all of the factors likely to affect electrode-measured redox potentials, that these potentials are unlikely to be quantitatively meaningful

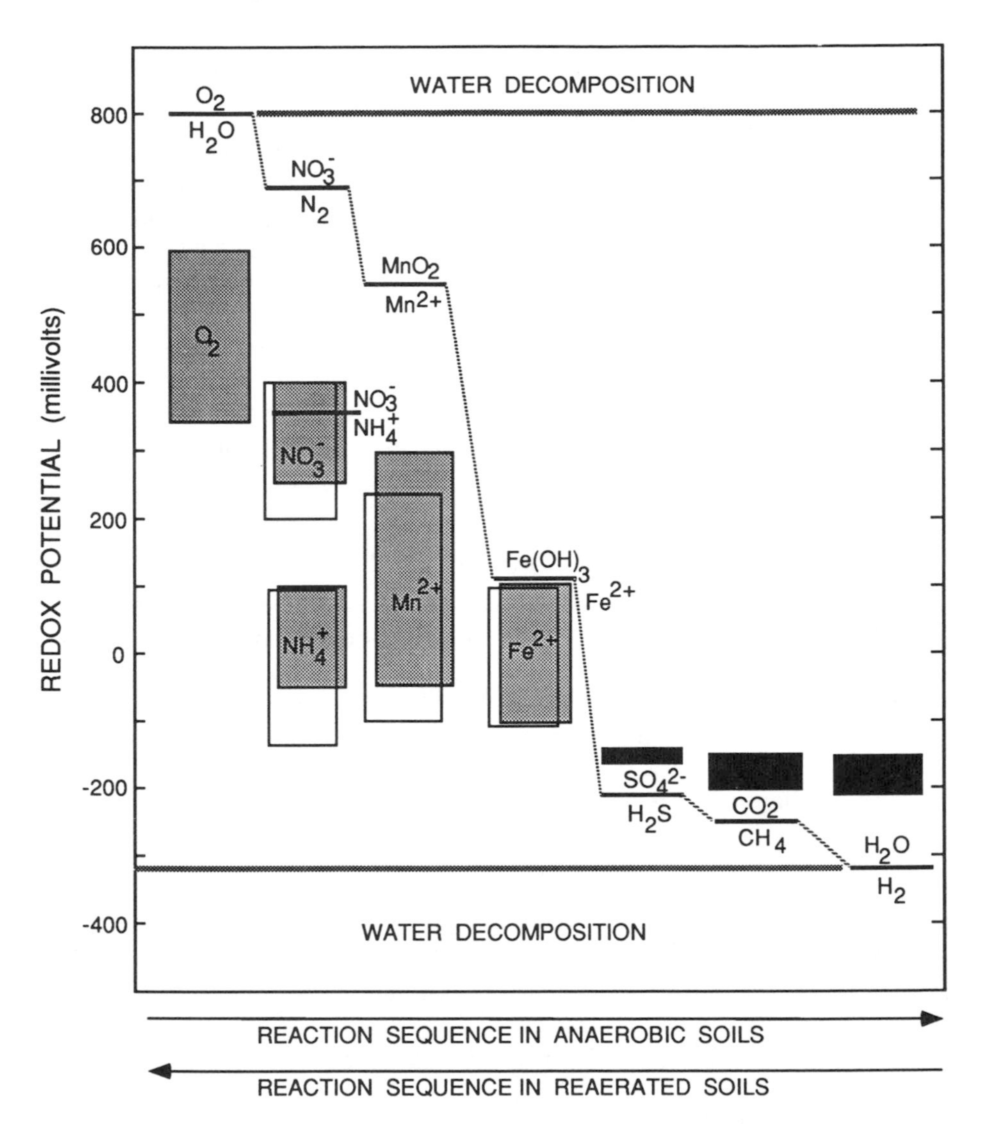

Figure 7.5. The reduction and oxidation sequence in soil solutions at pH 7. Theoretical potentials are indicated by solid lines, assuming equal activities of reduced and oxidized species unless otherwise noted (the pressure of H_2 is arbitrarily set at 10^{-3} atmosphere). Measured ranges of soil potentials over which the indicated species react (change concentration) during soil reduction and oxidation are specified by boxes (shaded for reduction, open for oxidation, black for initial appearance of the reduced form during reduction). (Data, in part, from W. H. Patrick and A. Jugsujinda. 1992. Sequential reduction and oxidation of inorganic nitrogen, manganese, and iron in flooded soil. *Soil Sci. Soc. Am. J.* 56:1071–1073.)

for soil solutions. Consequently, thermodynamic calculations of equilibrium states based only on these measurements should be viewed with suspicion.

7.1d. E_h Measurement by Redox Indicators

Given the problems inherent to the electrode measurement of E_h in soil solution, it would seem desirable to consider other methods of estimating E_h. Theoretically, one could analyze the soil solution for the reduced and oxidized species of a redox couple, say dissolved Fe^{2+} and Fe^{3+}, and use the Nernst relation (equation 7.2) to calculate E_h. Usually, however, there are difficulties with this approach. In the case of the Fe couple, Fe^{3+} solubility in all but very acid soil solutions is extremely low (below detection), so that an assumption must be made about the activity of the free Fe^{3+} ion. It might reasonably be assumed that the solubility product of Fe oxide limits this activity, but Fe oxides have a rather wide range of solubilities depending on oxide crystallinity, structure, and purity. A better approach to measuring E_h would be to use an "indicator" chemical that undergoes reversible electron transfer with natural redox couples in soil solution, that is,

$$Fe^{3+} + \text{indicator (reduced)} = Fe^{2+} + \text{indicator (oxidized)} \qquad (7.10)$$

The color of an ideal indicator would depend on the fraction of this chemical in the reduced and oxidized state, allowing a visual estimation of redox potential in much the same way that pH indicators are used to estimate pH. The quantity of indicator present must be sufficiently small that the electron transfer between indicator and solution does not in itself significantly alter the E_h of the solution.

The oxidized form of the indicator usually reacts with protons upon conversion to the reduced form, as is the case for quinone (Q) reduction to the phenol, hydroquinone (HQ):

$$Q + 2H^+ + 2e^- = HQ \qquad (7.11)$$

The involvement of H^+ ions in most reduction reactions of this kind means that the solution pH as well as E_h affects the fraction of indicator in the reduced and oxidized state. This is made evident by applying the Nernst equation to reaction 7.11:

$$E_h = E_h^0 - 0.059\,\text{pH} - (0.059/2)\log\left[\frac{(HQ)}{(Q)}\right] \qquad (7.12)$$

Since E_h^0, the standard-state potential for the Q—HQ redox couple, is a constant (0.699 volt), equation 7.12 reveals that the ratio of reduced to oxidized indicator at any particular pH is determined by the E_h of the solution. At a fixed pH, for example, this equation becomes

$$E_h = E_h^{0\prime} - (0.059/2)\log\left[\frac{(HQ)}{(Q)}\right] \qquad (7.13)$$

where $E_h^{0\prime}$ is the sum of the first two terms on the right side of equation 7.12. At pH 7, $E_h^{0\prime} = 0.699 - 0.413 = 0.286$ volt. In general, for redox indicators with unknown standard-state potentials, the value of $E_h^{0\prime}$ can be obtained by preparing mixtures containing known amounts of the oxidized and reduced indicator in pH buffers, measuring E_h by the Pt electrode method, and using an equation like 7.13 to calculate $E_h^{0'}$ for that particular pH.

The $E_h^{0'}$ values for a number of indicators useful in measuring redox potentials of soil solutions are listed in Table 7.2. The $E_h^{0'}$ value (adjusted for pH) represents the approximate solution E_h range for which that indicator is diagnostic. Outside of its range, any particular indicator is fully in the reduced or oxidized state; for the Q—HQ indicator, this would mean that the last term in equation 7.13 could not be quantified. In that event, the solution E_h could not be measured by this single indicator; it would be possible to state only that the E_h is *above* the $E_h^{0'}$ value (if the indicator is fully oxidized) or *below* the $E_h^{0'}$ value (if the indicator is fully reduced).

In practice, to measure the redox potential of a solution by an indicator, first the pH must be measured. The $E_h^{0'}$ value is then calculated for the indicator at that pH. Then E_h is estimated by an equation of the same form as equation 7.13 but specific for the indicator being used. This estimation is made by using the color of the solution to gauge the ratio of reduced to oxidized indicator. According to equation 7.13, when the indicator is equally in the reduced and oxidized form, $E_h = E_h^{0'}$.

Suppose that in a particular soil solution all indicators with $E_h^{0'}$ values above +200 millivolts converted to their reduced forms (which tend to be colorless), but all indicators with $E_h^{0'}$ below +200 millivolts remained in the oxidized (colored) form. The E_h of the solution must then have been very near +200 millivolts. Methylene blue, which is intensely colored in the oxidized state, has been used as an indicator of the onset of strongly reducing conditions in soil solutions. It becomes colorless at an E_h of about +11 millivolts, assuming a pH of 7 (see Table 7.2).

Other indicators of the redox state of the soil include complexing ligands (organic and inorganic) that produce different colors in the presence of Fe^{3+} and Fe^{2+}. Ammo-

Table 7.2. Redox Indicators Ranked by Their $E_h^{0'}$ Values at pH 7[a]

Indicator	$E_h^{0'}$ (millivolts)
Phenol blue	+224
o-Cresol Indophenol	+191
Thymol indophenol	+174
1-Naphthol-2-sodium sulfonate indophenol	+123
Toluylene blue	+115
Thionine	+63
Methylene blue	+11
Potassium indigo trisulfonate	−81
Potassium indigo disulfonate	−125
Diazine green	−255
Rosinduline	−281
Neutral red	−325
Benzyl viologen	−359

[a]Caution must be exercised in using these redox indicators because many of them are also pH indicators. This means that their color in the oxidized state may vary depending on the pH of the solution.

Source: P. R. Hesse. 1971. *A Textbook of Soil Chemical Analysis.* New York: Chemical Publishing.

nium thiocyanate, for example, forms a red color in the presence of Fe^{3+}, while o-phenanthroline forms a red complex with Fe^{2+} but not with Fe^{3+}. These are not true redox indicators because they specifically sense the oxidation state of only the iron redox pair. But if the Fe redox couple is in equilibrium with the other redox couples in soil solution (a questionable assumption), such Fe-specific indicators should report a redox status consistent with the E_h measured by true redox indicators.

Practical difficulties have limited the use of redox indicators in soils. In clear, colorless solutions that are well poised (i.e., have significant concentrations of reduced and oxidized species), redox indicators work well. In soils, however, indicators are much more difficult to use. They may adsorb at colloid surfaces or complex with metal cations, profoundly changing their redox properties, color, and detectability. Separating the soil solution from the soil solids by filtration or other methods solves some of these problems, but increases the chance that anaerobic solutions become aerated (unless extreme care is taken to exclude oxygen gas during the process). Fortunately, oxygen reacts rather slowly with many indicators in their reduced form, so that colorimetric detection of redox potential may be possible before aeration compromises the measurement. Nevertheless, the use of redox indicators has not been accepted in soil chemistry as a generally satisfactory method, even though clever experimental designs might circumvent the known drawbacks.

7.1e. The Concept of pϵ in Soil Solutions

Although free electrons do not exist in aqueous solutions, one can still quantify the degree of electron availability, sometimes termed the virtual electron activity, by the parameter pϵ:

$$\mathrm{p}\epsilon = -\log(e^-) \tag{7.14}$$

where (e^-) symbolizes electron activity measured in terms of the tendency of the solution to accept or donate electrons. Reducing solutions have a high value of (e^-), and consequently a low or negative pϵ. Conversely, oxidizing solutions have a low electron activity and a high (positive) pϵ.

For the general case of redox half-reactions:

$$\mathrm{ox} + n\,e^- \leftrightarrow \mathrm{red} \tag{7.15}$$

the virtual "equilibrium" expression[1] is

$$K = \frac{(\mathrm{red})}{(\mathrm{ox})\,(e^-)^n} \tag{7.16}$$

which in logarithmic form, considering the definition of pϵ, becomes

$$\log K - n\,\mathrm{p}\epsilon = \log\left[\frac{(\mathrm{red})}{(\mathrm{ox})}\right] \tag{7.17}$$

1. No such reaction can attain equilibrium in isolation, since at least one other redox pair must be present to donate or accept the electron(s). The "equilibrium constant" should then be thought of as a virtual or operational constant only.

Now the redox potential, E_h, is conventionally defined in terms of the potential of a cell composed of two half-cells: the half-cell of particular interest (given in general terms by reaction 7.15) and the standard hydrogen half-cell (with E_h and pϵ assigned values of zero). The reaction in the latter case is

$$\frac{n}{2} H_2 (g) \leftrightarrow n H^+ + n e^- \tag{7.18}$$

which, when combined with the second half-cell reaction (7.15), gives the overall cell reaction:

$$\frac{n}{2} H_2 (g) + \text{ox} \leftrightarrow n H^+ + \text{red} \tag{7.19}$$

The Nernst equation, applied to this reaction, produces the relationship:

$$E = E^0 - \frac{RT}{nF} \ln \frac{(H^+)^n (\text{red})}{(P_{H_2})^{n/2} (\text{ox})} \tag{7.20}$$

where R is the gas constant and F is the Faraday constant (F = 96,490 coulombs/mole). Because the hydrogen half-cell is defined to be at standard-state conditions, $(H^+) = 1.0$, $P_{H_2} = 1.0$ atmosphere, and the potential of this half-cell is 0.00 volt, meaning that E and E^0 in equation 7.20 can be taken as the measured and standard *half-cell* potential, respectively, of reaction 7.15. But the Nernst equation for half-cell reaction 7.15 is

$$E_h = E_h^0 - \left(\frac{RT}{nF}\right) \ln \left[\frac{(\text{red})}{(\text{ox})}\right] \tag{7.21}$$

Using equation 7.17, equation 7.21 can be reexpressed in terms of the equilibrium constant for the half-cell reaction:

$$E_h = E_h^0 - \left(\frac{2.303\ RT}{nF}\right) (\log K - n\ \text{p}\epsilon) \tag{7.22}$$

But the free energy of reaction, ΔG_r, is related to the reaction potential, E, by

$$\Delta G_r = -nFE \tag{7.23}$$

so that the standard-state free energy is given by

$$\Delta G_r^0 = -nFE^0 \tag{7.24}$$

Thermodynamics provides the relationship between ΔG_r^0 and the equilibrium constant (see Chapter 1):

$$\Delta G_r^0 = -RT \ln K \tag{7.25}$$

so it follows that

$$nFE^0 = RT \ln K \tag{7.26}$$

and consequently

$$E^0 = \frac{(2.303\ RT \log K)}{nF} \tag{7.27}$$

which at room temperature can be written as

$$E^0 \text{ (volts)} = \left(\frac{0.059}{n}\right) \log K \tag{7.28}$$

This provides the useful conversion between standard potentials and equilibrium constants of redox reactions.

Because the standard-state half-cell potential, E_h^0, is measured *relative* to the zero potential of the hydrogen half-cell, $E^0 = E_h^0$, and the definition of E^0 given by equation 7.27 is substituted into equation 7.22 to give

$$E_h = \left(\frac{2.303\ RT}{nF}\right) \log K - \left(\frac{2.303\ RT}{nF}\right) (\log K - n\ \text{p}\epsilon) = \left(\frac{2.303\ RT}{nF}\right) \text{p}\epsilon \tag{7.29}$$

At room temperature, this equation relating pϵ to E_h becomes simply

$$E_h \text{ (volts)} = 0.059\ \text{p}\epsilon \tag{7.30}$$

so that E_h and pϵ for half-reactions are seen to be easily interchangeable, related by the Nernst constant. The choice of E_h (volt units) or pϵ (unitless) in measuring redox status is a matter of preference and convenience.

In soils, pϵ values lie in the range of -6 (strongly reduced) to $+12$ (strongly oxidized). However, these pϵ values (and E_h values as well) are negatively correlated with pH. Such a correlation is expected from consideration of the individual redox reactions occurring in soils (see Figure 7.1).

Using the MnO_2-Mn^{2+} reaction as an example, equation 7.5, written in terms of pϵ, provides the pϵ-pH relationship:

$$\text{p}\epsilon = 20.8 - 0.5 \log (Mn^{2+}) - 2\ \text{pH} \tag{7.31}$$

If the Mn^{2+} activity in soil solution were more or less constant (perhaps controlled by ion exchange processes), then pϵ and pH would, as observed, be negatively correlated. Closer scrutiny of the pϵ-pH relationship in soils, however, reveals that the slope of the function tends to be near 1. This would be predicted if the controlling redox reaction(s) in the soil involved equal numbers of protons and electrons (that is, $m = n$ in reaction 7.1). This is clearly not the case for the MnO_2-Mn^{2+} reaction, whose pϵ-pH function has a slope of 2. For the O_2-H_2O and several other important redox couples in soils, however, the expected slope is at or near unity (Table 7.1 lists redox couples along with the number of protons and electrons consumed by reduction).

For the O_2-H_2O reaction in an aerated soil with $P_{O_2} = 0.2$ atmosphere, the Nernst equation gives

$$E_h = 1.219 - 0.059\ \text{pH} \tag{7.32}$$

or equivalently in terms of pϵ

$$\text{p}\epsilon + \text{pH} = 20.66 \tag{7.33}$$

Actual pϵ-pH correlations obtained by Pt electrode measurements in wet, aerobic soils give a somewhat different relationship:

$$\text{p}\epsilon + \text{pH} \approx 15 \tag{7.34}$$

plotted as the shaded band in Figure 7.6. The slope of this band, but not the intercept, is consistent with behavior of the O_2-H_2O and several other redox half-reactions. It is known that the O_2-H_2O redox couple does not behave in an ideal manner at the Pt electrode surface, so that the discrepancy between the expected and observed pϵ + pH value may be due to Pt electrode potential readings below the theoretical E_h value for aerated water. Despite this discrepancy, it is generally believed that the O_2-H_2O couple controls the pϵ-pH relationship in aerated soil systems.

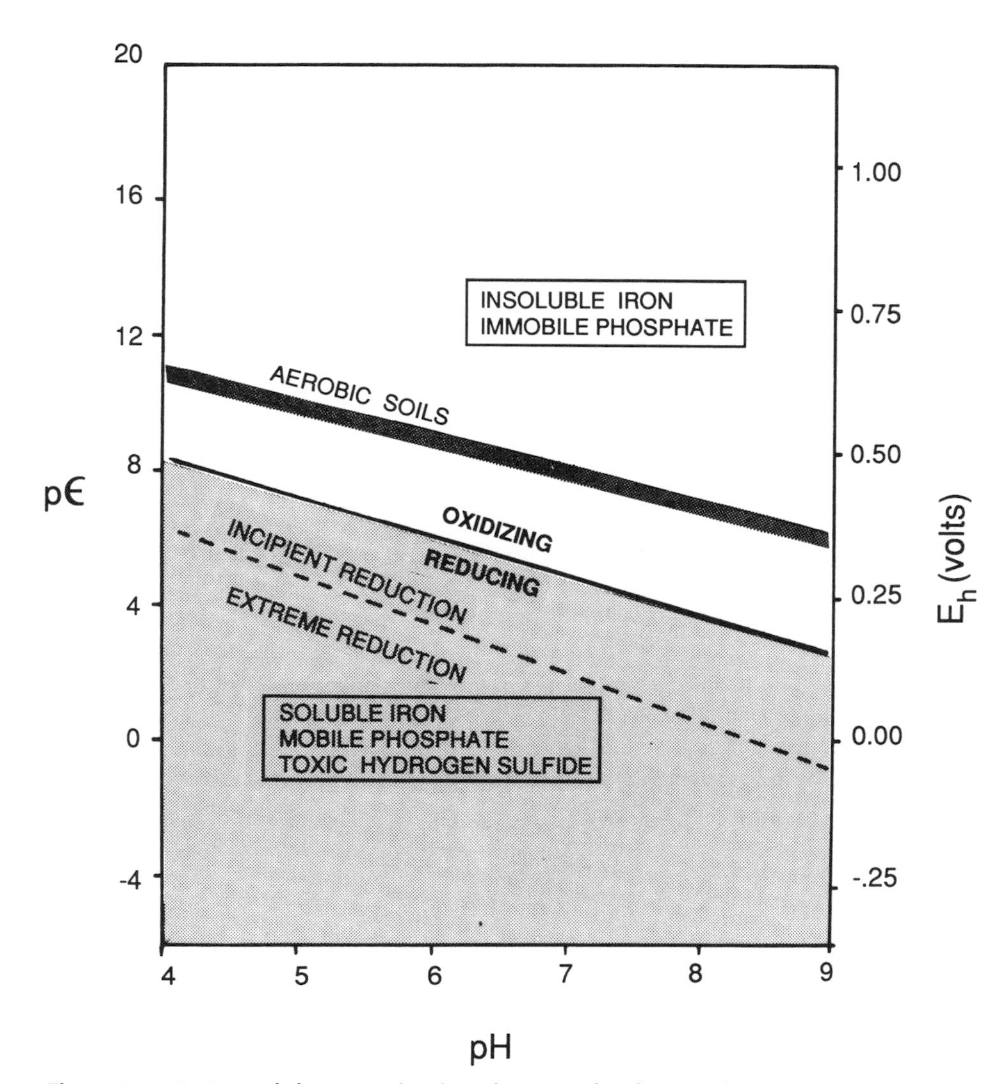

Figure 7.6. Estimated degrees of soil oxidation and reduction, based on pϵ-pH or E_h-pH values. The broad line labeled "aerobic soils" is the pϵ-pH relationship measured by electrode in aerated soils. (Data from P. R. Hesse. 1971. *A Textbook of Soil Chemical Analysis.* New York: Chemical Publishing; and W. L. Lindsay. 1979. *Chemical Equilibria in Soils.* New York: Wiley.)

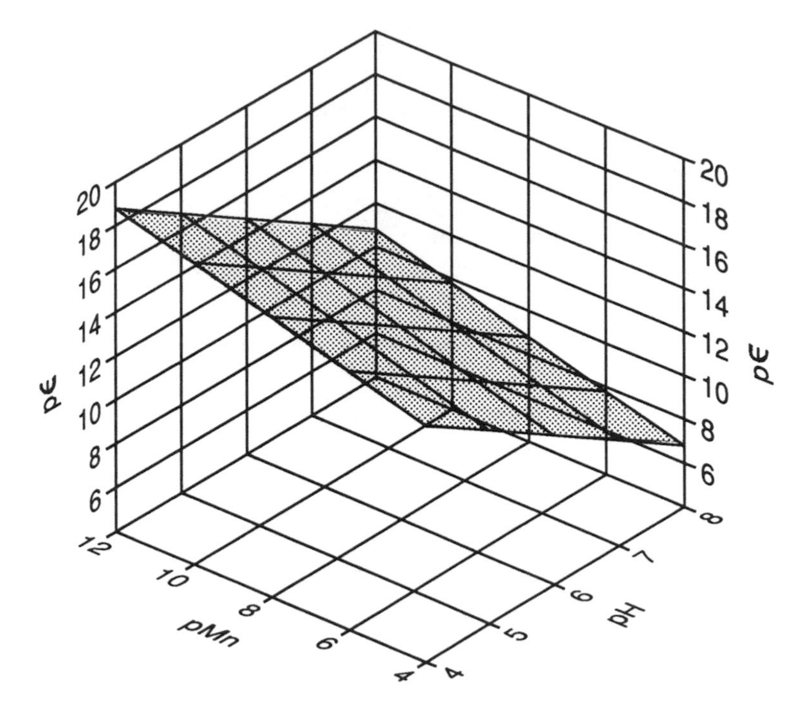

Figure 7.7. Theoretical dependence of pε on pH and dissolved Mn^{2+} activity (pMn = $-\log(Mn^{2+})$).

Holding the pH of a redox reaction system constant does not necessarily fix the pε of the system, although equations such as 7.33 and 7.34 suggest this. In the case of the MnO_2-Mn^{2+} couple, the Mn^{2+} activity influences the pε-pH relation according to equation 7.31. The Mn^{2+}-pε-pH relation is depicted graphically in Figure 7.7, indicating that, at a given pH, higher Mn^{2+} activity results in a lower pε of solution. Consequently, the oxidative strength of Mn oxides is lowered as Mn^{2+} is dissolved and builds up in the process of oxidizing other chemical species. Similarly, lower O_2 partial pressures in poorly aerated soils reduce the pε according to the equation:

$$p\epsilon = 20.8 - pH + 0.25 \log P_{O_2} \qquad (7.35)$$

In general, as reaction products of spontaneous redox reactions accumulate, they lower the pε of the redox couple that provides the oxidant and raise the pε of the redox couple that provides the reductant. When the two pε values converge, equilibrium is reached.

7.2. CHEMISTRY OF IMPORTANT REDOX COUPLES IN SOILS

7.2a. The H_2O-O_2 Couple

The four-electron reduction of O_2 to water has a very favorable potential of +1.229 volts (see Table 7.1), making O_2 a powerful oxidant, at least in principle. Two-electron reduction to hydrogen peroxide (H_2O_2) is much less favorable (E_h = +0.68

volt). As the subsequent two-electron reduction of H_2O_2 to H_2O is slow, the effective redox potential of O_2 may be only + 0.68 volt. In biologically mediated oxidations, enzymes catalyze what appears to be synchronous two- and four-electron reduction of O_2, and the O_2-H_2O and $O_2-H_2O_2$ reaction potentials may be relevant. In chemical oxidations, however, there is no evidence for reduction of O_2 in either four- or two-electron steps. Instead, oxidation of substrates by O_2 inevitably proceeds through a series of one-electron steps, which means that the full oxidative potential of O_2 is not necessarily exploited.

The one-electron reduction of O_2 to the superoxide radical ion, O_2^-, is unfavorable, with a potential of −0.56 volt. This means that single electron oxidation of molecules by O_2 itself is unlikely. The superoxide ion is a moderately effective *reducing* agent, but a poor oxidizing agent. However, the superoxide ion reacts with water to form peroxide ions, HO_2^- and O_2^{2-}. These are thought to be the operative oxidizing species in aerated water.

There are further reasons that O_2 demonstrates a surprisingly weak ability to oxidize organic molecules in spite of its high reduction potential. In its ground (unexcited) electronic state, O_2 has an electron spin of $S = 1$ (referred to as the triplet state, caused by two unpaired electrons). It can be excited by ultraviolet light to the $S = 0$ (singlet) state, in which there are no unpaired electrons. Most organic substrates are also in the singlet electronic state under normal conditions, and since electron transfer between molecules in different spin states is a spin-forbidden process, molecular oxygen in its normal (unexcited) electronic state is not expected to react with most organic molecules. In contrast, singlet oxygen, formed by photochemical excitation of triplet oxygen, is an extremely potent organic oxidant but is of very limited importance in soil chemistry because ultraviolet light is a factor only at the soil surface.

Molecular oxygen, if bonded to certain transition metal ions in solution or at oxide surfaces, can be more reactive than dissolved free oxygen. This property may contribute to the catalytic activity of metal oxide minerals in oxidation reactions.

All of these considerations of O_2 reduction mechanisms suggest that the four-electron reduction of O_2, with its very favorable potential, overestimates the oxidative power of O_2 in chemical systems. This should be kept in mind in the following discussions of redox reactions if O_2 is the oxidant.

Strong oxidation potentials can arise in aqueous solutions from sources other than dissolved O_2, at least if these solutions are irradiated with light. Low-energy ultraviolet light (wavelength $\approx$ 300 nm) induces the photoreduction of hydrolyzed Fe^{3+} species such as dissolved $FeOH^{2+}$ or colloidal Fe hydroxide by the $O^{2-} \rightarrow Fe^{3+}$ electron transfer. This reaction produces Fe^{2+} and $\cdot OH$, the hydroxy radical. The hydroxy radical is one of the most powerful and reactive oxidants known to occur in water, oxidizing most natural organics (carboxylic acids, phenols, etc.) and many metals.

7.2b. The Mn^{2+}—Mn Oxide System

Microorganisms can facilitate and use energy from redox reactions (without a net cost in energy) only if these reactions are favored thermodynamically. Microbes catalyze, via enzyme systems, the reduction of O_2 to H_2O. In the process, they may directly or indirectly promote Mn from oxidation state +2 to +3 or +4. In effect, the oxidative power of O_2 is converted into the form of Mn(+3,+4) oxides. Mn oxides are perhaps the most potent oxidizing solids in soils.

Chemically, oxidation of Mn^{2+} occurs spontaneously in alkaline aerated solutions, as the initially precipitated manganous hydroxide rapidly reacts with O_2 to produce a variety of oxide products depending on pH, the concentration of O_2, presence of cations, and other factors:

$$Mn^{2+}\text{ (soluble)} \xrightarrow{OH^-} Mn(OH)_2\text{ (s)} \xrightarrow{O_2} \begin{matrix} Mn_3O_4\text{ (s)} \\ MnOOH\text{ (s)} \\ MnO_2\text{ (s)} \end{matrix} \qquad (7.36)$$

Although Mn^{2+} in solution is thermodynamically unstable with respect to oxidation by O_2 at any pH higher than 4, the *rate* of uncatalyzed oxidation is not appreciable unless the pH is well above 8. The Mn oxidation reaction is *autocatalytic;* that is, the initial oxidation step:

$$Mn^{2+} + \tfrac{1}{2}O_2 \underset{\text{slow}}{\rightarrow} MnO_2\text{ (s)} \qquad (7.37)$$

is followed by selective adsorption of Mn^{2+} onto the freshly formed Mn oxide:

$$Mn^{2+} + MnO_2\text{ (s)} \underset{\text{fast}}{\rightarrow} Mn^{2+}\cdot MnO_2\text{ (s)} \qquad (7.38)$$

and the adsorbed Mn^{2+} is then oxidized relatively quickly:

$$Mn^{2+}\cdot MnO_2\text{ (s)} + \tfrac{1}{2}O_2 \underset{\text{fast}}{\rightarrow} 2\ MnO_2\text{ (s)} \qquad (7.39)$$

Consequently, as Mn oxides precipitate, oxidation accelerates in response to the increased surface available for selective adsorption of Mn^{2+}.

In soils, autocatalytic oxidation of Mn^{2+} by Mn oxides seems to be a mechanism by which the formation of Mn oxides can be explained at pH values well below 8. There is also evidence that Mn and Fe tend to co-precipitate in oxides, possibly because iron oxide surfaces also catalyze the oxidation of Mn^{2+}. In any event, the oxidizing agent for Mn^{2+} is molecular oxygen, either by a direct or indirect reaction path. The reason for this is clear from Figure 7.5. Except for NO_3^-, which tends to be kinetically inert, O_2 is the only sufficiently strong and common oxidant in soil to be able to oxidize Mn^{2+}.

Example Problem: What is the equilibrium concentration of Mn^{2+} in an aerated soil solution at pH 4, assuming that MnO_2 is the solid phase formed by Mn oxidation?

Solution: From Table 7.1, the relevant half-reactions might be:

$$\tfrac{1}{2}MnO_2\text{ (s)} + 2H^+ + e^- = \tfrac{1}{2}Mn^{2+} + H_2O \qquad E_h^0 = 1.23\text{ V}$$

$$\tfrac{1}{2}H_2O = \tfrac{1}{4}O_2\text{ (g)} + H^+ + e^- \qquad E_h^0 = -1.229\text{ V} \qquad (7.40)$$

since Mn^{2+} in aerated soil solutions can be assumed to be oxidized by dissolved molecular oxygen. If these two half-reactions are coupled (in equilibrium), the overall reaction is written as their sum:

$$\tfrac{1}{2}MnO_2\text{ (s)} + H^+ = \tfrac{1}{2}Mn^{2+} + \tfrac{1}{4}O_2\text{ (g)} + \tfrac{1}{2}H_2O \qquad E^0 = 0.001\text{ V} \qquad (7.41)$$

The Nernst relation for this reaction at room temperature is

$$E = E^0 - 0.059 \log \frac{(P_{O_2})^{1/4} (Mn^{2+})^{1/2}}{(H^+)} \quad (7.42)$$

which becomes

$$E = 0.001 - 0.01475 \log P_{O_2} - 0.0295 \log (Mn^{2+}) - 0.059 \text{ pH} \quad (7.43)$$

Now for aerated water at pH 4, $P_{O_2} \approx 0.2$ atmosphere, and equation 7.43 simplifies to

$$E = -0.225 - 0.0295 \log (Mn^{2+}) \quad (7.44)$$

Since this redox system is taken to be at equilibrium, there is no potential for the reaction to do work; that is, $E = 0$ (or equivalently, from equation 7.23, $\Delta G = 0$). Thus equation 7.44 can be used to solve for the equilibrium activity of Mn^{2+}:

$$(Mn^{2+}) = 10^{-7.63} \quad (7.45)$$

This means that the solution concentration of free Mn^{2+} ions at pH 4 should be less than 10^{-7} molar (neglecting corrections for activity coefficients). But much higher concentrations than this are usual in aerated soil solutions; it appears that equilibrium between Mn oxides and dissolved O_2 is not the usual situation. Disequilibrium could be attributed to the sluggish participation of O_2 in electron transfer reactions despite its potential to force reaction 7.41 strongly to the left. However, other explanations for the higher than expected Mn^{2+} concentrations in soils are possible, including:

1. Reduced zones within the generally aerated soil matrix
2. Control of redox reactions by forms of oxidized Mn that are quite different from MnO_2
3. Failure of reaction 7.40 to measure the operative chemical oxidation potential of molecular oxygen (see discussion about O_2 in section 7.2a)
4. Coupling of the Mn reduction to kinetically active reducing agents such as polyphenols.

The predominant stable forms of Mn in soils under different prevailing solution conditions of pH and E_h can be summarized in a pH-E_h or pH-pϵ diagram. For example, equation 7.31 provides a relationship between pϵ and pH for any particular activity of Mn^{2+} in solution. If a realistic Mn^{2+} activity, say 10^{-5}, is chosen to represent soil solution, equation 7.31 becomes simply

$$p\epsilon = 23.3 - 2 \text{ pH} \quad (7.46)$$

Plotted on a pϵ-pH axis system, this is a straight line specifying the redox and acidity conditions in the solution that allow MnO_2 and approximately 10^{-5} M Mn^{2+} to coexist in equilibrium. Below this line (lower pϵ), Mn^{2+} concentrations are higher than 10^{-5}; above this line (higher pϵ), Mn^{2+} concentrations are lower. In Figure 7.8, this line, along with other lines separating different chemical forms of Mn, is plotted. Areas enclosed by lines envelope pϵ and pH conditions favorable to the formation of specific chemical forms of manganese. It is apparent from the figure that the predom-

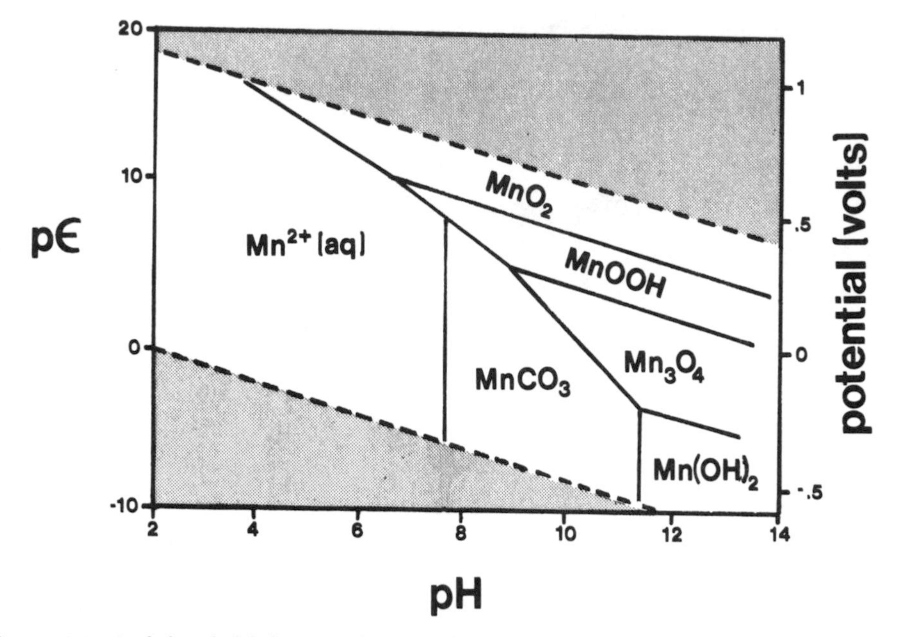

Figure 7.8. Stability field diagram for soluble and solid forms of Mn under a range of redox potential and pH conditions, assuming a dissolved CO_2 concentration of 10^{-3} *M*. Shaded areas delineate conditions that are unattainable because of water decomposition. (Adapted from W. Stumm and J. J. Morgan. 1981. *Aquatic Chemistry.* 2nd ed. New York: Wiley.)

inant form of Mn below pH 7 should be soluble Mn^{2+} unless the solution is strongly oxidizing (aerated). At higher pH, Mn^{2+} is expected to precipitate as $MnCO_3$ under reducing conditions, or to oxidize to an insoluble oxide under oxidizing conditions.

7.2c. The Fe^{2+}—Fe Oxide System

Ferrous iron (Fe^{2+}) appears later than Mn^{2+} in soils that have been subjected to prolonged waterlogging because, as Table 7.1 shows, the reduction potential of Fe^{3+} in oxides (and probably in many other soil minerals as well) is lower than that of Mn(+3,+4) in Mn oxides. Since Fe^{2+}, like Mn^{2+}, is rather soluble, it can reach appreciable concentrations in poorly aerated soil solutions. The introduction of dissolved oxygen causes rapid oxidation of Fe^{2+} and precipitation of ferric hydroxide if the solution pH is much higher than 6. The rate law of oxidation of dissolved Fe^{2+} is known to be

$$\frac{-d\,[Fe^{2+}]}{dt} = k[Fe^{2+}]\,[OH^-]^2\,P_{O_2} \tag{7.47}$$

where $k \approx 8.0 \times 10^{13}$ $min^{-1}atm^{-1}mole^{-2}liter^2$ at 20°C. The sensitivity of oxidation rate to pH is evident from this equation; Fe^{2+} persists for no more than a few minutes in aerated solutions of pH 7 or higher.

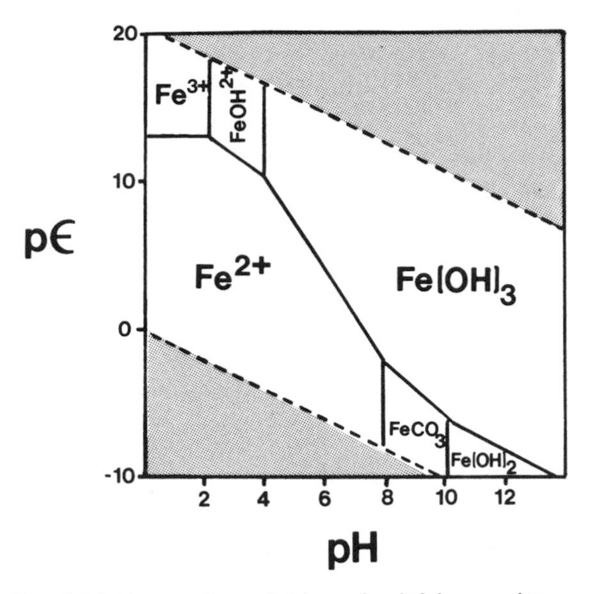

Figure 7.9. Stability field diagram for soluble and solid forms of Fe under a range of redox potential and pH conditions, assuming a dissolved CO_2 concentration of 10^{-3} *M*. Shaded areas delineate conditions that are unattainable because of water decomposition. (Adapted from W. Stumm and J. J. Morgan. 1981. *Aquatic Chemistry*. 2nd ed. New York: Wiley.)

The pϵ-pH diagram in Figure 7.9 depicts the thermodynamically stable forms of Fe under a range of conditions in solution. This shows that, if reducing conditions prevail in solution, soluble Fe^{2+} should be stable only if the pH is below about 8. If the pH is higher than 8, as may be the case in calcareous or sodic soils, solid $FeCO_3$ would be the predominant form of iron. Under oxidizing conditions, the highly insoluble ferric hydroxides and oxides are the most stable forms of iron over a wide pH range.

Example Problem: Since O_2 reacts directly with Fe^{2+} to limit iron solubility in aerated water, what is the equation relating Fe^{2+} activity to pH in aerobic soils?

Solution: From Table 7.1, the relevant half-reactions are

$$\mathrm{Fe(OH)_3\,(s) + 3H^+ + e^- = Fe^{2+} + 3H_2O} \qquad E_h^0 = 1.057\ \mathrm{V}$$

$$\tfrac{1}{2}\mathrm{H_2O} = \tfrac{1}{4}\mathrm{O_2\,(g) + H^+ + e^-} \qquad E_h^0 = -1.229\ \mathrm{V} \tag{7.48}$$

If these two half-reactions are coupled, the overall reaction is written as their sum:

$$\mathrm{Fe(OH)_3\,(s) + 2H^+ = Fe^{2+}} + \tfrac{1}{4}\mathrm{O_2\,(g)} + \tfrac{5}{2}\mathrm{H_2O} \qquad E^0 = -0.172\ \mathrm{V} \tag{7.49}$$

From the Nernst equation for this reaction at room temperature:

$$E = E^0 - 0.059 \log \frac{(P_{O_2})^{1/4}\,(\mathrm{Fe^{2+}})}{(\mathrm{H^+})^2} \tag{7.50}$$

which becomes

$$E = -0.172 - \tfrac{1}{4}(0.059) \log P_{O_2} - 0.059 \log (Fe^{2+}) - 2(0.059)pH \quad (7.51)$$

Since $E = 0$ at equilibrium, and aerated conditions allow P_{O_2} to be assigned a partial gas pressure of about 0.2 atmosphere, equation 7.51 can be simplified to give the desired relationship between Fe^{2+} activity and pH:

$$\log (Fe^{2+}) = -2.74 - 2\,pH \quad (7.52)$$

This result shows that, even in acid soil solutions, the equilibrium activity of Fe^{2+} would be far below detectable levels if the soil is aerated. Furthermore, as is easily shown by equation 7.51, traces of dissolved oxygen in solution would be sufficient to keep the Fe^{2+} concentration well below detection. That these predictions are not borne out in soils may suggest, as for the Mn case, that equilibrium is not usually attained with O_2 in soil solution and that soluble organics may reduce Fe^{3+} even when O_2 is present. Alternatively, reduced zones may exist in aerated soils because of slow O_2 diffusion into micropores, or the O_2-H_2O half-reaction may not be the appropriate one to express the oxidative potential of O_2 (see discussion in section 7.2a).

Various natural polyphenols in soils, including those in humic acids, can reduce Fe^{3+} to Fe^{2+}, for example:

$$Fe^{3+}\text{-humic complex} \rightarrow Fe^{2+} + \text{oxidized humic} \quad (7.53)$$

Processes such as this one maintain detectable concentrations of Fe^{2+} in many soil solutions despite the presence of dissolved O_2. Commonly, though, Fe^{2+} reacts *faster* with oxygen when these humics are present than when they are not, that is,

$$Fe^{2+} + \tfrac{1}{4}O_2 + \text{humic} \rightarrow Fe^{3+}\text{-humic complex} \quad (7.54)$$

When reactions 7.53 and 7.54 are combined, it is seen that no net change in the oxidation state of Fe has occurred. Instead, the $Fe^{3+}-Fe^{2+}$ system has merely acted as a catalyst for the oxidation of the humic acid by O_2.

Metal-complexing ligands, including humic and fulvic acids, by preferentially complexing with Fe^{3+} or Fe^{2+} actually shift the redox potential of soil solutions. This fact can be illustrated simply by considering the Nernst equation for the Fe^{3+}/Fe^{2+} redox couple (see Table 7.1):

$$E_h = E_h^0 (Fe^{3+}-Fe^{2+}) - 0.059 \log \left[\frac{(Fe^{2+})}{(Fe^{3+})}\right] \quad (7.55)$$

The redox potential of solution, E_h, is solely determined by the Fe^{2+}/Fe^{3+} activity ratio in this case. If the pH is kept very low so that dissolved Fe^{3+} does not hydrolyze to form Fe-hydroxy species, then this activity ratio in solution is approximated closely by the ratio of the total dissolved Fe^{2+} to Fe^{3+}. Once a complexing ligand is put in solution, the activity of one of the metal ions is lowered relative to the other, and the redox potential changes. For example, fluoride complexes much more strongly with Fe^{3+} than with Fe^{2+}, lowering the activity of Fe^{3+} relative to Fe^{2+}. According to equation 7.55, the consequence is a *decrease* in the redox potential. In chemical terms, this means that the tendency of Fe^{3+} to be reduced is diminished because complexation stabilizes the oxidized state of iron. Most natural organic complexing agents in soil solutions, being "hard" bases with oxygen ligands, would similarly stabilize oxidized iron and decrease the redox potential.

Realistically, it should be recognized that precipitated forms of Fe^{3+} are present in the soil in addition to soluble Fe^{3+} and Fe^{2+}. That is, the more appropriate half-reaction is the $Fe^{2+}-Fe(OH)_3$ couple (see Table 7.1). For this couple, the Nernst equation gives

$$E_h = E_h^0\,(Fe(OH)_3-Fe^{2+}) - 0.059 \log\left[\frac{(Fe^{2+})}{(H^+)^3}\right] \qquad (7.56)$$

Now we find that the redox potential is pH dependent because the Fe^{3+} activity is controlled by the solubility product of $Fe(OH)_3$ (or some other very insoluble mineral of Fe^{3+}). If the solution pH is increased, precipitation stabilizes the oxidized form of iron and the redox potential of the Fe system is lowered. If a ligand capable of forming soluble complexes with Fe^{3+} is added to the system, it will dissolve part of the $Fe(OH)_3$, increasing the total soluble Fe^{3+}. However, because this dissolution has no effect on the activity of the free Fe^{3+} and Fe^{2+} ions (free Fe^{3+} is controlled by solubility of the ferric oxide), the redox potential in this case is unaffected by the complexing ligand—at least, once equilibrium is established.

Fluctuating water tables are common in many soils, creating alternating aerobic and anaerobic conditions. During wet periods, iron oxides undergo reduction, with organic matter (directly or indirectly) supplying the electrons:

$$Fe(OH)_3 \underset{CO_2}{\overset{e^-}{\rightarrow}} Fe(HCO_3)_2 \qquad (7.57)$$

The Fe^{2+} ions may then occupy a significant fraction of exchange sites on the soil colloids (clays and humus):

$$Fe(HCO_3)_2 + Ca^{2+}\text{-colloid} = Fe^{2+}\text{-colloid} + Ca(HCO_3)_2 \qquad (7.58)$$

Because $Ca(HCO_3)_2$ is a soluble salt that can be leached out of the soil, there may be a tendency for exchangeable Fe^{2+} to accumulate as base cations are lost. Once the soil drains, aerobic conditions again prevail, and Fe^{2+} oxidation generates acid soils:

$$Fe^{2+}\text{-colloid} \underset{O_2}{\rightarrow} Fe(OH)_3\,(s) + 2H^+\text{-colloid} \qquad (7.59)$$

The overall process of soil acidification by alternating iron reduction and oxidation, diagrammed in Figure 7.10, is termed *ferrolysis.* A hardpan of iron oxide may build up at the interface of the aerobic and anaerobic zones if the water table tends to perch at a particular position in the soil profile. Magnetite, Fe_3O_4, a black oxide containing iron in both the reduced and oxidized state, is sometimes formed at this interface. In the Fe stability diagram (Figure 7.9), magnetite would appear above pH 6 over a narrow range of $p\epsilon$ at the interface of the reduced and oxidized forms of Fe.

The accumulation of acidity in the surface soil by ferrolysis is a localized process, enabled by the spatial separation of acid-generating Fe^{2+} from the alkaline bicarbonate ion. It occurs only where drainage permits bicarbonate to leach through the soil profile.

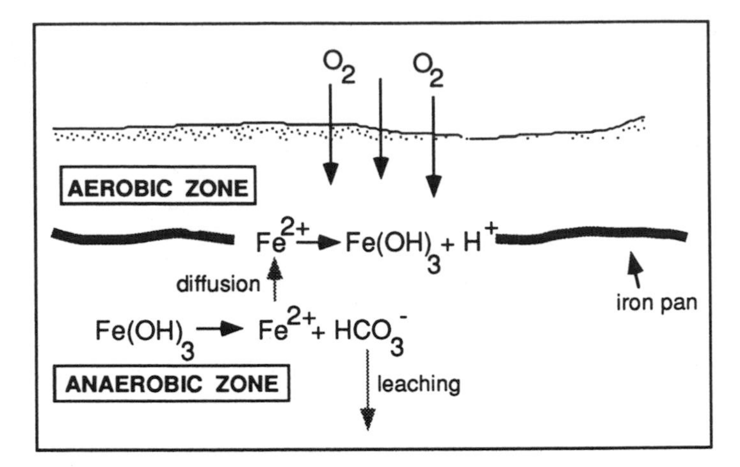

Figure 7.10. Schematic description of ferrolysis in a soil with a perched water table.

7.2d. The Carbon Cycle

Solar radiation, by the process of photosynthesis,[1] provides the energy to create simultaneously within plant cells both molecular oxygen (positive $p\epsilon$) and pockets of strongly reducing conditions (negative $p\epsilon$). This is a state of extreme disequilibrium at the subcellular level. The localized reducing conditions converts CO_2 to reduced organic compounds possessing high energy C—H (as well as organic N, S, and P) bonds. In effect, solar energy raises the oxidation state of oxygen atoms while lowering that of carbon atoms, as illustrated schematically in Figure 7.11. Nonphotosynthetic organisms then use O_2 to oxidize the energy-rich organics, returning the system to "equilibrium."

In aerobic soils, in fact, the only stable forms of carbon are CO_2, HCO_3^-, and CO_3^{2-}; all soil organic matter is potentially susceptible to oxidation by O_2. While the persistence of humus in soils for years, even centuries, may seem to belie this statement, the redox reaction moves slowly but inexorably in the direction of equilibrium. The reduced forms of carbon in the soil organic matter provide the energy (and electrons) that drives the "engine" of chemical reduction under water-saturated conditions.

7.2e. The Nitrogen Cycle

The most stable chemical form of nitrogen in aerated soil solutions is nitrate (NO_3^-), yet N_2 gas is prevalent in the air of soil pores. Oxidation of N_2 to nitrate, a thermodynamically favored reaction in aerated soils, is made difficult by the high bond order of N_2. The N—N triple bond has a formidably high dissociation energy of 225 kcal/mole, and biological organisms have not evolved enzymatic systems to

1. Photosynthesis is a complex sequence of catalyzed redox reactions, using the energy of photons absorbed by plant pigments. Clusters of Mn atoms in +3 and +4 oxidation states are now known to catalyze the water splitting reaction that generates O_2.

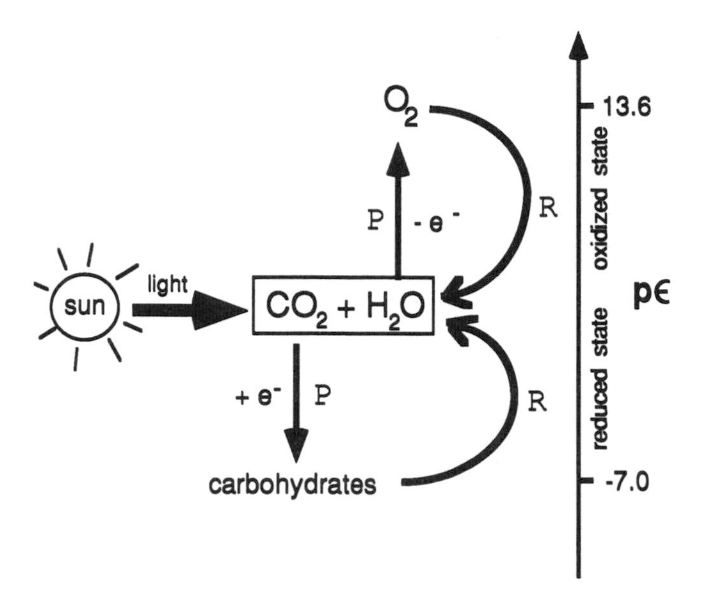

Figure 7.11. Schematic description of the carbon cycle, placing respiration (R) and photosynthesis (P) on a redox scale.

achieve the difficult oxidation to NO_3^-. On the other hand, the reverse reaction, reduction of NO_3^- to N_2, is favored under moderately reducing conditions ($-4 < p\epsilon < 12$). Termed *denitrification,* it proceeds by an indirect pathway through intermediates including nitrite (NO_2^-) and N_2O. Denitrification accounts for significant losses of nitrogen from wet, but not necessarily flooded, soils.

Under strongly reducing conditions ($p\epsilon < -4$), N_2 reduction to ammonium (NH_4^+) is thermodynamically possible, and when it occurs as an enzyme-catalyzed reaction in plants and algae is referred to as *nitrogen fixation.* In these N-fixing organisms, extreme reducing conditions are maintained at the subcellular level to permit the enzymatic system to catalyze N_2 reduction. Enabling this reaction is costly to the plant in terms of the energy invested. With the high energy required to break the triple N—N bond, only a few organisms have evolved with enzymes capable of reducing N_2.

Although chemical reduction of N_2 to ammonium by reaction with H_2 gas is thermodynamically possible at room temperature, the reduction proceeds only at very high temperature. Consequently, abiotic nitrogen fixation is not feasible even in strongly reduced soils.

The various nitrogen redox reactions are depicted in Figure 7.12, illustrating the oxidation state of the possible chemical forms.

7.2f. The Sulfur Cycle

In aerobic soils, sulfate (SO_4^{2-}) is the stable form of sulfur. Strongly reducing conditions cause sulfate to be reduced biologically to hydrogen sulfide (H_2S), a foul-smelling gas that dissolves readily in water according to the reaction:

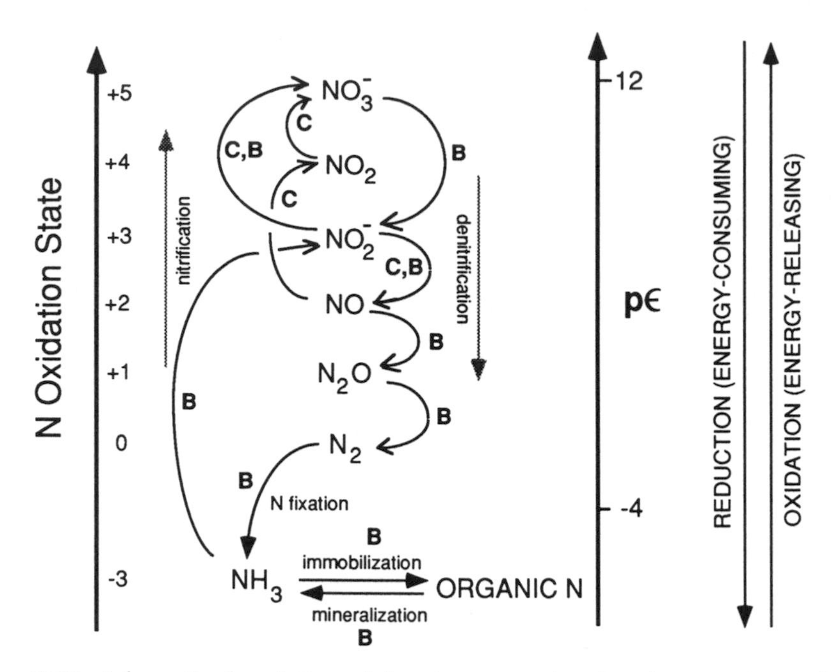

Figure 7.12. Schematic description of the nitrogen cycle, placed on a redox scale.

$$H_2S\ (g) = H_2S\ (aq) \qquad K = 10^{-0.99} \tag{7.60}$$

and dissociates in nonacid solutions:

$$H_2S\ (aq) = H^+ + HS^- \qquad K = 10^{-7.02} \tag{7.61}$$

A further dissociation step produces the sulfide anion:

$$HS^- = H^+ + S^{2-} \qquad K = 10^{-13.9} \tag{7.62}$$

which forms highly insoluble precipitates in anaerobic soils by reaction with Fe^{2+} and a number of other transition and heavy metals. For example, the reaction:

$$Fe^{2+} + S^{2-} = FeS\ (s) \qquad K = 10^{18.1} \tag{7.63}$$

limits Fe^{2+} solubility to very low levels in strongly reduced soils, whereas in moderately reduced soils, *siderite* ($FeCO_3$) may limit solubility to a degree determined by the CO_2 partial pressure.

Submerged soils commonly contain iron sulfides, or *pyrites,* with formulas ranging from FeS to FeS_2, which, once the soils are exposed to air by a fluctuating water table or drainage, begin to oxidize by biological and chemical means. The reaction pathway in soils and sediments submerged in sulfate-rich water, typically found on coastal margins of oceans, is illustrated in Figure 7.13. The acidity generated by sulfide oxidation, once these sediments are drained, can produce extremely low pH (3.0–3.5) along with sulfate minerals such as jarosite and gypsum. Soils showing evidence of acidifying weathering reactions and the associated sulfate minerals are called *acid sulfate soils.* Although usually located on coastal margins, these soils can occur

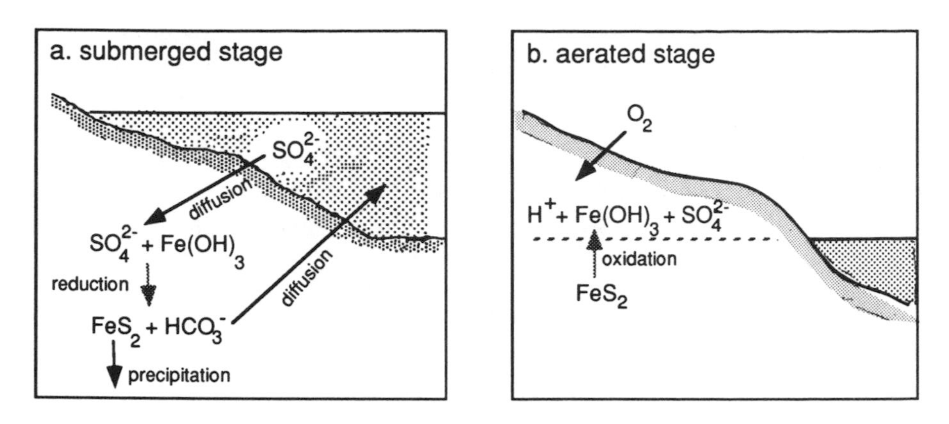

Figure 7.13. Process of sulfide accumulation (submerged stage) and acid sulfate generation (aerated stage) on a seacoast.

inland because pyrite-rich sedimentary deposits may be located along old coastlines. Furthermore, mining operations often bring pyritic tailings to the surface where oxidation generates extreme acidity in the form of sulfuric acid (see Figure 7.13b). Leachate from such sites poses a hazard because the acid dissolves and mobilizes toxic heavy metals in the rock (e.g., lead, cadmium, copper).

In semiarid climates, sulfate deposits in the soil are sometimes reduced under conditions of poor drainage. The conditions necessary for reduction are created in natural basins where water accumulates due to the very low water permeability of sodic soils. The sulfate reduction reaction:

$$Na_2SO_4 \rightarrow NaHCO_3 + H_2S\ (g) \tag{7.64}$$

allows hydrogen sulfide to escape to the atmosphere, while alkalinity in the form of $NaHCO_3$ accumulates as water evaporates at the soil surface. In this situation of poor drainage, the sulfate-sulfide reduction reaction has the overall effect of adding alkalinity to the soil. (The reverse reaction of iron sulfide oxidation adds acidity, but occurs in situations of better drainage.)

Wetlands of humid climatic zones often emit H_2S, as is evident from the rotten-egg odor of marshes and swamps. As long as the dominant exchangeable base cations are Ca^{2+} and Mg^{2+}, which is the case in most freshwater wetlands, H_2S formation should not cause the soils to become strongly alkaline. In these nonsodic soils, alkalinity generated by reduction forms precipitates of Ca (and Mg) carbonates. The low solubility of these carbonates prevents the pH from rising much above 8. In sodic soils, however, reaction 7.64 causes alkalinity to build up in the form of soluble Na carbonates (see Chapter 8, Section 8.1).

7.3. REDOX REACTIONS AT MINERAL SURFACES

Clays and oxides have the ability to promote certain redox reactions such as the oxidation of phenols and aromatic amines. These are surface reactions, involving an adsorption step and electron transfer step, unlike many of the chemical reactions discussed in the previous sections of this chapter which were assumed to be redox processes of homogeneous solutions. Soil minerals known to be reactive as electron acceptors include ferric oxides, manganese oxides, and Fe^{3+}-bearing layer silicate clays.

Surface-localized redox reactions can be viewed as the transfer of an electron between one particular surface metal ion and an adsorbed molecule, with a change in the oxidation state of the metal ion. This is a reasonable description if electrons have no mobility in the mineral; that is, if the mineral is an *insulator.* However, some minerals are *semiconductors* or *conductors,* in which case electron transfer might be better described as insertion of electrons into (or extraction of electrons from) the overlapping electronic orbitals of the solid. The resultant electron excess or deficit is then delocalized over the solid, not associated with one particular metal ion at one surface location.

In the *band theory* of solids, the overlapping atomic orbitals of the solid form "bands" of many energy levels, rather than a single energy level, as diagrammed in Figure 7.14. The orbitals occupied by the valence electrons constitute the *valence band,* while the first excited state of the electrons, normally unoccupied, is the *conduction band* of the solid. The bands of even higher excited states generally overlap the conduction band, creating a continuum of allowed electronic energy levels from E_c, the energy level at the bottom of the conduction band, out to infinite energy.

Most silicate clays and oxides are insulators and semiconductors, types of solids that possess a *band gap*—a nonallowed region of energy, E_g, between the filled valence band and empty conduction band (see Figure 7.14). This gap prevents electron flow in these minerals, so that they are nonconducting. In the solid, the *Fermi*

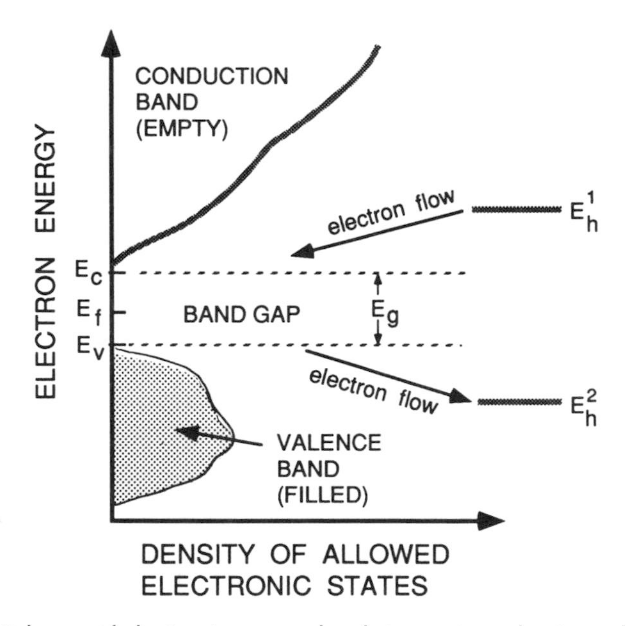

Figure 7.14. Scheme of electronic energy levels in semi-conducting solids. E_h^1 and E_h^2 refer to electronic energy levels of molecules in solution that may, respectively, donate electrons to or accept electrons from the solid.

energy, E_f, defines the electrochemical (redox) potential of the electrons; it is the energy level above which all levels are unoccupied and below which all levels are filled. For insulators and semiconductors, since the valence band is almost completely filled while the conduction band is essentially unoccupied, the Fermi energy falls within the bandgap region. This E_f of the solid is somewhat analogous to the E_h of redox couples in solution, so that the relative energy levels of E_f and E_h tend to determine the direction of electron flow between the solid and reducible or oxidizable molecules in solution. The electron flow, depicted schematically in Figure 7.14, must ultimately reach an equilibrium such that E_f and E_h are equalized. If, for example, the solid is initially able to oxidize molecules adsorbed from solution, electrons flowing into the solid from the molecules cause the Fermi energy to rise. Reduction of the solid continues until the Fermi energy rises enough to balance the solution redox potential.

Many soil minerals of interest, including Fe oxides, behave as insulators. However, minerals characterized by metals with mixed oxidation states, such as Mn oxides,[1] are often semiconductors or conductors, in which case electron flow through the conduction band of the solid is possible. Oxidation of dissolved molecules by these types of solids can be viewed as the transfer of electrons from solution and insertion into the conduction band. For example, Mn oxides can oxidize NO_2^- to NO_3^- without release of Mn^{2+} to solution. The electrons accepted by the oxide from NO_2^- are delocalized in the solid so that no Mn^{2+} is released.

Mn oxides are versatile oxidants. They can oxidize Co^{2+} to Co^{3+}, Cr^{3+} to Cr^{6+} (chromate), As^{3+} to As^{5+}, and phenols and aromatic amines to polymeric products. Some of the organic degradation reactions are discussed in more detail in Chapter 10.

In many cases in which structural or adsorbed Mn and Fe of minerals act as oxidizing agents, the mineral is only the catalyst, because O_2 ultimately reoxidizes the reduced Mn or Fe in the mineral, as follows:

$$\begin{array}{ccc} H_2O & Fe^{3+} & \text{substrate} \\ O_2 & Fe^{2+} & \text{oxidized product} \end{array} \qquad (7.65)$$

The reaction is much faster with the mineral present, although the reasons for this are not fully understood. After all, O_2 as an oxidant is theoretically more powerful than $Fe(OH)_3$ and as powerful as MnO_2 (see Table 7.1)

It is interesting to speculate about why the mineral surface should react much more rapidly than molecular oxygen with a substrate. One explanation might be that Mn^{3+} in oxides has a higher reduction potential than O_2 (see Table 7.1). However, such comparisons of reduction potentials should be made with caution. In the case of O_2, the half-cell reaction is that of a homogeneous solution-gas mixture. In the case of metals such as Fe^{3+} in the octahedral sites of silicate clays, the reduction is a solid-state reaction:

$$Fe^{3+}\text{ (structural)} + e^- = Fe^{2+}\text{ (structural)} \qquad (7.66)$$

1. Manganese in the structure of these oxides can have a mixture of oxidation states +2, +3, and +4.

which is quite different from the $Fe^{3+}-Fe^{2+}$ half-cell reaction in solution whose potential is given in Table 7.1. In fact, the reduction potential for reaction 7.66 is not known.[1] Similarly, the potentials for reduction of structural Mn^{4+} or Mn^{3+} in Mn oxides to lower oxidation states *within the solid* are unknown. The $Fe(OH)_3-Fe^{2+}$ and MnO_2-Mn^{2+} reaction potentials given in Table 7.1 are appropriate only for those particular solid forms of iron hydroxide and manganese oxide in which reduction releases Fe^{2+} or Mn^{2+} to solution. Electron transfer processes within the solid state are not described by standard-state potentials such as those listed in Table 7.1. They are complicated by the electron-conducting properties of the solids, and involve intermediate states of reduction or oxidation within the solids. The band theory of solids indicates that electron transfers in conducting and semiconducting minerals can occur without any particular atom changing valence by an integral amount. This property facilitates electron flow into and out of the mineral, and may account for the ease with which part of the structural Fe^{3+} in Fe-rich smectites is reduced to Fe^{2+}.

In solids, the immediate structural environment (crystal field) of a metal ion is likely to alter its reduction potential. For example, Fe^{3+} substituted in a silicate has a potential different from Fe^{3+} in $Fe(OH)_3$, and different again from Fe^{3+} in solution. Structures that stabilize Fe^{3+} relative to Fe^{2+} lower the reduction potential of Fe^{3+} in the solid. Solution properties such as pH are likely to affect these potentials in a manner different from the way they affect solution potentials. It is interesting that, in a recent study of biotite subjected to weathering in the presence of Cu^{2+}, the reduction of adsorbed Cu^{2+} to metallic copper (Cu^0) has been demonstrated (Earley et al., 1992). Reduction of Cu^{2+} by Fe^{2+} in soil clays is also known to occur. Although Cu^{2+} reduction by solution Fe^{2+} is not favorable, structural Fe^{2+} in the octahedral sheet of 2:1 silicate minerals seems to be a much more potent oxidant than aqueous Fe^{2+}. This is explained by the high stability of Fe^{3+} relative to Fe^{2+} in octahedral sites of silicates and oxides.

It appears from the above discussion that the reduction potentials of many solid phases in soils, such as Mn oxides of mixed oxidation state or Fe^{3+}-bearing layer silicates cannot be gauged by classical electrochemical concepts. The question of whether molecular oxygen or a particular mineral is the more potent oxidizing agent in soils is a complex one, since kinetic as well as thermodynamic factors come into play. That is, properties other than intrinsic reduction potential probably contribute to the effectiveness of Fe- and Mn-bearing minerals as oxidants relative to O_2. As opposed to dissolved O_2, which oxidizes molecules through chance encounters in three-dimensional solutions, mineral surfaces may attract substrate molecules electrostatically or by some other form of bonding, thereby concentrating them in two dimensions. The probability of electron transfer with adjacent structural Fe or Mn ions is then enhanced, as is the chance of direct reaction with dissolved O_2. Examples of reactions that seem to be promoted in this way include: (1) Benzidine oxidation by birnessite (a Mn oxide) and smectites, (2) phenol oxidation by Mn and Fe oxides, and (3) Cr^{3+} oxidation by Mn oxides. In each case, the substrate is adsorbed to the surface prior to electron exchange, either by electrostatic attraction (in the case of benzidine cations), complexation with surface metal cations (in the case of phenols), or coordination to surface oxyanions (in the case of Cr^{3+}). Intimate association at the surface is probably a necessary preliminary step to electron transfer between dissolved molecules and minerals.

1. Part of the structural Fe^{3+} in smectites can be reduced to Fe^{2+} by dithionite and other moderate reducing agents.

Mineral surfaces containing structural Fe^{3+} and Mn^{3+}, besides increasing the probability for reductant-oxidant encounter, are facile one-electron acceptors, unlike molecular oxygen.[1] Oxidation of organics at these mineral surfaces seems not to be constrained by the rule of spin parity that limits the reactivity of O_2 (see section 7.2a). Again, chemical factors besides the magnitude of reduction potentials appear to favor heterogeneous oxidations by soil colloids over homogeneous oxidations by molecular oxygen.

References

Bartlett, R. J. and B. R. James. 1992. Sunlight powered redox in soils and waters. *Agronomy Abstracts,* American Society of Agronomy, p. 233. Madison, WI.

Earley, D., E. S. Ilton, M. D. Dyar, and D. R. Veblen. 1992. Solid state redox chemistry of biotite during chemisorption of Cu^{2+} at 25°C and 1 bar. *Agronomy Abstracts,* American Society of Agronomy, p. 368.

Hesse, P. R. 1971. *A Textbook of Soil Chemical Analysis.* New York: Chemical Publishing Co., Chapter 17, 18.

Lindsay, W. L. 1979. *Chemical Equilibria in Soils.* New York: Wiley. Chapter 2.

Patrick, W. H. and A. Jugsujinda. 1992. Sequential reduction and oxidation of inorganic nitrogen, manganese, and iron in flooded soil. *Soil Sci. Soc. Am. J.* 56:1071–1073.

Stumm, W. and J. J. Morgan. 1981. *Aquatic Chemistry.* 2nd ed. New York: Wiley, Chapter 7.

Suggested Additional Reading

Drever, J. I. 1988. *The Geochemistry of Natural Waters.* 2nd ed. Englewood Cliffs, N.J.: Prentice Hall, chaps. 13, 14.

Wehrli, B. 1990. Redox reactions of metal ions at mineral surfaces. In *Aquatic Chemical Kinetics* (W. Stumm, ed.), pp. 311–336. Wiley, New York.

Questions

1. Why is the oxidation of most organic compounds energetically favorable? Which are the best electron acceptors for the biological oxidation of organic matter in soil?

2. Part of the soluble iron in soil solutions of spodosols is in the form of Fe^{2+}.
   (a) Calculate the expected "half-life" of 10^{-4} molar Fe^{2+}, assuming the solution is aerated and at pH 6. Repeat the calculation for pH 5 (use equation 7.47).
   (b) Explain the persistence of soluble Fe^{2+} in these well-drained soils, given that they contain significant dissolved organic matter.

3. Why are electrode-measured redox potentials and thermodynamically defined redox potentials not necessarily the same in soil solutions?

4. A soil solution at pH 5 contains 10 μg/ml of NO_3^- and 1 μg/ml of Fe as Fe^{2+}. No soluble Fe^{3+} is detected.
   (a) Are the ions in this solution at thermodynamic equilibrium?
   (b) Can a theoretical redox potential be defined for this solution?
   (c) Estimate the E_h that a Pt electrode would report for this solution.

1. The Mn^{4+} ion in pyrolusite (MnO_2) is rather inert in oxidation reactions, evidently because the one-electron $Mn^{4+}-Mn^{3+}$ reduction step is not a very favorable reaction.

5. For the following systems of waterlogged soils, estimate the pϵ assuming that the redox reactions are at equilibrium:
   (a) Soil solution with traces of dissolved O_2; pH is 7.
   (b) Soil containing both $Fe(OH)_3$ and $FeCO_3$; pH is 6.
   (c) Soil solution containing sulfate at 10^{-4} M and with the smell of hydrogen sulfide gas; pH is 6.
   (d) Soil containing MnO_2; solution has 10^{-5} M Mn^{2+} and pH is 7.
   (e) Soil air contains nearly equal amounts of carbon dioxide and methane.

6. The following reactions in some cases can control the solubility of Fe^{2+} and Mn^{2+} in soil solutions:

$$\gamma\text{-MnOOH (s)} + 3H^+ + e^- = Mn^{2+} + 2H_2O \qquad E^0 = 1.457\ \text{V}$$
$$Fe(OH)_3\ \text{(s)} + 3H^+ + e^- = Fe^{2+} + 3H_2O \qquad E^0 = 1.057\ \text{V}$$

   (a) Derive an equation that defines the relationship between soluble Fe^{2+} and Mn^{2+} in soil solution at equilibrium with MnOOH and $Fe(OH)_3$. What does this equation predict the activity of Fe^{2+} to be when the activity of Mn^{2+} is 10^{-5}?
   (b) Calculate the E_h and pϵ of this soil if the pH is 7 and Mn^{2+} activity is 10^{-5}.
   (c) If molecular oxygen were introduced into this anaerobic soil system, what would be the likely sequence of chemical processes?
   (d) Mn oxides are often intimately associated with Fe oxides in soils. Can you develop a plausible mechanism to explain this association?

7. Should Fe^{2+} in soil solution be able to reduce nitrate to nitrite? If so, what pH would be favorable to reduction?

8. A recent observation (Bartlett and James, 1992) was made that a solution at pH 4.2, containing hydrolyzed Fe^{3+} and Cr^{3+}, formed Fe^{2+} and chromate (CrO_4^{2-}) simultaneously on exposure to light. Dissolved O_2 did not seem to be necessary for the reaction.
   (a) Propose a mechanism for this reaction.
   (b) What would you expect to happen when the light is turned off?

9. Biotite is an Fe^{2+}-rich mica that weathers easily to release K^+ from the interlayer regions. When it is weathered in the presence of Cu^{2+}, the reduction of adsorbed Cu^{2+} to metallic copper (Cu^0) in the interlayer has been demonstrated (Earley et al., 1992).
   (a) Calculate the potential and free energy of the $2Fe^{2+} + Cu^{2+} = 2Fe^{3+} + Cu^0$ reaction *in solution* under standard-state conditions, using the known reduction equation:

$$Cu^{2+} + 2e^- = Cu^0\ \text{(s)} \qquad E_h^0 = +0.34\ \text{V}$$

   Is the reaction favorable in solution?
   (b) Since the electron transfer actually involves Fe in the solid state, the reduction equation:

$$Fe(OH)_3\ \text{(s)} + e^- = Fe(OH)_2\ \text{(s)} + OH^- \qquad E_h^0 = -0.56\ \text{V}$$

   seems to be an appropiate model for the biotite reaction. Calculate the potential and the free energy of the $2Fe(OH)_2 + Cu^{2+} = 2Fe(OH)_3 + Cu^0$ reaction at pH 7, assuming the solution activity of Cu^{2+} is maintained at 0.01. Is the formation of metallic Cu by $Fe(OH)_2$ a favorable reaction?
   (c) How does actual Fe^{2+} oxidation in biotite differ from the model $Fe(OH)_2 \rightarrow Fe(OH)_3$ reaction?

8

Salt-Affected and Swelling Soils

As discussed in Chapter 5, natural soil-forming processes of humid climates tend to acidify soils over the long term. However, in arid climates, because salts are not frequently leached from the soil by rainfall, salts and alkalinity tend to accumulate. This is because the potential evaporation and transpiration of water at the soil surface exceeds rainfall so that the net movement of water is *upward* in the soil profile. Consequently, salts accumulate at or near the surface of these arid-region soils. However, in localized cases, accumulation of salts and alkalinity arises from confinement, created by a physical barrier to water flow out of a depression in the landscape. Thus, salt problems are not restricted to soils of arid climates.

This chapter discusses the origin of salinity and alkalinity, how properties of soils are affected by these two conditions, and strategies for reclaiming these soils for agriculture or other purposes.

8.1. SOURCES OF SALINITY AND ALKALINITY

8.1a. Weathering and Alkalinity

Two questions naturally arise with respect to salt-affected soils: What is the origin of the accumulated salts in these soils, and why are these salts associated with alkalinity?

In response to the first question, localized sources are sometimes responsible. These include "fossil salts"; that is, salts deposited by ancient seas and later buried under sediments. The subsequent process of evaporation or lateral groundwater flow can move these salts to the surface, degrading the soil, as depicted in Figure 8.1. In coastal regions, rainfall can be "contaminated" by salts (principally NaCl) from ocean spray, so that it is not unusual to see elevated levels of exchangeable Na^+ in coastal soils. Human activities, including the application of poor quality irrigation water, fertilizers, and waste water to land, can accelerate the process of salt accumulation and soil degradation. Historically, salt-related problems of irrigated agriculture have been responsible for the loss of much productive land in semiarid parts of the world.

Even in the absence of these localized inputs, salts accumulate in soils that are

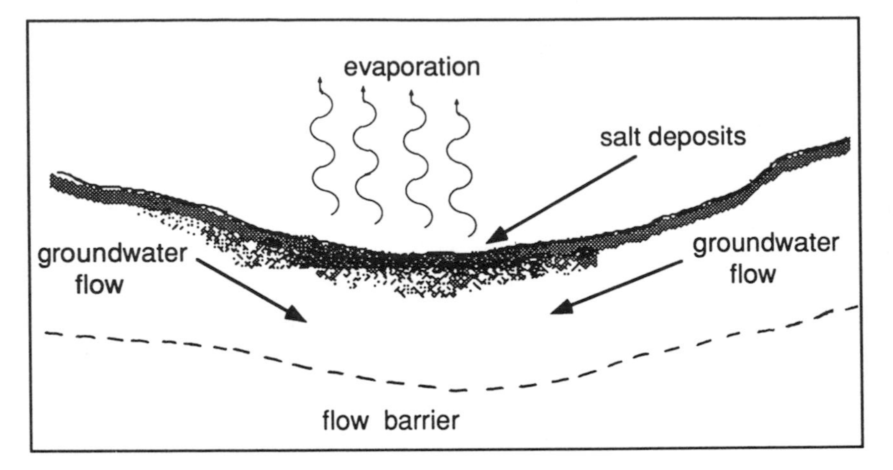

Figure 8.1. Localized situation conducive to the accumulation of salinity in a semiarid climate.

not subjected to frequent leaching. The source of these ubiquitous salts is mineral weathering, a process discussed in Chapter 6. Example weathering reactions, such as

$$\underset{\text{olivine}}{Mg_2SiO_4} + 4H_2O \rightarrow 2Mg^{2+} + Si(OH)_4^0 + 4OH^- \tag{8.1}$$

$$\underset{\text{feldspar}}{2NaAlSi_3O_8} + 11H_2O \rightarrow \underset{\text{kaolinite}}{Al_2Si_2O_5(OH)_4} + 4Si(OH)_4^0 + Na^+ + OH^- \tag{8.2}$$

reveal that, as a general rule, the hydrolysis of primary silicate minerals is a *non-reversible* process at earth-surface conditions that generates alkalinity in solution.[1] Conversely, oxidation of primary minerals containing elements in the reduced state (e.g., pyrite) can be acidifying, as shown below:

$$\underset{\text{pyrite}}{2FeS_2} + 7O_2 + 2H_2O \rightarrow 2Fe^{2+} + 4SO_4^{2-} + 4H^+ \tag{8.3}$$

Because silicate mineral hydrolysis and dissolution is the major weathering process in most soil-forming materials, weathering has the net effect of generating alkalinity, at least in the initial stages. This alkalinity can accumulate in soils of arid and semi-arid climates; it is neutralized by natural acid production in soils of humid climates (see Chapter 5). The process of silicate weathering, then, provides an answer to the second question that was posed above:

> *Mineral dissolution and release of cations and anions into solution are inseparably tied to the generation of alkalinity.*

The source of this alkalinity is the silicate anion liberated from the mineral by hydrolysis of the structure. This anion is a very strong base, reacting with protons from solution to form the weak acid, monosilicic acid:

1. It is useful to think of the primary silicate minerals as salts of weak acids (monosilicic) and strong bases (such as KOH), which therefore react with water to form alkaline solutions.

$$\left[\begin{array}{c} O \\ | \\ O-Si-O \\ | \\ O \end{array}\right]^{4-} + 4H^{+} = \left[\begin{array}{c} OH \\ | \\ HO-Si-OH \\ | \\ OH \end{array}\right]^{0} \tag{8.4}$$

silicate — monosilicic acid

If the protons for this reaction are provided by water dissociation (as would be the case if acidity were not generated by biological or other processes in the soil), then reaction 8.4 generates alkalinity in the form of OH^- ions. In natural soil systems, OH^- further reacts, with CO_2 dissolved in the water to convert the alkalinity into the carbonate or bicarbonate form:

$$\underset{}{OH^-} + \underset{\text{carbonic acid}}{H_2CO_3} = H_2O + \underset{\text{bicarbonate}}{HCO_3^-} \tag{8.5}$$

Thus the ubiquity of CO_2 in soils ensures that any alkalinity that accumulates does so largely in the form of carbonate or bicarbonate salts. Consequently, the cation associated with the alkaline anion becomes critical in limiting the severity of alkalinity in soil solution. As is shown in Table 8.1, Mg^{2+} and Ca^{2+} carbonates are not very soluble, while Na^+ and K^+ carbonates dissolve readily in water to form solutions with very high pH, according to the reaction:

$$\begin{array}{c} Na_2CO_3 \rightarrow 2Na^+ + CO_3^{2-} \\ CO_3^{2-} + H_2O \rightarrow HCO_3^- + OH^- \end{array} \tag{8.6}$$

This means that soils formed from primary minerals rich in Ca^{2+} and Mg^{2+} are less likely to become alkaline than soils formed from minerals rich in K^+ and Na^+. The alkalinity in Ca^{2+}-rich soils is precipitated in the form of $CaCO_3$ (calcite), which controls the solution pH at a value near 8.2 (assuming a CO_2 gas pressure of 0.3 millibar). These soils are termed *calcareous,* and are not considered to be alkaline despite their pH because they do not possess the negative attributes of the more strongly alkaline soils. If, on the other hand, alkalinity is present in the form of sodium and potassium bicarbonates and carbonates, the moist soil has a very high pH, and poses an "alkalinity hazard" to plants.

8.1b. Equilibrium in the CO_2-H_2O System

A detailed description of how dissolved CO_2 is involved in controlling the pH of alkaline and calcareous soils is useful at this point.

Table 8.1. Solubility of Carbonates in Cold Water

Carbonate	Solubility (g/liter)
$CaCO_3$	0.014
$MgCO_3$	1.76
Na_2CO_3	71
K_2CO_3	1120

Dissolved CO_2 in water exists in two forms, carbonic acid and hydrated CO_2:

$$CO_2 + H_2O = H_2CO_3 \qquad (8.7)$$
$$CO_2 + H_2O = CO_2 \cdot H_2O$$

When the water is in equilibrium with the atmosphere, the partial pressure of CO_2 in the air (P_{CO_2}) controls the concentration of total dissolved CO_2, written as $[H_2CO^*_3] = [CO_2] + [H_2CO_3]$, according to Henry's law:

$$[H_2CO_3^*] = K_H P_{CO_2} \qquad (8.8)$$

where $\log K_H = -1.46$. Therefore, at a normal atmospheric level of dissolved CO_2 (0.3 milliatmospheres), $[H_2CO_3^*] \approx 10^{-5}$ M.

The dissolved CO_2 dissociates to form bicarbonate and carbonate:

$$H_2CO_3^* = H^+ + HCO_3^- \qquad -\log K_1 = 6.35 \qquad (8.9)$$

$$HCO_3^- = H^+ + CO_3^{2-} \qquad -\log K_2 = 10.33 \qquad (8.10)$$

Combining equilibrium expressions for these two dissociation reactions with equation 8.8, the following two expressions are obtained (assuming that molar concentrations of dissolved ions can be approximated to activities):

$$-\log [HCO_3^-] = 7.81 - \log P_{CO_2} - pH \qquad (8.11)$$

$$-\log [CO_3^{2-}] = 18.14 - \log P_{CO_2} - 2\ pH \qquad (8.12)$$

These equations are graphed in Figure 8.2, assuming a fixed atmospheric P_{CO_2} pressure of 0.3 milliatmospheres. In soils, the P_{CO_2} pressure can be much higher than this because of biological activity, reaching 100 milliatmospheres or more in poorly aerated soils. As equations 8.11 and 8.12 predict, higher CO_2 levels in the soil atmosphere would increase dissolved H_2CO_3, HCO_3^-, and CO_3^{2-}, shifting the solubility lines upward in Figure 8.2.

For any chemical reaction system in solution, the charge of anions must be balanced by the charge of cations; this is the *electroneutrality requirement.* In the simple H_2O-CO_2 system with no inputs of bases or acids, the electroneutrality requirement is

$$[H^+] = [HCO_3^-] + 2[CO_3^{2-}] + [OH^-] \qquad (8.13)$$

since all of the cations and anions present in solution are accounted for in this equation. Equation 8.13 combined with equations 8.11 and 8.12 gives three equations with three unknowns ($[CO_3^{2-}]$,$[HCO_3^-]$, and pH) if P_{CO_2} is constant.[1] The pH of this system is then fixed and found to be 5.65 for $P_{CO_2} = 0.3$ milliatmospheres. This result is valid for a solution open to the atmosphere, where the atmosphere provides an "infinite" reservoir of CO_2. Note that as the CO_2 pressure in the system is increased, the equilibrium pH goes below 5.65.

1. Remember that $[OH^-]$ is fixed by pH because the dissociation constant of water requires that $[OH^-] = 10^{-14}/[H^+]$.

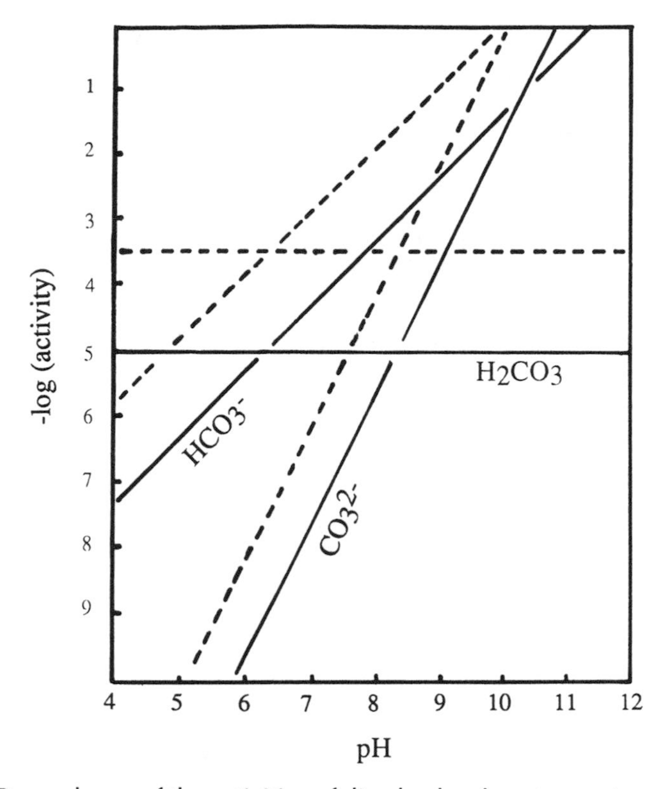

Figure 8.2. Dependence of the activities of dissolved carbonate species on solution pH, assuming atmospheric pressure (0.3 milliatmospheres) of CO_2 (solid lines). The effect of higher CO_2 (10 milliatmospheres) on each species is displayed as a broken line parallel to the original line.

8.1c. Equilibrium in the $CaCO_3-CO_2-H_2O$ System

In cases where Ca^{2+} is present to influence the carbonate system by the dissolution/precipitation reaction of calcite:

$$CaCO_3\,(s) = Ca^{2+} + CO_3^{2-} \qquad \log K_{SO} = -8.35 \tag{8.14}$$

an additional constraint is added to the system of carbonate equilibrium, described by the equation:

$$\log\,[Ca^{2+}] + \log\,[CO_3^{2-}] = -8.35 \tag{8.15}$$

In this system, the electroneutrality requirement is

$$2[Ca^{2+}] + [H^+] = [HCO_3^-] + 2[CO_3^{2-}] + [OH^-] \tag{8.16}$$

If one of the ionic concentration variables is fixed, it can be shown that only one solution is possible from combining equations 8.8, 8.11, 8.12, 8.15, and 8.16. By again setting the P_{CO_2} pressure at that of the atmosphere (0.3 milliatmospheres), the equations are solved to give a pH of 8.3, $[Ca^{2+}] = 10^{-3.3}$ M, $[HCO_3^-] = 10^{-3.0}$ M, and $[CO_3^{2-}] = 10^{-5.0}$ M.

The general relationship among CO_2 pressure, soluble Ca^{2+}, and pH, obtained by combining equations 8.8, 8.11, 8.12, and 8.15, is:

$$-\log [Ca^{2+}] = 2\,pH - 9.79 + \log P_{CO_2} \tag{8.17}$$

which applies to the case of a soil system in which $CaCO_3$ is present. This equation reveals that, if CO_2 in the soil air is increased by biological activity, there will be a resulting *decrease* in the pH and an increase in soluble Ca^{2+}, with the soluble Ca^{2+} concentration obeying the approximate relationship that $2\,[Ca^{2+}] = [HCO_3^-]$, which is obtained from a simplification of equation 8.16.

These calculations predict that calcareous soils should have pH values near 8.3; however, the actual pH of these soils in the field is often well below 8.3, in the range of 7.5 to 7.8. This may be a consequence of the higher CO_2 pressure in soil air compared with the atmosphere, or it may reflect the presence of carbonate solids other than pure crystalline calcite (Mg carbonates, for example). Calcareous soil solutions are commonly supersaturated with respect to calcite, perhaps because dissolved molecules and ions inhibit crystallization.

8.2. MEASURES OF SALINITY AND ALKALINITY

8.2a. The Residual Sodium Carbonate (RSC) Value

The alkalinity of water is measured as the residual sodium carbonate (RSC) value, given by:

$$RSC = [HCO_3^- + CO_3^{2-}] - [Ca^{2+} + Mg^{2+}] \tag{8.18}$$

where all concentrations are expressed in millimoles of *charge* per liter.[1] This parameter is useful because, as evaporation of water from soils proceeds, Ca^{2+} and Mg^{2+} precipitate the alkaline anions as insoluble and relatively innocuous carbonates, but any excess of carbonate or bicarbonate not precipitated by Ca^{2+} and Mg^{2+} poses an alkalinity hazard (not to mention the physical problems created in the soil by the Na^+ and K^+ carbonate salts, discussed in section 8.4). The RSC is equal to the quantity of acidity (in units of mmoles H^+/liter) that is needed to neutralize the solution alkalinity in excess of the alkalinity associated with Ca^{2+} and Mg^{2+}.

The RSC is a property of solutions, and has been used to quantify the alkalinity hazard of irrigation water applied to soils. The accepted scale, based on experience is

RSC $>$ 2.5	Hazardous
RSC = 1.25–2.50	Potentially Hazardous
RSC $<$ 1.25	Generally Safe

1. Ideally, the alkalinity of water is given by $[HCO_3^- + CO_3^{2-}] + [OH^-] - [H^+]$, but the concentrations of OH^- and H^+ are not generally significant in alkaline soil solutions exposed to air.

This means that the repeated application of water having an RSC above 2.5 will result in the soil becoming alkaline.

8.2b. Electrical Conductivity (EC)

Salinity, or the concentration of dissolved salts, is most conveniently measured by electrical conductivity (EC). This is based on the principle that the conductivity, or ease with which an electric current is carried through a solution, is proportional to the quantity of ions (actually, the quantity of ionic charge) in solution.

To measure soil conductivity, the soil is mixed with water until it has a pastelike consistency. A pair of electrodes, commonly made of platinum metal, is inserted into this "saturated paste." The electrodes are characterized by their surface area, A cm^2 (usually about 1 cm^2), and their separation distance of L cm (usually about 1 cm), as illustrated in Figure 8.3. An alternating current is then applied to the electrode pair, and the conductance of the paste is measured in units of reciprocal ohms (mho), or Siemens (1 Siemen = 1 mho). But this conductance depends on A and L and is therefore a characteristic of the measuring device as well as the solution. It is more useful to define a specific conductance, termed the *electrical conductivity* or EC, by the equation:

$$\text{EC} = \text{conductance (Siemens)} \times \frac{L}{A}\,(\text{cm}^{-1}) \tag{8.19}$$

EC, unlike conductance, is strictly a property of the solution, whereas L/A is a characteristic of the electrode geometry and is called the cell constant. For a saturated paste, an EC value greater than 4 milliSiemens/cm (mS cm^{-1}) is diagnostic of a saline soil.

The current-carrying ability of a mole of cation or anion charge in solution is termed the *equivalent conductance.* This value is measured in very dilute solutions to ensure that the salts are fully dissociated into individual cations and anions. The measured conductance is then assumed to result from the summation of conductances contributed by each of the ions. As a result, each ion can be assigned an equiv-

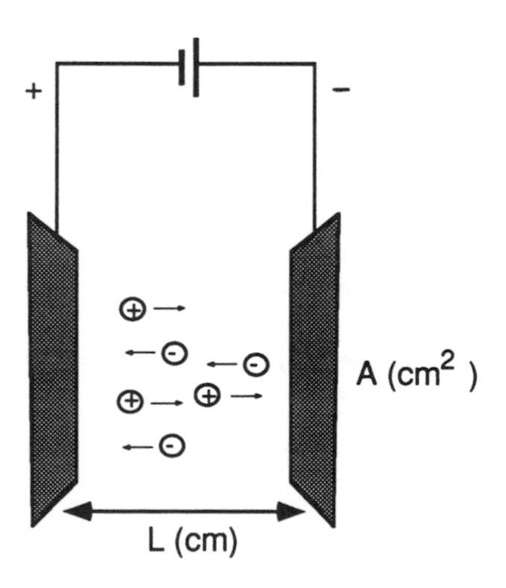

Figure 8.3. Two-electrode cell used to measure electrical conductivity of solutions and saturated soil pastes.

Table 8.2. Equivalent Conductances of Ions at Infinite Dilution

Anion	Conductance	Cation	Conductance
OH^-	198	H^+	350
Cl^-	76	Na^+	50
Br^-	78	K^+	74
ClO_4^-	74	Ag^+	62
NO_3^-	71	$Cu^{2+}/2$	57
CH_3COO^-	41	$Zn^{2+}/2$	56
$SO_4^{2-}/2$	80	$Fe^{3+}/3$	68

alent conductance, listed in Table 8.2, that represents its contribution (per mole of charge) to the measured (total) conductivity. It is apparent that H^+ and OH^- have anomalously high conductances, while most cations and anions contribute about equally *per mole of ionic charge* to the conductivity of solutions. This fact allows simple but approximate empirical relationships to be established between the concentration of electrolyte solutions and the conductivity, irrespective of ionic composition. For example, the total molarity of cations or anions can be estimated:

$$\text{Total cations (mmoles(+)/liter)} \approx \text{EC (mS cm}^{-1}) \times 10 \tag{8.20}$$

$$\text{Total anions (mmoles(−)/liter)} \approx \text{EC (mS cm}^{-1}) \times 10 \tag{8.21}$$

or, the quantity of "total dissolved solids" (TDS) in the form of soluble salts can be approximated:

$$\text{TDS (mg/liter)} \approx \text{EC (mS cm}^{-1}) \times 640 \tag{8.22}$$

These equations are likely to apply reasonably well to soil solutions of different ionic compositions unless the pH indicates the solution to be very acid or very alkaline.

8.3. SODICITY

From the mass-action principles of cation exchange, one expects the percentage of soil exchange sites occupied by Na^+ ions, termed the *exchangeable sodium percentage* (ESP), to be a function of the relative concentrations of Na^+ and competing cations (Ca^{2+}, Mg^{2+}, K^+) in solution. Exchangeable Al^{3+} can be disregarded in this relationship, because acidity and aluminum are not usually associated with soils possessing significant exchangeable Na^+.[1] Consequently, the assumption can be made that the exchange cations complementing Na^+ in the soil are Ca^{2+} and Mg^{2+}. The K^+ ion rarely occupies a significant fraction of the exchange sites because of its low solubility in most soils.

1. One exception is acid sulfate soils formed on coastal sediments due to the influence of sulfate in seawater (see Chapter 7).

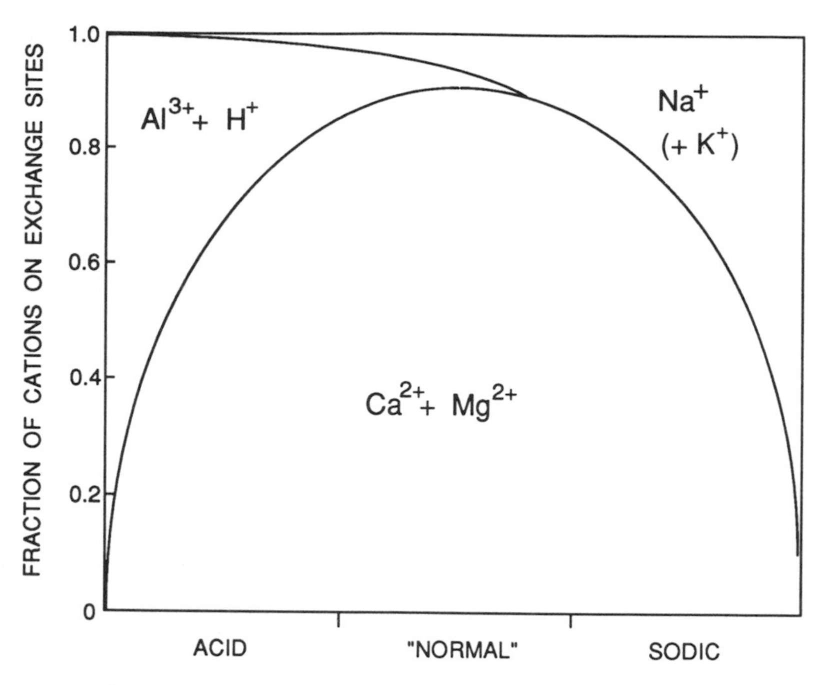

Figure 8.4. Relative cation composition of exchange sites for sodic soils compared with nonsodic ("normal" and acid) soils. (Adapted from G. H. Bolt et al. 1976. Adsorption of cations by soil. In G. H. Bolt and M.G.M. Bruggenwert (eds.), *Soil Chemistry*. New York:Elsevier.)

The typical compositional range of exchange sites in soils, shown in Figure 8.4, reveals that *sodic* soils have higher than normal levels of exchangeable Na^+. The important ionic concentrations in soil solution that affect the level of exchangeable Na^+ are $[Na^+]$, $[Ca^{2+}]$, and $[Mg^{2+}]$. The *sodium adsorption ratio,* or SAR, is a solution property defined by the equation

$$\mathrm{SAR} = \frac{[\mathrm{Na^+}]}{\sqrt{([\mathrm{Ca^{2+}}] + [\mathrm{Mg^{2+}}])/2}} \tag{8.23}$$

where all solution concentrations are expressed in units of millimoles(+) per liter.[1] The SAR is related to ESP through the process of cation exchange, and it has been found empirically that the equation:

$$\frac{\mathrm{ESP}}{100 - \mathrm{ESP}} = 0.015\ \mathrm{SAR} \tag{8.24}$$

predicts reasonably well the composition of the exchange sites of the soil (ESP) based on a knowledge of the soil solution composition (SAR).

1. These units are millimoles of cationic charge (that is, milliequivalents) per liter.

The complexity of the overall ESP-SAR relation, which is seen by substituting the definition of SAR (equation 8.23) into equation 8.24, may not be justified in practice. A partitioning factor, K_p, defined as

$$K_p = \frac{\{Na\} \cdot a_{\mathrm{Ca}}}{\{Ca\} \cdot a_{\mathrm{Na}}} \tag{8.25}$$

or in logarithmic form:

$$\log\left(\frac{\{Na\}}{\{Ca\}}\right) = \log\left(\frac{a_{\mathrm{Na}}}{a_{\mathrm{Ca}}}\right) + \log K_p \tag{8.26}$$

where $\{Na\}$ and $\{Ca\}$ are equivalents of exchangeable Na^+ and Ca^{2+} per unit weight of soil and a_{Na} and a_{Ca} are the soil solution activities of Na^+ and Ca^{2+}, provides a reasonable first approximation for relating solution to soil composition in salt-affected soils. Fitting of actual soil data to a function of the general form of equation 8.26 produces the relation (Kittrick, 1976):

$$\log\left(\frac{\{Na\}}{\{Ca\}}\right) = 0.79 \log\left(\frac{a_{\mathrm{Na}}}{a_{\mathrm{Ca}}}\right) - 1.41 \tag{8.27}$$

8.4. CLAY SWELLING AND DISPERSION

Exchangeable Na^+ in soils is strongly associated with poor structural characteristics. Soils high in exchangeable Na^+ are predisposed toward swelling, surface crusting, sealing, erosion, and other undesirable consequences of colloidal dispersion. The amount of clay that will disperse by shaking a soil in water, simply termed the water-dispersible clay, is taken as a measure of the potential for structural degradation. Because clay expansion and dispersion are the precursors of degradation, the causes and properties of these two colloidal phenomena need to be understood. This section describes models that have had some success in explaining expansion and dispersion.

8.4a. Concept of Colloidal Flocculation and Dispersion

Stable colloidal suspensions remain suspended in water because of Brownian motion, which is the random movement of particles arising from collisions with water molecules in thermal motion, and particle-particle repulsion. The agglomeration of individual particles in water suspension to form visible aggregates, which then settle out by gravity, is called *flocculation.* Both high-charge exchange cations and high electrolye concentrations in solution promote flocculation. The minimum concentration of electrolyte (salt) that causes a dispersed clay to flocculate in a specified time is called the *flocculation value* or *critical coagulation concentration* (CCC). This value, sensitive to the magnitude of charge of the electrolyte cation (for the case of negatively charged colloids), is about 0.025 to .150 M for monovalent cations, 0.0005 to 0.002 M for divalent cations, and 0.00001 to 0.0001 M for trivalent cations. In fact, a general rule, now called the Schulze–Hardy rule, was proposed about a century ago to predict the CCC for colloidal suspensions:

The critical coagulation concentration for a colloid suspended in an aqueous solution of electrolyte is determined by those ions with charge opposite in sign to that of the colloidal particles, and is proportional to an inverse power of the valence of the ions.

A low CCC is desirable because low levels of salts in soils are then sufficient to prevent dispersion of aggregates and the consequent structural degradation. At the critical coagulation concentration, there are enough salts in solution to induce an attractive particle-particle interaction, and the clay-sized particles associate into aggregates composed of a sufficient number of particles that gravity supersedes Brownian motion. The aggregates then settle out of suspension.

Na^+-saturated layer silicate clays, when in suspension, have the greatest viscosity at very low and at very high electrolyte concentrations, as illustrated by the graph in Figure 8.5. This seemingly odd behavior may be explained in terms of three stages of clay particle interaction, depending on the electrolyte (NaCl) concentration. The suggested arrangement of individual clay sheets at each stage is sketched in Figure 8.6.

Stage 1: At very low NaCl concentrations, the individual platelets do not interact (i.e., are dispersed) unless the clay concentration is high enough to force overlap between the *co-volumes* of the particles. The co-volume is roughly equivalent to the volume arced out by Brownian motion of the particle. The thickness of the very asymmetrical particles of smectite is about 10 Å, while the diameter, d, is as large as 2 μm. The co-volume of a plate-like particle would then be about $\frac{1}{2}d^3$, the spherical volume arced out during rotation. It is easy to calculate that a 1 percent clay suspension would be sufficiently concentrated to force co-volume overlap. This hinders free rotation of the particles, and platelets align with one another as a result. The high

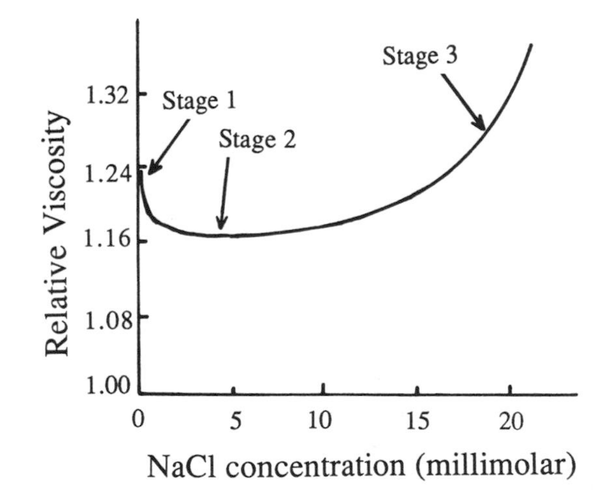

Figure 8.5. Viscosity of a 0.23-percent suspension of Na^+-smectite in water as a function of the amount of NaCl added to solution. (Adapted from H. van Olphen. 1977. *An Introduction to Clay Colloid Chemistry*. 2nd ed. New York: Wiley.)

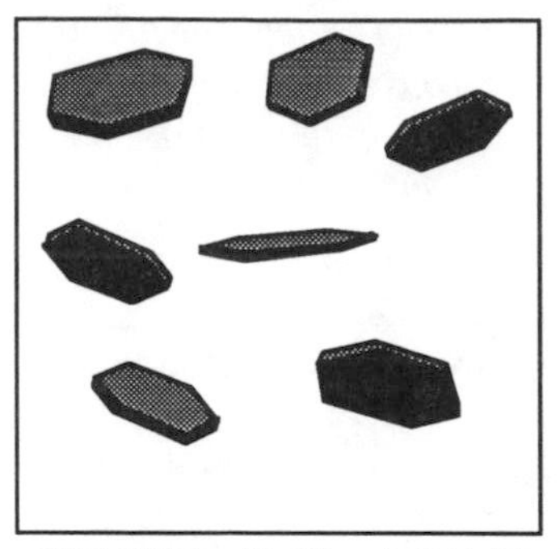

STAGE 1 : NaCl < 0.001 M

STAGE 2 : NaCl > 0.001 M

STAGE 3 : NaCl > 0.01 M

Figure 8.6. Suggested arrangement of individual clay platelets at stage 1 (dispersed), stage 2 (aggregated), and stage 3 (both aggregated and flocculated). Platelets are represented schematically, with their thickness exaggerated relative to their diameter.

suspension viscosity evident at low NaCl concentration (see Figure 8.5) may reflect this strong particle interaction. Furthermore, the co-volume model seems to explain interactions of colloidal particles over long distances (> 1000 Å) without invoking "long-range forces" of mysterious origin. Average center-to-center interparticle distances in these rather dilute suspensions is on the order of micrometers (10^{-6}m). With the possible exception of electrostatic forces, the known types of forces do not operate over such long distances.

Stage 2: At intermediate electrolyte concentrations, clay platelets may begin to approach closely enough to form tactoids, groups of aligned platelets. This produces

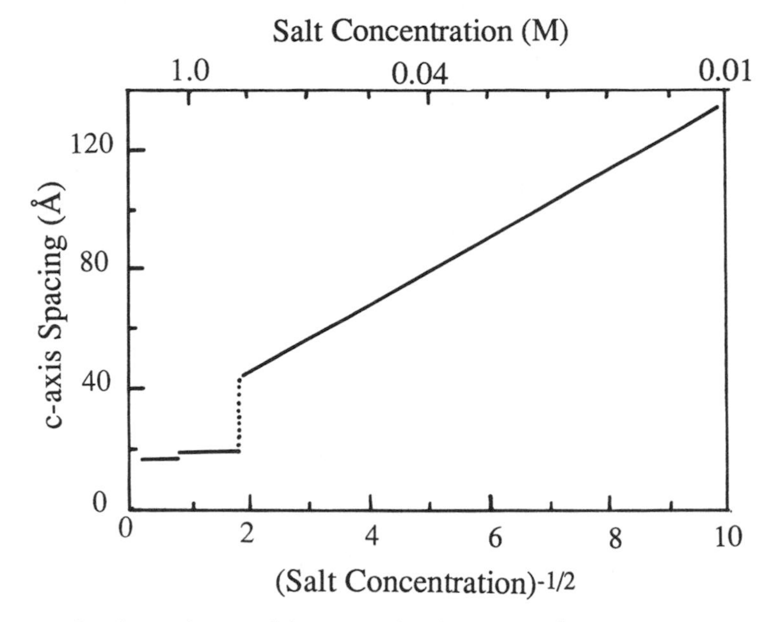

Figure 8.7. The dependence of the c-axis (001) spacing of smectite in suspension on the NaCl concentration. (Adapted from K. Norrish. 1954. The swelling of montmorillonite. *Dis. Faraday Soc.* 18:120–134.)

a smaller number of particles in the same volume, so that particle interaction and viscosity are reduced. For example, a NaCl concentration of 0.01 *M* causes individual platelets to align in parallel orientation with interlayer distances of about 100 to 150 Å, and as the relationship of Figure 8.7 shows, this separation decreases at even higher NaCl concentrations. An explanation for this behavior, based on an osmotic model of clay swelling, will be discussed later in this chapter.

Stage 3: Finally, at very high electrolyte concentrations, interlayer spacings diminish even further and extensive flocculated structures build up, again raising the viscosity (Figure 8.6).

Traditional models of layer silicate particle interaction in water suspension or in soil are based on the idea that these platelike particles could, in principle, display three kinds of association: edge-to-face (EF), edge-to-edge (EE), and face-to-face (FF). In actual fact, direct evidence exists to support only the FF arrangement, which can form particles from packets of aligned platelets, referred to as quasi-crystals. The irregular edges of many layer silicates, particularly smectites, are not amenable to significant EE, or even EF, contact. Even where contact is possible, there is some question about the magnitude and sign of the charge on the edge surfaces. If this charge is positive at ambient pH, then an argument for electrostatic EF attraction can be made. Nevertheless, studies of smectite particle arrangements under the high magnification allowed by electron microscopy have shown only FF interaction. Apparent EF contacts, where individual platelets meet nearly at right angles, are revealed to be FF contacts permitted by the bending of the flexible silicate sheets into a parallel alignment. Such arrangements "cross-link" the clay gel, allowing semirigid structures to be built.

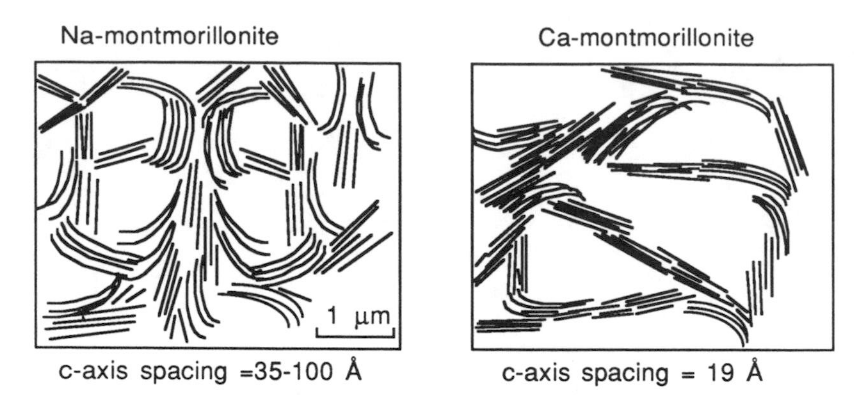

Figure 8.8. Stacking arrangement of the individual smectite layers (edge view) in the Ca^{2+} and Na^{+} exchange forms of clay gels formed in 10^{-3} *M* solutions of $CaCl_2$ and NaCl, respectively. Note the larger micropores and more extended stacking of the Ca^{2+} clay. (Adapted from H. Ben Rhaiem et al. 1985. Factors affecting the microstructure of smectites. In L. G. Schultz et al. (eds.), *Proceedings of International Clay Conference,* Denver. Bloomington, Ind.: Clay Minerals Society.)

8.4b. Role of Salinity and Sodicity in Swelling and Dispersion

Clay swelling is the initial process that enables the individual platelets of layer silicate clay aggregates to separate and form a stable dispersion. While this swelling can be suppressed by high concentrations of salt, such a situation is unusual except in saline soils. One might then expect swelling to be the natural condition of many soils; this is not the case because multivalent exchange cations, namely Ca^{2+}, Mg^{2+}, and Al^{3+}, counteract the tendency toward swelling by forming electrostatic bridges between adjacent platelets, creating larger particles as platelets stack together into "quasi-crystals." Consequently, in sodic soils, the $Na^{+}/(Ca^{2+} + Mg^{2+})$ ratio on exchange sites, defined by ESP, controls particle size, arrangement, and dispersibility. The diagram in Figure 8.8 reveals the very different microstructure of stacked platelets of Ca^{2+}-saturated and Na^{+}-saturated smectite gels.

The ESP has long been used as a measure of the dispersibility (aggregate instability) of a soil. As the ESP of a soil is increased, the CCC increases and the soil aggregate structure deteriorates. This in turn can be the cause of reduced permeability and drainage of the soil, poor aeration, and enhanced surface crusting and shrink-swell behavior under cycles of wetting and drying. ESP does not provide a complete description of clay dispersibility; clays with a given ESP are found to be more dispersible at high pH than at low pH. This pH tendency is less pronounced for smectites, which have dominantly permanent charge, than for kaolinites or illites. In the typical pH range of soils (5–9), smectite dispersibility is fairly constant. At extreme pH values (above 9, below 5), increased dispersion of silicate clays is sometimes seen, but this should be interpreted with caution because dissolution reactions could alter the mineral surface charge. Actually, layer silicate clays tend to flocculate at low pH because Al^{3+} is dissolved from the clay structure and adsorbs on remaining exchange sites.[1] This pattern of behavior is described qualitatively in Figure 8.9. Oxides of Fe

1. Acid mineral soils are commonly the most dispersion-resistant, stabilized by exchangeable Al^{3+}.

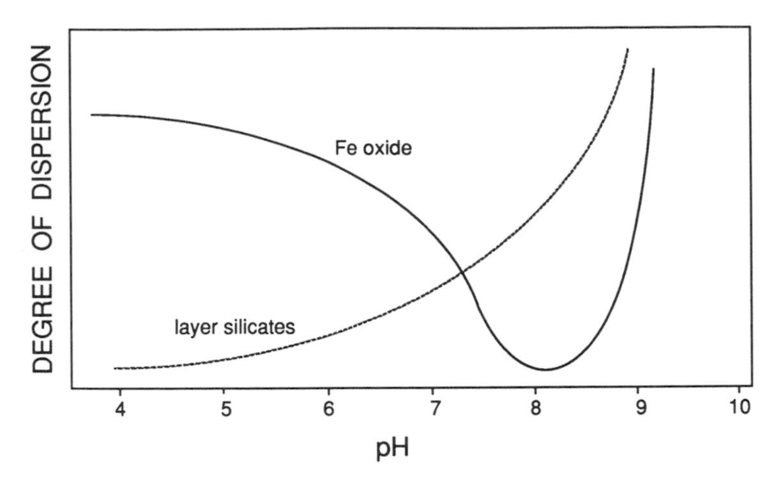

Figure 8.9. Dependence of the dispersibility of permanent-charge layer silicate clays and variable-charge oxides on solution pH.

and Al, because they possess no permanent charge, disperse at both low and high pH, tending to flocculate at a pH near the PZC of the oxide (Figure 8.9). This observation illustrates an important principle of colloidal dispersion, that:

> *Maximum stability of aggregated particles (minimum dispersion) occurs at the pH of lowest surface charge.*

In summary, soil colloids (including organic matter as well as silicate clays) disperse readily at low salt concentration, high exchangeable Na^+ levels, and high pH. Figure 8.10 describes graphically the boundary conditions that separate dispersion from flocculation. This graph indicates that, for any particular pH and ESP in the soil, there should be a defined critical coagulation concentration (CCC) above which the soil colloids flocculate. For many soils, the CCC is near 10^{-3} M (monovalent salt), and soils subjected to severe leaching may disperse as salts are removed from soil solution.

8.4c. The Osmotic Model of Clay Swelling

From the observations described in sections 8.4a and 8.4b, a simple description of layer silicate clay swelling based on water activity can be proposed. Electrolytes (salts) dissolved in water lower the free energy (or activity) of water relative to pure water according to the relation

$$\Delta G = RT \ln \frac{P}{P_0} \tag{8.28}$$

where P and P_0 are the vapor pressure of the solution and pure water, respectively, and ΔG is the change in free energy of water. Although the free energy of water in the interlayer region between silicate sheets must also be lower than that of free water, there is no direct way of evaluating the free energy change caused by interaction with the clay surfaces and exchange cations.

When a dry clay is immersed in a dilute salt solution, the free energy of water in

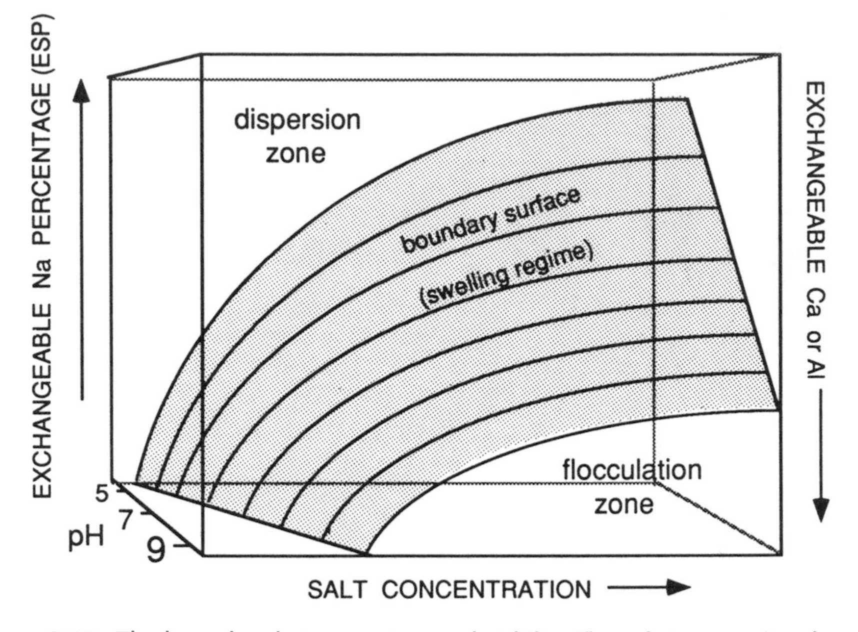

Figure 8.10. The boundary between structural stability (flocculation zone) and instability (dispersion zone) as influenced by chemical factors in soils.

the clay phase is initially lower than that in the solution. Since exchange cations cannot diffuse out of the clay phase, water molecules must diffuse inward (as depicted in Figure 8.11a) in order to equalize the free energy of water in the solution and clay phases. If the concentration of salt in solution is then increased, water moves out of the clay phase in response to the lowered free energy of water in the solution. As a result, the interlayer spacing begins to collapse, as shown in Figure 8.11b. This model is analogous to one in which a semipermeable membrane (permeable to water only) separates salt solutions of different concentration, creating an *osmotic pressure.* No physical membrane actually separates the clay from the solution, but semipermeability is nevertheless created by the capacity of colloidal clays to electrostatically constrain cations to a region near the surface. Water is free to move in and out of the clay phase, but these cations are not.

For dilute solutions, the osmotic pressure, π, is a function of the molar volume of water, V, the mole fraction of water in the solution, x_A, and the temperature, T:

$$\pi = -\left(\frac{RT}{V}\right) \ln x_A \tag{8.29}$$

An approximate form of this equation is $\pi = n_B RT/v$, where v is the volume of solution containing n_B moles of ions.

If a similar equation could be applied to the clay phase, π would represent the osmotic pressure or *swelling pressure* of the clay. It is possible, by monitoring the spacing between clay layers (using x-ray diffraction methods), to determine the external pressure that must be applied to a clay gel to exactly counterbalance the swelling pressure. If the applied pressure is greater than the clay's swelling pressure, the clay

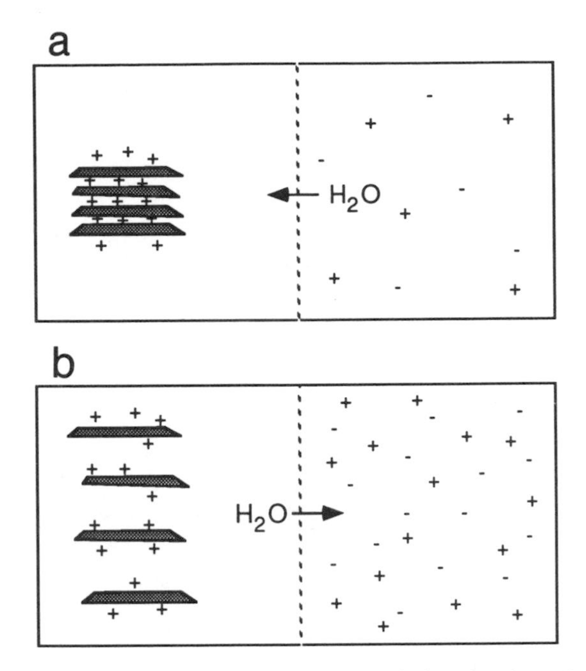

Figure 8.11. Osmotic model of clay swelling (a) and clay shrinking (b). The left chamber contains clay and is separated from the right chamber (containing salt solution) by a membrane permeable only to water.

layers will be pushed together as water is squeezed out of the interlayer. The result, plotted in Figure 8.12 for two freely swelling forms of silicate clays, Li^+-saturated vermiculite and Na^+-saturated montmorillonite, shows that two expanded states of the clays coexist at certain pressures. A minimum c-axis spacing ($\approx$40 Å for montmorillonite) is reached as increasing pressure is applied: the $>$ 40 Å phase is referred to as the "gel" phase of the clay. A second stable c-axis spacing ($\approx$20 Å for montmorillonite) is also apparent, becoming more dominant at higher pressure, and is termed the "hydrated-solid" phase.

The results, then, suggest that expanding clays swell from the dry state in two stages: first, hydration swelling that adds several molecular layers of water between the clay layers, and second, osmotic swelling that takes the layers out to large separation distances. This two-stage behavior is clear in the data of Figure 8.7, and the osmotic stage of swelling is typified by a linear relationship between interlayer spacing and the reciprocal of the square root of the solution salt concentration, $1/\sqrt{C}$. Hydration swelling, on the other hand, is much less sensitive to the concentration of salts in the external solution.

This swelling model does *not* apply to clays with mostly divalent and polyvalent cations on their exchange sites: these clays show the hydration stage of expansion but not the osmotic one. The explanation for their behavior is found in the next section.

To put the pressures plotted in Figure 8.12 into perspective, we can estimate that a 100-Å-thick interlayer would have an *average* Na^+ concentration of about 0.25 *M*

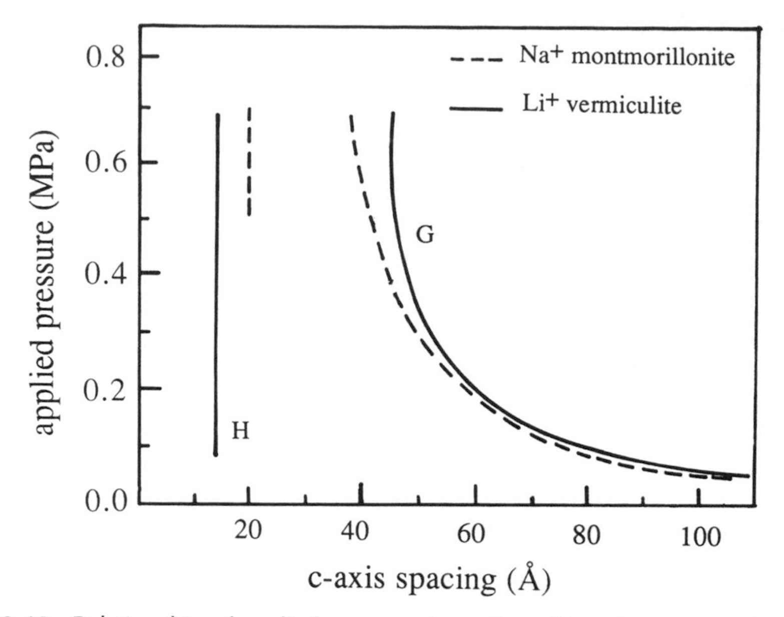

Figure 8.12. Relationship of applied pressure (megaPascals) to the measured c-axis spacing of expanded Li^+-vermiculite and Na^+-montmorillonite in 10^{-4} *M* LiCl and NaCl, respectively. The symbols G and H refer respectively to the two states, gel and hydrated solid, that appear to coexist. (Adapted from B. E. Viani et al. 1985. Direct measurement of the relation between swelling pressure and interlayer distance in Li^+ -vermiculite. *Clays Clay Min.* 33:244–250.)

(assuming a CEC of 0.9 mmole/g). Equation 8.29 can be used to estimate the osmotic pressure of this "solution," giving a pressure of about 0.6 megaPascal (MPa). As revealed by Figure 8.12, this is much more pressure than is actually needed to prevent the clay platelets from separating beyond 100 Å. Failure of this simple osmotic model to *quantitatively* explain clay swelling has put the model in disfavor with some clay scientists. The model predicts the exchange cations to generate a considerably higher osmotic pressure than is actually measured in the swelling pressure. It should be remembered, however, that equation 8.29 quantitatively applies only to dilute solutions (which the interlayer regions of silicate clays clearly are not). Furthermore, it is really the concentration of cations at the midpoint between the clay plates that determines the incremental osmotic swelling pressure, a concentration that is less than the average concentration of interlayer cations. A large fraction of some exchange cations may be localized at the surfaces, not dissociated from the surface. Such cations would contribute little to the osmotic pressure. One of the most common ways to calculate predicted swelling pressures of smectites is to use the mid-plane cation concentration obtained from the diffuse double-layer equations (see Appendix of Chapter 3 for a discussion of diffuse double-layer theory and near-surface ion distribution), and obtain the predicted pressure from equation 8.29. As long as this pressure is greater than the osmotic pressure of the external solution, the clay will continue to swell. Finally, the swelling pressure, which is the expression of the imbalance of all operative forces, may not be a direct reflection of osmotic pressure because attractive electrostatic forces counter the repulsive osmotic force to some (unknown) degree.

An osmotic model of clay swelling can be defended by the fact that the surface O atoms of layer silicates are weak electron donors, and cations such as Li^+ (and to a considerable extent Na^+) dissociate from the silicate surface on hydration. Osmotic pressure develops from the presence of these cations in the interlayer region.

8.4d. The Hydration States of Expanding Layer Silicate Clays

Clay expansion from the dry state up to several monolayers of adsorbed water molecules is driven by powerful hydration forces, not osmotic forces. This is evident when comparing swelling of the Na^+-saturated forms of smectites with that of the Ca^{2+}-saturated forms in Figure 8.13. At these small interlayer distances, hydration of the exchange cations and the clay surface are the dominant driving forces for expansion. With either Na^+ or Ca^{2+} as the exchange cation, the 12-Å phase, which is the one-layer hydrate (diagrammed in Figure 8.14), is stable at very low relative humidity (low free energy of water). The 15-Å phase, which is the two-layer hydrate (also diagrammed in Figure 8.14) is stable at higher humidity. However, the transition from the one-layer to the two-layer hydrate occurs at much lower relative humidity in the Ca^{2+}-smectites, a result of the greater hydration energy of Ca^{2+} compared with Na^+.

To get some idea of how much expansion pressure could be generated by the one- to two-layer hydration of Ca^{2+}-smectite, we can use equation 8.28 to calculate the free

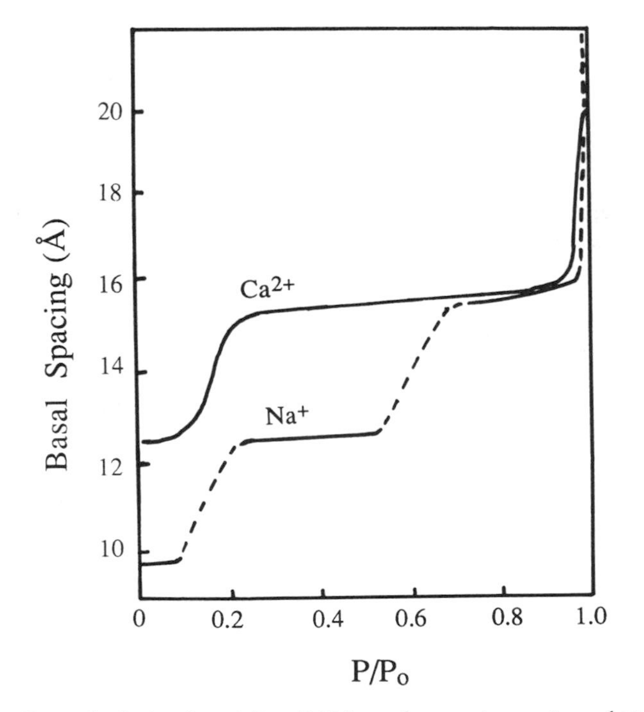

Figure 8.13. Effect of relative humidity (P/P_o) on the c-axis spacing of Na^+ and Ca^{2+} exchange forms of smectite. (Adapted from H. Suquet et al. 1975. Swelling and structural organization of saponite. *Clays Clay Min.* 23:1–9.)

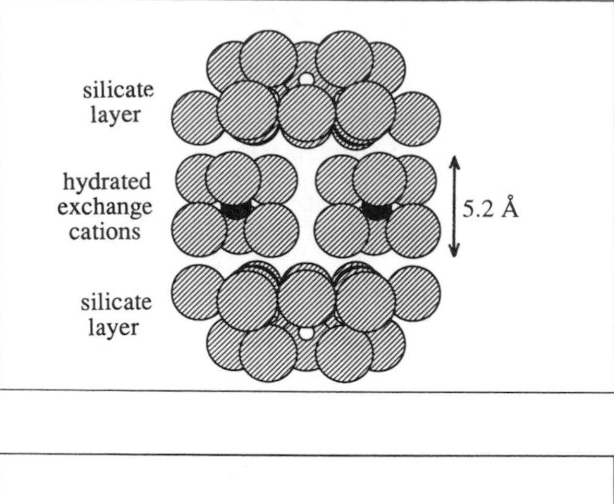

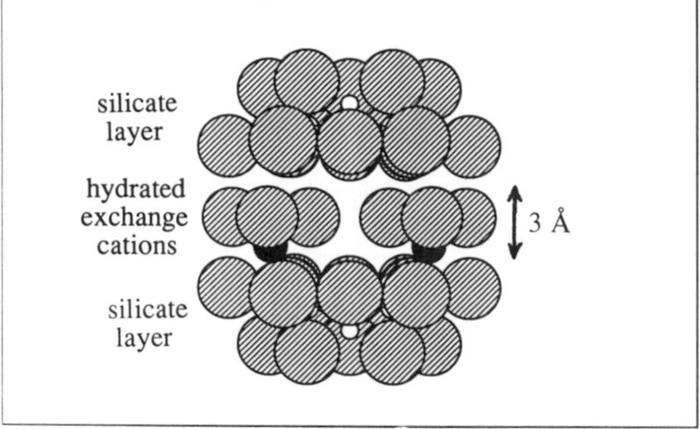

Figure 8.14. Cross-sectional view of the interlayer region in the one-layer and two-layer hydrate of layer silicate clays. Water molecules and silicate oxygens are represented as large shaded circles; exchange cations are shown as smaller black circles. The c-axis is oriented in the direction of the arrows.

energy of water at a relative humidity of 20 percent ($P/P_0 = 0.2$), which is about -950 cal/mole of water (relative to free water). Using the fact that 1 cm^3 of water contains 0.056 mole of H_2O, the energy units can be converted to pressure units (energy per unit volume is dimensionally equivalent to pressure), and it can be shown that free water adsorbed by Ca^{2+}-smectite at 20 percent relative humidity (the clay is about 25% water by volume in this dry state) would generate more than 500 atmospheres (bars) of pressure. On an area basis, this is about 0.5 metric ton/cm^2! It is not difficult to see how clay expansion could heave roads, buildings, and other structures that are built on dry soils containing smectites or other clays capable of hydrating and undergoing volume change.

When the relative humidity is raised to 100 percent, Na^+-smectites seem to expand freely, although Figure 8.12 indicates a discontinuity between a 20-Å phase ($\approx$four-layer hydrate) and a 40-Å gel phase. This discontinuity separates the "clay hydration" mode of swelling from the "osmotic" mode, described in Figure 8.15 as stage 1 and stage 4, respectively, of declining system energy. Coexistence of two apparently stable expanded states of Na^+-smectite suggests that the potential energy

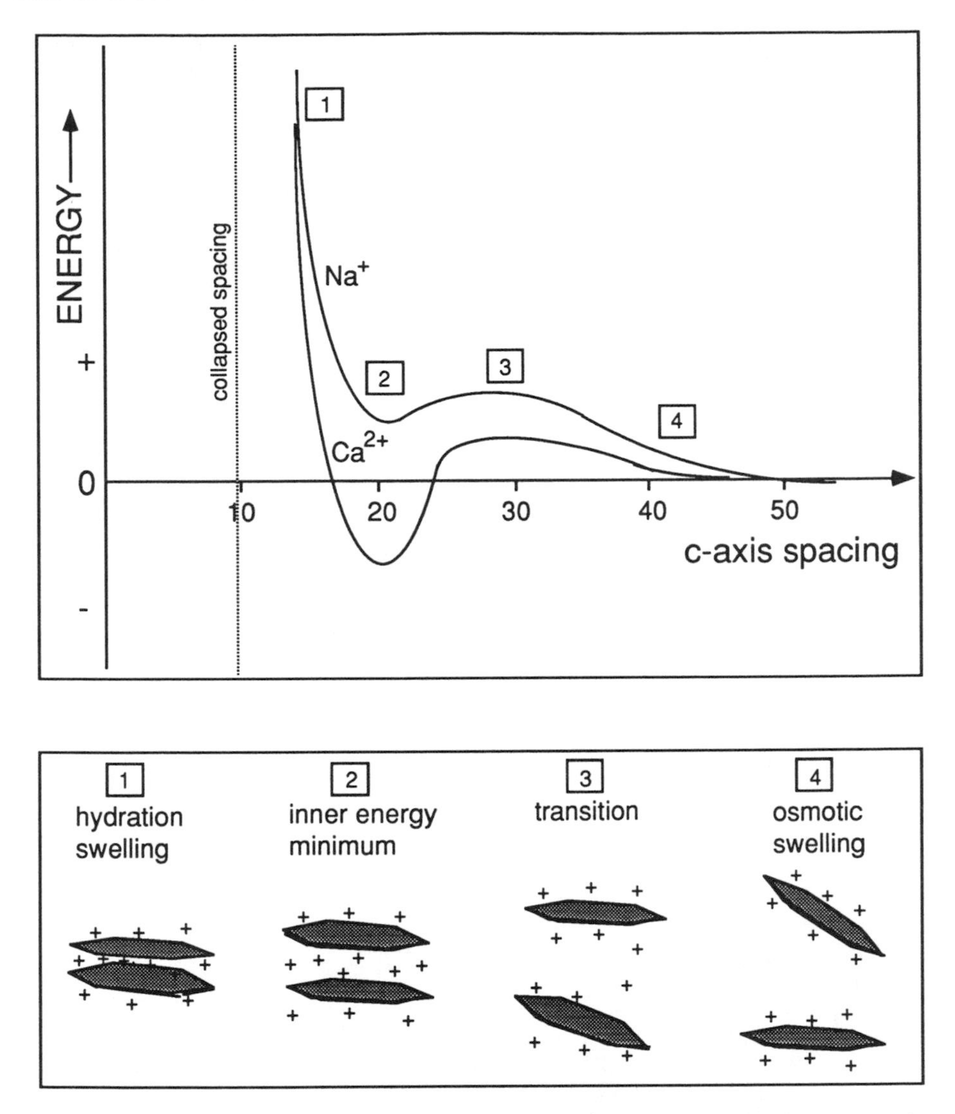

Figure 8.15. Suggested dependence of the energy of a smectite-water system on the c-axis spacing. The stages of clay expansion are numbered and diagrammed to show interparticle and counterion (+) arrangement.

curve of the clay-water system has two energy minima. The inner minimum (stage 2 in Figure 8.15) corresponds to the situation in which the counterions of adjacent aligned plates are coalesced, while the outer minimum (stage 4) follows the division of the counterions as they associate with one particular surface. The energy barrier (stage 3) separating these two minima may represent the energy required to overcome the electrostatic attraction in the 20-Å "quasi-crystal" phase and separate the layers with their associated counterions. Once the counterion populations are separated with their respective particles, weak electrostatic or osmotic repulsion maintains the dispersed state.

Measurements of the forces between hydrated layer silicate sheets (micas) have shown several energy minima at sheet separations that are integral multiples of the molecular diameter of water. Thus, the hydration force is an oscillating function of sheet separation at distances of less than 20 Å between silicate surfaces, as shown graphically in Figure 8.16. It arises from the energetically favorable packing of water molecules between layers that are separated by an integral multiple of the molecular diameter of water, and therefore the force has a periodicity of about 2.5 to 3.0 Å. Nonintegral separations necessarily result in voids, inefficient packing, and a higher energy state of the system.

The oscillating hydration force is superimposed on a longer range repulsive force between the layers, which is predicted to be an osmotic force from the diffuse double-layer theory (but may still contain a hydration component). Thus, as Figure 8.16 shows, closer separations between the silicate sheets result in the oscillating interlayer force becoming more repulsive on average.

Unlike Na^+-smectite, Ca^{2+}-smectite expands to a c-axis spacing of 19 to 20 Å at 100 percent relative humidity, and expands no further. Since the osmotic force is repulsive at all separation distances, this limited expansion implies the existence of

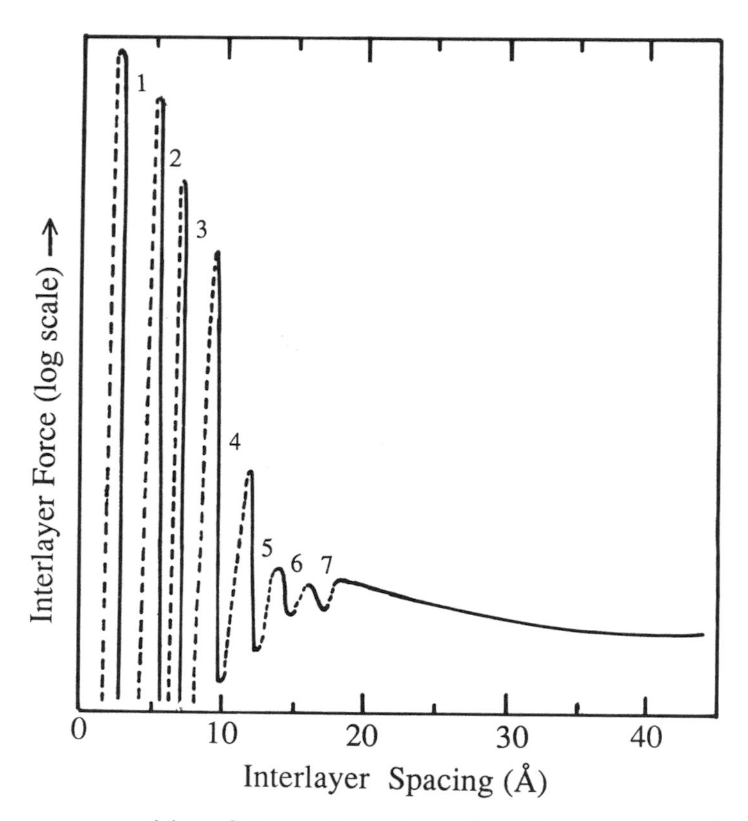

Figure 8.16. Measured force between two mica surfaces immersed in 10^{-3} *M* KCl solution. The force oscillates with a periodicity of 2.5 to 3.0 Å, about equal to the diameter of a water molecule. The numbers count how many molecular layers of water are expected to be between the surfaces at each oscillation. (Adapted from J. N. Israelachvili. 1985. Measurements of hydration forces between macroscopic surfaces. *Chemica Scripta* 25:7–14. Used with permission.)

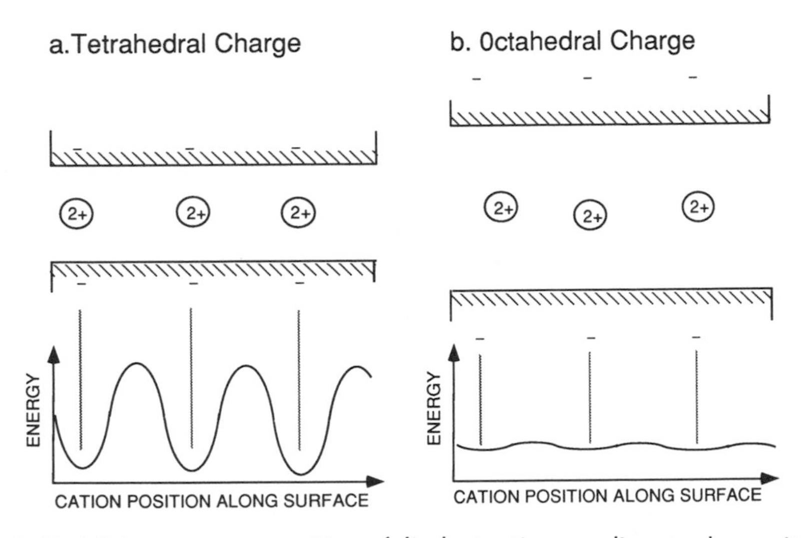

Figure 8.17. Minimum energy position of divalent cations on discrete charge sites (−) of tetrahedrally charged clays (a) and smeared charge sites (−) of octahedrally charged clays (b). This diagram is idealized in that charge positions on clays are not likely to be evenly spaced.

an attractive force to counteract the tendency toward osmotic swelling. The resultant of these two opposing forces must produce an energy minimum to account for the 20-Å spacing, as depicted in Figure 8.15. However, this energy "well" is apparently deeper for Ca^{2+}-smectite than for Na^{+}-smectite, because the ≈20-Å phase is much more stable when Ca^{2+} (or any other multivalent metal cation) is the exchange cation. The attractive force that limits swelling in clays with predominantly divalent and polyvalent exchange cations is electrostatic in nature, and is particularly important in clays with most of their layer charge localized in the tetrahedral sheet (such as vermiculite and certain types of smectites). A diagrammatic explanation for the strong electrostatic attraction between clay layers occupied by Ca^{2+} ions, enhanced by the more localized layer charge of clays with tetrahedral isomorphous substitution, is shown in Figure 8.17.

The deeper energy "well" noted with divalent, as opposed to monovalent, exchange cations needs an explanation. Diffuse double-layer theory, while explaining satisfactorily the swelling pressure of Na^{+}-smectites, fails to predict the much lower swelling pressure (i.e., stronger attractive force) of Ca^{2+} smectites. The attractive force is probably dominantly electrostatic, with an attractive energy given roughly by

$$E = \frac{z^{+} z^{-} e^{2}}{\epsilon r_{A}} \tag{8.30}$$

where z^{+} is the cation charge, z^{-} is the local surface charge, e is the unit of electron charge, ϵ is the dielectric constant of the solvent (water) separating the two charges by a distance r_A. At small interlayer spacings, most of the repulsive force arises from cation hydration, which is dominantly an ion-dipole interaction with energy given by

$$E = \frac{-z^{+} e\mu}{r_{B}^{2}} \tag{8.31}$$

where μ is the dipole moment of the water molecule and r_B is the cation-dipole distance. As the clay layers move apart during swelling, both r_A and r_B increase, but the electrostatic attractive energy diminishes with increasing layer separation distance r, according to $1/r$ while the repulsive energy diminishes according to $1/r^2$. This means, as shown in Figure 8.18, that the electrostatic force is more long range than the hydration force, and that an energy minimum, or "well," exists that permits a stable hydration state to form with several layers of water molecules in the interlayer space.

Of course, this model implies that a similar stable hydration state should exist for all layer silicate clay systems, and indeed it does for vermiculites (with the exception of Li^+-vermiculite, which evidently has an unusually strong short-range repulsive energy because of the high hydration energy of the very small Li^+ ion) and for smectites saturated with divalent (e.g., Ca^{2+}, Mg^{2+}) and trivalent (e.g., Al^{3+}) cations. Why then do Ca^{2+}- and Na^+-smectite have such different swelling behavior? The answer must be that the Ca^{2+} clay has a larger attractive electrostatic force tending to hold the layers together, consistent with the diagrams in Figure 8.18. Contributing to this attractive force may be the fact that high-charge cations strongly order the dipoles of water molecules around them, creating a phenomenon called *dielectric saturation.* The effective dielectric constant of water in the vicinity of high-charge, small-radius cations is much *lower* than the bulk water value of about 80, approaching 3 to 6 within a few Ångstroms of the ion. Small, high-charge cations are associated with a larger number of water molecules, which make up the so-called hydration sphere. For example, the hydration sphere of Na^+, a fairly large low-charge cation, consists of about four water molecules, while that of the smaller Mg^{2+} ion comprises about sixteen water molecules. Consequently, the dielectric constant, ϵ, in equation 8.30 is smaller for Ca^{2+}-smectite than for Na^+-smectite. This may account at least in part for the fact that the electrostatic (attractive) component of the total energy is larger in the Ca^{2+}-smectite. The result is the limited expansion described by Figure 8.18a. With Na^+-smectite, the lower attractive energy means that any energy "well," if it does exist, is not very deep, as diagrammed in Figure 8.18b. There is some evidence, however, for a quasi-stable 20-Å hydration state of Na^+-smectite, as described earlier (see Figures 8.12 and 8.15).

Recent statistical mechanical analyses of Ca^{2+}- and Na^+-smectites indicate that another fundamental difference exists between them (Kjellander et al., 1988). Ca^{2+} in interlayers interacts sufficiently strongly with other interlayer Ca^{2+} ions that instantaneous relative positions of these ions are correlated to some degree. As a result, a net electrostatic attraction between clay layers is generated by these "correlation forces." The Na^+-Na^+ interactions are weaker so that correlation forces in Na^+-smectites are too weak to counter the repulsive osmotic and hydration forces. However, because these correlation forces were calculated for clays in which the layer charge was assumed to be evenly "smeared" over the surface, the analysis failed to show a "configurational" electrostatic attraction force of the type suggested by Figure 8.17. An attractive force between layers can arise from the arrangement of cations relative to layer charge sites depending on the extent to which the charge sites are localized. On octahedrally charged smectites, the layer charge is rather delocalized, and "configurational" attraction may be weak.

The presence of localized layer charge would affect the behavior of Ca^{2+} and Na^+ exchange ions on clays differently. Monovalent cations would approach these charge sites to the extent permitted by hydration forces and thermal motion, thereby minimizing electrostatic energy. Divalent cations, however, could not closely approach the charge sites, not only because of hydration forces, but also because this approach would overcompensate the layer charge locally and induce a repulsion. Therefore divalent exchange cations tend to reside in the middle of the interlayer where they hold the layers together.

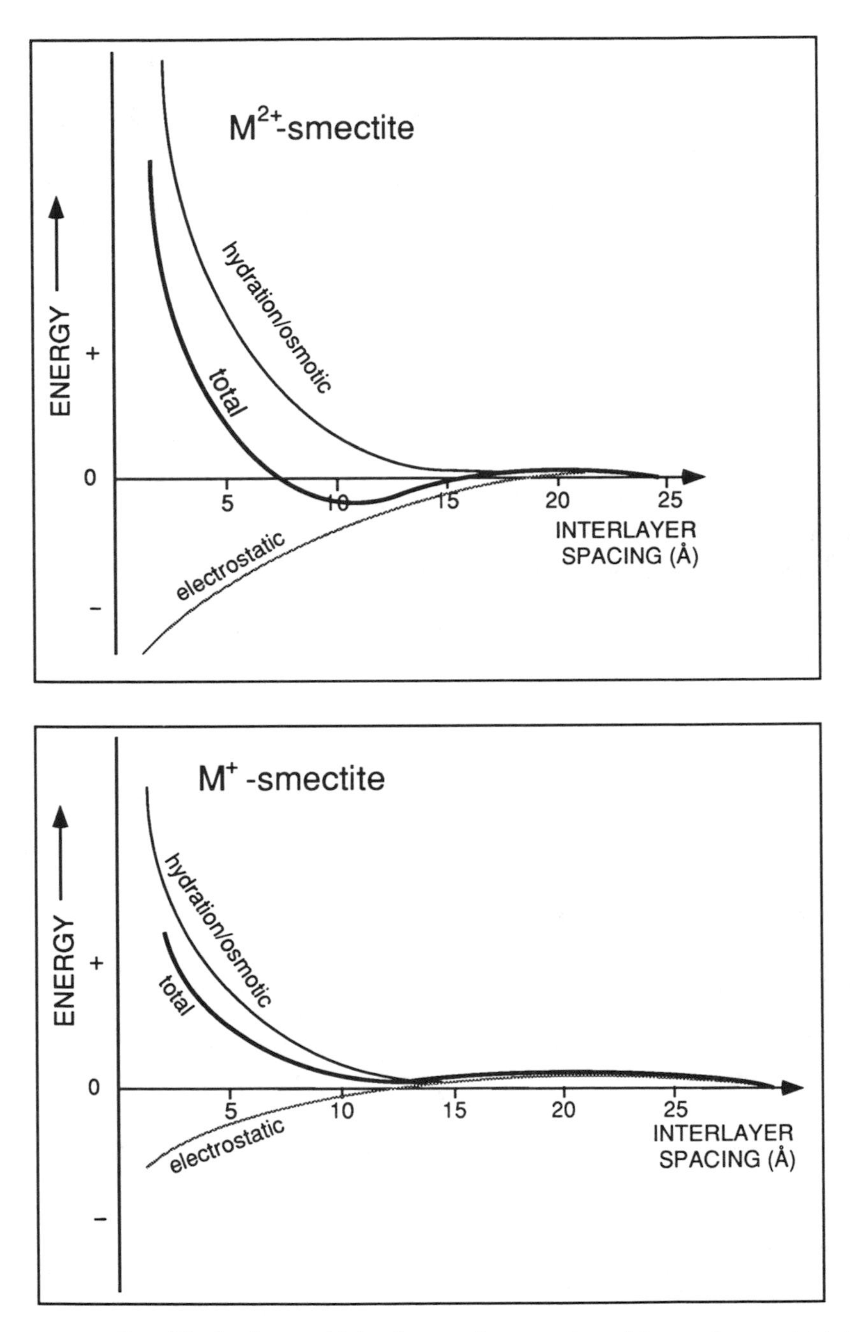

Figure 8.18. Simplified scheme of the forms of energy (repulsive and attractive) that combine to give the total energy of M^{2+}-saturated and M^{+}-saturated smectite as a function of the interlayer spacing in water.

It should be pointed out that the electrostatic force in the classical diffuse double-layer theory (described in Appendix of Chapter 3), referred to as the DLVO (Derjaguin–Landau–Verwey–Overbeek) theory, is assumed to be *repulsive* at *all* separation distances because it arises from the overlap of static double layers. The interparticle attraction is then attributed to van der Waals forces. However, there is uncertainty about the magnitude and range of the van der Waals forces in clay-water systems. Furthermore, double layers would repel only as long as they were far enough apart to prevent a mingling of the counterion "swarms" around individual particles. Once the particles approached closely enough, these swarms would coalesce and the electrostatic force would become attractive. For these and other reasons, the model presented here assumes that the electrostatic force is attractive, at least at short range ($<$ 20-Å separation), and that repulsion is largely due to hydration (short range) and osmotic forces (long range).

With Ca^{2+}-smectite, the stable two-layer hydrates (of air-dry clay) and four-layer hydrates (of fully wet clay) are described by the arrangement of water molecules pictured in Figure 8.19. Such ordered arrays of water dipoles stabilize the clay-water system by providing "dipolar links" to the surface, in effect transferring positive charge from the cations to surface O atoms. In the case of vermiculites, the structural negative charge is nearer the surface, and distributed over fewer surface O atoms, so that the two-layer hydrate is more stable than the four-layer hydrate. Because smectites typically have lower charge than vermiculites, and much of this charge is often buried deeper in the structure (in octahedral as opposed to tetrahedral sites), water molecules bond less strongly on smectite surfaces.[1] The more diffuse distribution of negative charge on smectites favors an extended array of "dipolar links" to the surface, which is more possible in the four-layer than the two-layer hydrate. However, this four-layer structure should be viewed as dynamic. Experimental evidence shows that hydrated exchange cations tumble rather freely in the $\approx$10-Å space created by these four layers of water molecules.

The impression should not be given that the division between clays that freely swell in water and those that do not is a clear one. Some vermiculites have sufficiently low charge that they freely expand if Li^+ ions occupy the exchange sites. Some smectites, if most of their charge resides in tetrahedral sites, do not expand beyond an 18 to 20-Å c-axis spacing in water even when saturated with Na^+ ions. The criteria that should be used to distinguish smectites from vermiculites are likely to be debated for some time. It is now clear that some clays have structural and swelling properties that place them in the nebulous zone between "well-behaved" smectites and vermiculites (see Figure 2.14).

The tendency of clays to expand in water is a function of both the *magnitude* and *location* of layer charge. In Figure 8.20, the hydration swelling transition of Na^+-smectites from the $\approx$15-Å to the 19 to 20-Å spacing is seen to occur at some critical concentration of NaCl in the external solution, a concentration that depends on the charge properties—both magnitude and location—of the smectite. Those smectites with high tetrahedral charge tend not to expand at all in the concentrated NaCl solutions, whereas those with lower or primarily octahedral charge undergo expansion once NaCl concentrations are lowered below 0.5 to 2.0 molal.

1. This fact will be seen in Chapter 10 to be very important in determining the ability of different clays to adsorb organic molecules from solution.

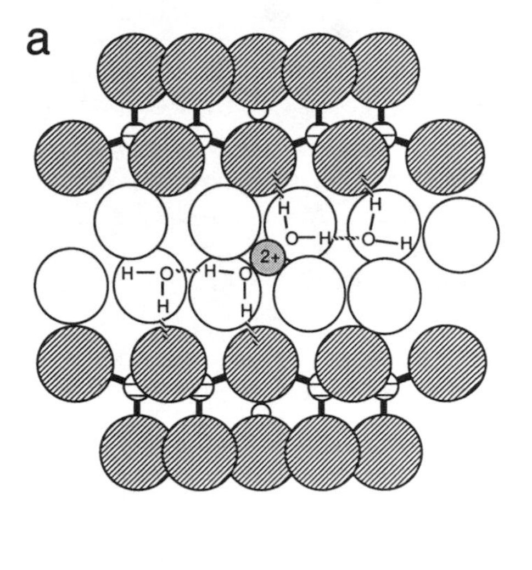

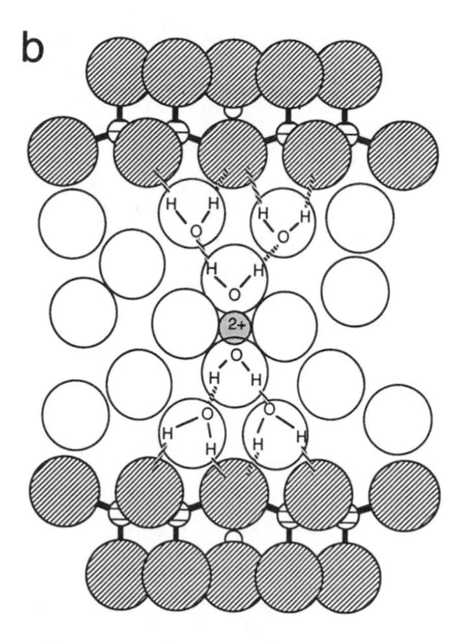

Figure 8.19. Arrangement of H_2O chains linking divalent exchange cations to the surfaces of the 15-Å two-layer hydrate (a), and the 20-Å hydrate (b). (Adapted from V. C. Farmer. 1978. Water on particle surfaces. In D. J. Greenland and M.H.B. Hayes (eds.), *The Chemistry of Soil Constituents.* New York: Wiley.)

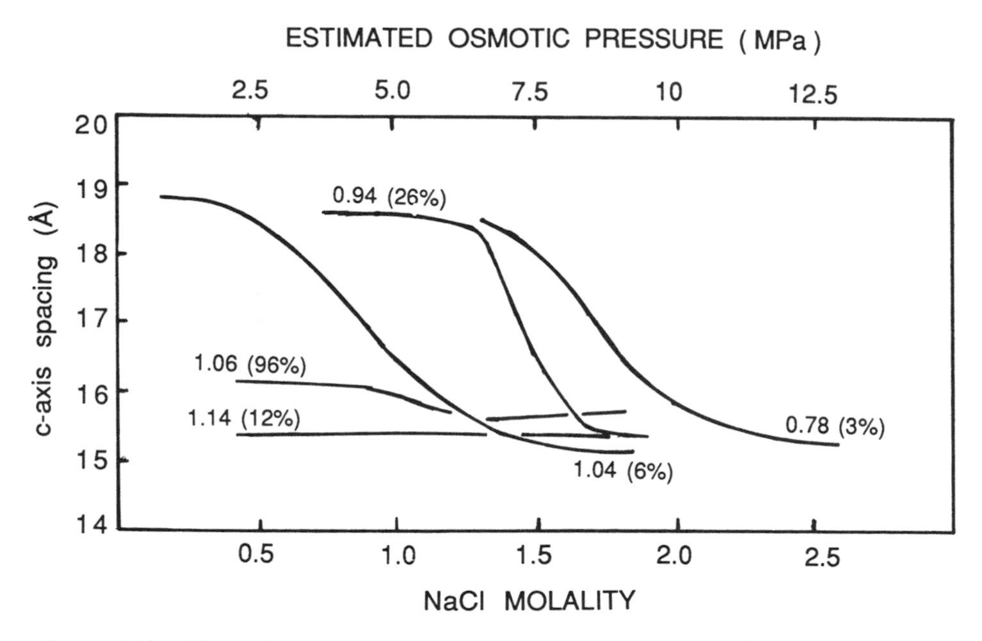

Figure 8.20. Effect of NaCl concentration on the c-axis spacing of Na^+-saturated smectites having different total layer charge (number labeling curves is the layer charge per unit cell) and percentage of charge in tetrahedral sites (value in brackets). (Adapted from P. G. Slade et al. 1991. Crystalline swelling of smectite samples in concentrated NaCl solutions in relation to layer charge. *Clays Clay Min.* 39:234–238.)

The osmotic pressures of the NaCl solutions in which the clays expand can be estimated from equation 8.29. The mole fraction of water, x_A, in, say, a 0.5 molal NaCl solution, is calculated from the fact that this solution contains 1.0 mole of Na^+ plus Cl^- ions dissolved in 1000 ml (55.5 moles) of water. The mole fraction, x_A, is then $55.5/(55.5 + 1.0) = 0.982$. Equation 8.29 is then solved for room temperature conditions to give $\pi = -[(82.05 \times 297)/18] \ln 0.982 = 24.6$ atmospheres. This is about 2.5 MPa of osmotic pressure, against which the smectite must develop an equivalent swelling pressure in order to expand from the ≈15-Å to the 19 to 20-Å spacing. The more freely swelling Na^+-smectites, such as hectorite (layer charge = 0.78), can generate swelling pressures of 100 atmospheres or more during this expansion. Such smectites are often used as "models" of soil clays, but it is clear from Figure 8.20 that they represent one extreme of swelling behavior. Expansion of many layer silicate clays occurs only in dilute salt solutions if at all. Vermiculites and illites as well as high-charge smectites belong to a large and abundant group of clays that do not expand freely, even when exchange sites are saturated by Na^+.

It can be inferred from the osmotic model of clay swelling (discussed in the previous section) that any influence on the clay-water system that lowers the free energy of water in the aqueous phase will reduce clay swelling. As seen earlier, high salt

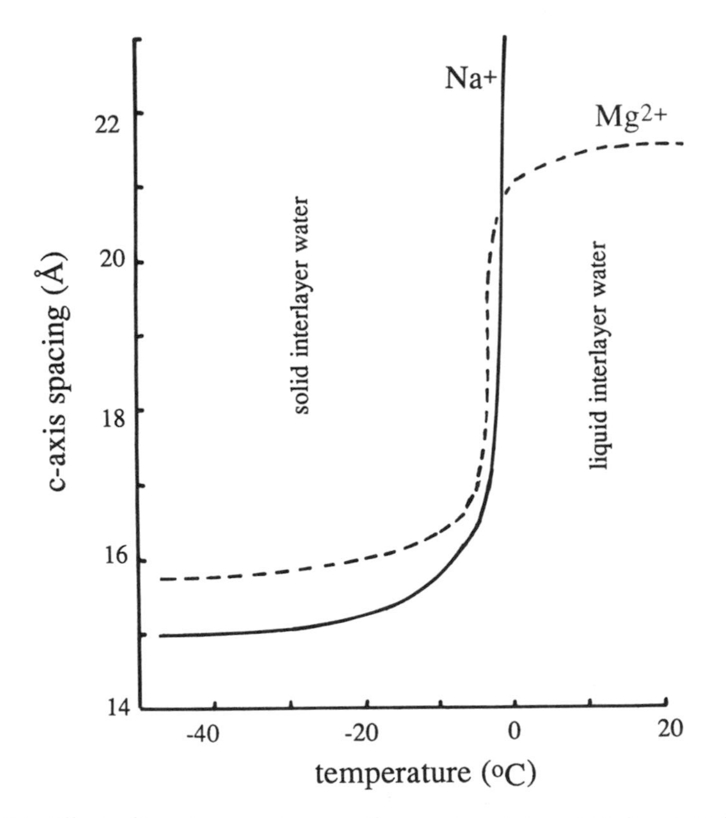

Figure 8.21. Effect of low temperature on the c-axis spacing of Na^+ and Mg^{2+} exchange forms of smectite gels.

Table 8.3. Swelling Volumes of Metal-Exchanged Smectites in Water (cm^3/g)

H^+	Li^+	Na^+	K^+	Ca^{2+}	Ba^{2+}
(2.20)[a]	10.8	11.1	8.6	2.5	2.5

[a]Clay has probably decomposed to the Al^{3+}-saturated form.

concentrations reduce the interlayer spacing (see Figure 8.7), as do low water vapor pressures created by air drying (see Figure 8.13) or freezing (Figure 8.21). In the last case, subzero temperatures cause ice crystals to form outside the clay interlayers, as water is drawn out of the collapsing interlayer regions. Nevertheless, Figure 8.21 shows that, even at temperatures as low as $-50°C$, at least two monolayers of "unfrozen" water remain in the interlayer. Soil freezing, then, has an effect on clay dehydration that is comparable to the effect of air drying. This enhances quasicrystal formation and improves soil structure.

The increase of soil volume from clay swelling in the field has important practical consequences, particularly in soils containing smectites. The extent of this problem depends very much on the nature of the exchange cation, as illustrated by the swelling volumes of smectites reported in Table 8.3. These results emphasize the management problems likely to be encountered with soils that are sodic.

8.5. EFFECTS OF SALT-DEGRADED SOILS ON PLANTS

There are basically three potential hazards of salt-degraded soils to plants—salinity, sodicity, and alkalinity. The first two hazards are used to classify soils as *saline* or *sodic,* or both, as described in Table 8.4. The alkalinity hazard (measured by RSC) is not usually applied directly to the classification of soils, but the pH values reported in Table 8.4 indicate that sodic soils are the most likely to have a problem of high alkalinity. This arises partly from the hydrolytic exchange reaction, described in Chapter 3, which is enabled by the presence of appreciable exchangeable Na^+ and low salt concentrations, conditions that are diagnostic of sodic soils. Hydrolytic exchange can be viewed as a two-step reaction:

$$Na^+\text{-clay} + H_2O = H^+\text{-clay} + NaOH \tag{8.32}$$

$$NaOH + H_2CO_3 = NaHCO_3 + H_2O \tag{8.33}$$

Table 8.4. Classification of Saline and Sodic Soils

	EC (mS/cm)	ESP (%)	Typical pH	Structure
Saline	>4	<15	<8.5	Good
Sodic	<4	>15	>9.0	Poor
Saline-sodic	>4	>15	<8.5	Fair-good

that produces alkaline salts. It should be apparent that reaction 8.32 does *not* actually *create* alkalinity—instead, the mechanism separates alkaline salts from acidified clays, permitting the soluble forms of alkalinity to accumulate by evaporation (i.e., at the soil surface). This process is suppressed by high salt levels in saline and saline-sodic soils, since excess soluble cations inhibit the adsorption of H^+ on exchange sites. As a result, saline and saline-sodic soils do not typically have severe alkalinity problems.

The three hazards of salt-degraded soils can create a problem for plant growth in several ways. High salinity lowers the free energy of water in soil solution, thereby reducing the ability of the plant roots to extract water from the soil. This is an osmotic effect that becomes significant only in extremely saline soils. The osmotic pressure generated by salts in solution can be estimated from the electrical conductivity (EC) of the solution using the empirical equation

$$\text{Osmotic pressure (atmospheres)} = \text{EC (mS/cm)} \times 0.36 \qquad (8.34)$$

For a soil that is barely saline (EC $\approx$ 4 mS/cm), an osmotic pressure of about 1.5 atmospheres would result. In other words, the water in this moist soil would not be freely available to plant roots, but would be retained in a lowered free energy state equivalent to 1.5 atmospheres of water tension. This, however, is insufficient tension to seriously impede water uptake by plants.[1] More important effects of soil salinity are likely to include particular ion toxicity effects (e.g., Na^+, Cl^-) and nutritional imbalances (e.g., excessive Na^+ or K^+ uptake relative to Ca^{2+} and Mg^{2+}).

Beyond the potentially toxic accumulation of Na^+ in plant tissue, sodicity presents a physical limitation to plant growth due to the dispersing effects of exchangeable Na^+ on clays and organic colloids. Although dispersion is suppressed in saline-sodic soils by high salt concentrations, if the excess salt is removed by leaching, clay swelling and dispersion result (see section 8.4). The consequences of dispersion are colloid migration, the clogging of soil pores, and dramatic reduction in the rate at which water can percolate through the soil. Dispersed soils have very poor structure, with the result that surface crusting, cementation, and soil erosion become problems. Since humus becomes dispersed in sodic soils, it can deposit at the surface during the evaporative loss of water from the soil. This process gives some sodic soils a characteristically dark appearance; the name "black alkali" has been used historically to describe these soils.

Alkalinity of soil solutions can create additional toxicity for plants. Typically, the alkalinity is in the form of HCO_3^- and CO_3^{2-}, anions known to reduce the availability of Fe to plants. The high pH is likely to reduce the availability of numerous other micronutrients (e.g., Zn, Mn) as well. At high pH, Al solubility actually increases because of the reaction:

$$Al(OH)_3\,(s) + OH^- = Al(OH)_4^- \qquad (8.35)$$

and the dissolution of Al-organic complexes. There is some evidence that Al toxicity might contribute to poor plant growth in alkaline soils. Certain micronutrients, notably Cu and Mo, also become more soluble in alkaline soils.

1. Most plants reach their permanent wilting point at about 15 atmospheres of water tension.

8.6 AMELIORATION OF SALT-DEGRADED SOILS

The above discussions have stressed that there are three quantifiable measures of degradation in salt-affected soils: salinity, sodicity, and alkalinity. In reclaiming the soils for agriculture, all three must be diminished to acceptable levels. Since salt-degraded soils are typically associated with semiarid and arid climates, irrigation water is often a factor in their formation or reclamation. Water of poor quality can degrade soils; water of good quality can improve them. Water quality is measured by the degree of salinity, sodicity, and alkalinity, as shown in Table 8.5.

In dry land water management, the assumption is usually made that continuous application of irrigation water to the soil will ultimately cause the soil to inherit any undesirable chemical characteristics of the applied water. That is, although the soil is to some degree "buffered" against compositional change, it must gradually come to a new equilibrium that is set by the properties of the irrigation water. For example, irrigation water with an SAR value of 13, if applied continuously to a nonsodic soil, would ultimately raise the exchangeable sodium percentage (ESP) to a level estimated by equation 8.24:

$$\frac{\mathrm{ESP}}{100 - \mathrm{ESP}} = 0.015 \times 13 = 0.195 \qquad (8.36)$$

The solution to equation 8.36 is ESP $=$ 16.3 percent, so that the soil is predicted to eventually exceed 15 percent exchangeable Na^+ and acquire sodic features.

Besides the sodicity hazard, irrigation water can also pose a salinity hazard for soils. This danger is magnified by the process of evapotranspiration, as can be understood by applying a simple mass balance to the salt content of the soil. Because the EC of drainage and irrigation water estimates fairly well the salt concentration of this water, then EC multiplied by the water volume (V) is roughly proportional to total salt in the drainage or irrigation water. In a situation of steady state, the soil is neither accumulating nor losing salt (plant uptake of salts is assumed negligible), so that the quantity of salt coming into the soil dissolved in irrigation water, Q_{IW}, must equal the quantity of salt escaping in drainage water, Q_{DW}:

$$Q_{\mathrm{IW}} = Q_{\mathrm{DW}} \qquad (8.37)$$

or equivalently:

$$EC_{\mathrm{IW}} \times V_{\mathrm{IW}} = EC_{\mathrm{DW}} \times V_{\mathrm{DW}} \qquad (8.38)$$

Table 8.5. Characterization of Water Based on Its Potential to Degrade Soil Properties

Salinity Hazard (EC)	Sodicity Hazard (SAR)	Alkalinity Hazard(RSC)
	← Low (safe) →	
0.25	7	1.25
	← Medium (marginal) →	
0.75	13	2.5
	← High (unsuitable) →	
2.25	20	-
	← Very high →	

It is assumed for this most simple scenario that rainfall is very low; therefore, irrigation water provides virtually all of the water that enters the soil. The steady-state situation, along with the situations in which salts are accumulating or being lost from the soil profile, is diagrammed in Figure 8.22.

It is apparent from equation 8.38 that the salinity of the irrigation water determines whether salt buildup in the soil can be avoided by the application of a sufficient volume of irrigation water. Evapotranspiration concentrates salt in the soil. It might be found, for example, that 80 percent of the applied irrigation water is subsequently lost at the soil surface by evaporation and transpiration, meaning that the volume of drainage water is only 20 percent of the volume of irrigation water (in a steady-state situation where the soil moisture level is not changing). By rearranging the steady-state equation 8.38, it becomes apparent that for this situation:

$$\frac{EC_{IW}}{EC_{DW}} = \frac{V_{DW}}{V_{IW}} = 0.2 \tag{8.39}$$

and the EC (or salt concentration) in the drainage water is expected to be five times that in the irrigation water. The value of V_{DW}/V_{IW}, in this case equal to 0.2, is termed the *leaching ratio* (or leaching requirement) because it specifies the *minimum* volume of irrigation water needed to prevent salt buildup. Application of suitable irrigation water greater than indicated by this ratio would actually decrease salt levels in saline soils, and would provide a margin of safety, but this may not be feasible because of the lack or cost of good quality irrigation water.

Amelioration of salt-degraded soils in arid climates requires that salts be removed by irrigation and drainage. The irrigation water must have acceptable values of EC, SAR, and RSC as described in Table 8.5. Even so, saline-sodic soils are likely to disperse as the excess soluble salts are leached out, and sodic soils may be impermeable to such a leaching treatment from the start. In practice, flocculating agents such as gypsum ($CaSO_4$) should be applied to saline-sodic and sodic soils before the salts are leached out. Lime ($CaCO_3$) is not recommended as a source of Ca^{2+} to promote flocculation; it is only slightly soluble in nonacid soils and it contributes to the

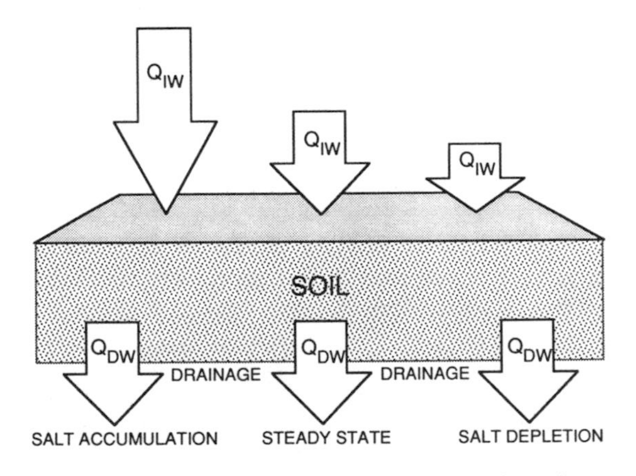

Figure 8.22. The three possible situations of mass balance for salts in soils of arid and semiarid climates.

total alkalinity of the soil. In fact, acidifying agents such as elemental sulfur or sulfuric acid can be beneficial amendments for the more alkaline salt-degraded soils. These agents neutralize alkalinity and bring the pH into a more acceptable range for plant growth.

References

Ben Rhaiem, H., C. H. Pons, and D. Tessier. 1985. Factors affecting the microstructure of smectites: Role of cation and history of applied stresses. In L. G. Schultz, H. van Olphen, and F. A. Mumpton (eds.), *Proceedings of the International Clay Conference,* Denver. Bloomington, In: Clay Minerals Society, pp. 292–297.

Bolt, G. H., M.G.M. Bruggenwert, and A. Kamphorst. 1976. Adsorption of cations by soil. In G. H. Bolt and M.G.M. Bruggenwert (eds.), *Soil Chemistry. Part A.* New York: Elsevier, pp. 54–90.

Farmer, V. C. 1978. Water on particle surfaces. In D. J. Greenland and M.H.B. Hayes (eds.), *The Chemistry of Soil Constituents.* New York: Wiley, pp. 405–448.

Israelachvili, J. N. 1985. Measurements of hydration forces between macroscopic surfaces. *Chemica Scripta* 25:7–14.

Kittrick, J. A. 1976. The separation factor applied to some soil ion exchange equilibria. *Soil Sci. Soc. Am. J.* 40:147–148.

Kjellander, R., S. Marcelja, R. M. Pashley, and J. P. Quirk. 1988. Double-layer ion correlation forces restrict calcium-clay swelling. *J. Phys. Chem.* 92:6489–6492.

Norrish, K. 1954. The swelling of montmorillonite. *Dis. Faraday Soc.* 18:120–134.

Slade, P. G., J. P. Quirk, and K. Norrish. 1991. Crystalline swelling of smectite samples in concentrated NaCl solutions in relation to layer charge. *Clays Clay Min.* 39:234–238.

Suquet, H., C. de la Calle, and H. Pezerat. 1975. Swelling and structural organization of saponite. *Clays Clay Min.* 23:1–9.

van Olphen, H. 1977. *An Introduction to Clay Colloid Chemistry.* 2nd ed. New York: Wiley.

Viani, B. E., C. B. Roth, and P. F. Low. 1985. Direct measurement of the relation between swelling pressure and interlayer distance in Li^+-vermiculite. *Clays Clay Min.* 33:244–250.

Suggested Additional Reading

Baveye, P. and M. B. McBride (editors). 1994. Clay Swelling and Expansive Soils. Kluwer, Dordrecht, the Netherlands.

Bohn, H., B. McNeal, and G. O'Connor. 1985. Soil Chemistry, 2nd Edition. Wiley, New York. Chapter 9.

Kamphorst, A. and G. H. Bolt. 1976. Saline and sodic soils. In G. H. Bolt and M.G.M. Bruggenwert (eds.), *Soil Chemistry. Part A.* New York: Elsevier, pp. 171–191.

Parker, J. C. 1986. Hydrostatics of water in porous media. In D. L. Sparks (ed.), *Soil Physical Chemistry.* Boca Raton, Fl: CRC Press, pp. 209–296.

Problems

1. Soils of arid climates commonly contain smectite minerals with Na^+ as a predominant exchange cation, yet Ca^{2+} can be present in the form of $CaCO_3$. Explain this separation of Ca^{2+} and Na^+ into separate solid phases.

2. If Na^+-saturated smectite is added to a suspension of $CaCO_3$, the pH is observed to rise from about 8 to 10 or higher. Explain the pH change.

3. (a) Using the solubility of Na_2CO_3 (Table 8.1), calculate the pH of a saturated solution of Na_2CO_3 in CO_2-free water.
   (b) Write the chemical reaction that occurs as CO_2 from the air is allowed to dissolve into this solution. Calculate an equilibrium pH for this solution exposed to the atmosphere.

4. Calculate the equilibrium pH for a calcareous soil in which the partial pressure of CO_2 is 0.1 atmosphere.

5. (a) Write the chemical reaction for the dissolution of calcium carbonate in a calcareous soil by carbonic acid from rainwater.
   (b) Using the reaction from part (a), estimate the quantity of Ca^{2+} (kg/ha) that could be leached out of a calcareous soil in one year by reaction with the carbonic acid in pristine rainfall. Assume a moist temperate climate with 90 cm of rainfall per year.
   (c) Calculate the equilibrium concentration of Ca^{2+} that would be present in the solution of a calcareous soil with a pH of 7.8 and a CO_2 gas pressure of 10 milliatmospheres. How much annual Ca^{2+} loss by leaching could this dissolution reaction cause under the same conditions described in part (b)? (Assume rapid kinetics of calcite dissolution.)

6. The leaching of saline-sodic soil with "good" water (low salinity, low SAR) causes the pH to rise. This can result in the mobilization of trace elements such as selenium in some soils (selenium exists in the soil as an oxyanion, selenite or selenate), elements that adsorb better at low pH than high pH. How would you prevent this potential environmental risk to groundwater during the reclamation of the saline-sodic soil?

7. Calculate the maximum possible swelling pressure of Na^+-smectite during expansion from the monolayer hydrate to the two-layer hydrate based on the data in Figure 8.13.

8. Consider the diagrammed system below to be at equilibrium:

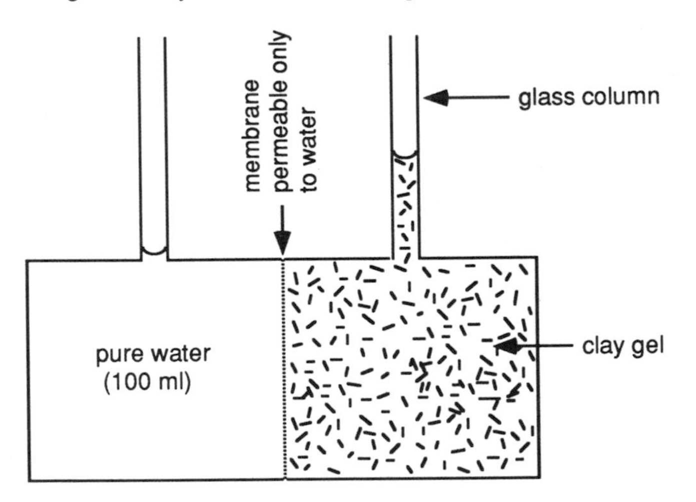

   (a) If the suspension column attains a level 10 cm higher than the water column, calculate the swelling pressure of the clay suspension (in atmosphere units).
   (b) What would happen if 10 g of NaCl were added to the water (left side)?

9. An irrigation water contains 100 mg/liter of total dissolved salts and is applied at a rate of 60 cm/year. Evapotranspiration accounts for the loss of 50 cm/year of this water during and after application.

(a) Calculate the leaching ratio.
(b) Estimate the EC of the irrigation and drainage water.
(c) Would the salt levels in the irrigation or drainage water create problems for plants?

10. A clay soil of predominantly smectite mineralogy is irrigated with water containing 3 millimoles/liter NaCl, 2 millimoles/liter $CaCl_2$, and 0.5 millimole/liter $MgCl_2$. The quantity of irrigation water applied is 15 percent greater than the the quantity consumed by evapotranspiration. Compared with irrigation, rainfall is negligible.
(a) Classify the irrigation water with respect to its potential to increase soil salinity and sodicity.
(b) Calculate the percent exchangeable Na^+ (ESP) expected in the surface soil after long-term use of this irrigation water.
(c) Determine the concentration of NaCl, $CaCl_2$, and $MgCl_2$ present in the drainage water once steady-state conditions are reached. Classify this drainage water for its salinity and sodicity hazard.
(d) Could the drainage water be reused for irrigation?

9

Trace and Toxic Elements in Soils

Fertile soils supply plants with all of the trace elements essential for growth, believed at the present time to be Fe, Mn, Zn, B, Cu, Mo, and Cl.[1] These seven elements are called the *micronutrients,* a term that indicates the small quantities needed by plants but not necessarily the concentrations found in soils. Deficiencies can occur in soils either because they contain extremely low concentrations of these elements or because the elements are present in very unavailable (insoluble) forms. Conversely, many trace elements, including all of the micronutrients, can reach concentrations in soils that are toxic to plants and microorganisms. Some of the most toxic are mercury (Hg), lead (Pb), cadmium (Cd), copper (Cu), nickel (Ni), and cobalt (Co). The first three are particularly toxic to higher animals. The last three are more toxic to plants than animals and are termed *phytotoxic.* From the standpoint of potential hazard to human health, an extended list of "priority metals" has been established. This list presently consists of:

As	Hg
Be	Ni
Sb	Se
Cd	Ag
Cr	Tl
Cu	Zn
Pb	

It is no exaggeration to state that modern analytical methods can detect most of the natural elements in soils at some level of concentration. The specific elemental composition of each particular soil reflects, to a degree modified over time by weathering, the chemical composition of the parent material from which the soil formed. However, knowledge of a soil's composition in terms of *total elemental content* is usually not very useful when it comes to understanding the processes and dynamics of element availability and cycling. Nevertheless, if elemental concentrations are

1. Some plant species need Na and Si, legumes require Co to fix nitrogen, and there is evidence that Ni is a plant micronutrient as well. Essential elements for humans and animals are not the same as those for plants.

greatly in excess of those expected for a particular soil type, this may be a sign of pollution from human activity or accumulation from natural biogeochemical processes.

Whether an element is present naturally in the soil or has been introduced by pollution, a measure more useful than total elemental content for most purposes is an estimation of "availability" or "lability" of the element, since it is this property that can be related to mobility and uptake by plants and extractability by chemical treatments. Chemical soil tests are designed to extract a quantity of the element from the soil solids that correlates statistically to the size of the "available pool" in the soil, defined by the quantity of element taken up by plants. Thus, soil tests are empirical, giving little insight into the chemical mechanisms in the soil that control availability. They estimate the potential for toxicity (or deficiency) to plants and animals. However, the extractability of different elements depends on their properties, such as their tendency to:

1. Complex with organic matter
2. Chemisorb on minerals
3. Precipitate as insoluble sulfides, carbonates, phosphates or oxides
4. Co-precipitate in other minerals

Consequently, some elements are at least partially extracted from many soils with solvents such as water or salt solutions, whereas others may resist extraction even by chemically aggressive solutions such as concentrated acids or powerful chelating agents.

Generally speaking, neither the total quantity of an element in the soil nor the quantity extracted by aggressive reagents is closely correlated to the plant-available "pool" of the element. The concentration of the element at any one time in soil solution often seems a better measure of availability. However, concentrations of trace elements *in natural soil solutions* commonly fall in the range of 1 to 1000 μg/liter, with some elements falling below the 1 μg/liter level. Consequently, the solution concentration is often below the detection limit of standard analytical methods. Furthermore, the solution concentration reveals only part of the situation regarding availability, providing a "snapshot" of a very dynamic and complex equilibrium, represented in a much-simplified form by Figure 9.1. Since prediction of plant availability is such a critical (and elusive) goal in soil and environmental science, a more detailed discussion of the meaning of availability follows.

9.1. AVAILABILITY OF ELEMENTS

9.1a. Controlling Processes

There are five steps that must be considered in getting an element from soil solids to plant tops. These are depicted in Figure 9.2, and a brief description of each of them is given here.

1. **Desorption or dissolution.** This step can be fast or slow depending on the element (see Chapter 4), and may limit availability to plants if desorption is particularly difficult or if the dissolution of a very insoluble solid is involved.

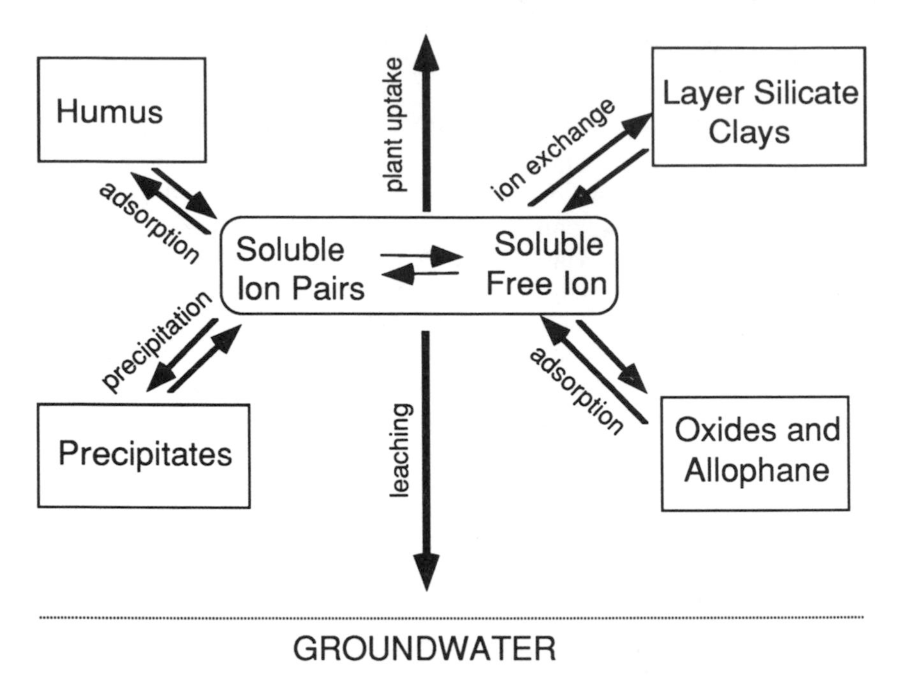

Figure 9.1. Dynamic interactive processes governing solubility, availability, and mobility of elements in soils.

2. **Diffusion and convection.** This step can be very slow for trace elements because extremely low solution concentrations of these elements are common, resulting in small quantities of ions moved by diffusion. Convection is important for non-trace elements, such as Ca^{2+}, that tend to be found at fairly high concentrations in soil solutions. Plants transpire water to create a flow of water in soil pores toward the roots, carrying amounts of these elements sufficient for plant growth.

3. **Sorption or precipitation at new sites.** After desorption, the element in question has some probability of resorbing before reaching the root. This can greatly limit movement of certain elemental forms in soils that have high levels of humus or clays. For example, phosphate ions move extremely slowly through oxidic soils because, even when these anions desorb from oxide surfaces, their probability of readsorbing on nearby sites is high. In contrast, Cd^{2+} can move fairly rapidly through the soil matrix because it tends to adsorb in exchangeable form.

4. **Absorption by roots.** Absorption of elements by roots may be passive or active, but in either case uptake depends on the concentration of the element in the soil solution near the root. The root may modify the solution chemistry of the root zone (creating the "rhizosphere effect"), locally changing soil properties such as pH or redox potential. This rhizosphere effect can be useful to plants by decreasing availability of toxic elements or increasing availability of deficient elements. However, some toxic metals (e.g., Pb) may become more soluble in the rhizosphere. One way the plant creates this effect is to exude protons and organic chelating agents that tend

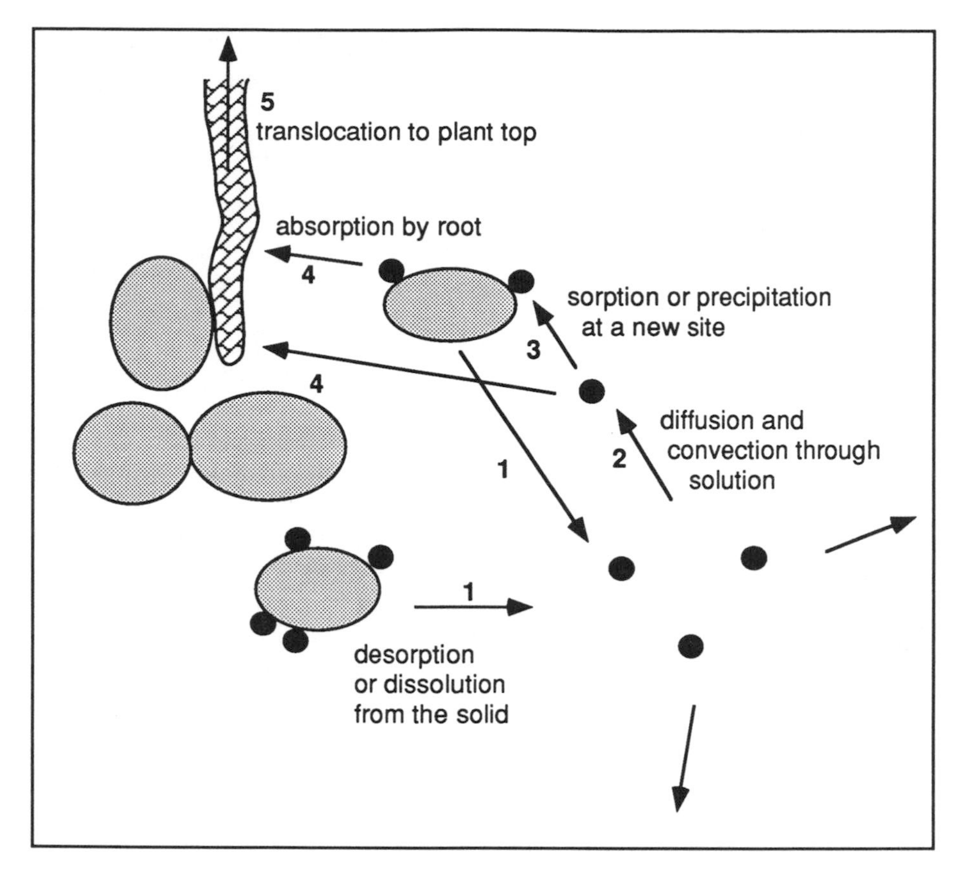

Figure 9.2. The five steps necessary for an element to move from soil solids to plants.

to increase total metal cation solubility and uptake. This is in spite of the fact that chelating agents *decrease* the concentrations of *free* (uncomplexed) metal cations in solution; plant roots either absorb the chelated metals or extract the metals from the chelate. In the particular case of Fe uptake, for example, Fe^{3+} chelators exuded by roots dissolve Fe^{3+} from minerals, thereby increasing Fe availability, because plant roots are able to reduce the strongly complexed Fe^{3+} to Fe^{2+}. The Fe^{2+} chelate is much less stable than the Fe^{3+} chelate, so that metal extraction from the chelate is facilitated by chemical reduction at the root. It is worth noting that the addition of metal chelating agents to soils raises metal mobility by influencing steps 1, 2, and 3 in Figure 9.2.

5. **Translocation in plant.** The last step in elemental availability is translocation from roots to tops, a biological process that would seem outside the control of soil chemistry. A number of trace elements, especially those that take the form of cations in soils, do not readily translocate to plant tops. Therefore it is common for metals such as Cu, Pb, and Cd, after absorption from the soil, to accumulate in (or on) roots. Translocation behavior of plants is complex and not understood for many of the elements of environmental concern, but soil chemical factors (alkalinity, phosphate

level, base cation concentration, etc.) and environmental factors can influence movement into plant tops.

The first three of these five steps combine to describe the *mobility* of the element in the soil, a soil property that may be directly related to plant availability, at least as long as the plant root itself does not complicate the situation further by altering the chemistry in the rhizosphere. A well-designed soil test could be expected to measure mobility, and perhaps availability as well.

From this outline of the factors that determine availability, we see that the soil controls elemental availability to the extent that it limits mobility (steps 1–3). The soil may also influence absorption by roots (step 4) because it has some control over the chemical forms (speciation) that elements take in solution. The important effect of speciation is discussed in more detail in the next section. The soil may even affect translocation of elements *within the plant,* since there is evidence that mobility in plants is sensitive to the specific chemical form absorbed from soil solution. For example, it appears that iron absorbed from soil solution as a bicarbonate salt is immobile within the plant root and not translocated to the top. Another example is the immobility of zinc and lead in plant roots that are well supplied with phosphate. It may be that chemical precipitation reactions within (or possibly on) roots are limiting translocation. The questions raised here about the chemical forms of elements within plants fall into the realm of plant physiology and will not be pursued further.

9.1b. Speciation and Availability

Elements, both metallic and nonmetallic, may occur in one of several oxidation states and in soluble complexes with different organic or inorganic ligands. A description of the chemical form(s) that an element takes in solution is termed its *speciation.* Each element has unique speciation tendencies. Soil solutions provide great opportunity for variety in speciation, as they contain organic ligands (fulvic acid), HCO_3^-, CO_3^{2-}, OH^-, and numerous other anions that are capable of forming soluble complexes with metal cations. The organic ligands in particular increase the "carrying capacity" of soil solutions for strongly complexing metals such as Cu^{2+}, increasing total metal solubility. Soluble complexing ligands usually enhance metal mobility, as described in Figure 9.1, influencing at least the first four steps that ultimately lead to elemental accumulation in plant tissues.

Dissolved hydroxyl (OH^-), HCO_3^-, CO_3^{2-}, and organic matter all increase in concentration as the soil pH is raised. At the same time, metal cation adsorption at mineral and organic surfaces is favored at higher pH (see Chapter 4). Consequently, total metal solubility in soil solutions often displays a two-stage trend, decreasing up to pH 6 or 7, but increasing again at higher pH as soluble ligands bring the metals into solution. This trend, shown for copper in Figure 9.3, is typical behavior for metal cations that form stable and soluble complexes with hydroxyl, carbonate, or fulvic acid.

The speciation behavior of an element in soils profoundly affects its bioavailability. Sometimes this is not evident until a soil property is changed. For example, absorption of metal ions such as Cu^{2+} and Cd^{2+} by plant roots is correlated to the

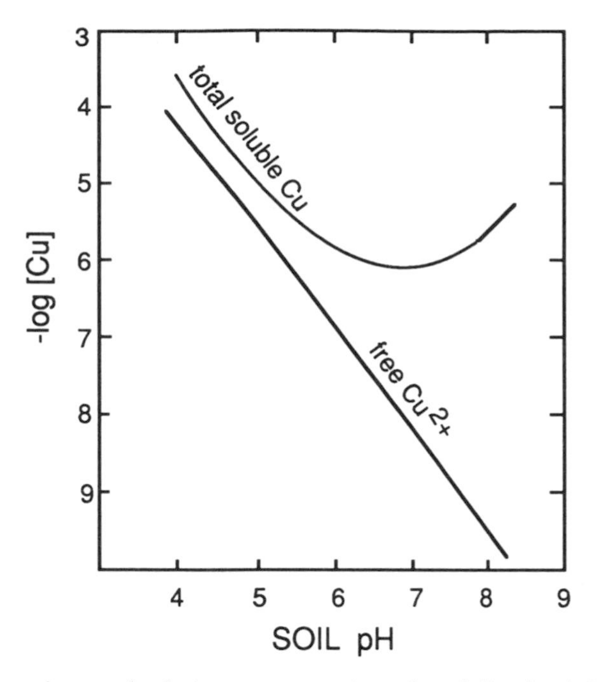

Figure 9.3. Dependence of solution concentration of total dissolved Cu and free (uncomplexed) Cu^{2+} on pH in a soil contaminated with 320 ppm of Cu.

free metal ion concentration in soil solution. Once high levels of organic matter are added to soils, however, a greater uptake of Cu^{2+} by roots occurs at any particular solution concentration of free (uncomplexed) Cu^{2+}. It is necessary to conclude that, at least for copper, soluble organic matter acts as a "cation carrier," facilitating metal diffusion from soil particles to roots. As Figure 9.3 illustrates, the carrier effect is readily explained by the much higher solution concentration of the Cu-organic complex compared with the free Cu^{2+} ion. As the roots take up copper, perhaps necessitating dissociation of the complex, the soluble complex replenishes the copper supply at the root surface at a much faster rate than diffusion of free Cu^{2+} ions could. For metals prone to forming soluble complexes with bicarbonate or carbonate, such as Cd^{2+}, the concentration of the "carrier" increases with both CO_2 level and pH in the soil. The latter effect arises from the greater total dissolved bicarbonate and carbonate in water at higher pH, so that even as free Cd^{2+} concentrations decrease at higher pH, total soluble cadmium may increase. Nevertheless, liming soils to raise the pH and enhance sorption or precipitation is generally found to be a successful practice for limiting cadmium uptake by plants.[1]

From these observations it appears that both free metal activity (or concentration), representing an *intensity* factor, and total sustained soluble metal concentration, representing a *capacity* (or buffering) factor, are important to biological systems.

1. It may be that part of the positive effect of lime is to increase competition from Ca^{2+} for plant uptake and translocation.

In fact, for metal cations in soils, the following general rules relate bioavailability to speciation:

> *Short term toxicity to plants and microorganisms is most closely related to* free *metal cation concentration in solution.*
>
> *Metal uptake over the long term depends to a large extent on the* total *metal concentration in solution and on the ability of the soil to maintain this concentration.*

9.1c. Soil Tests and Plant Availability

Soil tests attempt to predetermine, by a rapid procedure, the longer term availability of elements to crops. For some metals, such as Zn^{2+}, Co^{2+}, and Cd^{2+}, long-term uptake by plants may correlate fairly well with the slopes of measured sorption curves. That is, if the sorption curve of a soil, measured as the quantity of metal sorbed as a function of the concentration of the metal in solution, is steep, this indicates strong metal retention by soil particles and correspondingly low long-term plant uptake. This behavior arises from competition between roots and soil surfaces for metal ions. A soil test that is based on this correlation between uptake and sorption strength has the advantage that it automatically considers all the soil properties that influence metal sorption. Nevertheless, because the test requires that the element in question be added in increments to the soil to establish a sorption curve, it may fail in cases of elemental deficiency because it does not establish whether the soil initially contains the minimum level of the element to sustain growth. A soil test based on sorption curves may also fail because it does not measure the kinetics of desorption (dissolution) of the element, a necessary step for uptake. For example, the slopes of Cu sorption curves in soils do not generally correlate (negatively) with plant uptake, perhaps because desorption kinetics have not been accounted for.

For metal cations such as Cu^{2+} and Fe^{3+}, migration to the root is almost certainly in the form of organic complexes. As these forms of the metals are essentially nonadsorbing, factors (such as higher pH) that increase adsorption and decrease the solubility of the *free* metal cations may not markedly change metal uptake by plants. Thus it is not uncommon to observe little or no effect of raising the soil pH on Cu concentration in plants. For metal cations such as Zn^{2+} and Cd^{2+}, there is less tendency to form soluble organic complexes, and factors that promote adsorption and precipitation (notably, higher pH) inhibit uptake. Ideally, soil tests could be tailored to consider these differences in behavior among groups of elements.

Current soil-testing procedures for most trace elements bathe the soil in a solution of complexing, acidifying, or reducing chemicals. This alters soil properties that control the solubility of the element; redox potential, pH, mineral solubility, and organic matter solubility may all be modified. Information about elemental speciation and availability is lost in the process of attempting to extract part or all of the "bioavailable pool." Success in predicting availability has been limited with tests of this sort. More widely useful soil tests for trace elements, when they are developed, will probably measure activities or concentrations of the elements in undisturbed soil solutions (the intensity factor) and the capability of the soil solids to replenish the solution (the capacity factor). Ion exchange resins, when mixed with soils, act as a sink for ionic forms of elements. They are known to be useful in predicting the capacity of soils to supply ionic forms of elements to plants.

9.2. MOBILITY OF ELEMENTS IN SOILS

9.2a. The Controlling Factors

As described earlier, movement of adsorbed elements in soils generally requires that a sequence of processes occurs, beginning with desorption or dissolution followed by diffusion and convection. Readsorption or precipitation can then immobilize the element at another location in the soil. Relative mobility of elements depends on several important factors, including:

1. **The chemical form and nature of the element.** Most trace metal cations have a low mobility in soils because they adsorb strongly on minerals and organic matter, or form insoluble precipitates (e.g., oxides, carbonates, sulfides). Some elements that take the form of anions in soils, such as boron, are relatively mobile. Other elements that form anions, like phosphorus, are considered to be immobile because they form insoluble precipitates and bond strongly with mineral surfaces. In Figure 9.4, many of the elements of interest in soils are classified on the basis of their ionic radii and valence. The elements tend to fall naturally into one of four groups:

Group 1. Soluble weakly hydrating cations
Group 2. Soluble, strongly hydrating cations
Group 3. Easily precipitated amphoteric hydroxides
Group 4. Soluble anions

The valence/radius ratio, or *ionic potential,* is a key parameter in determining which of these four chemical forms the element takes in water. Figure 9.4 can be used in a general way to predict chemical form for elements, but an assessment of elemental mobility requires knowledge of soil properties as well. In broad terms, the mobility of the four groups of elements in soils can be classified as follows:

Group 1. Strongly held by clays and humus
Group 2. Exchangeable and somewhat mobile, excepting strong chemisorbers such as Pb^{2+} and Cu^{2+}
Group 3. Immobile as insoluble oxides
Group 4. Mobile, excepting strong chemisorbers such as phosphate

2. **The chemical and mineralogical nature of the soil.** Elements are less mobile in those soils that provide a large quantity of sorption sites or a chemical environment favorable to precipitation of the element. Oxides of Fe, Al, and Mn provide chemisorption sites for both cation and anion forms of elements. Layer silicate minerals provide exchange sites for cations, and a few chemisorption sites (at edges, defects, etc.) for both cations and anions. Noncrystalline aluminosilicates (allophanes) possess large quantities of chemisorption sites for both cations and anions. Soil organic matter is principally involved in metal cation adsorption, although the borate anion (and perhaps several other anions) can bond covalently to organic mat-

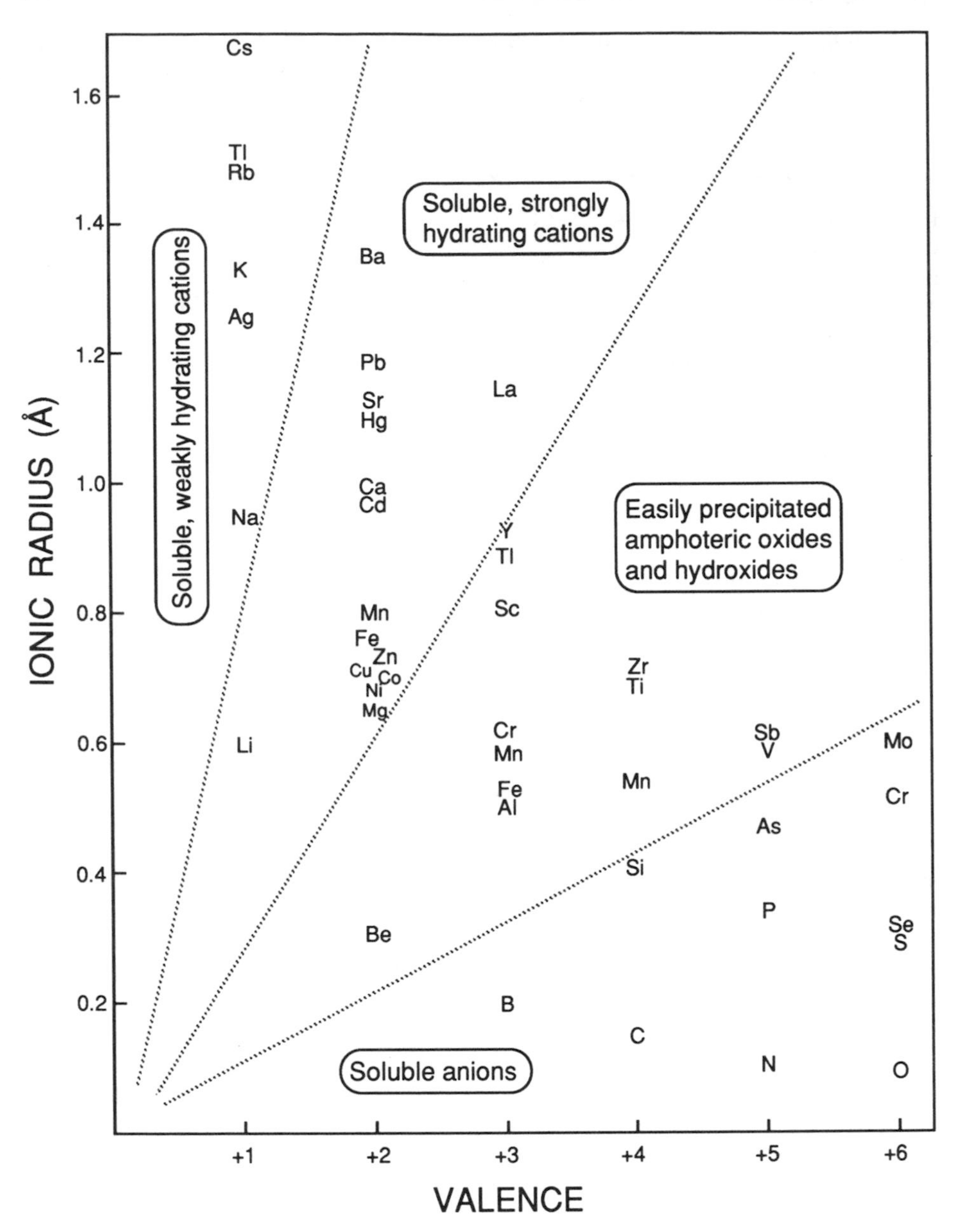

Figure 9.4. Classification of elements into four groups on the basis of ionic charge (valence) and radius. (Adapted from G. Millot. 1970. *Geology of Clays.* New York: Springer-Verlag.)

ter. Overall, soils high in clays, oxides, or humus strongly retain most of the trace metals.

The chemical conditions in the soil can also be critical in element retention. For metal cations, high pH favors sorption and precipitation as oxides, hydroxides, and carbonates (see Chapter 4). For many anions, such as molybdate and selenite, low pH favors sorption and precipitation. Alkaline conditions in soils are generally associated with elevated Na^+ and K^+ levels relative to Ca^{2+} and Mg^{2+}. Because many anions form much more soluble compounds with Na^+ and K^+ than with Ca^{2+} or Mg^{2+}, soil alkalinity is usually accompanied by a high degree of anion mobility. For example, borate and molybdate are easily leached from alkaline soils of arid regions. Alkalinity may mobilize some metal cations (e.g., Cu) as well because it favors the formation of soluble metal complexes of organic matter and hydroxyl. Salinity in soils is associated with high levels of halide ions, particularly chloride, and this anion is able to mobilize certain heavy metals by forming soluble metal-chloride ion pairs. Mercury (Hg^{2+}) is especially prone to chloride-induced mobilization, but Cd^{2+} and Pb^{2+} are affected to some degree as well.

Soil redox potential is also critical in controlling elemental mobility. Some elements are much more soluble and mobile in one oxidation state than another (examples include Cr, Mn, Se, and others). The elements classified as *chalcophiles* (e.g., Hg, Cu, Pb, Cd, Zn, As, Se) form insoluble sulfide minerals in reducing environments where sulfide (S^{2-}) is generated from sulfate reduction (see Chapters 4 and 7). Mobility for chalcophiles is then extremely low unless oxidizing conditions are restored in the soil. Those elements that, in the sulfide form, have the very lowest solubility products (notably mercury, copper, lead, and cadmium) are the most likely to become highly immobile and unavailable in reduced soils.[1]

3. **The physical and biological environment of the soil.** High solubility of elements is not manifested as significant migration unless there is water movement through the soil. In arid climates, net water movement in the soil profile is upward, and the mobile elements that are carried to the surface become concentrated by evaporation. Conversely, in wet climates, mobile elements leach downward as long as there is free drainage. However, roots can intercept mobile elements and extract immobile elements, accumulating them in the plant. As the plant material decays to humus (which retains ions selectively), particular elements *bioaccumulate* in the surface soil. Some of these elements are not required nutrients and may even be phytotoxic.

Experience with many elements that have found their way into soils as pollutants from agriculture or industry suggests that mobility is so low that the presence of the undesired element is virtually permanent. This is certainly true for the strongly sorbing metals such as Pb and Cu. Agricultural sites that were subjected to the use of lead arsenate and copper sulfate as pesticides many decades ago still retain Pb and Cu in the soil surface today, although in some cases the arsenate has moved deeper in the soil profile. Even in wet climates where leaching is more or less continuous, removal of a large portion of these less mobile elements by natural processes could take thousands of years.

1. Unfortunately, mercury forms volatile organomercury compounds under reducing conditions, so that its bioavailability is not as low as the solubility of HgS would suggest.

Depending on the controlling factors for mobility, the elemental composition of soils may or may not reflect the composition of the soil's parent material. However, it is generally observed that, all other factors being equal:

a. Soils high in clay-sized minerals (particularly silicates and oxides) tend to have a higher concentration of most trace elements than coarse-textured soils such as sandy and gravelly soils.

b. Soils rich in humus tend to have higher concentrations of most trace elements than soils that are not. This observation applies to comparisons between sites as well as within a single soil profile, as many elements bioaccumulate in the organic-enriched surface horizon.

c. Soils of moderate to high pH tend to have higher concentrations of most trace elements than soils that have been acidified naturally (or otherwise) by acid water leaching.

9.2b. The Case of Mobile Elements

It may be instructive to analyze in some detail the situation of a reversibly adsorbed ion, whose mobility in the soil is relatively high. A case in point is Cd^{2+}, a toxic metal that adsorbs by cation exchange in acid and near-neutral soils, but may precipitate into carbonates and phosphates in calcareous and alkaline soils. The theory needed for the analysis of ion leaching through soils, in which the main retention mechanism is ion exchange, is developed in Chapter 3 (section 3.5) using Cs^+ as the example ion.

The problem posed is the following: Suppose a slightly acid clay soil, with a CEC of 50 mmoles/kg, has its exchange sites occupied largely by Ca^{2+}. A small amount of the soluble salt, $Cd(NO_3)_2$, is accidentally spilled on the surface of this soil in the field. An attempt is then made to leach the cadmium out of the soil surface using 0.01 M $Ca(NO_3)_2$, relying on the fact that Ca^{2+} and Cd^{2+} have nearly equal affinity for clay exchange sites. The practical question that arises is: What volume of $Ca(NO_3)_2$ solution must be surface applied to move the cadmium to 1 meter depth in the soil?

To address this question, we must first define K_S for $Ca^{2+}-Cd^{2+}$ exchange (see Chapter 3 for a description of K_S). The exchange equation that describes K_S is

$$K_S = \frac{N_{Cd}[Ca^{2+}]}{N_{Ca}[Cd^{2+}]} \approx 1 \tag{9.1}$$

The measured value of K_S is about 1 so long as the pH is relatively low and those anions most inclined to form soluble complexes with Cd^{2+} (e.g., chloride) are not present. A low pH inhibits both chemisorption and precipitation reactions of Cd^{2+}; otherwise the soil's apparent selectivity for Cd^{2+} would greatly exceed unity. Furthermore, only if the total soluble Cd can be approximated by the free Cd^{2+} concentration, $[Cd^{2+}]$, would equation 9.1 be expected to provide the basis for estimating cadmium retention and mobility in the soil. Soluble complexes of Cd, unless they happen to be cations, bypass cation exchange sites and are mobilized.

For this example of exchange between cations of equal charge, the fraction of exchange sites occupied by Cd^{2+} and Ca^{2+} are given by $N_{Cd} = m_{Cd}/\text{CEC}$ and $N_{Ca} = m_{Ca}/\text{CEC}$, where m_{Cd} and m_{Ca} are the equivalents (moles of charge) of these metals on the soil's exchange sites. Equation 9.1 then becomes

$$K_S = \frac{m_{Cd}[Ca^{2+}]}{m_{Ca}[Cd^{2+}]} = 1 \tag{9.2}$$

Since $[Ca^{2+}]$ is made nearly constant by the introduction of 0.01 M $CaCl_2$ into the soil, and $m_{Ca} \gg m_{Cd}$, then $[Ca^{2+}]$ and m_{Ca} can be replaced by N (solution normality) and the CEC, respectively. Equation 9.2 then becomes

$$m_{Cd} = \text{CEC}\left(\frac{K_S}{N}\right)[Cd^{2+}] \tag{9.3}$$

The specific values of variables in equation 9.3 for the present case, being careful to keep units consistent, are as follows:

$$\begin{aligned} \text{CEC} &= 0.050 \text{ moles/kg} \\ K_S &= 1.0 \text{ (unitless)} \\ N &= 0.02 \text{ moles (of cation charge)/liter} \end{aligned}$$

When these values are entered into equation 9.3, a simple relation is found:

$$m_{Cd} = 2.5[Cd^{2+}] \tag{9.4}$$

where 2.5 is now identified as the value of the distribution coefficient, K_d, for this particular system because it quantifies the relation between the concentration of Cd^{2+} in soil solution and the quantity of Cd^{2+} adsorbed on the soil exchange sites. K_d has units of liters per kilogram in this example as a result of the choice of units for the variables.

The K_d value determined in this way is then used in equation 3.102 from Chapter 3, which estimates the velocity of a chemical's movement under saturated flow, v_{Cd}, relative to the leaching velocity of water, v. For cadmium, that equation is

$$\frac{v_{Cd}}{v} = \frac{1}{[1 + (\rho_B/\o)K_d]} \tag{9.5}$$

where ρ_B and $\o$ are the bulk density and fractional porosity of the soil. Reasonable values for bulk density and porosity are 1.5 g/cm^3 and 0.42. These units remain consistent with those of equation 9.4 because 1.5 g/cm^3 equals 1.5 kg/liter. Equation 9.5 is then solved:

$$\frac{v_{Cd}}{v} = \frac{1}{[1 + (1.5/0.42)2.5]} = 0.10 \tag{9.6}$$

to reveal that the leaching velocity of the cadmium should be about 10 percent of the leaching velocity of water (or nitrate) when 0.01 M $Ca(NO_3)_2$ is applied to the surface.

From this result, the surface-applied $Ca(NO_3)_2$ solution must leach to a 10-meter depth to force the Cd^{2+} ions down 1 meter (on average). Because the soil's porosity is 0.42, a 10-meter depth of solution in the soil matrix is equivalent to a 4.2-meter depth of applied water. It would take a number of years of rainfall in a temperate humid climate to provide this extent of leaching, and since the concentration of exchanging cations in soil solutions (Ca^{2+}, Mg^{2+}, etc.) is typically well below the 0.01 M Ca^{2+} concentration used in this calculation, cadmium movement out of the surface horizon of soils is likely to take several decades *even if no chemisorption of the metal occurs*. Nevertheless, if the soil is somewhat acid, cadmium movement can be detected over time spans of years to decades, unlike more strongly bound metals for which movement can take centuries.

Accelerated leaching experiments using 0.01 *M* $CaCl_2$ to leach a column of acid mineral soil have produced some movement of trace metal ions initially applied to the column surface. Results are shown for Cd^{2+}, Zn^{2+}, Ni^{2+}, and Cu^{2+} in Figure 9.5a. Cd^{2+}, Zn^{2+}, and Ni^{2+} moved readily, so that most of these ions migrated below 30 cm depth after seven hours leaching. The mobility of Cu^{2+} was obviously lower than that of the other three metals, consistent with its strong chemisorption on clays, oxides, and humus. The same soil, after liming with $Ca(OH)_2$ to pH 6.5, retained all four metals more strongly (Figure 9.5b), so that even after seven hours of leaching, Cd^{2+}, Zn^{2+}, and Ni^{2+} remained within the soil column. Liming reduces both mobility and plant availability of many of the toxic heavy metals.

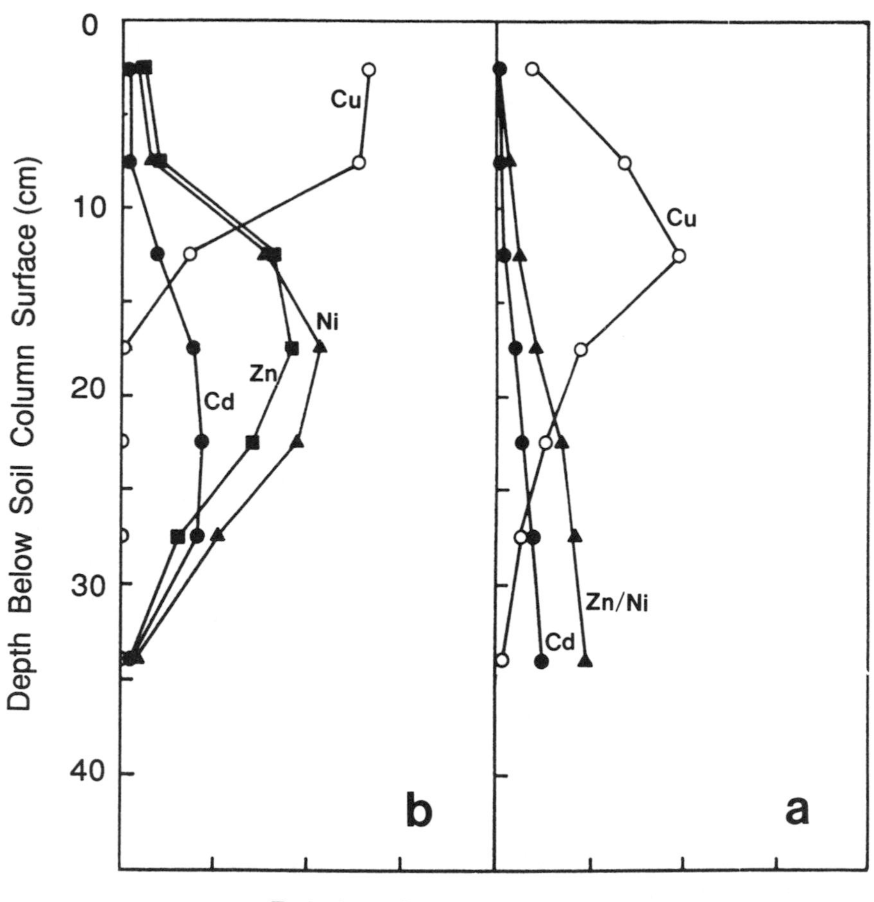

Figure 9.5. Profiles of adsorbed Cu^{2+}, Ni^{2+}, Zn^{2+}, and Cd^{2+} concentrations in a soil column to which these metals were initially surface-applied and then leached with 0.01 *M* $CaCl_2$. The acid mineral soil was either left unlimed (a) or was limed to pH 6.5 (b).

While the attenuation of Cd^{2+}, Zn^{2+}, and Ni^{2+} migration is largely attributable to cation exchange in these columns, and could be modeled by equation 9.5, Cu^{2+} retention is by nonreversible chemisorption and is not subject to the same approach. A further danger in using such a simple model as that represented by equation 9.5, even for Cd^{2+}, is that it does not consider the importance of metal "load" in the soil. If a very small quantity of any of these heavy metals is applied to soil, selective retention processes such as chemisorption and chelation are likely to be active, and mobility will be much lower than that predicted from models based on fairly non-selective cation exchange reactions. For cases of strong and seemingly irreversible sorption of elements, a different approach is needed.

9.2c. The Case of Immobile (Fixed) Elements

Consider now adsorbed molecular or ionic species that are, practically speaking, immobilized in the soil. Unless the soil is extremely acid, metals such as Cu^{2+}, Cr^{3+}, and Pb^{2+} fall into this category. Also, certain anions such as phosphate bond so strongly on minerals that they too behave as immobile elements. The property that all of these ions have in common is that their sorption isotherms are not reversible within a time scale relevant to soil processes; the adsorption (forward) isotherm is usually approximated closely by a Langmuir function of the strong-affinity type,[1] but the desorption (backward) isotherm deviates markedly from the adsorption isotherm. This kind of nonequilibrium behavior, depicted in Figure 9.6, is sometimes referred to as *hysteresis*. Possible reasons for hysteresis in chemisorption are discussed in Chapter 4.

The important consequence of the irreversibility illustrated by Figure 9.6 is that not many of the initially sorbed ions are able to desorb once the concentration of these ions is lowered in solution. This is behavior inconsistent with the ion exchange model of leaching described in the last section. That model would overpredict mobility of elements that are retained by chemisorption or precipitation reactions.

Taking phosphate for our example of an immobile ion, we know from equation 4.41 in Chapter 4 that this anion should adsorb according to the Langmuir function, having the form:

$$Q = \frac{Q_M[P]}{C + [P]} \qquad (9.7)$$

where Q is the quantity of phosphate sorbed by the soil at equilibrium with the solution phosphate concentration, [P]. Q_M is the maximum quantity of phosphate sorbed, determined in principle by the number of available bonding sites in the soil. C, the bonding constant, is a function of pH and is described in more detail in Chapter 4.

In soils, phosphate sorption data usually take the shape of a high-affinity (H-type) isotherm, like that of Figure 9.6. Because of this and the nonreversible behavior of sorption, the downward movement of phosphate and other strongly bonded elements can be approximated by the "tipping bucket" model. This model visualizes the layers

1. See Chapter 10 for a description of this shape of adsorption isotherm, classified as L-type or H-type.

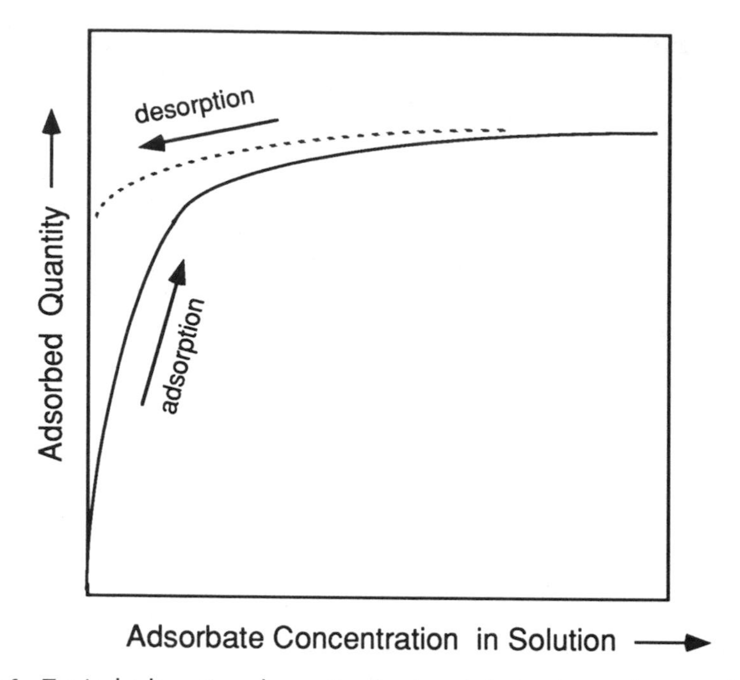

Figure 9.6. Typical adsorption-desorption hysteresis in isotherms for strongly bonding ions such as phosphate.

of a soil column as a stack of empty "buckets," each bucket symbolizing the sorption capacity of one soil layer. As phosphate is applied to the surface layer, the top bucket must be filled (the sorption capacity satisfied) before phosphate can "spill" into the soil layer immediately below. The changing status of the isotherms for each soil layer as phosphate moves deeper is shown schematically in Figure 9.7.

This model is most appropriate for adsorbates with high-affinity isotherms, because, as can be seen in Figure 9.7, very little of the adsorbate resides in soil solution until practically all of the sorption capacity in that layer has been satisfied. The opportunity for significant leaching into the next layer below arises only when sorption nears the sorption capacity ($Q \approx Q_M$), and a significant concentration of soluble adsorbate builds up. This is in contrast to the leaching model for mobile ions based on ion exchange, where mobility can be appreciable without saturating the exchange sites with the absorbate.

The tipping bucket model provides a straightfoward way to estimate at what soil loading level the chemical of concern is expected to "break through" a specific depth of soil into groundwater, drainage pipes, or surface waters. The only measurement needed for this estimate is a value for Q_M, the sorption capacity of the soil for that chemical. Because sorption of ions such as phosphate is a complex process in soils, involving both chemisorption and precipitation reactions, several mechanisms contribute to Q_M. Nevertheless, equation 9.7 is still useful as an empirical function describing overall sorption, even though use of this Langmuir function implies that a single chemisorption reaction is being described. The difficulty with this equation and the tipping bucket model, as they are applied to the sorption of chemicals, is the

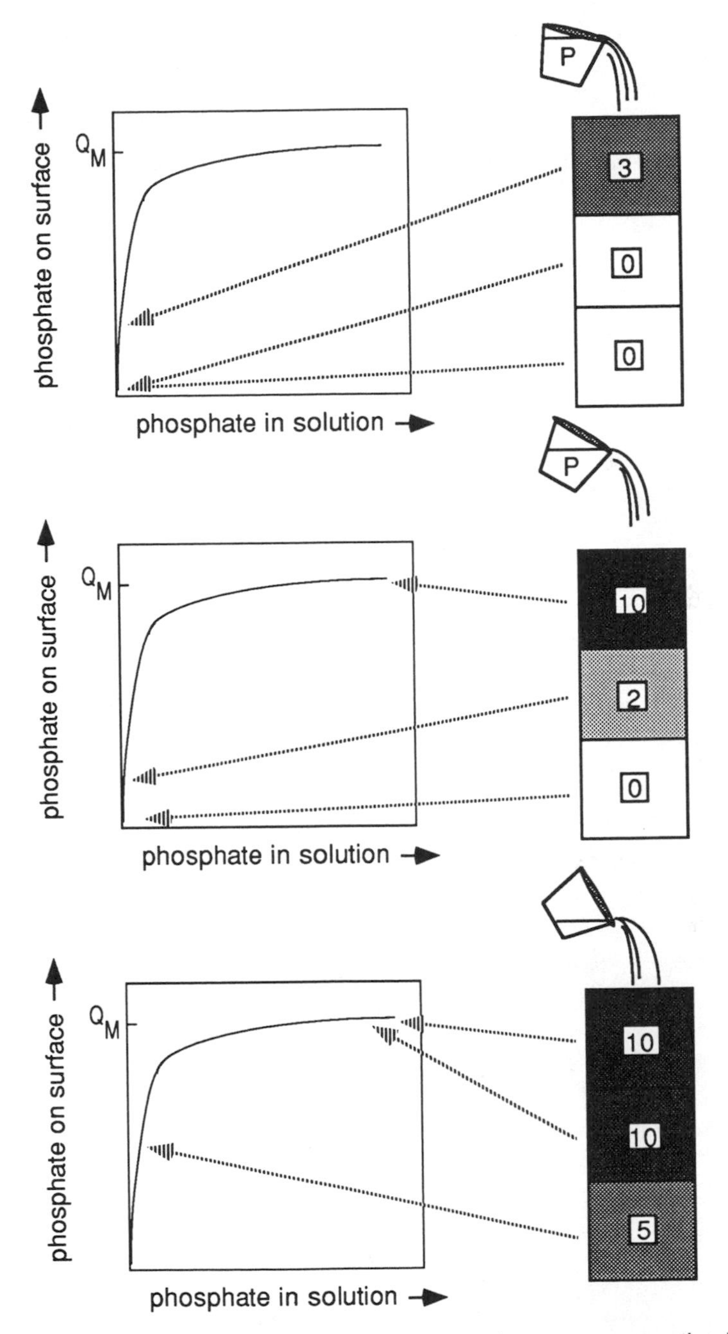

Figure 9.7. Progression of phosphate retention in a homogeneous soil column as increasing amounts of phosphate are applied to the surface. The position on the sorption isotherm, and the relative quantity of phosphate sorbed, is illustrated for each soil layer (Q_M, the sorption maximum, is 10 units).

fact that slow chemical reactions and diffusion processes in the soil tend to gradually increase the measured value of Q_M as time goes on, a phenomenon referred to as "regeneration" of sorption capacity. That is, sorption capacity is not so well determined and fixed as the model would suggest. Regeneration has practical consequences for the mobility of chemicals in soil. For phosphate, it means that the tipping bucket model overestimates mobility, especially in cases where phosphate movement in the soil is measured over a long time. In fact, phosphate migration through mineral soils and into groundwater has not been detected very often, even when phosphate loadings have exceeded the short-term sorption capacity of the soil (Q_M) by a wide margin.

It is perhaps worthwhile to use a simple problem to illustrate the application of the tipping bucket model.

Example Problem: Suppose that a deep and fairly homogeneous soil is going to be used to dispose of septic tank waste water that contains 10 mg/liter of soluble phosphorus in the form of phosphate. A sorption isotherm is measured for the soil, from which the sorption capacity of the soil is estimated at $Q_M = 200$ mg P/kg soil. The 600 liters of waste water that is generated each day is evenly disposed of over a 70 square meter area. Estimate the depth of phosphate movement in the soil after ten years, and make a decision whether a shallow well (10 meters deep) at the edge of the disposal area could become contaminated by phosphate.

Solution: This problem is approached by first calculating the total "P load" on the disposal area from the information given. A discharge of 600 liters per day containing 10 mg P/liter amounts to:

$$600 \text{ liters/day} \times 10 \text{ mg/liter} = 6000 \text{ mg P/day}$$

which, over ten years, accumulates a "P load" of

$$6000 \text{ mg P/day} \times 365 \text{ days} \times 10 = 21.9 \times 10^6 \text{ mg P}$$

Because the soil's average sorption capacity is 200 mg P/kg, the weight of soil that can sorb the entire P load is

$$\frac{(21.9 \times 10^6 \text{ mg P})}{(200 \text{ mg P/kg soil})} = 1.10 \times 10^5 \text{ kg soil}$$

This weight has to be converted to a volume using the soil's bulk density. A reasonable estimate is 1.5 g/cm^3 (1.5 kg/liter), so that the soil volume needed to retain all of the phosphate is

$$\frac{(1.10 \times 10^5 \text{ kg soil})}{(1.5 \text{ kg/liter})} = 7.3 \times 10^4 \text{ liters} = 73 \text{ m}^3$$

Since the waste water is spread over a 70 m^2 area, the depth of phosphate penetration below the surface is determined by dividing the "phosphate-saturated" volume of soil by the disposal area:

$$\frac{73 \text{ m}^3}{70 \text{ m}^2} \approx 1 \text{ meter}$$

The solution reached predicts that phosphate should move no deeper than 1 meter after ten years, and the shallow well is at no immediate risk from phosphate contamination.[1] Nitrate contamination of the well would, however, be likely because NO_3^- is a very mobile anion.

9.3. PROPERTIES OF INDIVIDUAL ELEMENTS IMPORTANT IN SOIL

A description of the most important properties of elements of interest, because they are essential for (or toxic to) life, is given in this section. A preliminary outline of biological function and level of toxicity is provided in Table 9.1. Those elements listed in the table that are more toxic to animals than to plants present the most insidious hazard to human health because the elements may accumulate in apparently healthy plants to levels that are poisonous to humans and animals. That is, for such elements the plant fails to play the role of a biological alarm that would warn of toxicity. The rating of elemental phytotoxicity in Table 9.1 should not be confused with the actual likelihood or frequency of toxicity in the field. For example, manganese is rated to have a fairly low intrinsic toxicity to plants, yet its toxicity is commonly seen because very high Mn^{2+} concentrations can develop in wet soils. In contrast, chromium and lead are rated as more phytotoxic, but they are generally so insoluble in soils that toxicity from these metals is rarely seen.

The chemistry of the elements listed in Table 9.1 is now discussed in more detail, excluding Fe and Al, which have been dealt with in some detail throughout the text. The parts per million (ppm) unit is used to report total elemental content by weight in the dry soil, and is equivalent to micrograms of the element per gram of soil ($\mu g/g$). These reported contents are averages given by Kabata-Pendias and Pendias (1984) for soils that were selected to try to avoid polluted sites. Being averages, they indicate typical contents but do not reveal the actual wide variation of elemental concentrations that can be encountered in some regions of the world. The following approximations may be useful in converting the reported contents to practical units:

$$\begin{aligned} 1 \text{ ppm} &\approx 2 \text{ pounds of element per acre of soil (6 inches deep)} \\ &\approx 2 \text{ kilograms of element per hectare (15 cm deep)} \end{aligned}$$

9.3a. Silver

The only stable oxidation state of Ag in aerated solutions is +1. The Ag^+ cation, having a very large radius, does not hydrolyze to AgOH until quite high pH. Ag_2O precipitates at high pH, reaching minimum solubility near pH 12. The important soluble form of Ag in the pH range of most soils is, therefore, the Ag^+ cation. At very low pH, the element has medium mobility in soils, sorbing selectively on clay exchange sites, humus, and oxides. At neutral or alkaline pH, Ag has very low mobility because chemisorption on minerals and organic matter is favored at higher pH.

1. Possible complicating factors in this analysis include the regeneration phenomenon, which would restrict P movement even further than predicted, and channel flow of water, which might allow some P to reach the well before all of the soil volume above had encountered P.

Table 9.1. Biological Function and Toxicity[a] of Important Trace Elements

Element	Biological Function	Phytotoxicity[b]	Mammalian Toxicity
Ag Silver	None known	H (5–10)	H
Al Aluminum	May activate succinic dehydrogenase	M (50–200)	L
As Arsenic	None known in animals. Constituent of phospholipid in algae and fungi	MH (5–20)	H
B Boron	Essential to plants. Phosphogluconate constituent	M (50–200)	L
Ba Barium	None known	L (500)	H (soluble forms)
Be Beryllium	None known	MH (10–50)	H
Cd Cadmium	None known	MH (5–30)	H Cumulative poison
Co Cobalt	Essential for mammals. Cofactor in numerous enzymes. Role in symbiotic N_2 fixation	MH (15–50)	M
Cr Chromium	May be involved in sugar metabolism in mammals	MH (5–30)	H(Cr^{6+}) M(Cr^{3+})
Cu Copper	Essential to all organisms. Cofactor in redox enzymes, O_2-transport pigments	MH (20–100)	M
F Fluorine	Strengthens teeth in mammals	LM (50–500)	M
Fe Iron	Essential to all organisms. Cofactor in many enzymes, heme proteins	L (>1000)	L
Hg Mercury	None known	H (1–3)	H (soluble or volatile forms). Cumulative poison.
Mn Manganese	Essential to all organisms. Cofactor in numerous enzymes. Involved in H_2O-splitting reaction of photosynthesis.	LM (300–500)	M
Mo Molybdenum	Essential to almost all organisms. Enzyme cofactor in N_2 fixation and NO_3^- reduction	M (10–50)	M
Ni Nickel	None known in mammals. May be essential to plants. Found in urease enzyme	MH (10–100)	M
Pb Lead	None known	M (30–300)	H Cumulative poison
Sb Antimony	None known	M (150)	H
Se Selenium	Essential to mammals and some plants	MH (5–30)	H
Tl Thallium	None known	MH (20)	H
V	Required by green algae; may be involved	H	H

Table 9.1. (continued)

Element	Biological Function	Phytotoxicity[b]	Mammalian Toxicity
Vanadium	in N_2 fixation. Porphyrin and heme constituent	(5–10)	
Zn Zinc	Essential to all organisms. Cofactor in numerous enzymes	LM (100–400)	LM

[a]Letters rate the elemental toxicity as low (L), moderate (M), and high (H).

[b]Numbers in parentheses denote the concentration of element in leaf tissue (μg/g dry weight or ppm) that shows toxicity in plants that are neither highly sensitive or tolerant (data from Kabata-Pendias and Pendias, 1984).

Source: Adapted from J. E. Huheey. 1972. *Inorganic Chemistry.* New York: Harper & Row.

The Ag^+ ion is a "soft" acid, displaying an extremely high affinity for "soft" bases, particularly S-containing ligands such as sulfhydryl and sulfide. To the extent that soil organic matter contains sulfhydryl groups, Ag^+ should associate strongly with the organic fraction of soils. In anaerobic soil environments, sulfide precipitates Ag^+ into highly insoluble solids, and Ag mobility is very low.

Peaty soils can have Ag contents exceeding 1 ppm. Much higher levels of Ag (> 10 ppm) are generally an indication that the soil has been polluted by human activities such as mining.

Soil Ag (range of means, worldwide) = 0.03 to 8 ppm

9.3b. Arsenic

The most probable oxidation states of As in soil environments are +3 and +5, although the −3 and 0 oxidation states are at least possible in strongly reduced soils and sediments. Arsenite (+3), which takes various forms such as $As(OH)_3$, $As(OH)_4^-$, $AsO_2(OH)^{2-}$, and AsO_3^{3-}, is the reduced state of As that is most likely to be found in anaerobic soils. Arsenate (+5), AsO_4^{3-}, the oxidized state, is stable in aerobic soils.

Arsenate has chemical behavior similar to that of phosphate in soils; it is chemisorbed by Fe and Al oxides, noncrystalline aluminosilicates, and, to a smaller extent, layer silicate clays. Being the anion of the strong acid, H_3AsO_4, with pK_a values (2.24, 6.94, and 11.5) similar to those of phosphoric acid, arsenate adsorbs most effectively at low pH. Consequently, its mobility is fairly low in acid soils with high clay or oxide content. In neutral to alkaline soils, especially those that are sodic, As may be mobile in the soluble Na arsenate form. Soil microbes and Mn oxides are able to promote the oxidation of arsenite to arsenate under aerobic conditions.

Based on chemical arguments, arsenite might be expected to adsorb more strongly on clays and oxides than arsenate (see Chapter 4). However, $As(OH)_3$ behaves like boric acid, forming the anion only at high pH:

$$As(OH)_3 + H_2O = As(OH)_4^- + H^+ \qquad \log K = -9.29$$

The weak acidity of $As(OH)_3$ causes adsorption of arsenite on oxides and soils to be weaker than that of arsenate, at least when the pH is below 7. Like borate, arsenate appears to adsorb most effectively in the pH 7 to 9 range.

In soils that are poorly drained, once anaerobic conditions are established, both arsenate and arsenite can be released into solution by the dissolution of Fe and Mn oxides. Subsequently the desorbed arsenate is reduced to arsenite, which eventually is converted to insoluble forms (McGeehan and Naylor, 1992). Thus, As solubility may first increase and then decrease if anaerobic conditions are maintained. This reaction sequence explains the fact that the flooding of soils has been reported to *increase* As solubility, whereas increasing the redox potential of flooded soils has generally *reduced* As availability to plants.

Several competing processes are evidently involved when soils remain anaerobic for extended periods. Sulfides formed under anaerobic conditions may co-precipitate As in its lower oxidation state, a consequence of the chalcophilic tendencies of this element. However, volatile alkylarsene compounds may also form under strongly reducing conditions, causing loss of some As from the soil. The various biological and chemical transformations that interact to control solubility, mobility, and availability are diagrammed in Figure 9.8.

Amendments of ferrous salts, which oxidize to $Fe(OH)_3$ in soils, and calcite (lime), which sorbs arsenate and supplies Ca^{2+} to form Ca arsenate precipitates, seem to diminish the level of toxicity of As in soils.

Soil As (range of means, worldwide) = 2.2 to 25 ppm
Soil As (range of means, U.S.) = 3.6 to 8.8 ppm

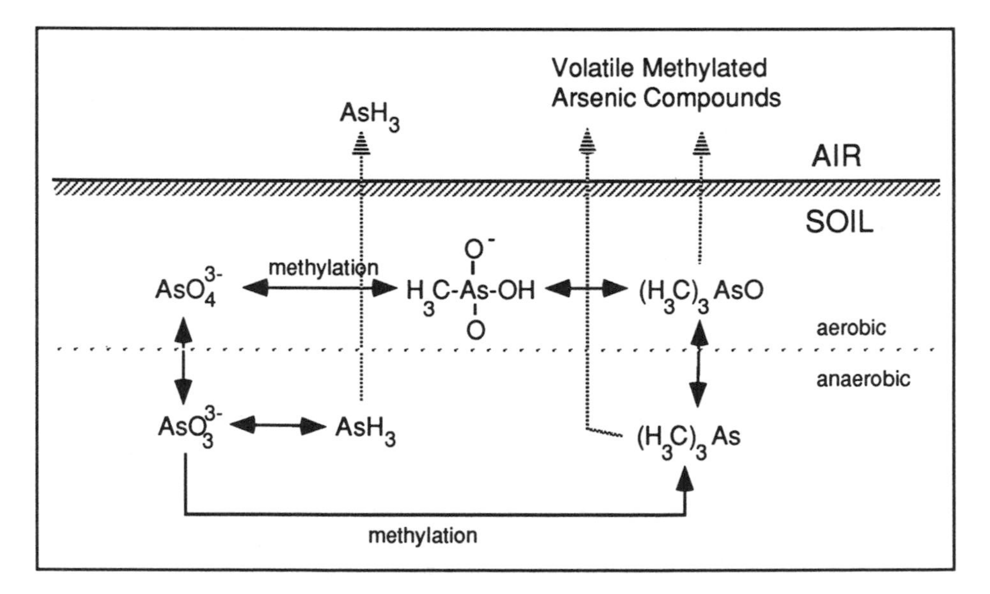

Figure 9.8. Biological and chemical transformations of arsenic in the soil. Broken arrows denote the loss of volatile forms of As to the atmosphere or the air-filled pores of the soil. (Modified from B. E. Davies. 1980. Trace element pollution. In B. E. Davies (ed.), *Applied Soil Trace Elements*. New York: Wiley.)

9.3c. Boron

Boron occurs in solution as boric acid, $B(OH)_3$, which is a very weak acid, but instead of donating a H^+ like most acids, it accepts an OH^- to convert to borate at high pH:

$$B(OH)_3 + OH^- = B(OH)_4^- \qquad K = 10^{-9.0}$$

Thus, $B(OH)_3$ is a Lewis acid rather than a Brønsted acid (see Chapter 1). Because boron adsorbs most effectively in the pH 8 to 9 range on Al and Fe oxides and silicate minerals, its availability is generally low in coarse-textured, acid-leached soils and in calcareous soils. Deficiency in acid soils is the result of boron depletion by leaching, while deficiency in calcareous soils is caused by strong adsorption and precipitation as relatively insoluble Ca borate salts. In contrast, B toxicity is most commonly found in alkaline soils of arid regions; these soils often contain high levels of Na^+ which forms quite soluble borate salts. A lack of rainfall allows soluble borate to accumulate to phytotoxic levels.

Boron is rated as a quite mobile element, leaching out of soils in humid climates and concentrating in the surface soil in arid and semiarid climates. In some soils, a large fraction of the total B is extractable by water.

Soil B (range of means, worldwide) = 9 to 85 ppm
Soil B (range of means, U.S.) = 20 to 55 ppm

9.3d. Barium

Barium occurs only in the +2 oxidation state, associating geochemically with feldspars and biotite micas in soils. It substitutes readily for K^+ in these structures because of the similar ionic radii of the large Ba^{2+} cation and K^+. As soils weather, the released Ba^{2+} can be immobilized by precipitation with sulfate or carbonate, or by "fixation" on high-charge layer silicate clays such as vermiculites. Clay exchange sites and humus show a high cation exchange selectivity for Ba^{2+} over more strongly hydrating cations such as Ca^{2+} and Mg^{2+}. Consequently, barium is rated as a rather immobile element in soils. Its average concentration in soils is higher than that of most of the trace elements.

Soil Ba (range of means, worldwide) = 84 to 838 ppm
Soil Ba (range of means, U.S.) = 265 to 835 ppm

9.3e. Beryllium

Beryllium in soils exists in the +2 oxidation state only. Its chemistry in soil solution parallels that of Al^{3+}. The free cation, Be^{2+}, is predominant below pH 6, but as the pH is raised, soluble Be-hydroxy species form (e.g., $BeOH^+$, $Be(OH)_2^0$, $Be_3(OH)_3^{3+}$) as $Be(OH)_2$ begins to precipitate. $Be(OH)_2$ reaches its minimum solubility near pH 9, because at higher pH the soluble anions, $Be(OH)_3^-$ and $Be(OH)_4^{2-}$, raise the solubility of total Be.

In soils of low pH, Be chemisorbs on surfaces as the cation, and is known to complex particularly strongly with organic matter. Since Be^{2+} is the smallest of all metal cations, it has the potential to form a very strong ionic bond with the abundant negatively charged functional groups in organic matter. In soils of neutral to slightly alkaline pH, hydrolysis reactions of Be^{2+} form hydroxy cations and solid $Be(OH)_2$, further reducing Be solubility. Mobility of Be in neutral to alkaline soils is consequently low. In very alkaline soils, soluble anionic Be-hydroxy and Be-carbonate complexes form, and the element may be mobilized to some degree.

$$\text{Soil Be (range of means, U.S.)} = 1.2 \text{ to } 2.1 \text{ ppm}$$

9.3f. Cadmium

Cadmium is a chalcophile, associating geochemically with Zn in the sulfide minerals of rocks. The oxidizing conditions of weathering in soils release Cd as the soluble and mobile Cd^{2+} ion. This cation is even more soluble than Zn^{2+} in acidic oxidizing solutions, and is rated as having medium to high mobility in well-drained acid soils. The high mobility is attributable to the fact that Cd^{2+} adsorbs rather weakly on organic matter, silicate clays, and oxides unless the pH is higher than 6. Above pH 7, Cd^{2+} can co-precipitate with $CaCO_3$ or precipitate as $CdCO_3$, and Cd phosphates may limit solubility as well. Therefore, mobility and bioavailability of Cd in neutral to alkaline soils is low. Liming acid soils is an effective means of limiting uptake of Cd by plants.

In continuously waterlogged soils, the low solubility of CdS results in low mobility. This fact can be used to advantage in severely contaminated soils to limit bioavailability. Well-drained soils polluted with Cd often allow hazardous levels of Cd uptake by crops; the same soils flooded for rice production allow much less uptake by that crop.

Generally, Cd concentrations in soils exceeding 0.5 $\mu g/g$ are considered to be evidence of soil pollution from one of a number of possible sources: industry (mining, metallurgy, etc.), proximity to highways, high-Cd phosphate fertilizers, or sewage sludge application. Nevertheless, natural geochemical processes have been known to concentrate Cd in surface soils. Like Hg, Cd tends to accumulate in peaty soils by precipitation in sulfide minerals, and Cd levels may also be high in poorly drained soils or soils of arid and semiarid climates. This has created a problem in some irrigated farming regions where the climate is too dry for leaching to deplete naturally high levels of Cd in the soil. The combination of generally high bioavailability in soils and very high toxicity to animals and humans has made Cd the element of greatest concern in considering the value of sewage sludge as a soil amendment.

$$\text{Soil Cd (range of means, worldwide)} = 0.06 \text{ to } 1.1 \text{ ppm}$$

9.3g. Cobalt

Cobalt occurs in two oxidation states in soil, +2 and +3, but Co^{2+} is the dominant form in soil solution. This metal associates preferentially with Fe and Mn oxides because of chemisorption and co-precipitation. There is evidence that, on Mn oxides, Co^{2+} is oxidized and strongly bound as Co^{3+}. Consequently, strongly oxidizing conditions in the soil are likely to favor the adsorption of cobalt. As the soil pH is raised, Co solubility decreases because of increased chemisorption on oxides and silicate clays, complexation by organic matter, and possibly precipitation of $Co(OH)_2$. Organic matter complexes with Co^{2+} are fairly labile, so that organically bound Co^{2+}

is bioavailable. Cobalt is considered to be somewhat mobile in acid soils, but less so as the pH approaches neutrality. Under strongly reducing conditions, the formation of Co sulfides can restrict mobility.

Cobalt toxicity is occasionally found in high-Co soils formed from serpentinite and other ultrabasic rocks. Deficiency is most likely in coarse-textured, acid-leached soils; alkaline or calcareous soils; and humus-rich soils. Extractability by strong acids can range from very little ($<1\%$) to a large fraction ($>30\%$) of the total Co, depending on the forms of Co in the soil.

Soil Co (range of means, worldwide) = 1.6 to 21.5 ppm
Soil Co (range of means, U.S.) = 1 to 17 ppm

9.3h. Chromium

Chromium in soils potentially occurs in the +3 (chromic) oxidation state as the Cr^{3+} cation, and in the +6 oxidation state as chromate, CrO_4^{2-}. However, soil conditions generally favor the Cr^{3+} form, a very immobile cation that complexes strongly with organic matter and chemisorbs on oxides and silicate clays, even at quite low pH. Furthermore, Cr^{3+} readily substitutes for Fe^{3+} in mineral structures, and precipitates as insoluble $Cr(OH)_3$ at higher pH. The chromic form is, therefore, very immobile in most soils and generally unavailable to plants, at least if the soil is not exceedingly acid.

At higher pH, a small fraction of the Cr^{3+} in soils can be oxidized to chromate, CrO_4^{2-}, a very toxic form of chromium. This oxidation is promoted by Mn oxides. Chromate adsorbs less strongly than Cr^{3+}, and the mobility and bioavailability of this anion is consequently higher. Generally, however, if pollutants containing chromate are applied to soils, most or all of the chromate is spontaneously reduced to Cr^{3+}, especially under acid conditions and with organic matter present. Organic matter supplies reducing agents and complexing groups, stabilizing the chromic form. The soil therefore has the ability to detoxify chromate and immobilize the element.

Chromium is rated as an immobile element, most of which is difficult to extract from soils even by aggressive chemical agents. Toxicity of Cr to plants is occasionally seen in unusually Cr-rich soils formed from the parent rock, serpentinite, or under high pH conditions favorable to Cr^{3+} oxidation.

Soil Cr (range of means, worldwide) = 7 to 221 ppm
Soil Cr (range of means, U.S.) = 20 to 85 ppm

9.3i. Copper

Copper occurs in soil solids and solutions almost exclusively as the divalent cation Cu^{2+}. However, reduction of Cu^{2+} (cupric) to Cu^{+} (cuprous) and Cu^{0} (metallic copper) is possible under reducing conditions, especially if halide or sulfide ions ("soft" bases) are present to stabilize Cu^{+} (a "soft" acid). Copper is classified as a chalcophile, owing to its tendency to associate with sulfide in the very insoluble minerals, Cu_2S and CuS. In reduced soils, then, copper has very low mobility. Most of the colloidal material of soils (oxides of Mn, Al, and Fe, silicate clays, and humus) adsorb Cu^{2+} strongly, and increasingly so as the pH is raised. For soils with high Cu accumulations, precipitation as cupric hydroxide, oxide, or hydroxy-carbonates (malachite and azurite) is possible above pH 6. The trace quantities of Cu found naturally in most soils are probably widely dispersed on sorption sites and within structural (octahedral) sites of oxides and silicates. Organically complexed Cu^{2+} is bound more tightly than any other divalent transition metal; this fact is most evident at low metal loadings in humus when very selective complexing groups (amines or polyphenols) are involved. Lability of these complexes is rather low, limiting bioavailability. For this reason, farmers have been able to apply large amounts of Cu salts to organic soils over time without causing toxicity to crops.

Because of the high affinity of Cu^{2+} for soil colloids, copper is rated a low-mobility element in near-neutral soils. It builds up in the surface of contaminated soils, showing virtually no downward migration. In more alkaline soils, while free Cu^{2+} solubility is exceedingly low, soluble complexes of Cu^{2+} (most importantly hydroxy, carbonate, and organic matter complexes) form and increase the total copper solubility. Consequently, mobility may be significant under high pH conditions. Most of the total dissolved copper in surface soils over a fairly wide range of pH, and particularly at higher pH, is in the form of Cu^{2+}-organic complexes.

Copper deficiencies are most common in soils that are peaty, acid-leached, coarse-textured, or calcareous. Soils with less than 8 ppm total Cu may be deficient for crops.

Toxicity of natural origin is found in soils formed from Cu sulfide-rich parent rocks, especially when the soil is acid. Bioaccumulation of Cu in humus followed by episodes of reduction can concentrate the element in sulfide form in natural wetlands. Because copper is not only phytotoxic but also a commonly abundant metal pollutant in waste materials, Cu in wastes such as sewage sludges is often the first element to limit land application.

Soil Cu (range of means, worldwide) = 6 to 80 ppm
Soil Cu (range of means, U.S.) = 14 to 29 ppm

9.3j. Fluorine

Fluorine occurs exclusively as the fluoride anion, F^-, in soils, where it complexes strongly with metals such as Al^{3+} and Fe^{3+}. It is found in structures of hydrous minerals, isomorphously substituting for structural OH^-. Thus, F^- can be found in micas, amphiboles, layer silicate clays, apatite (rock phosphate), and numerous other minerals. Because it is associated with clay structures, the natural concentrations of fluorine in fine-textured mineral soils and sedimentary rocks can be high.

The fluoride ion chemisorbs on clays and oxides by ligand exchange of surface OH^-, a reaction favored at low pH and on oxide and silicate minerals of low crystallinity. Fluoride, a "hard" base, has a particular affinity for Al^{3+}, a "hard" acid. Soluble Al^{3+}-fluoride cationic and anionic complexes are quite stable, and can dominate the speciation of dissolved aluminum in low-humus soils. The mobility of Al can be increased by the presence of F^-; soluble complex formation with Al may explain the rather high solubility and mobility of F^- in acid soils.

In calcareous soils, F^- mobility is low. Perhaps its solubility is limited by incorporation into insoluble Ca minerals such as hydroxyapatite. In sodic soils, F^- mobility is enhanced by the fact that NaF is a very soluble salt.

Naturally occurring F^- associated with hydrous minerals has low mobility because it is occluded in structures. Airborne fluoride pollutants (from smelters, rock phosphate fertilizer factories, etc.) are, in contrast, easily dissolved on contact with the soil. These forms of fluoride can be bioaccumulated by plants before leaching, sorption, or precipitation processes have a chance to lower solubility.

Soil F (range of means, worldwide) = 73 to 566 ppm
Soil F (range of means, U.S.) = 205 to 465 ppm

9.3k. Mercury

This element is a chalcophile, and in unweathered rocks is most commonly found as the mineral cinnabar (HgS). In soil environments, the cationic form, Hg^{2+}, is most common, as the reduced oxidation state (+1) has a limited stability range. Reduction to the metallic elemental form, Hg^0, is easily achieved in soils by both biological and chemical reactions. Elemental mercury is somewhat volatile, and the vapor is extremely toxic to organisms. Under anaerobic conditions at least, soil microbes methylate mercury, forming volatile organomercury compounds that are bioavailable and present a health hazard. At the same time, however, anaerobic conditions can convert Hg^{2+} into the exceedingly insoluble sulfide, HgS. Some of the more important transformations possible for mercury in soil are summarized in Figure 9.9.

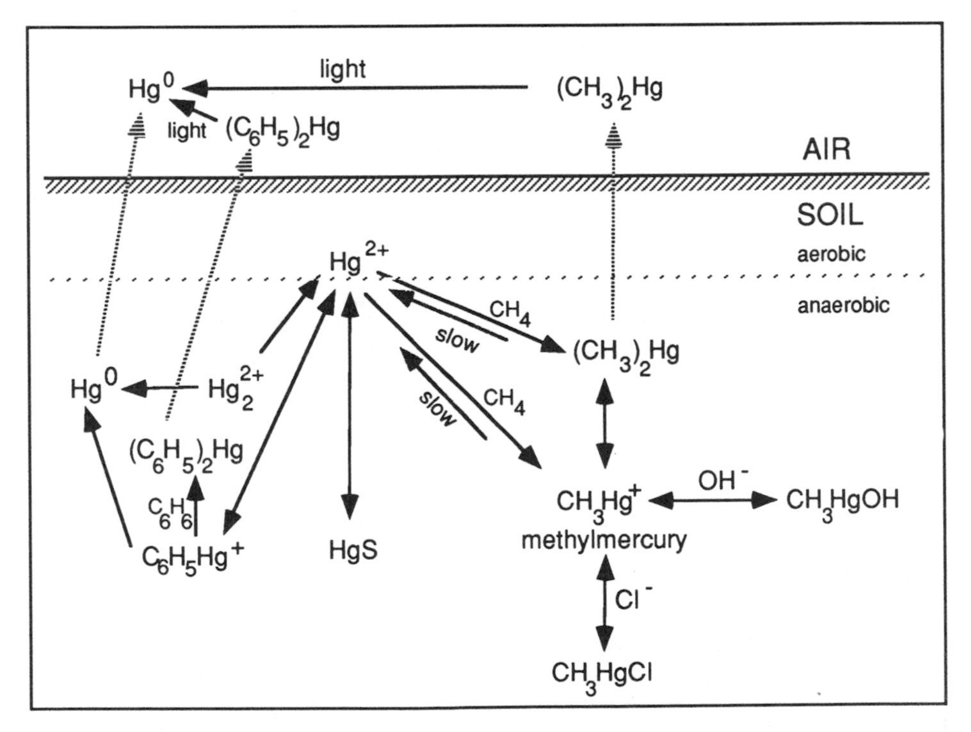

Figure 9.9. Biological and chemical transformations of mercury in the soil. Broken arrows denote the loss of volatile forms of Hg to the atmosphere or the air-filled pores of the soil. (Modified from B. E. Davies. 1980. Trace element pollution. In B. E. Davies (ed.), *Applied Soil Trace Elements*. New York: Wiley.)

Because of the complex chemistry of Hg, general statements about mobility are difficult to make. Acidic oxidizing conditions in soils tend to stabilize the +2 oxidation state (Hg^{2+}), which complexes only moderately strongly with the predominant functional groups of organic matter, phenolic and carboxylate groups. However, the Hg^{2+} cation is a "soft acid," bonding very strongly with "soft" ligands such as sulfhydryl and sulfide. The sulfhydryl ligand (−SH) is a minor functional group in humus (which typically contains less than 1% total sulfur), and yet may bind chalcophilic metal cations such as Hg^{2+}, Cd^{2+}, and Pb^{2+}. Mobility of Hg^{2+}, then, is probably very sensitive to loading, with traces of Hg^{2+} being very immobile and higher levels having medium mobility.

Adsorption of Hg^{2+} on silicate clays and oxides is more favorable at higher pH, and hydrolysis of Hg^{2+} causes the dominant soluble form to be $Hg(OH)_2^0$ above pH 4. The solubility of this neutral molecule is high enough to prevent precipitation from lowering the solubility of mercury in Hg-contaminated soils to trace levels. Precipitation of $Hg(OH)_2$ does, however, limit the activity of the free Hg^{2+} cation to extremely low values in the neutral to alkaline pH range.

In reducing soil environments, the strong association of Hg with sulfide creates a situation of low mobility, although there is the danger that volatile forms of the element can lead to some mobilization.

Mercury accumulation in soils tends to correlate with the organic matter level. The highest natural Hg concentrations have been reported in peaty and waterlogged soils. Toxicity to organisms is usually attributable to soil conditions that favor the production of volatile forms of Hg; evidently, Hg^{2+} cations interact strongly enough with humus and clays to limit availability of this form to plants and animals.

Soil Hg (range of means, worldwide) = 0.02 to 0.41 ppm
Soil Hg (range of means, U.S.) = 0.04 to 0.28 ppm

9.3l. Manganese

Manganese has three possible oxidation states in soils, +2, +3, and +4. The most reduced form of Mn, the Mn^{2+} ion, is the only stable form in soil solution. The Mn^{3+} ion is a powerful oxidant that either disproportionates to Mn^{2+} and Mn^{4+} or oxidizes water to liberate O_2:

$$2Mn^{3+} + H_2O = 2Mn^{2+} + 2H^+ + \frac{1}{2}O_2$$

Needless to say, Mn^{3+} has a very short life in solution, but it can be detected in some cases if complexing anions such as pyrophosphate, oxalate, or sulfate are present to stabilize it for a time.

Both Mn^{3+} and Mn^{4+} are stable only in the solid phase of soils, where they form insoluble oxide and hydroxide minerals of variable structure and oxidation state (e.g., MnO_2, $MnOOH$, Mn_3O_4, $Na_4Mn_{14}O_{27}$). The Mn^{2+} ion is released from these solids by spontaneous dissolution or cation exchange, especially under acidic or reducing conditions. Release of Mn^{3+} can also be detected in the presence of pyrophosphate.

Manganese solubility is controlled by the redox potential and pH of the soil. The Mn^{2+} ion is a very soluble species in water, forming hydroxide and carbonate precipitates only at high pH (>7). However, as the pH is raised above 6 in soils,

Mn^{2+}, despite being the most weakly complexing transition metal, bonds with organic matter, oxides, and silicates and its solubility decreases. Small changes in the soil redox potential or pH can shift the Mn^{2+} − Mn oxide reaction. Low pH or low E_h (see Chapter 7) favors the reduction of insoluble Mn oxides and an increased solubility of Mn^{2+}. As a result, Mn solubility within any particular soil can fluctuate tremendously over time, sometimes ranging from deficient to toxic levels.

The mobility of Mn defies classification because it is extremely sensitive to soil conditions (acidity, wetness, biological activity, etc.). Toxicity to plants is most likely in waterlogged soils or acid soils with low humus content. Deficiency is most often found in soils that are saline and alkaline, calcareous, peaty, coarse textured, or acid leached.

Total Mn in soils is highly variable and not closely related to the Mn content of the parent material, probably because of the tendency for cyclical reduction-oxidation to rapidly mobilize, then reprecipitate Mn as oxides in nodules and other deposits. Mn tends to associate with ferric iron in these deposits.

Soil Mn (range of means, worldwide) = 80 to 1300 ppm
Soil Mn (range of means, U.S.) = 260 to 840 ppm

9.3m. Molybdenum

Molybdenum is primarily in the +6 oxidation state in soils, taking the form of the molybdate anion, MoO_4^{2-}. Protonation of this oxyanion, described by the reaction:

$$MoO_4^{2-} + H^+ = HMoO_4^- \qquad K = 10^{3.7}$$

occurs at low pH. Molybdate is chemisorbed in soils by oxides, noncrystalline aluminosilicates, and to a lesser extent, layer silicate clays, with sorption on all these minerals increasing at lower pH. Consequently, Mo is least soluble in acid soils, particularly those containing iron oxides. The availability of adsorbed molybdate diminishes with time. Phosphate additions to soil may release molybdate from bonding sites, increasing bioavailability of the element. Precipitation of Mo into pure phases is considered highly unlikely because of the very low concentration of the element in almost all soils.

Molybdenum has a high mobility in neutral or alkaline soils and medium mobility in acid soils. Liming soils is known to increase availability of the element. Factors controlling mobility in anaerobic soils are not understood. In one respect, the association of Mo in lower oxidation states with sulfide should reduce its solubility under poor drainage. However, poor soil drainage also leads to organic matter accumulation, and there is circumstantial evidence that molybdate bonds with organic matter by the same mechanism that borate does—via two adjacent OH groups such as would be found in the polyphenols of humus. Humus-rich soils then accumulate Mo as long as they remain poorly drained, because little leaching is occurring.

Molybdenum is most commonly deficient in coarse-textured acid soils and low-humus soils. Plant toxicity is associated with poor drainage and alkaline, calcareous, clayey, or peaty soils.

Soil Mo (range of means, worldwide) = 1 to 3 ppm
Soil Mo (range of means, U.S.) = 0.35 to 5.8 ppm

9.3n. Nickel

The +2 oxidation state is the only stable form of nickel in soil environments. The Ni^{2+} cation is almost as electronegative as Cu^{2+}; this fact and its electronic structure favor the formation of complexes with organic matter that are comparable in stability to those of Cu^{2+}. Bioaccumulation of Ni in humus is pronounced, and like Cu^{2+}, Ni^{2+} favors bonding to "softer" organic ligands containing nitrogen and sulfur. As the smallest of the divalent transition metal cations, Ni^{2+} fits easily into octahedral sites, co-precipitating readily into Mn and Fe oxides in soils. Chemisorption on oxides, noncrystalline aluminosilicates, and layer silicate clays is favorable above pH 6, but lower pH favors exchangeable and soluble Ni^{2+}. Because solubility decreases markedly at higher pH, mobility of Ni, rated as medium in acid soils, becomes very low in neutral to alkaline soils. Under reducing conditions, Ni^{2+} is incorporated into sulfides that restrict mobility to very low levels.

Toxicity of Ni to plants is found in acid soils formed from serpentinite or other ultrabasic rocks. High organic matter levels in Ni-rich soils can solubilize Ni^{2+} as organic complexes, at least at higher pH. Nickel is a strongly phytotoxic element, being several times more toxic than copper. Like copper, it occurs commonly in industrial wastes and sewage sludges at appreciable levels, and may reach levels toxic to plants in waste-treated soils.

Soil Ni (range of means, worldwide) = 4 to 55 ppm
Soil Ni (range of means, U.S.) = 13 to 30 ppm

9.3o. Lead

Lead exists principally in the +2 oxidation state in soils. It is strongly chalcophilic, occurring primarily as PbS in rocks and becoming very insoluble in reduced soils because of its precipitation by sulfide generated from sulfate reduction. Sulfhydryl groups, if present in humus, strongly complex the Pb^{2+} cation. Under oxidizing conditions, the Pb^{2+} ion becomes less soluble as soil pH is raised. Complexation with organic matter, chemisorption on oxides and silicate clays, and precipitation as the carbonate, hydroxide, or phosphate are all favored at higher pH. In alkaline soils, solubility may increase by formation of soluble Pb-organic and Pb-hydroxy complexes. The Pb^{2+} ion has a particularly high affinity for Mn oxides, a fact perhaps explained by Mn oxidation of Pb^{2+} to Pb^{4+}, a very insoluble ion. Lead is the least mobile heavy metal in soils, especially under reducing or nonacid conditions.

As expected from the strong complexation of Pb^{2+} by organic matter, lead bioaccumulates in the humus-rich surface layer of soils. Soils polluted at the surface with lead deposited from aerial contaminants show little indication of metal leaching over many years. Lead levels in sewage-sludge-treated soils are considerably higher than levels in most natural soils because of the generally high concentration of Pb in wastes contaminated by old plumbing and by various industries.

Most of the lead in soils appears to be unavailable to plant tops. Plant-absorbed Pb^{2+} concentrates in the roots, translocating very little from roots to tops as long as the plant is actively growing.[1] Toxic effects of lead on plants have not often been

1. There is some evidence that Pb translocates readily from roots to tops as plants senesce and die, perhaps due to breakdown of membrane barriers to ion transport.

observed, but a hazard to animals exists because of the inherently higher toxicity of this element to animals. The health concern with lead-polluted soils arises mostly from the contamination of plants by soil particles, and ingestion of soil by humans and grazing animals. The risk of lead movement from soils to edible plant parts is believed to be low.

Soil Pb (range of means, worldwide) = 10 to 84 ppm
Soil Pb (range of means, U.S.) = 17 to 26 ppm

9.3p. Antimony

Antimony has geochemical behavior similar to arsenic, as it occurs most commonly in the +3 (antimonite) and +5 (antimonate) oxidation states and also tends to associate with sulfides in rocks, sediments, and soils. In soil solutions, the Sb^{3+} and Sb^{5+} oxidation states are stable under reducing and oxidizing conditions, respectively.

In reducing soil solutions, Sb^{3+} is likely to have the form of the uncharged $Sb(OH)_3^0$ molecule, except at very acid and alkaline pH where $Sb(OH)_3^0$ converts into the $Sb(OH)_2^+$ cation and $Sb(OH)_4^-$ anion, respectively. The oxide of Sb^{3+}, Sb_2O_3, is too soluble to limit solubility of the element except in highly polluted soils. The oxide becomes more soluble below pH 3 and above pH 10.

In oxidizing soil solutions, Sb^{5+} is likely to form the anionic molecule $Sb(OH)_6^-$, above pH 4, and $Sb(OH)_5^0$ in more acid solution. As an anion, $Sb(OH)_6^-$ may adsorb by ligand exchange on oxides and silicate clays. Antimony associates with ferric hydroxide in soils, perhaps a result of chemisorption of the anion on this mineral.

Antimony is a common pollutant of soils in industrial and mining sites. Its mobility in soils is rated as medium, with reducing conditions associated with poor drainage probably lessening mobility.

Soil Sb (range of means, worldwide) = 0.19 to 1.77 ppm

9.3q. Selenium

Selenium, a chalcophile, tends to be associated with sulfide minerals in rocks. Weathering processes in the soil oxidize these very insoluble reduced forms, including elemental Se (Se^0), the selenides (Se^{2-}), and selenium sulfides, to the more soluble selenites (SeO_3^{2-}) and selenates (SeO_4^{2-}). With numerous oxidation states possible for Se in soils, redox potential is a critical factor in Se behavior.

In alkaline, oxidized soils, selenates are the dominant form. These oxyanions bond rather weakly to oxides and other minerals, so that Se mobility in neutral to alkaline soils is rated high. In slightly acid soils that are oxidized, selenites prevail, showing lower mobility than selenates because of the ability of SeO_3^{2-} to chemisorb strongly on oxides and aluminosilicates and to precipitate as the insoluble ferric selenite. In wet, acid, or humus-rich soils, the insoluble reduced forms of Se tend to predominate, so that Se mobility and bioavailability are very low. However, under reducing conditions, biological methylation of Se is possible, forming volatile compounds that may mobilize the element.

The solubility and total content of Se in many soils is quite low, so that crops often contain Se levels that could produce deficiencies in animals and humans. In contrast, Se often concentrates as the soluble and highly available selenate form in the surface soils of arid and semiarid regions. Plants can bioaccumulate this available Se, creating a potential toxicity hazard to foraging animals. The geographic distribution of Se in crop and forage plants of North America is, broadly speaking, low in the humid regions of the northwest, northeast, and southeast and moderate to high (sometimes toxic) in the arid and semiarid western plains. Irrigation of arid-region alkaline soils transports natural selenate into drainage water; this can contaminate bodies of water with levels of Se toxic to aquatic animals.

High Se levels are often associated with poorly drained sites, and yet, as described above, Se solubility is generally low under reducing conditions. Poor drainage allows dissolved Se salts to accumulate at the surface by evaporation of water.

A large fraction of Se in soils can be plant available and easily extractable. This is the case for nonacid soils, particularly calcareous ones, that often contain Se in the relatively soluble selenate form.

Soil Se (range of means, worldwide) = 0.05 to 1.27 ppm
Soil Se (range of means, U.S.) = 0.19 to 1.05 ppm

9.3r. Thallium

In soil solutions, both the +1 and +3 oxidation states of Tl are possible as Tl^{3+} is easily reduced to Tl^{+}. The Tl^{3+} cation behaves somewhat like Al^{3+} but hydrolyzes even more readily than Al^{3+}, and the insoluble oxide, Tl_2O_3, forms at lower pH ($<$ 2) and remains insoluble to higher pH ($>$10) than Al_2O_3. Thus, mobility of the oxidized form of Tl (+3) should be very low due to precipitation and adsorption reactions. The reduced form of Tl (+1) has very different behavior, as the Tl^{+} ion acts more like an alkaline metal such as K^{+} or Rb^{+}. This reduced form should have higher mobility than Tl^{3+} under well-drained conditions, although fixation of Tl^{+} on clays may be possible. In poorly drained soils, the reducing conditions may limit solubility and mobility, since Tl^{+} is readily incorporated into sulfide minerals.

Soil Tl (range of means, U.S.) = 0.02 to 2.8 ppm

9.3s. Vanadium

Vanadium in soil solutions exists predominantly in the +5 and +4 oxidation states, as the vanadate anionic forms ($VO_3(OH)^{2-}$, $VO_2(OH)_2^-$, $VO(OH)_3^0$, and more complex polymeric anions), and as the vanadyl cation, VO^{2+}. However, the V^{3+} ion, which is not stable in aerated solutions, may substitute readily for Fe^{3+} in minerals such as Fe oxides. The soil redox potential determines the oxidation state, and therefore solubility, with V having very high mobility and bioavailability in oxidized soils that are neutral to alkaline. It is likely that vanadate anions bond on oxides and silicates most effectively at low pH, following the pattern of phosphate and many other oxyanions. Consequently, V solubility should be quite high at high pH, but lower if the soil is more acid. Contributing to this trend is the fact that lower pH favors chemical reduction of vanadate to vanadyl, a cation that behaves much like Cu^{2+}, complexing tightly with organic matter and chemisorbing on oxides and aluminosilicates. As a result, the mobility of V under reducing or acid conditions is probably moderate to low. Since there is evidence that humus promotes the rather easy reduction of vanadate to VO^{2+}, a low redox potential may not be necessary to immobilize V in organically bound form. In any event, it is observed that the element tends to associate with humus in soils.

Vanadium is found substituted into the structures of layer silicate clays as the VO^{2+} (V^{4+}) ion, a form that must have very low bioavailability. Its tendency to substitute into oxides and layer silicates as V^{4+} (and perhaps V^{3+}) may account for the correlation between clay content and total V content of soils.

Soil V (range of means, worldwide) = 5 to 190 ppm
Soil V (range of means, U.S.) = 38 to 136 ppm

9.3t. Zinc

As a chalcophile, zinc tends to occur as the sulfide mineral, sphalerite (ZnS) in rocks, but weathers to the soluble Zn^{2+} ion in the oxidizing environment of soils. The +2 oxidation state is the only one possible in the soil. In acid, aerobic soils, Zn has medium mobility, held in exchangeable forms on clays and organic matter. At higher pH, however, chemisorption on oxides and aluminosilicates and complexation with humus lower the solubility of Zn^{2+} markedly. Consequently, Zn mobility in neutral soils is very low. If soils are slightly alkaline, even though the activity of the free Zn^{2+} ion is extremely low, Zn-organic complexes can become soluble and raise mobility. In strongly alkaline soils, Zn-hydroxy anions may form to increase solubility.

In soils contaminated by high levels of Zn, precipitation of Zn oxide, hydroxide, or hydroxycarbonate may limit Zn^{2+} solubility at pH 6 or higher. Co-precipitation of Zn^{2+} into octahedral sites of oxides and silicates is theoretically possible, although the importance of this in soils is not known.

In the reducing environment of flooded soils, release of Zn^{2+} from dissolving Fe oxides may initially increase availability, but Zn mobility is ultimately restricted by the extreme insolubility of ZnS.

Under acidic, oxidizing conditions, Zn^{2+} is one of the most soluble and mobile of the trace metal cations. It does not complex tightly with organic matter at low pH. Acid-leached soils often have Zn deficiency because of depletion of this element in the surface layer. Calcareous and alkaline soils also commonly have Zn deficiency, but the cause is low solubility.

Toxicity of Zn to plants is most likely to appear in acid soils that have not been subjected to prolonged acid leaching. The rather high potential solubility of Zn^{2+} in acid soils, and the fact that Zn^{2+} is typically a high-concentration pollutant of industrial wastes and sewage sludges, combine to create a significant potential for phytotoxicity from land application of wastes. This is in spite of the fact that Zn toxicity to plants is inherently fairly low.

Soil Zn (range of means, worldwide) = 17 to 125 ppm
Soil Zn (range of means, U.S.) = 34 to 84 ppm

References

Davies, B. E. 1980. Trace element pollution. In B. E. Davies (ed.), *Applied Soil Trace Elements*. New York: Wiley, pp. 287–351.

Huheey, J. E. 1972. *Inorganic Chemistry.* New York: Harper & Row.
Kabata-Pendias, A. and H. Pendias. 1984, 2nd ed. 1992. *Trace Elements in Soils and Plants.* Boca Raton, Fl.: CRC Press.
McGeehan, S. L. and D. V. Naylor. 1992. In *Agronomy Abstracts.* Madison, Wis.: Soil Science Society of America. p. 242.
Millot, G. 1970. *Geology of Clays.* New York: Springer-Verlag.

Suggested Additional Reading

Alloway, B. 1990. *Heavy Metals in Soils.* Halsted Press, New York.
Davies, B. E. (ed.). 1980. *Applied Soil Trace Elements.* New York: Wiley.
Swift, R. S. and R. G. McLaren. 1991. Micronutrient adsorption by soils and soil colloids. In G. H. Bolt, M. F. DeBoodt, M.H.B. Hayes, and M. B. McBride (eds.), *Interactions at the Soil Colloid-Soil Solution Interface.* Dordrecht, Netherlands: Kluwer, pp. 257–292.

Questions

1. The powerful metal chelator, EDTA (ethylenediaminetetraacetic acid), is an efficient extracting agent for organically bound trace metal cations in soils at pH 6 and higher. The formation constants, K_{ML}, for some metal-EDTA complexes in water are given below:

Metal	log K_{ML}
Cu^{2+}	19.7
Ni^{2+}	19.5
Zn^{2+}	17.2
Cd^{2+}	17.5
Ca^{2+}	11.5
Al^{3+}	17.6
Fe^{3+}	26.5

   (a) Why should EDTA be more efficient at extracting trace metals from neutral or alkaline soils than from acid soils?
   (b) The EDTA-extractable quantities of trace metals in soils amended with sewage sludges sometimes increase over a period of years. Can you propose a mechanism that would account for this change in the soil?
   (c) How would EDTA additions to a Ni-contaminated soil at pH 7 affect Ni^{2+} activity, total soluble Ni, and Ni availability to plants?

2. The availability of arsenic in soils could possibly be limited by the solubility product of Ca or Fe arsenate, given below:

$$Ca_3(AsO_4)_2 \quad K_{SO} = 6.8 \times 10^{-19}$$
$$FeAsO_4 \quad K_{SO} = 5.7 \times 10^{-21}$$

   Assuming that the soils have been contaminated by the use of arsenate pesticides, calculate the upper limit of solubility of AsO_4^{3-} in:
   (a) an acid mineral soil at pH 4.5, if the solution activity of Fe^{3+} is controlled by the solubility of $Fe(OH)_3$ ($K_{SO} = 10^{-39}$).
   (b) a calcareous soil at pH 8, if the solution activity of Ca^{2+} is maintained at about 2×10^{-3} *M*.

   In which soil is arsenic likely to be more mobile and bioavailable? (Ignore activity corrections for solution concentrations in this problem.)

3. Ferrous chloride, $FeCl_2$, is known to be an effective soil amendment for lowering soluble levels of selenium in alkaline soils. Propose a reaction sequence that would explain this.

4. Explain the phenomenon of bioaccumulation, that is, the gradual buildup over time of certain trace metals in the humus fraction of the soil surface.

5. Given that Al^{3+} is not intrinsically a highly phytotoxic element (see Table 9.1), how can the widespread and common toxicity of Al to crops, especially in soils of the humid tropics, be explained?

6. A calcareous soil, contaminated with a high level of copper, was found to contain the Cu hydroxycarbonate mineral, malachite. The solubility of malachite is defined by the dissolution reaction:

$$Cu_2(OH)_2CO_3(s) + 4H^+ = 2Cu^{2+} + 3H_2O + CO_2(g) \qquad \log K = 14.16$$

   (a) If the pH of this soil is 7.4, and the gas pressure of CO_2 in the soil air is assumed to be $P_{CO_2} = 10^{-2}$ atmospheres, calculate the activity of the Cu^{2+} ion in soil solution. (Ignore activity corrections for solution concentrations in this problem.)
   (b) Severe symptoms of Cu toxicity were observed in a corn crop grown in this soil. Does the answer in (a) provide a reliable indication that toxicity is likely to develop over a growing season?
   (c) How can the rhizosphere manipulate the activity of Cu^{2+}?

7. The fallout from nuclear tests in the 1950s resulted in widespread deposition of the radioactive metal, ^{90}Sr, in trace amounts on soils. This metal takes the form of the exchangeable cation, Sr^{2+}, in soils. Consider the question of the downward migration of Sr^{2+} by saturated flow (leaching) through a well-drained nonacid clayey soil with exchange sites occupied largely by Ca^{2+}. The soil has a porosity of 0.42 and a bulk density of 1.5 g/cm^3.
   (a) Calculate the value of the distribution coefficient, K_d, to be used in equation 9.5 if the soil has a CEC of 150 mmoles/kg and a solution Ca^{2+} concentration of 10^{-3} M. (Sr^{2+} shows some preference for exchange sites over Ca^{2+}, so that $K_S \approx 2$ is a reasonable estimate for the cation exchange selectivity coefficient.)
   (b) Use equation 9.5 to determine how many years it would take for the radioactivity to leach out of the reach of the root systems of most annual crops (about 1 meter). Assume a humid climate with about 40 cm of rainfall that actually leaches through the soil.

8. In Chapter 7, Table 7.1, the reduction potentials of MnO_2 and $MnOOH$ are listed. Use the Nernst equation to determine if either of these Mn oxides in a soil solution at pH 7 would be able to generate chromate from precipitated $Cr(OH)_3$. Assume that ion exchange limits Mn^{2+} solubility to 10^{-5} M. What would be the significance of this reaction to the environment?
   (Note: The relevant chromate-chromic half-cell reaction is:

$$CrO_4^{2-} + 4H_2O + 3e^- = Cr(OH)_3(s) + 5OH^- \qquad E_h^\circ = -0.13V)$$

10

Organic Pollutants in Soil

Soils function as chemical and biological "filters" that lessen the environmental impact of organic chemicals introduced into the biosphere by design or accident. Soils thereby form a first line of defense against leakage of these compounds into groundwater and the biosphere.

Soils curtail mobility of organic chemicals in two ways: by sorption and by biological or chemical degradation. Consequently, the basis for predicting the behavior of organic pollutants in soil, and for evaluating the risk that a chemical might leach into groundwater, follows from an understanding of the nature and extent of these sorption and degradation processes. This chapter will describe some important aspects of sorption theory and use it to explain the behavior of organic molecules at mineral and humic surfaces. Some of the nonbiological mechanisms of organic molecule degradation will also be discussed. Biological degradation, although very important, is too complex a subject to be covered adequately in this book. Detailed treatment of the subject is available in a number of texts and review articles.

10.1. CHEMICAL CLASSIFICATION OF PESTICIDES AND OTHER ORGANIC POLLUTANTS

Adsorption and desorption of organic molecules in soils is controlled by the chemical properties of the molecules and the surface properties of the particular soil. The relevant properties of the organic adsorbate are:

1. Identity of functional groups attached to the molecule
2. Acidity or basicity of the functional groups
3. Molecular size and shape
4. Polarity and charge of the molecule
5. Polarizability of the molecule

These properties determine the water solubility of the molecule, and as will be demonstrated later, the tendency of the molecules to adsorb on soil surfaces.

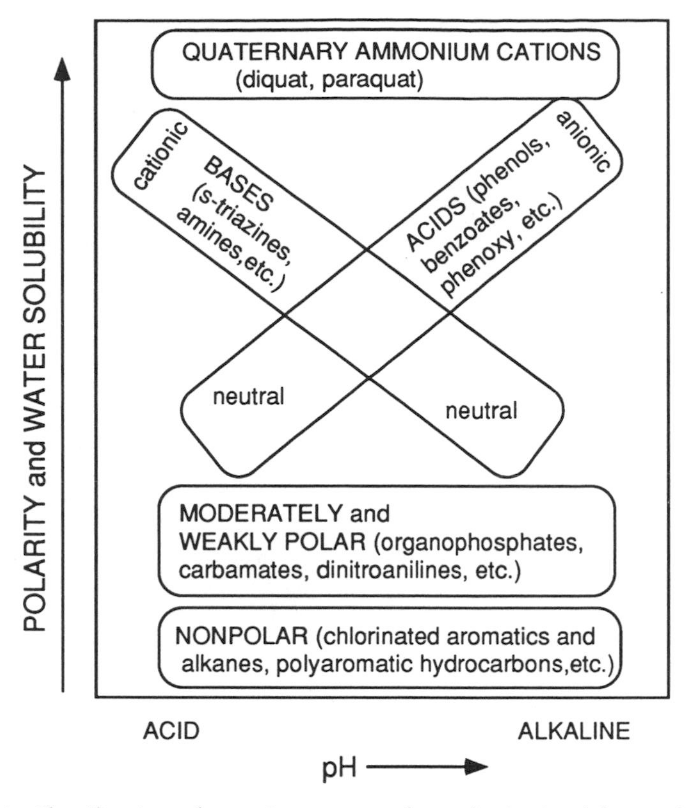

Figure 10.1. Classification of organic compounds on the basis of their polarity and charge.

The polarity and charge is a sufficiently important property that it is used to broadly classify organic molecules, as shown in Figure 10.1.

10.2. THE NATURE OF PHYSICAL AND CHEMICAL ADSORPTION

The mineral and organic surfaces of soils (referred to here as the adsorbent) may adsorb organic molecules (the adsorbate) weakly or strongly depending on the strength of the adsorbate-adsorbent interaction. Strong interaction is indicative of chemical adsorption, or *chemisorption,* in which a covalent or short-range electrostatic bond forms between the molecule and the surface. Weak adsorption, on the other hand, is characteristic of *physical adsorption,* in which the bonding interaction is not very energetic (typically <10 kcal/mole of adsorbate). The contrasting features of chemisorption and physical adsorption are listed in Table 10.1.

Adsorption data are most commonly represented by an *adsorption isotherm,* which is a plot of the quantity of adsorbate retained by a solid as a function of the concentration of that adsorbate in the bathing gas or solution phase that is *at equilibrium with the solid.* The shape of this isotherm line suggests (but does not confirm) infor-

Table 10.1. Characteristics of Physical and Chemical Adsorption

Property	Physical	Chemical
Heat of adsorption	<10 kcal/mole	>20 kcal/mole
Temperature range of adsorption	Only below boiling point of adsorbate	Both low and high temperature
Slope of adsorption isotherm	Greater at higher adsorbate concentration	Less at higher adsorbate concentration
Dependence on properties of adsorbent	Relatively little	Great
Dependence on properties of adsorbate	Great	Great
Activation energy for adsorption	Low or none	May be high
Number of layers of adsorbed molecules	Multiple (at most)	Single (at most)

mation about the adsorbate-adsorbent (organic-surface) interaction; to this end, isotherms have been classified into four types, diagrammed in Figure 10.2:

1. The *L-type* (Langmuir) isotherm reflects a relatively high affinity between the adsorbate and adsorbent, and is usually indicative of chemisorption (see Table 10.1 for the features of chemisorption).

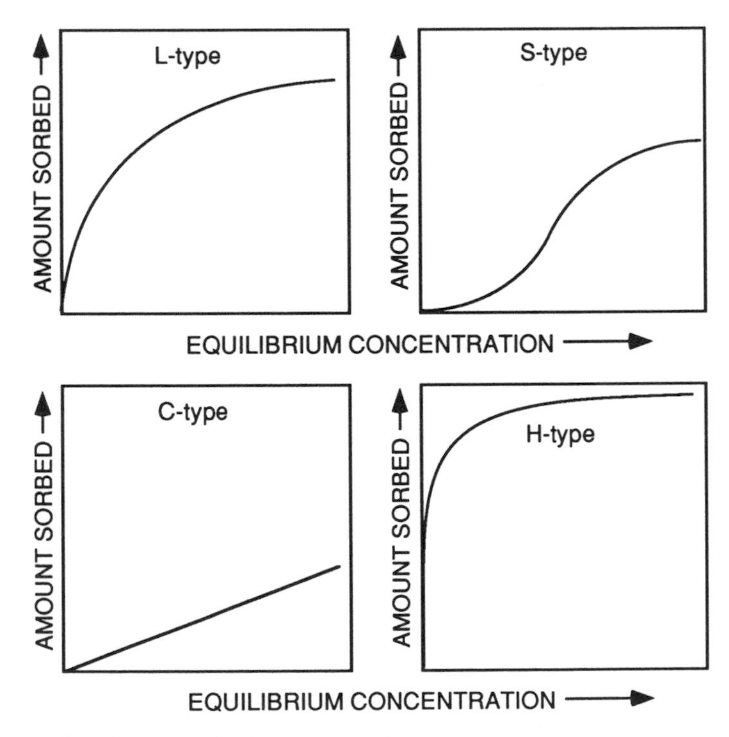

Figure 10.2. Classification of adsorption isotherms. (After Giles et al., 1960.)

2. The *S-type* isotherm suggests "cooperative adsorption," which operates if the adsorbate-adsorbate interaction is stronger than the adsorbate-adsorbent interaction. This condition favors the "clustering" of adsorbate molecules at the surface because they bond more strongly with one another than with the surface.
3. The *C-type* (constant-partitioning) isotherm, which suggests a constant relative affinity of the adsorbate molecules for the adsorbent, is usually observed only at the low range of adsorption. Deviation from the linear isotherm is likely at high adsorption levels. Nevertheless, because many nonpolar organic compounds of interest in soils are adsorbed at quite low concentrations, the linear C-type isotherm is often a reasonable description of adsorption behavior.
4. The *H-type* isotherm, indicative of very strong adsorbate-adsorbent interaction (i.e., chemisorption), is really an extreme case of the L-type. This isotherm is not often encountered with organic molecules because few of them form strong ionic or covalent bonds with soil colloids.

More complex isotherm shapes are possible, but these can be generally recognized as hybrids of the four main types listed here.

It must be cautioned that isotherm features can never prove the adsorption mechanism involved. They can only point to reasonable mechanisms that must ultimately be confirmed by more direct methods such as molecular spectroscopy.

10.3. ADSORPTION EQUATIONS

10.3a. Kinetic Derivation of the Langmuir Equation

Any valid chemical model of adsorption should predict an equation that accurately describes the experimental adsorption isotherm. In 1918, Langmuir developed a kinetic model of adsorption that described vapor adsorption on a homogeneous surface. The surface is assumed to possess a certain number of identical sites, S. Of these sites, S_A are occupied by adsorbate molecules and $S_0 = S - S_A$ are vacant. The rate of evaporation of molecules from the surface is reasoned to be proportional to S_1 because all molecules are adsorbed with the same bond strength on identical sites (adsorbate molecules on adjacent sorption sites are assumed not to interact). Thus, each molecule has an equal chance of desorbing from the surface in any particular time period. Mathematically, this means that

$$\text{Rate of evaporation} = k_1 \cdot S_A \tag{10.1}$$

where k_1 is a reaction rate constant. Conversely, the rate at which molecules condense on the surface is taken to be proportional to both the number of vacant surface sites, S_0, and the gas pressure of the adsorbate, P. In mathematical terms:

$$\text{Rate of condensation} = k_2 \cdot P \cdot S_0 \tag{10.2}$$

This equation is based on the argument that each gas molecule colliding with a vacant surface site has a finite and fixed probability of "sticking" on this site, while molecular collisions at occupied sites have no net effect on condensation. Furthermore, gas pressure is a measure of the number of molecular collisions at a surface per unit of surface area, so the rate of condensation should increase with pressure in a linear fashion.

At equilibrium, steady state is reached and the evaporation and condensation rates are equal. That is:

$$k_1 S_A = k_2 P S_0 = k_2 P(S - S_A) \qquad (10.3)$$

If θ is defined as the fraction of surface sites occupied, then $\theta = S_A/S$ and equation 10.3 can be reexpressed in terms of θ:

$$\theta = \frac{bP}{(1 + bP)} \qquad (10.4)$$

where $b = k_2/k_1$. This is the famous *Langmuir equation,* describing the fraction of surface covered by a monolayer of adsorbate, θ, as a function of the partial pressure of the adsorbate, P.

10.3b. The Langmuir Equation Derived from Mass Action

Because equation 10.4 describes an equilibrium state, it stands to reason that it should be possible to derive the equation from arguments that are not based on reaction kinetics. In fact, this is the case, and a more general derivation of equation 10.4 from statistical mechanics, or more simply from the mass-action principle, is possible.

A derivation based on the mass-action law produces equation 10.4 as follows:

Consider the reversible adsorption of an adsorbate, A, onto sites given the label S:

$$S + A = S - A \qquad (10.5)$$

The equilibrium constant for this reaction, K_1, takes the form:

$$K_1 = \frac{(S - A)}{(S)(A)} \qquad (10.6)$$

where the parentheses denote activities of surface and solution species. The activity of the vacant sites, (S), and of the site-adsorbate complex, (S − A), are not accessible to measurement, but it is assumed from ideal mixture theory that activities are proportional to mole fractions. The activity terms in equation 10.6 are then replaced by mole concentrations, denoted by square brackets.

The fraction of sites occupied by A, θ_A, is defined as

$$\theta_A = \frac{[S - A]}{[S] + [S - A]} \qquad (10.7)$$

Similarly, the fraction of unoccupied sites is

$$1 - \theta_A = \frac{[S]}{[S] + [S - A]} \qquad (10.8)$$

The ratio of occupied to unoccupied sites is then found to be

$$\frac{\theta_A}{(1 - \theta_A)} = \frac{[S - A]}{[S]} \tag{10.9}$$

and equation 10.6 can then be rewritten in the form

$$\frac{\theta_A}{(1 - \theta_A)} = K_1(A) \tag{10.10}$$

which becomes, on rearrangement, the standard Langmuir equation:

$$\theta_A = \frac{K_1(A)}{1 + K_1(A)} \tag{10.11}$$

The case of the S-shaped isotherm, representing cooperative adsorption, is a special situation with a similar derivation. Suppose that there is a tendency for the adsorbate, A, to adsorb in pairs on the surface. Then, in addition to reaction 10.5, the following adsorption reaction must be considered:

$$S + 2A = S - A_2 \tag{10.12}$$

along with the equilibrium constant:

$$K_2 = \frac{(S - A_2)}{(S)(A)^2} \tag{10.13}$$

Again converting from activities to fraction of sites occupied by dimers, θ_{A_2}, equation 10.13 becomes

$$K_2 = \frac{\theta_{A_2}}{(1 - \theta_T)(A)^2} \tag{10.14}$$

where θ_T is the fraction of sites occupied by both monomers, A, and dimers, A_2. That is, $\theta_T = \theta_A + \theta_{A_2}$. Equations 10.10 and 10.14 rearrange to

$$\theta_A = K_1(A)(1 - \theta_T) \tag{10.15}$$

$$\theta_{A_2} = K_2(A)^2(1 - \theta_T) \tag{10.16}$$

and summing θ_A and θ_{A_2} gives θ_T:

$$\theta_T = K_1(A)(1 - \theta_T) + K_2(A)^2(1 - \theta_T) \tag{10.17}$$

which becomes on rearrangement:

$$\theta_T = \frac{K_1(A) + K_2(A)^2}{1 + K_1(A) + K_2(A)^2} \tag{10.18}$$

However, adsorption isotherms are not generally expressed in terms of site occupancy, but rather in terms of how much adsorbate is removed from solution. If the adsorbed quantity of A is given the symbol Q_A, it is clear that for a given number of adsorption sites, N, each adsorbed dimer removes *two* adsorbate molecules from solution, and Q_A is given by

$$Q_A = N(\theta_A + 2\theta_{A_2}) \tag{10.19}$$

The final expression for Q_A is then obtained by multiplying equation 10.16 by two and adding it to equation 10.15:

$$Q_A = N\left[\frac{K_1(\mathrm{A}) + 2K_2(\mathrm{A})^2}{1 + K_1(\mathrm{A}) + 2K_2(\mathrm{A})^2}\right] \tag{10.20}$$

This is an S-shaped function for which the degree of cooperative adsorption is measured by the relative magnitude of the equilibrium constants, K_1 and K_2, for the monomer and dimer.

10.3c. The Linearized Langmuir Equation

The general shape of the Langmuir isotherm (equation 10.4) is shown in Figure 10.3 for several selected values of b, the bonding constant. This constant can be shown from the statistical mechanical derivation to be a function of the bonding energy, Q, as follows:

$$b = b_0 e^{Q/RT} \tag{10.21}$$

where R is the gas constant and b_0 is a constant for any given adsorbate at a particular temperature.

A linear form of the Langmuir equation can be obtained if θ in equation 10.4 is replaced by v/v_m, where v_m is the maximum volume of gas adsorbed per unit weight of adsorbent (representing a monolayer of adsorbed molecules, with all S surface sites occupied), and v is the volume of gas (measured at STP[1]) adsorbed per unit weight of adsorbent at a pressure, P. Equation 10.4 then becomes

$$v = \frac{v_m bP}{(1 + bP)} \tag{10.22}$$

which can be rearranged into a linear form:

$$\frac{P}{v} = \frac{1}{bv_m} + \frac{P}{v_m} \tag{10.23}$$

A plot of P/v versus P is then expected to produce a straight line as shown in Figure 10.3, allowing the two constants, b and v_m, to be determined from the slope and intercept of the line. (The line has a slope of $1/v_m$ and an intercept of $1/bv_m$.)

Commonly, surface areas of solids are measured by this method of estimating v_m, using the equation

$$A = \left(\frac{v_m}{v_0}\right) N_0 \mathring{a} \tag{10.24}$$

where A is the specific surface area of the solid (m^2/g), $\mathring{a}$ is the area of one adsorption site, v_0 is the volume of a mole of the adsorbing gas (22.4 liters at 273 K), and N_0 is Avogadro's number. The value of $\mathring{a}$ must be estimated in some way before A can be calculated from the experimentally determined value of v_m. Commonly, the estimate is based on the dimensions of an atomic model of the adsorbate, assuming that the

1. STP stands for conditions of standard temperature and pressure, 298 K and 1 atmosphere (760 mm Hg).

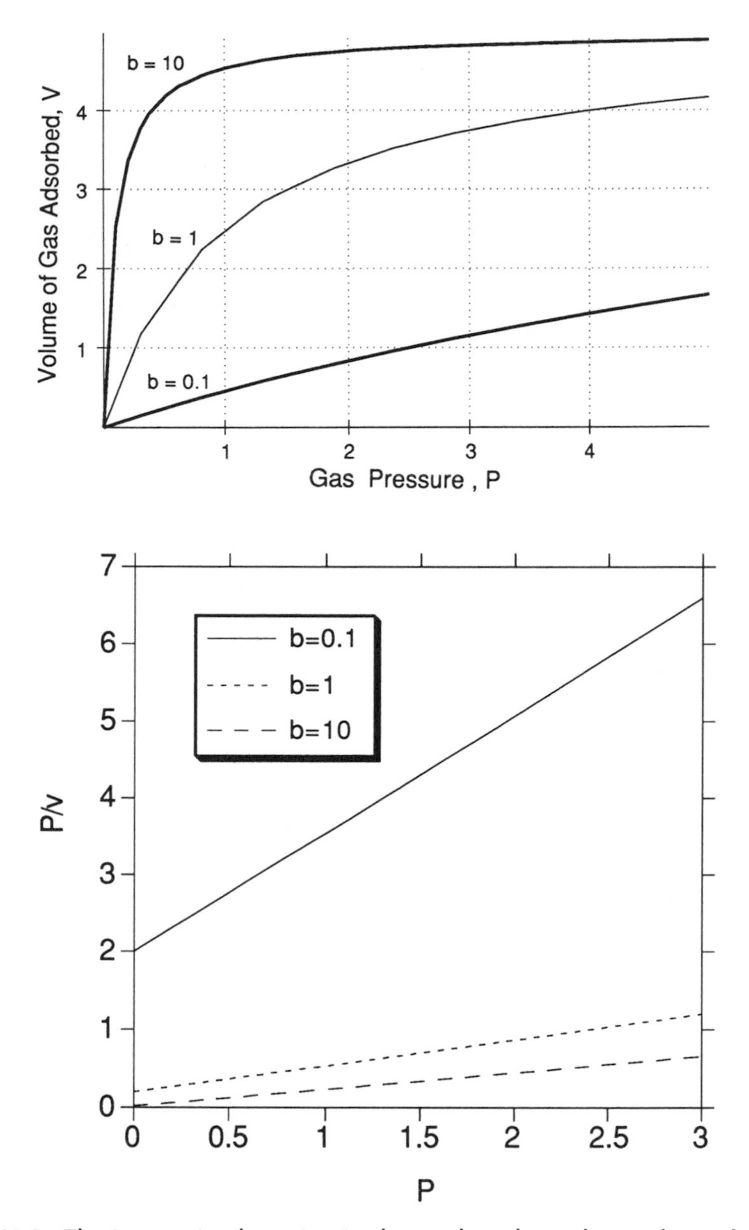

Figure 10.3. The Langmuir adsorption isotherm plotted as volume of gas adsorbed (v) against the equilibrium partial pressure, P (nonlinear form), and as P/v against P (linear form). The isotherms are shown for three values of b, assuming an arbitrary maximum adsorption volume, v_m, of 5 units.

adsorbed molecules form a nearly close-packed monolayer on the surface when $v = v_m$.

10.3d. The BET Equation and Surface Area Measurement

The Langmuir isotherm often describes chemisorption quite well because chemical bond formation to surface atoms necessitates that adsorption be limited to a monolayer at most. On the other hand, this isotherm is not usually appropriate to describe physical adsorption. In that case, multilayer adsorption is possible because physical adsorption is typified by a small energy of adsorption, Q_1, of the first layer of molecules. Succeeding layers often adsorb with energy comparable to Q_1, making multilayer adsorption quite possible. This process is illustrated in Figure 10.4a. Note that multilayer adsorption begins before monolayer adsorption is complete, forming the shape of isotherm depicted in Figure 10.4b. Multilayer physical adsorption is described by a function known as the BET equation:

$$\frac{v}{v_m} = \frac{c(P/P_0)}{(1 - P/P_0)[1 + (c - 1)P/P_0]} \tag{10.25}$$

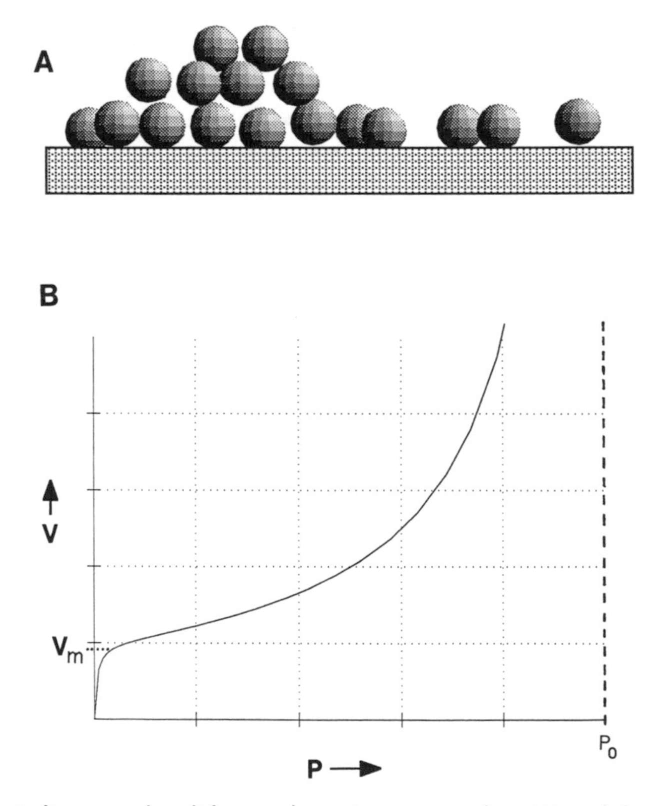

Figure 10.4. A diagram of multilayer adsorption on a surface (A) and the resulting BET adsorption isotherm (B). P_o is the saturation vapor pressure of the adsorbing gas.

where P/P_0 is the partial pressure of the gaseous adsorbate, c is a constant, and v and v_m are defined as before. Here, however, v_m, the volume of gas adsorbed at monolayer coverage, is not the maximum volume adsorbed (see Figure 10.4b). Like the Langmuir equation, the BET equation can be transformed into a linear form:

$$\frac{P/P_0}{v(1 - P/P_0)} = \frac{1}{cv_m} + \frac{(c-1)(P/P_0)}{cv_m} \qquad (10.26)$$

so that a plot of $(P/P_0)/[v(1 - P/P_0)]$ versus P/P_0 produces a straight line, from which v_m and c can be obtained using the values of the intercept and slope.

Physical adsorption of N_2 gas on surfaces at 77 Kelvin is used as a standard method of measuring surface areas; the data for N_2 adsorption in the partial pressure range of $P/P_0 = 0.05$ to 0.3 are fitted to equation 10.26. The value of v_m is then calculated, and the specific surface area of the adsorbent is determined from equation 10.24. It has been estimated that $\mathring{a}$, the adsorption site area, is 16.2 $Å^2$ for N_2.

Example Problem: The adsorption of N_2 gas onto a sample of dry manganese oxide, weighing 0.3144 g, was measured volumetrically at three partial pressures of the N_2 gas: 0.098, 0.196, and 0.294. The volumes adsorbed at these pressures were 1.550, 1.904, and 2.22 ml, respectively, measured under ambient conditions of 22°C and 747 mm Hg of atmospheric pressure. Calculate the specific surface area of this mineral in units of square meters per gram.

Solution: The ideal gas equation allows the gas volumes to be converted to molar quantities:

$$n \text{ (moles of } N_2) = \frac{PV}{RT} \qquad (10.27)$$

where:

P = pressure (in atmosphere units) = 747/760 = 0.983 atm
V = gas volume (cm^3)
R = ideal gas constant = 82.054 cm^3 atm/deg-mole
T = temperature = 273 + 22 = 295 K

The molar quantities of N_2 adsorbed by the oxide at the three P/P_0 values are then calculated using equation 10.27, and are listed in the column titled X below:

P/P_o	V (ml)	X (moles)	$(P/P_o)/[X(1 - P/P_o)]$
0.098	1.550	6.295×10^{-5}	1726
0.196	1.904	7.733×10^{-5}	3152
0.294	2.220	9.015×10^{-5}	4619

Since the gram molecular weight of N_2 is 28, the values of X are multiplied by 28 to obtain grams of N_2 adsorbed. A plot of $(P/P_0)/[X(1 - P/P_0)]$ versus P/P_0 should give a straight line with slope $S = (c-1)/cX_m$ and intercept $I = 1/cX_m$, where X_m is the mole quantity of N_2 adsorbed at monolayer surface coverage, based on the conversion of equation 10.26

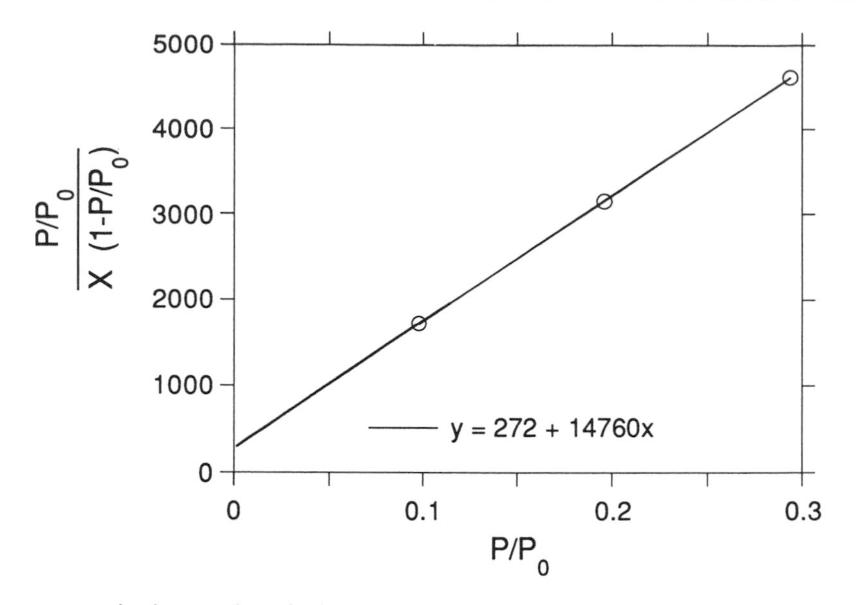

Figure 10.5. The linear plot of adsorption data according to the BET equation.

to mole from volume units for the adsorbed gas. This plot is shown in Figure 10.5, with a calculated slope and intercept of 14,760 and 272, respectively. It is readily shown that

$$S + I = \frac{(c-1)}{cX_m} + \frac{1}{cX_m} = \frac{1}{X_m} \tag{10.28}$$

Thus, $14760 + 272 = 15032 = 1/X_m$, and $X_m = 6.65 \times 10^{-5}$ moles of N_2. The area of one adsorption site, $\mathring{a}$, is estimated to be 16.2 Å^2 for a N_2 molecule. A mole of N_2 would then cover $16.2 \times N_0 = 16.2 \times 6.023 \times 10^{23} = 9.757 \times 10^{24}\ \text{Å}^2 = 9.757 \times 10^4\ m^2$. For the calculated value of X_m, the monolayer would cover a surface of 6.65×10^{-5} moles $\times\ 9.757 \times 10^4\ m^2/\text{mole} = 6.49\ m^2$. But this amount of N_2 was adsorbed on 0.3144 g of oxide. The specific surface area is then $6.49\ m^2/0.3144\ g = 20.6\ m^2/g$.

The BET equation actually describes a more general situation than the Langmuir equation. The BET constant, c, is sensitive to the bonding energy of the first monolayer (Q_1) relative to that of succeeding monolayers ($Q_2, Q_3, \ldots, Q_n$). If c is very large (that is, $Q_1 \gg Q_2, Q_3, \ldots, Q_n$), the BET equation reduces to the Langmuir equation. Thus, Langmuir-type adsorption is seen to be a special case of BET-type adsorption. If c is small ($Q_1 \approx Q_2, Q_3, \ldots, Q_n$), a low-affinity isotherm (similar to S-type at low vapor pressures) results, a consequence of the low energy of attraction of the first monolayer for the surface relative to that of subsequent layers. The range of isotherm shapes possible for $c = 0.1$ to $c = 1000$ is shown in Figure 10.6.

The BET equation is useful in describing water vapor adsorption by soils and clays, as shown in Figure 10.7, and may also be applicable to the adsorption of organic vapors (including volatile pesticides) on dry soils and clays.

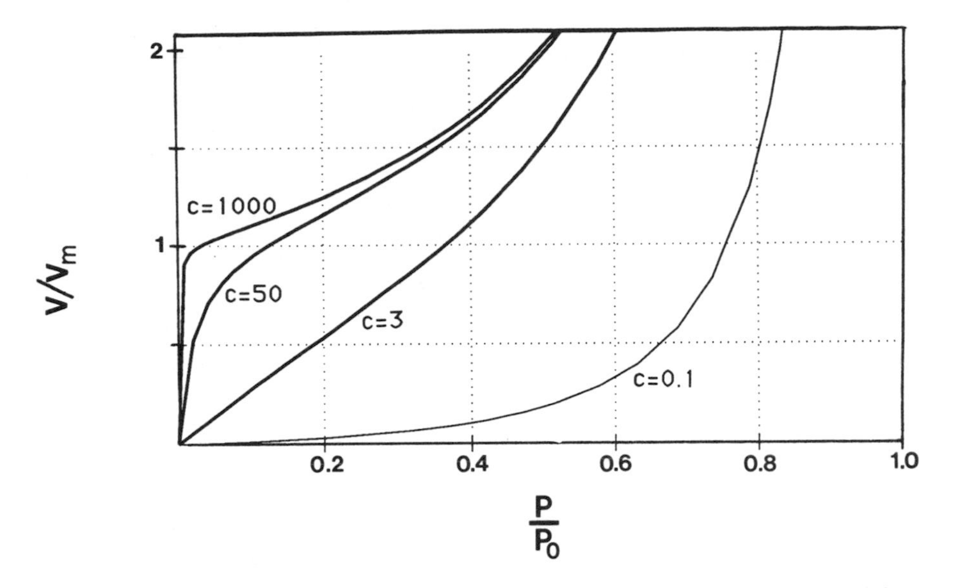

Figure 10.6. The dependence of BET isotherm line shape on the magnitude of the c parameter.

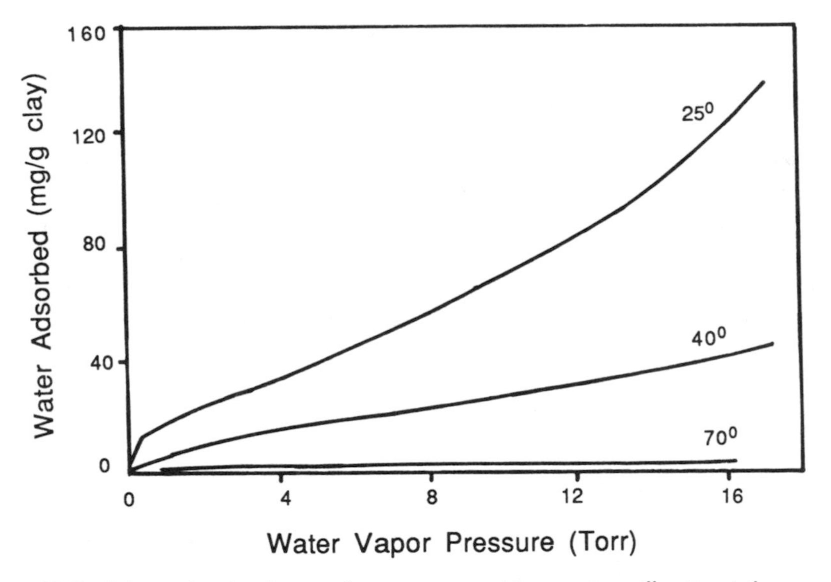

Figure 10.7. Adsorption isotherms for water on a Na-montmorillonite at three temperatures. (Adapted from P. L. Hall and D. M. Astill. 1989. Adsorption of water by homoionic exchange forms of Wyoming montmorillonite. (SWy-1). *Clays Clay Min.* 37:355–363.)

10.3e. The Clausius–Clapeyron Equation and Temperature Effect on Adsorption from Gas Phase

For the adsorption of a mole of any gaseous molecule onto an "inert" surface (one that is not changed by the adsorption itself), it can be shown from thermodynamic principles (see Chapter 1) that this vapor-adsorbate phase change is described by the Clausius–Clapeyron equation:

$$\frac{d(\ln P)}{d(1/T)} = -\frac{\Delta H}{R} \tag{10.29}$$

This equation relates the vapor pressure, P, at which adsorption occurs, to the temperature of adsorption, T. ΔH is then the net enthalpy change (per mole) associated with the phase change from vapor to adsorbed state, and is a differential value. This enthalpy can be viewed as the sum of the enthalpies of condensation and adsorption of the vapor. If adsorption isotherms are collected at different temperatures, then a plot of ln P (where P is the vapor pressure at which a particular quantity of adsorption occurs) against $1/T$ should be linear with a slope of $-\Delta H/R$. Consequently, if the ΔH calculated in this way from equation 10.29 were just equal to the enthalpy of condensation of the vapor, then the phase change is simple condensation on surfaces or in pores, and does not involve true adsorption.

If adsorption from the gas phase were insensitive to temperature, this would mean that ΔH is close to zero. However, because adsorption to an inert surface from the vapor phase requires that molecules lose degrees of freedom of motion in bonding at the surface, the entropy change for the process, ΔS, is inevitably negative. Consequently, it would seem that ΔH must always be negative in order to drive the spontaneous adsorption process (which by definition has a negative free energy change). As is suggested by Figure 10.7, this means that: *physical adsorption from the vapor phase should always be less at higher temperature.*

This general rule of temperature effect on gas-phase adsorption is valid so long as the adsorbent is "inert." However, soil colloid surfaces are not likely to qualify as "inert," both because the adsorbing molecules may have to displace other molecules already present on the surface and because conformational changes, such as swelling, may result from adsorption. Such changes in the state of the adsorbent invalidate the assumption of the Clausius–Clapeyron equation that ΔH measures the change in enthalpy of only the adsorbing phase. For example, in the case of water vapor adsorption on a swelling clay (see Figure 10.7), the measured ΔH is affected by the energy involved in separating the clay layers and is a property not of the water but of the entire system.

10.3f. Organic Adsorption from Aqueous Solution

It is common in soil chemistry for adsorption of organics from solution to be of more interest than adsorption from vapor. This usually does not involve multilayer adsorption because of the tendency of water molecules to compete effectively for adsorption sites, at least if physical adsorption is the main process. In such cases, the Langmuir equation may be more accurate in describing adsorption results than the BET equation. The Langmuir equation is readily converted from the linear form used in gas phase adsorption (equation 10.23) to one relevant for the liquid phase, by replacing gas pressure P with its solution-phase analogue, adsorbate concentration C_A. At the same time, adsorbed molecules are quantified more conveniently by weight, x_A, instead of volume. The resulting linear form of the Langmuir equation is

$$\frac{C_A}{x_A} = \frac{1}{bx_{A(m)}} + \frac{C_A}{x_{A(m)}} \tag{10.30}$$

where $x_{A(m)}$ is the weight of adsorbate A retained by the solid adsorbent at full (monolayer) surface coverage. This form of equation often describes rather well the adsorption of organics (and metal ions, as described in Chapter 4) by soils and soil materials, especially if the mechanism of adsorption involves a fairly strong and specific bond with the surface.

For describing weak physical adsorption from solution, which is commonly the case for nonpolar and weakly polar organic molecules, the BET equation is useful once it is modified to account for adsorption from solution rather than from the gas phase. Nonpolar organics have low solubility in water and are adsorbed onto soil solids from very dilute solutions. This would be analogous to restricting gas phase adsorption to very low pressure, P, of the organic vapor. The BET isotherm (equation 10.25) is essentially linear for $P/P_0 \ll 1$ and small values of c ($c < 1$), taking the form:

$$\frac{v}{v_m} = c\left(\frac{P}{P_0}\right) \tag{10.31}$$

If this equation is converted to a form applicable to adsorption from solution as was done for the Langmuir equation (equation 10.30), the result is

$$\frac{x_A}{x_{A(m)}} = cC_A \tag{10.32}$$

where the symbols are defined as before but $x_{A(m)}$ is best described as the monolayer adsorption level rather than maximum adsorption. However, as long as C_A, the equilibrium concentration of the organic in solution, is much less than the solubility of the organic in water (analogous to keeping $P/P_0 \ll 1$ in the vapor adsorption case), surface adsorption is likely to be much below the monolayer level (i.e., $x_A \ll x_{A(m)}$).

In summary, equation 10.32 suggests that weak physical adsorption by soils of nonpolar organics at low concentrations should display a near-linear isotherm when the quantity of adsorption, x_A, is plotted against solution concentration, C_A. This is termed *constant partitioning* of the adsorbate, because at each level of the organic in the soil, the relative quantity found in the soil solid and solution phases remains the same. The linear isotherm is sometimes given its own designation as the C-type isotherm, but as shown here it can be thought of as a special case of the BET equation. The BET bonding constant, c, and the quantity of adsorption sites in the monolayer, $x_{A(m)}$, determine the slope of this isotherm. A high capacity of soil for adsorption (for example, a high surface area of soil solids) or a stronger attraction between the soil surfaces and the organic adsorbate (that is, a higher c value) would produce a steeper isotherm, according to equation 10.32.

In actual fact, the adsorption of nonpolar and weakly polar organics on soils, soil organic matter, and soil minerals usually *does* produce a linear isotherm, at least at low adsorbate concentrations. A linear isotherm would *not* be expected for (1) high adsorbate concentrations in solution (high relative to the solubility limit of the adsorbate in water), (2) polar or charged adsorbates, which typically display large values of c, the bonding constant. As Figure 10.6 shows, a large value of c produces an isotherm that is initially steep, much like the L-type isotherm described earlier.

The effect of temperature on adsorption of molecules from solution is described by the van't Hoff equation, which has the form

$$\frac{d(\ln K)}{dT} = \frac{\Delta H^0}{RT^2} \tag{10.33}$$

Thus, at two temperatures, T_1 and T_2, the equilibrium constants would be related by the expression:

$$\ln\left(\frac{K_2}{K_1}\right) = \frac{\Delta H^0}{R}\left[\frac{1}{T_1} - \frac{1}{T_2}\right] \tag{10.34}$$

In this equation, K_1 and K_2 are the equilibrium constants for the adsorption reaction at the two temperatures T_1 and T_2, respectively. The magnitude of K measures the degree of partitioning of molecules between the solution and surface. This equation shows that a reaction that is exothermic under standard state conditions (one with $\Delta H^0 < 0$) is more favorable at low temperature, while an endothermic reaction (with $\Delta H^0 > 0$) is more favorable at high temperature.

10.4. ADSORPTION BEHAVIOR OF IONIC ORGANIC MOLECULES

10.4a. Organic Cations

Although the adsorption of this class of organics on soils and soil minerals has been studied in some detail, there are actually few examples of organic cations that are used as pesticides or occur as pollutants in soils. The best known examples are the two bipyridilium quaternary salts, diquat and paraquat, herbicides that are very soluble in water because of their ionic nature. The structure of the chloride salt of paraquat is diagrammed below:

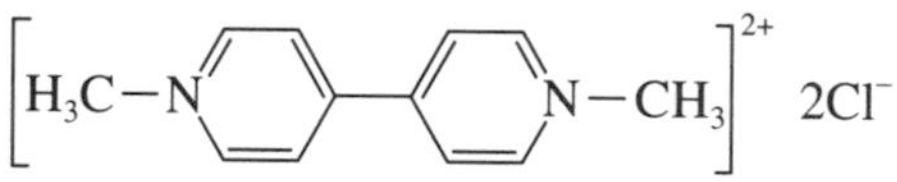

$$\left[H_3C-N\langle\bigcirc\rangle-\langle\bigcirc\rangle N-CH_3\right]^{2+} \; 2Cl^-$$

These molecules are strongly adsorbed by cation exchange on smectites, and somewhat less strongly adsorbed on vermiculites and kaolinite. The smectite adsorption isotherm is H-type, revealing a strong selectivity of the clay for the organic relative to the displaced metal exchange cations. In fact, very little desorption of paraquat and diquat from smectites occurs, even with the help of high concentrations of competing metal cations. The exchange reaction, written for a Ca^{2+}-saturated smectite is

$$PQ^{2+} + Ca^{2+}\text{-smectite} = PQ^{2+}\text{-smectite} + Ca^{2} \tag{10.35}$$

where PQ symbolizes paraquat. This reaction has a selectivity constant, K_s, defined as

$$K_s = \frac{\{PQ^{2+}\} \cdot [Ca^{2+}]}{\{Ca^{2+}\} \cdot [PQ^{2+}]} \tag{10.36}$$

with { } brackets denoting mole quantities of cations on exchange sites and square brackets indicating solution concentrations at equilibrium. The H-type isotherm is evidence that the value of K_s for adsorption is large ($\gg 1$), but because strongly bonded organics such as paraquat seem not to adsorb reversibly, the K_s value may severely overestimate desorption of PQ^{2+} when its activity in solution is lowered.

Cationic molecules adsorb on soil organic matter, which is negatively charged even at low pH. However, organic matter adsorbs diquat and paraquat less strongly than do layer silicate clays, with L-type isotherms typical for organic matter. How strongly organic cations are adsorbed by organic matter depends on the cation that initially occupies the exchange sites. The weakly held Na^+ is easily displaced by organocations, and K_s for paraquat-Na^+ exchange is large. On the other hand, Ca^{2+} is less easily exchanged from humic materials by paraquat, in part a result of stronger association of higher charge cations with carboxylate groups (the main site of negative charge in humus), and a lower K_s value is observed. However, higher charge cations such as Ca^{2+} and Al^{3+} also electrostatically interconnect the macromolecules of soil organic matter, restricting diffusion of paraquat to exchange sites that become "buried" in the aggregated structures (Hayes and Mingelgrin, 1991).

As the pH is lowered, organocations such as paraquat adsorb to a lesser extent on humus, in part because the H^+ ion bonds more strongly than metal cations with the functional groups in organic matter that generate negative charge, that is,

$$R{-}COO^- \cdots Na^+ + H^+ = R{-}COOH + Na^+ \tag{10.37}$$

The charge-neutralizing effect of H^+ on humic polymers reduces the tendency of the polymer structures to hydrate and swell in water, and accessibility of internal exchange sites is limited.

The adsorption behavior of organic cations on soil colloids explains the immediate loss of herbicidal activity once paraquat and diquat contact soil. The potential for leaching and groundwater pollution should be extremely low for this class of compounds in soils containing layer silicate clays and reasonable levels of organic matter. On the other hand, strong adsorption in interlayer regions of the clays or micropores of humic materials could protect the organic cations from microbial attack and degradation, increasing the lifetime of these organics in soil.

Application of thermodynamics to ion exchange (ideal) produces an equation relating the selectivity constant, such as that defined by equation 10.36, to the free energy of the exchange reaction, ΔG_{ex}:

$$-\Delta G_{ex} = RT \ln K_s \tag{10.38}$$

The properties of the organic cation that determine the magnitude of ΔG_{ex}, and therefore the degree of selectivity of clays and organic matter for that particular cation are:

1. **Molecular weight.** In most cases, organocations of higher molecular weight are adsorbed on exchange sites with higher selectivity. For example, the order of preference for a series of quaternary ammonium cations on smectite is

$$(CH_3CH_2CH_2)_4N^+ > (CH_3CH_2)_4N^+ > (CH_3)_4N^+$$

This order does not arise so much from a difference in energy of attraction for the surface as from an entropy contribution to ΔG_{ex}. Larger molecules must displace more water molecules from the colloid surface during exchange, that is,

$$R_4N^+ + Na^+\text{-smectite} = R_4N^+\text{-smectite} + Na^+ + n\,H_2O \qquad (10.39)$$

In this reaction, as the organic group R becomes larger, n increases, and the reaction entropy, ΔS_{ex}, assumes a larger positive value because more water molecules gain freedom of motion by displacement into the solution. At the same time, removal of the larger (organic) cation from solution lowers the energy of the solution phase, so that energy as well as entropy may drive the overall reaction.

2. **Molecular hydration energy.** Organocations that do not very effectively shield their positive charge from solution interact strongly with water by ion-dipole forces. They are then hydrated in solution and, like strongly hydrating metal cations, adsorb with lower selectivity on exchange sites. For example, ethylammonium, $CH_3CH_2NH_3^+$, is readily displaced from smectites by Ca^{2+} or the quaternary ammonium ion, $(CH_3CH_2CH_2)_4N^+$. The latter cation has its charge shielded from solution by propyl groups, and has a very low hydration energy, which enables it to adsorb selectively on exchange sites. Furthermore, the absence of a strong hydration force, which normally keeps the interlayer region of swelling clays expanded (see Chapter 8), can at high adsorption levels collapse the interlayer space and flocculate the clay. The surface properties of smectites loaded with weakly hydrating organocations change from *hydrophilic* to *organophilic*, and the smectites become much better adsorbents of nonpolar organic molecules in solution.[1]

The properties of the adsorbing surfaces are critical as well in determining how strongly any particular organocation is adsorbed on exchange sites. These properties include:

1. **The nature of the cation occupying the exchange sites prior to exchange.** Charge and hydration energy are key properties of the cation.

2. **The charge density of the clay mineral.** Clays with high permanent charge density, such as vermiculites, normally expand in water to no more than a 14.5 Å c-axis spacing, leaving less than 5 Å of space for organocations to diffuse between the silicate layers. This is too narrow for larger organocations such as $(CH_3CH_2CH_2)_4N^+$, and a molecular "sieving" effect results in which small cations are adsorbed in preference to large cations. Furthermore, since silicate clays with high charge localized in the tetrahedral sheet (i.e., vermiculites) hydrate more strongly than those with lower charge localized mainly in the octahedral sheet (i.e., many smectites), organocations adsorb less energetically on vermiculite surfaces because of

1. This property of layer silicate clays has been used to modify smectites so that they can adsorb and retain non-polar pollutants in water, such as chlorinated benzene.

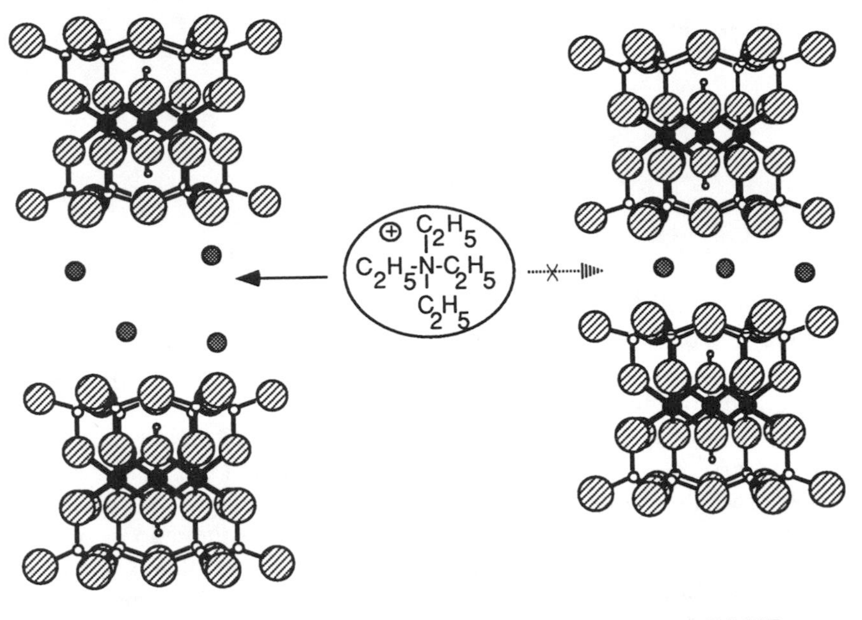

Figure 10.8. Sketch of the molecular "sieving" action of layer silicate clays for relatively bulky organic cations.

the greater energy cost in displacing water from the adsorption site. The "sieving" effect for quaternary ammonium cations is diagrammed in Figure 10.8.

10.4b. Basic Molecules

These organics become positively charged only when they accept a proton.[1] For any basic molecule B, the protonation reaction can be written:

$$B + H^+ = BH^+ \tag{10.40}$$

where BH^+ is the conjugate acid of B. The dissociation constant of the conjugate acid, K_a, is defined by the equation

$$K_a = \frac{(B)(H^+)}{(BH^+)} \tag{10.41}$$

This equation shows that, as the solution is made more acidic, a greater fraction of the base is converted to the cationic BH^+ species.

One of the most important groups of basic compounds used in agriculture is the s-triazine group, which includes common herbicides such as atrazine, simazine, and

1. Although some of the amines considered in the above section are basic compounds, they are charged over the entire pH range likely to be encountered in soils, and are therefore treated as organic cations.

prometone. The protonation reaction of atrazine, with a pK_a ($-\log K_a$) of 1.68, is given below:

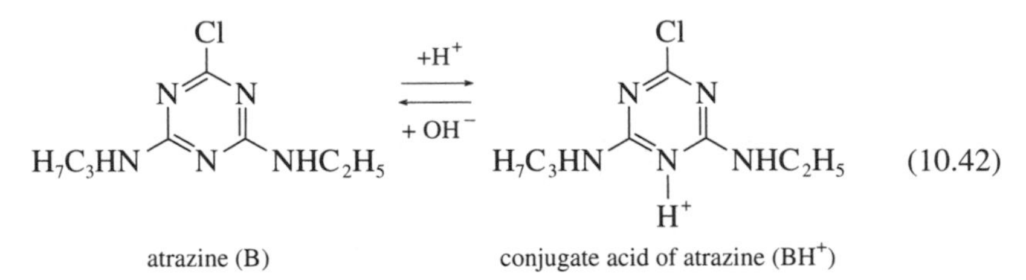

On the basis of the earlier discussion of the adsorption characteristics of organic cations, the conjugate acid is expected to adsorb strongly on permanent-charge clays and organic matter, while the base should adsorb weakly by physical forces. This expected behavior is confirmed by experiment. The model adsorbent, Na^+-montmorillonite, adsorbs the s-triazines to the greatest extent at low pH, as shown in Figure 10.9a. The adsorption isotherms *at any given pH* obey the Langmuir equation, as shown in Figure 10.9b. The pH effect confirms that, as the ratio of BH^+ to B in solution increases, adsorption increases. The decrease in adsorption at extremely low pH can probably be attributed to partial decomposition of the montmorillonite, which releases Al^{3+} from the clay structure. The Al^{3+} cations then occupy exchange sites and limit adsorption of the organic cation. Suppression of triazine adsorption by Al^{3+} would only be expected in the most acid of mineral soils.

Figure 10.9 also demonstrates the importance of the pK_a value of the conjugate acid in determining the pH of maximum adsorption. Those s-triazines with higher pK_a values tend to adsorb more readily at high pH than those with lower pK_a. This simply reflects the ease with which the base protonates, as those molecules with the greatest tendency to form the conjugate acid cation are the most adsorptive. From equation 10.41, the relationship between pK_a and ratio of the concentration of organic cation to neutral molecule, $(BH^+)/(B)$, is found to be

$$pK_a = \log\left\{\frac{(BH^+)}{(B)}\right\} + pH \tag{10.43}$$

which reveals that, at a given pH, the relative amount of organic in the cation form depends on the pK_a (acidity) of the organic cation. Yet Figure 10.9a reveals that some s-triazines do adsorb at high pH even though a very small fraction of the molecules in solution are charged. For example, at pH 8, prometryne adsorbs to some extent, but equation 10.43 reveals that

$$\log\left\{\frac{(BH^+)}{(B)}\right\} = pK_a - pH = 4.05 - 8.00 = -3.95 \tag{10.44}$$

This means that only about one in 10,000 molecules in solution is in the cation form at pH 8, but significant adsorption still occurs. Results of this sort have been used to argue that clay surfaces have a much lower "pH" than the solution in contact with the clay, and that this low "pH" enhances the protonation of organic bases at the clay-solution interface. It is probably more correct to say that negatively charged clays adsorb the cation, BH^+, in strong preference over the neutral molecule, B, with

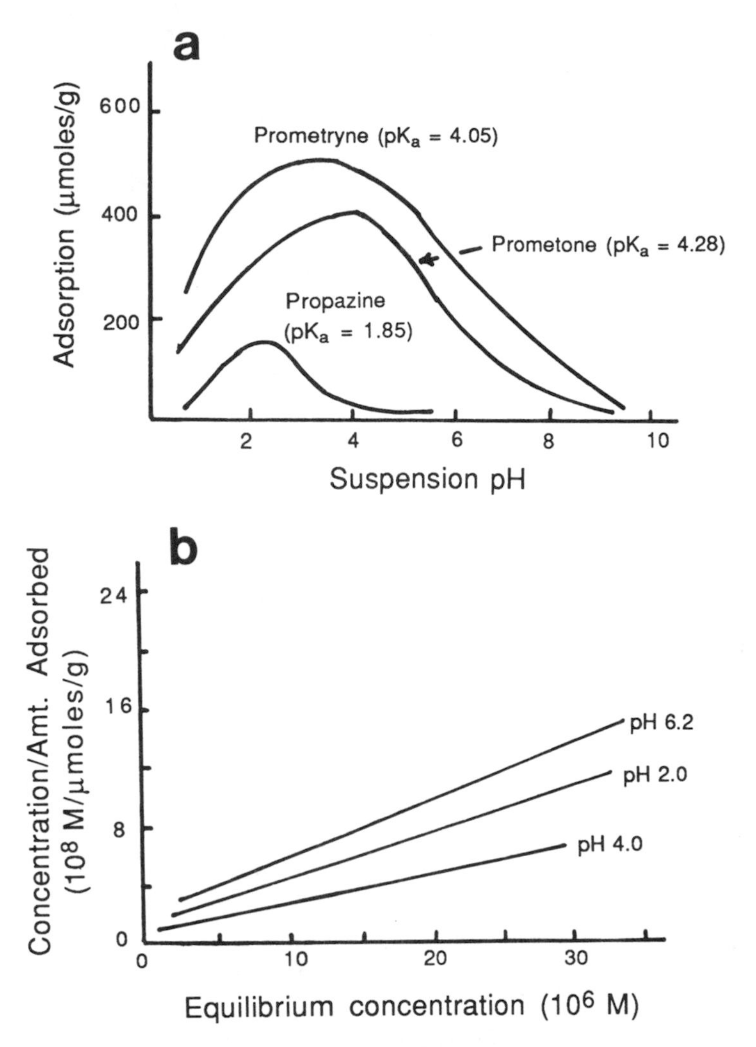

Figure 10.9. Adsorption of s-triazines by Na^+-montmorillonite as a function of pH (a) and as a function of s-triazine (prometone) concentration (b). (Adapted from J. B. Weber. 1970. Mechanisms of adsorption of s-triazines by clay colloids and factors affecting plant availability. In *Residue Reviews.* Vol. 32: *The Triazine Herbicides.* New York: Springer-Verlag.)

the result that the BH^+/B ratio is much higher on the clay surface than in the solution that is in equilibrium with the clay. This situation is depicted in Figure 10.10, emphasizing that no matter how many organic cations are adsorbed by the clay, the BH^+/B ratio in solution is controlled only by the solution pH. This phenomenon of "enhanced protonation" of organic bases in colloidal suspensions can simply be attributed to electrostatic attraction between the conjugate acid cation and the clay.

Extreme surface acidity is expected only on clay surfaces that have been dried, as will be discussed later in this chapter. Air-dry layer silicate clays, especially when saturated

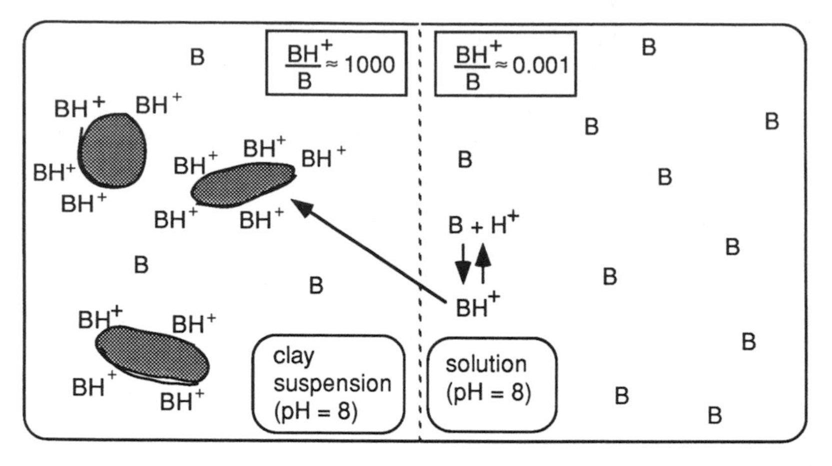

Figure 10.10. Schematic diagram of the enhanced protonation of an organic base (prometryne) at pH 8 in the presence of layer silicate clays.

with moderately acidic exchange cations such as Mg^{2+} or Al^{3+}, are able to protonate extremely weak bases such as urea. In dry systems such as these, pH is a questionable concept, although "surface acidity" can be calibrated using a series of very weak bases with different pK_a values. These are called Hammett indicators, as they change color on the clay surface on conversion from the base to the conjugate acid form.

Organic matter can also adsorb basic molecules, and s-triazines in particular, by cation exchange. Adsorption is sensitive to pH, which not only determines the BH^+/B ratio in solution, but also the fraction of carboxylic acid groups that dissociate and thereby become potential adsorption sites for the organic cation. Adsorption isotherms are L-type or H-type, depending on the acidity of the soil organic matter. The steps involved in adsorption are:

1. **Organic matter dissociation.**

$$R{-}COOH = R{-}COO^- + H^+ \tag{10.45}$$

2. **Protonation of the organic base.**

$$B + H^+ = BH^+ \tag{10.46}$$

3. **Adsorption of the conjugate acid.**

$$R{-}COO^- + BH^+ = R{-}COO^- \cdots BH^+ \tag{10.47}$$

Because reaction step 1 is favored at higher pH and step 2 is favored at lower pH, there exists an intermediate pH at which adsorption attains a maximum. This optimum pH is usually near the pK_a of the conjugate acid, for much the same reasons that bases were seen to adsorb to the greatest extent on clays when pH $\approx pK_a$ (see Figure 10.9a). We have, then, the general rule that:

Optimal adsorption of an organic base on negatively charged soil colloids occurs near the pH numerically equal to the pK_a *of the organic conjugate acid.*

However, because pK_a values of basic compounds are often below the soil pH, as soils become more acid, adsorption of bases such as s-triazines increases. This means that s-triazine herbicides tend to be inactivated once they come into contact with acid soils. On the other hand, residual herbicidal activity can persist in nonacid soils where adsorption is relatively weak, controlled largely by physical adsorption rather than electrostatic attraction to soil surfaces.

10.4c Acidic Molecules

These organic compounds possess acidic functional groups, such as carboxylic or phenolic, that dissociate to form anions:

$$\begin{aligned} R{-}COOH &= R{-}COO^- + H^+ \\ \phi{-}OH &= \phi{-}O^- + H^+ \end{aligned} \tag{10.48}$$

where R is any organic group and ϕ is the phenyl group. The acid dissociation constant of these acids is defined as it is for the conjugate acid of bases in the previous section. For example, the equation defining the pK_a for carboxylic acids is

$$pK_a = pH - \log\left\{\frac{(R{-}COO^-)}{(R{-}COOH)}\right\} \tag{10.49}$$

The pK_a for phenolic groups, because they are typically much weaker acids than carboxylic groups, is usually quite high (> 9).

In general, the anion forms of these acidic molecules adsorb little, if at all, on soil organic matter and layer silicate clays because of electrostatic repulsion between the molecules and the negative charge of these soil particles. Conversely, the neutral form is retained, albeit weakly, by physical adsorption (on organic matter in particular). The usual pattern of behavior in soils of temperate climates, then, is that acidic organic compounds are adsorbed most effectively when the soil pH is low and organic matter content is high. Otherwise, this class of organics is considered highly mobile and susceptible to leaching in soils.[1] Bonding of these compounds to soil organic matter is weak, with adsorption isotherms being the C- or S-type. As a result, acidic organic compounds do not adsorb on soils (with permanent-charge minerals) to the extent that cationic and basic compounds do.

A comparison of the adsorption of several acidic herbicides by an acid (pH = 4) muck soil is useful in beginning to generalize the factors that control retention in soil organic matter. The common names, structures, and significant chemical properties of these herbicides are listed in Table 10.2. Since adsorption by the muck soils follows the order of preference:

$$\text{dinoseb} > \text{picloram} > \text{2,4-D} = \text{dicamba}$$

it is concluded that the solubility of these organics in water is inversely related to adsorption. That is, the least soluble compounds tend to adsorb to the greatest extent, a trend that might be expected if the surface interaction is weak.[2] The compounds of

1. The exception is found in soils of variable charge, where allophanic and oxidic minerals are able to adsorb the dissociated form of acid organic molecules, as will be described later in this section.

2. As will be seen in the section on nonpolar molecules, a low polarity is generally associated with low solubility in water and greater tendency for the molecule to "escape" the solution phase by condensing at interfaces, particularly colloid-water interfaces.

Table 10.2. Structure and Properties of Several Acidic Pesticides

	Dinoseb	Picloram	2,4-D	Dicamba
Structure	NO_2, NO_2, C_4H_9, OH	NH_2, Cl, Cl, Cl, N, COOH	Cl, Cl, O, CH_2, COOH	Cl, OCH_3, Cl, COOH
pK_a	4.40	3.6	2.80	1.93
Water solubility (ppm)	45	430	650	4,500

Source: S. B. Weed and J. B. Weber. 1974. Pesticide-organic matter interactions. In W. D. Guenzi (ed.), *Pesticides in Soil and Water.* Madison, Wis.: Soil Science Society of America.

Table 10.2 probably adsorb as the undissociated neutral molecules, or even as organic anion-metal cation ion pairs on humus. As the solution pH is raised, the neutral molecules dissociate to a greater extent, and adsorption is further suppressed by electrostatic repulsion (and the fact that the charged form of the molecule is more soluble than the uncharged form). Thus, pK_a values of organic acids are correlated positively with tendency to adsorb and negatively with solubility in water.

From this description of adsorption properties, it is reasonable to expect these acidic organics to desorb readily from soils that are leached by water. This is found to be the case in many soils, with little indication that the organics can be immobilized by adsorption. There is, however, evidence that variable-charge minerals (e.g., oxides and allophanes) can strongly retain certain of these molecules by a chemical adsorption (ligand exchange) process. For example, 2,4-D is adsorbed by Fe oxides, as shown in Figure 10.11, probably by coordination of the carboxylate group to a surface Fe^{3+} ion, to form the complex:

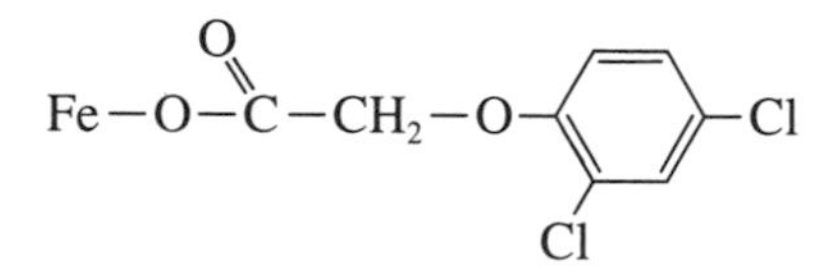

The fact that there is a pH at which adsorption is optimized, which happens to coincide approximately with the pK_a of the organic acid (2.80), suggests the following reaction mechanism:

1. **Protonation of the oxide surface.**

$$Fe{-}OH + H^+ = Fe{-}OH_2^+ \tag{10.50}$$

2. **Dissociation of the organic acid.**

$$R{-}COOH = R{-}COO^- + H^+ \tag{10.51}$$

3. **Surface bonding by ligand exchange.**

$$R{-}COO^- + Fe{-}OH_2^+ = Fe{-}OOC{-}R + H_2O \tag{10.52}$$

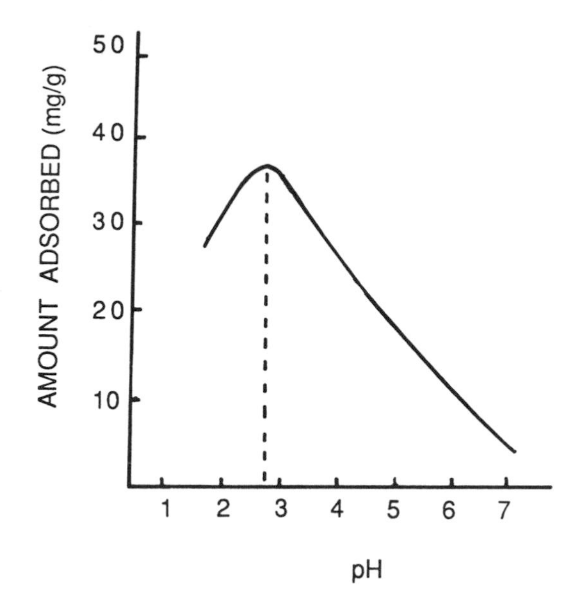

Figure 10.11. Adsorption of 2,4-D on goethite (FeOOH) as a function of pH. (Adapted from J. R. Watson et al. 1973. Adsorption of the herbicide 2,4-D on goethite. *J. Soil Sci.* 24:503–511.)

This scheme explains reduced adsorption at very *low* pH on the basis of reaction 2, since it is the dissociated molecule, $R-COO^-$, that actually adsorbs. At pH < 2.8, most of the organic in solution is in the weakly adsorbing nonionic form. The scheme further explains reduced adsorption at *high* pH based on reaction 1, which indicates that the oxide has fewer reactive (i.e., protonated) sites[1] as the pH is raised. As a result, there is an optimal pH for adsorption that is determined by the "pK_a rule," described earlier for organic bases on negatively charged surfaces, but valid as well for organic acid adsorption on oxides and other variable-charge minerals:

> *Optimal adsorption of an organic acid on variable-charge soil colloids occurs near the pH numerically equal to the* pK_a *of the organic.*

The effect of surface charge of the adsorbing mineral should be, in principle, to shift the pH of optimal adsorption away from the pK_a and toward the PZC of the mineral.

Besides carboxylic acids, phenolic compounds can also bond to variable-charge minerals by direct coordination between the phenolic oxygen and the surface. This means that pesticides such as dinoseb (see Table 10.2) may be retained to some degree by allophane and oxide minerals of soils.

A more detailed and quantitative analysis of organic acid bonding on variable-charge mineral surfaces is possible. Representing all organic acids by the chemical symbol RH,

1. The $Fe\text{-}OH_2^+$ sites are believed to be active in adsorption, while the Fe-OH sites are not, because water molecules, unlike OH^- groups, are easily displaced by organic ligands.

bonding of the dissociated acid with a metal atom, M, at the surface of a variable-charge mineral can be described by the reaction scheme:

$$RH = R^- + H^+ \tag{10.53}$$

$$>M-OH_2^+ + R^- = >M-R + H_2O \tag{10.54}$$

Reaction 10.53 is fully described by the acid dissociation constant, K_a:

$$K_a = \frac{(H^+)(R^-)}{(RH)} \tag{10.55}$$

Reaction 10.54 is the ligand exchange of metal-coordinated water by the conjugate base, R^-, and its tendency to proceed is determined by the basicity of the organic ligand relative to that of water. This reaction can be described by a surface complexation constant, K_c:

$$K_c = \frac{S_R}{S_+(R^-)} \tag{10.56}$$

in which S_+ and S_R symbolize the quantities of reactive surface sites initially protonated and complexed with the organic, respectively. It is assumed that organic adsorption requires a two-step process in which some of the sites first protonate to give S_+ reactive sites, and a fraction of these sites then undergo ligand exchange with R^- to form S_R organic complexes. This reaction model is depicted in Figure 10.12.

S_+ is related to the total potentially reactive sites S by the fact that $S_+ = S\theta_+$, where θ_+ is the fraction of these sites that are actually protonated before adsorption of the organic. The expression for θ_+, given in equation 3.85 in Chapter 3, can be simplified by ignoring H^+-OH^- competition for sites (which is only significant in the pH range near the PZC of the oxide), so that

$$S_+ = S\theta_+ = S\left[\frac{B_{HX}(H^+)(X^-)}{1 + B_{HX}(H^+)(X^-)}\right] \tag{10.57}$$

By substituting this expression for S_+ into equation 10.56, we obtain

$$K_c = \frac{S_R}{S\{B_{HX}(H^+)(X^-)/[1 + B_{HX}(H^+)(X^-)]\}\,(R^-)} \tag{10.58}$$

From equation 10.55:

$$(R^-) = K_a\frac{(RH)}{(H^+)} \tag{10.59}$$

and the fact that the total organic acid concentration in solution, (R_T), must equal the sum of dissociated and nondissociated forms:

$$(R_T) = (R^-) + (RH) \tag{10.60}$$

an expression for the dissociated organic acid concentration in solution is derived:

$$(R^-) = \frac{K_a\,(R_T)}{(H^+) + K_a} \tag{10.61}$$

Substituting this expression for (R^-)into equation 10.58, the final expression for the quantity of adsorbed organic acid, S_R, is found:

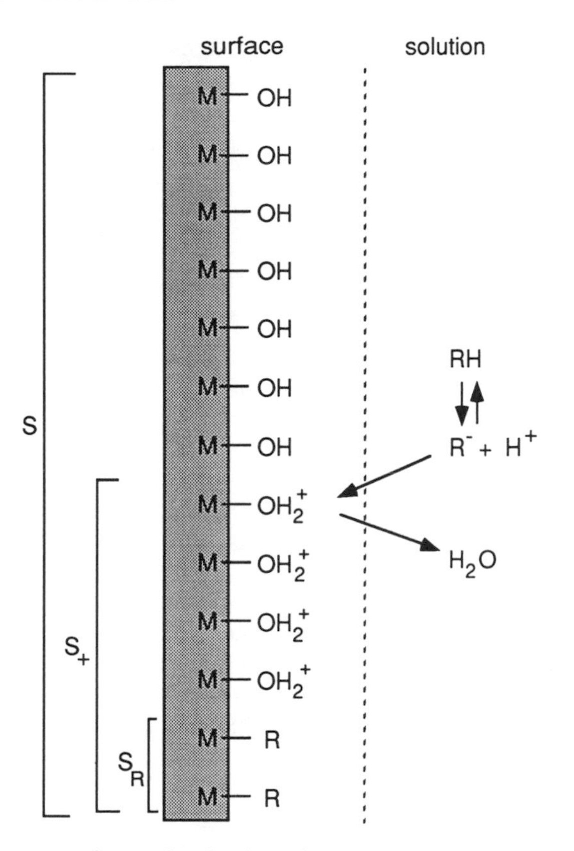

Figure 10.12. Reaction scheme for the ligand exchange reaction of organic acid, RH, at protonated sites of a variable-charge mineral surface.

$$S_{\mathrm{R}} = S \cdot K_{\mathrm{c}} \left[\frac{B_{\mathrm{HX}}(\mathrm{H}^{+})(\mathrm{X}^{-})}{1 + B_{\mathrm{HX}}(\mathrm{H}^{+})(\mathrm{X}^{-})} \right] \cdot \left[\frac{K_{\mathrm{a}}(\mathrm{R}_{\mathrm{T}})}{(\mathrm{H}^{+}) + K_{\mathrm{a}}} \right] \tag{10.62}$$

This equation, while unlikely to be quantitatively correct, reveals the variables that are predicted by the model to influence adsorption of organic acids on a given weight of variable-charge mineral. These variables are:

- S the potential sites for adsorption, determined by the specific surface area (particle size) and crystallinity of the mineral.
- K_c bonding constant, determined by the intrinsic strength of the surface metal (M) bond with the organic ligand, R^-, and sensitive to the electron-donating properties (Lewis basicity) of the ligand.
- K_a dissociation constant of the organic acid, a measure of the ease of proton removal from the organic acid. Easy proton removal facilitates adsorption.
- B_{HX} measure of the relative acidity/basicity of the surface sites, indicated by the PZC. For Al and Fe oxides, PZC values are in the range of 7.5 to 9.5, denoting a high B_{HX} and strong proton bonding by surface sites.

Table 10.3. Acid Dissociation Constants of Substituted Benzoic Acids

Substituent	pK_a
$p-NO_2$	3.44
$p-CH_3$	4.38
$p-OCH_3$	4.48
$p-NH_2$	4.85

(X^-) concentration of the "bathing" indifferent[1] anion. High concentration of X^- increases the activity of the mineral acid, HX, in solution at any given pH, with the result that greater surface protonation and organic adsorption occur.

(H^+) the solution activity of H^+, affecting both surface protonation and organic acid dissociation.

(R_T) total concentration of soluble organic.

Two limiting situations for adsorption can now be considered to illustrate the utility of this equation:

1. Adsorption of p-substituted carboxylic acids on $Fe(OH)_3$. These acids have dissociation constants, K_a, that depend on the electron-withdrawing ability of the substituted groups. In Table 10.3, the pK_a values for several p-substituted benzoic acids are listed. In this adsorption experiment, the pH was controlled at 5.3, so that the dominant form of the organics in solution was R^- rather than RH. The adsorption isotherm becomes, in effect, a plot of S_R (amount of organic adsorbed) versus (R^-), the solution activity of dissociated organic acid, because $(R_T) \approx (R^-)$ at pH 5.3. Since $(H^+) < K_a$, the last term in equation 10.62 is approximately equal to (R_T). In this particular experiment, the "bathing" indifferent electrolyte was 0.05 *M* $NaClO_4$, so that (X^-) can be assumed to be constant along with (H^+). Equation 10.62 then takes on a simplified form for the adsorption isotherm of the benzoic acids on any particular variable-charge mineral:

$$S_R = S \cdot K_c \cdot K_L \cdot (R_T) \tag{10.63}$$

where K_L is a constant whose value depends on both the bonding strength of protons at surface groups and the bathing electrolyte concentration. Thus we see that the quantity of adsorption of organic S_R, plotted against organic acid concentration in solution should be a straight line whose slope depends on the number of reactive sites, *S*, the bonding affinity of sites for protons, B_{HX}, the electrolyte concentration, (X^-), and the strength of surface bond formed by the particular organic ligand. The acidity constant, K_a, is found to be of little consequence as long as the pK_a values of the organic acids are well below the value of the solution pH.

In comparing the adsorption of the substituted benzoic acids listed in Table 10.3, the only variable of equation 10.63 that is not fixed by the experimental conditions is K_c. The isotherms, shown in Figure 10.13, reveal that p-aminobenzoic acid adsorbs the most, a fact that is explained by a stronger surface bond (larger K_c) for the p-aminobenzoate ligand. In fact, adsorption is in the same order as pK_a values: the *least* acidic benzoic acid

1. Indifferent salts do not strongly compete with the organic anion for adsorption sites on variable-charge mineral surfaces. Competing anions, such as sulfate and phosphate, are not considered "indifferent" and would have to be explicitly accounted for in the model, since their presence in solution would suppress the adsorption of the organic acid.

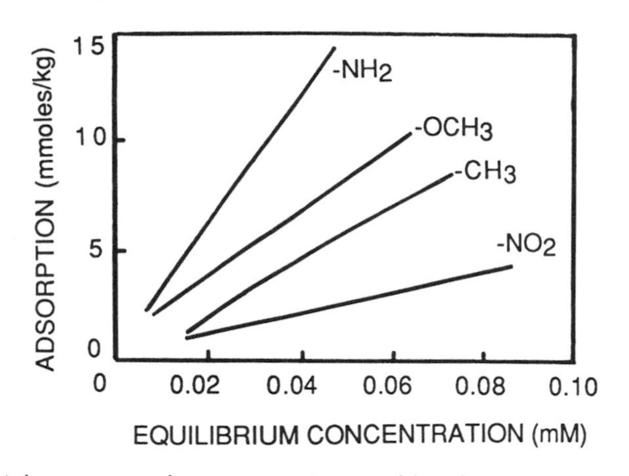

Figure 10.13. Adsorption isotherms on $Fe(OH)_3$ of four benzoic acids substituted at the para position with the functional groups shown. (Adapted from K.-H. Kung and M. B. McBride. 1989. Adsorption of para-substituted benzoates on iron oxides. *Soil Sci. Soc. Am. J.* 53:1673–1678.)

adsorbs to the greatest extent at the fixed pH of 5.3. This relationship stems from the fact that the most strongly electron-donating groups, when substituted on the aromatic ring, cause the carboxylate group to become more strongly basic. Consequently, the basicity of these carboxylate ligands follows the order determined by the inductive electronic effects of the substituents:

$$p{-}NH_2 > p{-}OCH_3 > p{-}CH_3 > p{-}NO_2$$

Since the more basic ligands form stronger covalent bonds with metal ions, it is reasonable to expect K_c, the complexation constant for bonding between the carboxylate anion and the surface metal, to be largest for p-aminobenzoate. The coordination complex formed is diagrammed below:

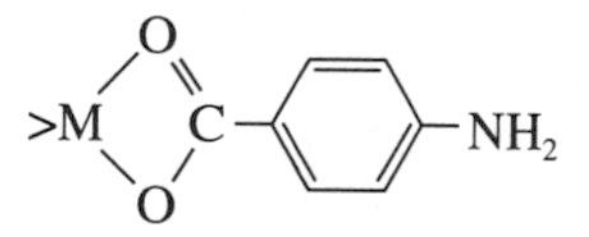

K_c, which is in general a property of the conjugate base R^-, is therefore useful in predicting adsorption of carboxylic acids on oxides, at least at pH values that are not too low. K_a, a property of the acid RH, would actually have predicted the reverse order of adsorption to that observed, since those p-substituted benzoic acids with the weakest basicity are also those most easily dissociated.

2. Adsorption of chlorinated phenols on oxides. The second limiting case to consider is phenols, which as a group have much higher pK_a values than carboxylic acids, and solution pH values at which adsorption is studied are typically *below* these pK_a values. This means that, in equation 10.62, $(R_T) \approx (RH)$, and $(H^+) > K_a$. For fixed solution pH and bathing electrolyte level, equation 10.62 then simplifies to

$$S_R = S \cdot K_c \cdot K_L \left[\frac{K_a(R_T)}{(H^+) + K_a} \right] \qquad (10.64)$$

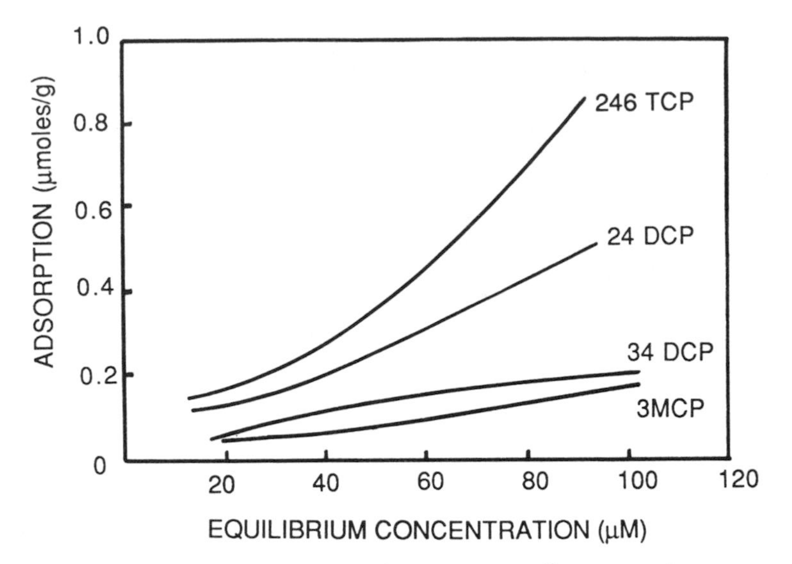

Figure 10.14. Adsorption of chlorophenols on noncrystalline Fe oxide at pH 5.4. (246 TCP = 2,4,6-trichlorophenol; 24 DCP = 2,4-dichlorophenol; 34 DCP = 3,4-dichlorophenol; 3 MCP = 3-monochlorophenol.) (Adapted from K.-H. Kung and M. B. McBride. 1991. Bonding of chlorophenols on iron and aluminum oxides. *Environ. Sci. Tech.* 25:702–709.)

and again the isotherm is predicted to be a straight line, in this case relating adsorbed phenol to concentration of phenol in solution. Now, however, both the acidity of the phenol (K_a) and the basicity of the phenolate ligand (measured by the bonding constant K_c) can influence the observed adsorption. More acidic phenols and more basic phenolates should adsorb most readily. But there is a negative correlation between these two properties of organic acids; that is, phenolic groups that are strongly acid dissociate to form phenolate ligands that are generally weak bases. For any group of variously substituted phenols, then, K_c tends to be inversely related to K_a. Since the product, $K_a \cdot K_c$, appears in equation 10.64, it is not clear what the *net* effect of these counteracting variables will be on adsorption. However, the adsorption data in Figure 10.14, for the chlorophenols with pK_a values listed in Table 10.4, are a clear indication that K_a has more influence than K_c. As a result, the most acid phenols adsorb in greater quantities on variable-charge minerals (if the solution pH is not high). The surface coordination bonds formed by these acid phenols are weak, however, because these same phenols form weak Lewis bases upon dissociation.

Table 10.4. Acid Dissociation Constants of Selected Chorophenols

Chlorophenol	pK_a
3-monochloro	9.10
3,4-dichloro	8.63
2,4-dichloro	7.85
2,4,6-trichloro	6.15

Equation 10.62 provides an explanation for the observation that organic acids tend to show an adsorption maximum at a solution pH near the pK_a value of the acid (the "pK_a" rule). We see that the far right term in this equation increases with increasing pH, while the adjacent term *decreases* with increasing pH. The product of the two terms, then, may reach a maximum at one particular pH. If the $K_a(R_T)/[(H^+) + K_a]$ term in equation 10.62 is differentiated with respect to $-\ln(H^+)$, the resulting function is $K_a(R_T)(H^+)/[(H^+) + K_a]^2$, which has a maximum near $(H^+) \approx K_a$. At lower and higher pH, the function's value decreases as the pH is adjusted away from pH $= pK_a$. This means that the most *abrupt* change in adsorption should occur when the pH is adjusted through the pH range that corresponds to the pK_a of the organic acid. In fact, inflections in adsorption of organic (and inorganic) anions are usually observed when the pH is close to the pK_a value.

The model developed here makes the assumption that the anion of the bathing electrolyte, X^-, is "indifferent"; that is, X^- does not compete with the organic anion for adsorption sites. But observation does not bear out this assumption. Even very weakly coordinating anions such as perchlorate, ClO_4^-, or nitrate, NO_3^-, can exchange part of the adsorbed organic acids from variable-charge mineral surfaces. This means that the model can be improved by allowing for competitive adsorption from inorganic anions found in soil solution (e.g., Cl^-, NO_3^-, HCO_3^-). It suggests that part of organic acid adsorption involves electrostatic attraction of the outer-sphere type:

$$>M-OH_2^+ + R^- = >M-OH_2^+ \cdots R^- \tag{10.65}$$

in addition to direct metal-ligand coordination. Nevertheless, the selectivity that is inherent in organic acid adsorption cannot be explained by outer-sphere electrostatic attraction.

10.4d Chelating Molecules

Some pesticides of the acidic type have several functional groups that can simultaneously bond to metal ions at the oxide surface, forming a *chelate.* A good example of this is the widely used herbicide, glyphosate (Roundup), which has amine, carboxylate, and phosphonate groups, as shown below:

$$HOOC-CH_2-NH-CH_2-\overset{\overset{\large O}{\|}}{\underset{\underset{\large OH}{|}}{P}}-OH$$

glyphosate

This molecule, which has basic as well as acidic character, is able to form a terdentate (three-bond) or tetradentate (four-bond) chelate, with several of the coordination positions on surface metal ions being occupied by ligand groups:

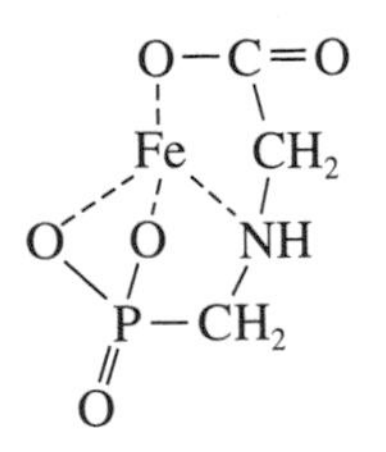

The multiple bond is expected to be very strong, in addition to being favored by the positive entropy change associated with chelation reactions (see discussion on chelation in Chapter 1). Glyphosate is readily and selectively adsorbed by Fe oxides, and its herbicidal function is rapidly inactivated on contact with soils, evidence of a strong surface interaction.

How generally important ligand exchange and chelation reactions are to organic interaction with variable-charge minerals in soils has not been evaluated. Certainly, the highly weathered soils, with their often large component of Fe and Al oxides, are likely to display some ability to immobilize organic acids. Also, many soils, especially those formed from volcanic ash, contain microcrystalline and noncrystalline allophanes or oxides as constituents of the clay mineral fraction. These minerals, especially in noncrystalline form, could contribute importantly to the capacity of soils to bond organic acids. It must be kept in mind, however, that these reactive minerals exist in soils with numerous natural organic and inorganic ligands already adsorbed at their surfaces.

10.5. ADSORPTION PROPERTIES OF NONIONIC AND NONPOLAR ORGANICS

As pointed out in section 10.4, the neutral forms of basic and acidic molecules are adsorbed sparingly by weak physical forces on soil particle surfaces. These molecules have polar[1] functional groups that may allow them to be attracted to surfaces by hydrogen bonding or ion-dipole attraction (see Chapter 1 for a description of these forces). For example, amines and alcohols can adsorb on clay minerals by coordinating directly with exchange cations on the surfaces or by hydrogen bonding to hydration water of the cations, as diagrammed in Figure 10.15. However, these mechanisms are not likely to be effective under moist conditions in soils because the excess water can easily displace physically adsorbed organic molecules from the surfaces. Water molecules have the dual competitive advantage of usually being present greatly in excess of the organic, and having a greater polarity than most organics. Therefore, even organics that can form fairly strong bonds of the type depicted in Figure 10.15 are readily moved through soils by water. This is in part because polar molecules tend to be highly soluble in soil solution.

When the huge group of organic molecules classified as nonionic, nonpolar[2] is considered, bonding on soil materials is even weaker than that of the polar nonionic molecules. Mechanisms such as hydrogen bonding and ion-dipole interaction are not operative for nonpolar molecules. In fact, adsorption of these molecules on *wet* layer silicate and oxide minerals is often very low, if detectable at all. Adsorption of nonpolars *does* appear to occur on soil organic matter, although there is some debate about the nature of this process. Is it actually adsorption, or is it the partitioning of the molecule into an organophilic phase in the humic material? Because of this

1. Molecules or molecular fragments in which the centers of positive and negative charge do not coincide are termed polar.

2. In this context, non-polar will also refer to weakly polar molecules, since few organic molecules are strictly non-polar (i.e., have zero dipole moment).

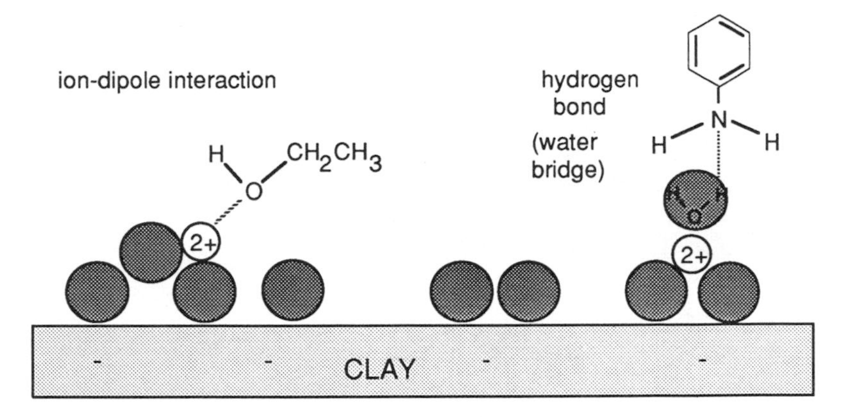

Figure 10.15. Possible mechanisms for the adsorption of ethanol and aniline by Mg^{2+}-saturated layer silicate clay. Water molecules are depicted as shaded circles. (Adapted from B.K.G. Theng. 1974. *The Chemistry of Clay-Organic Reactions*. London: Adam Hilger.)

uncertainty, retention of nonpolar organics on soil materials is usually referred to as "sorption" rather than "adsorption." The term "sorption" is inclusive of adsorption (a two-dimensional process) as well as the partitioning mechanism that might be more accurately called "absorption" or "dissolution" (a three-dimensional process). In any case, it will be shown later that the sorption process is driven more by the low solubility of the nonpolar molecule in solution than by a significant force of attraction to the organic matter.

Two important variables that have been used to characterize and predict sorption of nonpolar organics by soils are:

1. The octanol/water partition constant, K_{OW}, a property of the adsorbate (or sorbate)
2. The fraction of soil that is organic carbon, f_{OC}, a property of the adsorbent (or sorbent)

The first variable, K_{OW}, is measured by allowing the nonpolar compound of interest, R, to distribute between the two immiscible liquid phases, octanol and water, placed in contact with one another:

$$\text{R (water)} \leftrightarrow \text{R (octanol)} \tag{10.66}$$

The value of K_{OW} is measured at equilibrium:

$$K_{OW} = \frac{[\text{R}]_{\text{octanol}}}{[\text{R}]_{\text{water}}} \tag{10.67}$$

where the brackets denote concentrations (μmoles/liter) in the respective liquid phases. K_{OW} is one measure of the hydrophobicity[1] of the compound R. The utility

1. A compound is hydrophobic if its molecular attraction to water is much weaker than the attraction of water to itself.

of this variable can be illustrated for the sorption reaction of nonpolar organics, R, described by

$$R_{aq} \leftrightarrow R_s \tag{10.68}$$

where *aq* denotes the molecule in aqueous solution and *s* denotes the molecule in the solid phase. The partitioning coefficient, K_P, for sorption is then defined as

$$K_P = \frac{[R_s]}{[R_{aq}]} \tag{10.69}$$

where the brackets denote concentration units. The magnitude of K_P for sorption in soils is correlated with K_{OW}, as is clearly evident from the data plotted in Figure 10.16. Furthermore, K_P is correlated to the organic carbon content of the sorbent, f_{OC}, as indicated by the data plotted in Figure 10.17. The sorbents represented in Figure 10.17 include soils, various sediments, and even a processed sewage sludge. The same correlation holds for all of these materials, despite their chemical disparity, an indication that the physical adsorption of nonpolar molecules is rather insensitive to the specific nature of the sorbent, as long as the sorbent has some organophilic character.

On the basis of the correlations shown in Figures 10.16 and 10.17, the partitioning coefficient of any nonpolar organic sorbate is predicted by the empirical equation:

$$\log K_P = a \log K_{OW} + \log f_{OC} + b \tag{10.70}$$

where a and b are only slightly dependent on the type of organic sorbent. This equation can be very useful in predicting the sorption behavior *within* groups of chemically related organics, such as Cl-substituted benzenes or polyaromatic hydrocarbons (PAHs). However, the good correlations displayed in Figures 10.16 and 10.17 dete-

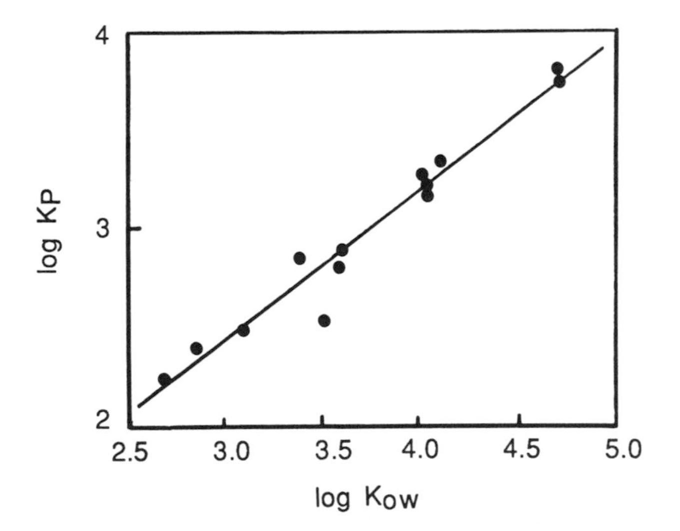

Figure 10.16. Partition constant, K_P, of substituted benzenes sorbed on soil material as a function of the K_{OW} of the sorbate. (From J. C. Westall. 1987. Adsorption mechanisms in aquatic surface chemistry. In W. Stumm (ed.), *Aquatic Surface Chemistry*. New York: Wiley. Used with permission.)

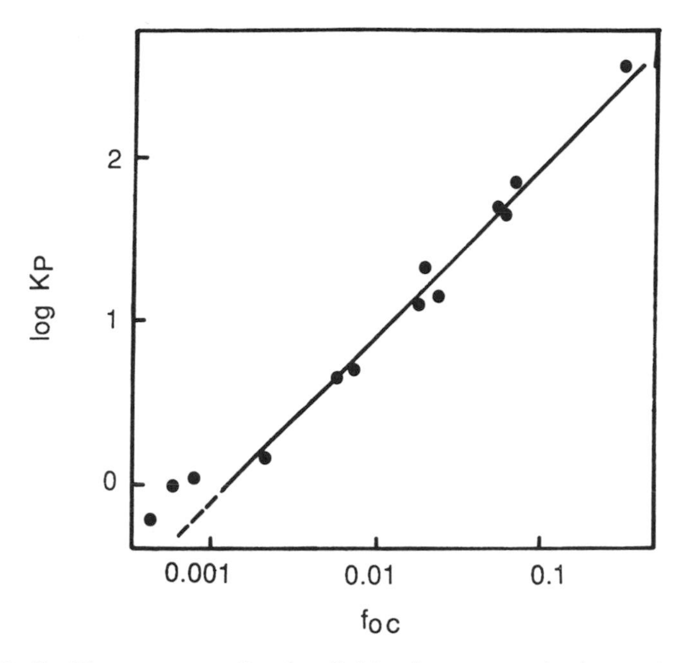

Figure 10.17. Partition constant, K_P, of p-dichlorobenzene sorbed on soil materials as a function of the f_{OC} of the sorbent. (From J. C. Westall. 1987. Adsorption mechanisms in aquatic surface chemistry. In W. Stumm (ed.), *Aquatic Surface Chemistry*. New York: Wiley. Used with permission.)

riorate when a broader range of chemical structures is included as sorbates. Equation 10.70 may then provide inaccurate estimates of K_P for types of nonpolar organics that were not employed to generate the empirical equation in the first place. A cautionary rule is:

Do not extend the $K_P - K_{OW}$ or $K_P - f_{OC}$ empirical relationships beyond the class of chemical structures from which they were derived.

The partition coefficient, K_P, has units of milliliters per gram of soil, but because equation 10.70 implies that the only sorbate of significance in soils for nonpolar organics is organic matter, it is common practice to express K_P in units of milliliters per gram of organic matter. It then becomes the partitioning coefficient for organic matter, K_{OM}, a value that is fairly constant in different soils for any particular nonpolar organic. K_{OM} is more a property of the organic sorbate than the soil. A very large K_{OM} (>1000) is indicative of strong sorption and immobility of a specific organic sorbate, whereas a small K_{OM} (<100) denotes mobility of the sorbate.

The fact that nonpolar molecules have a stronger attraction for soil organic matter than for mineral surfaces is ascribed to the existence of hydrophobic (organophilic) surfaces or phases within the organic matrix. The phenomenon of *hydrophobic attraction,* defined as the enhanced attraction between the surface and sorbate in a solvent because the solvent-sorbate attractive interaction is weaker than the solvent-solvent attractive interaction, explains the "affinity" of nonpolar organic molecules for humus. Since water is the solvent in soils, hydrophobic attraction becomes a driving force for sorption if the organic sorbate has less affinity for water than water has for itself. The disruption of water structure by the hydrophobic surfaces of dissolved nonpolar molecules, especially large ones, is energetically unfavorable. Consequently, nonpolar organics are forced out of the solution and onto weakly hydrating surfaces. This mechanism is diagrammed in Figure 10.18. The bond formed between the molecule and adsorbing surface is typically quite weak.

The apparent attraction of nonpolar molecules for humus surfaces can be visualized as analogous to the attraction between two droplets of oil immersed in water. They coalesce into a single droplet because this reduces the total oil-water interfacial area, thereby lowering the energy of the system. Although organic molecules are much smaller than these droplets, they occupy a specific volume in aqueous solution, creating an organic-water interface that disrupts the normal hydrogen-bonded arrangement of water molecules. Because a large molecule of high molecular weight destabilizes the aqueous solution to a greater extent than a small molecule, larger nonpolar molecules are "pushed" out of solution onto surfaces more completely than small ones.

The importance of molecular volume in driving hydrophobic adsorption is apparent in a comparison of the sorption of several chlorinated hydrocarbons by a muck soil, shown in Figure 10.19. These C-type (linear) isotherms, described by

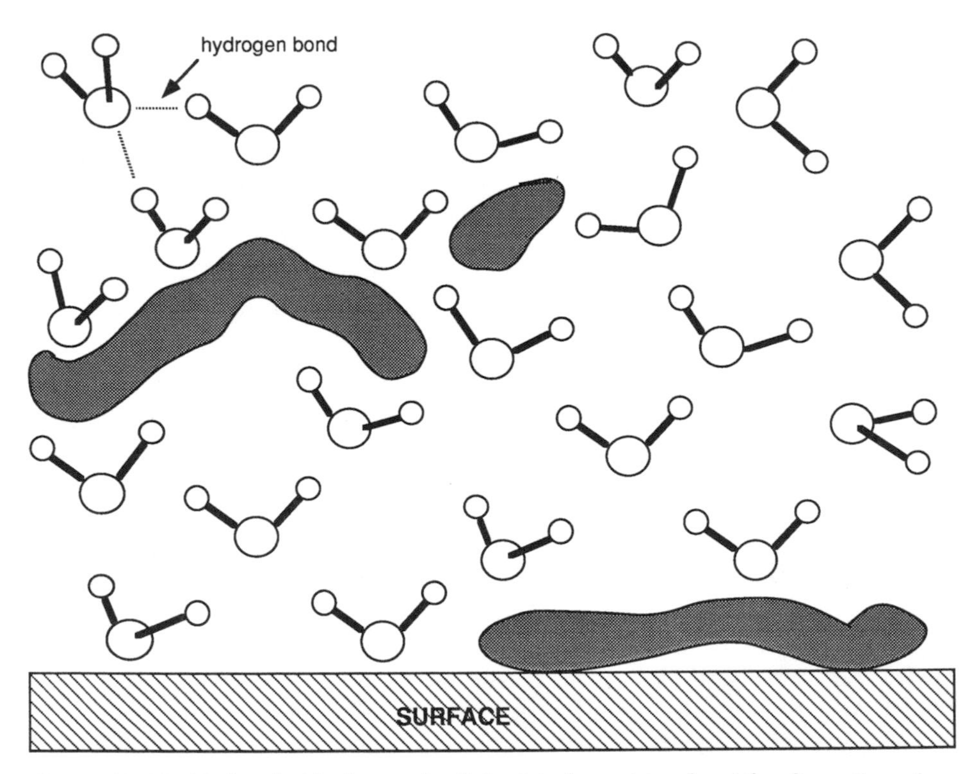

Figure 10.18. Hydrophobic "attraction" depicted as arising from the disruption of hydrogen-bonded water structure by molecules.

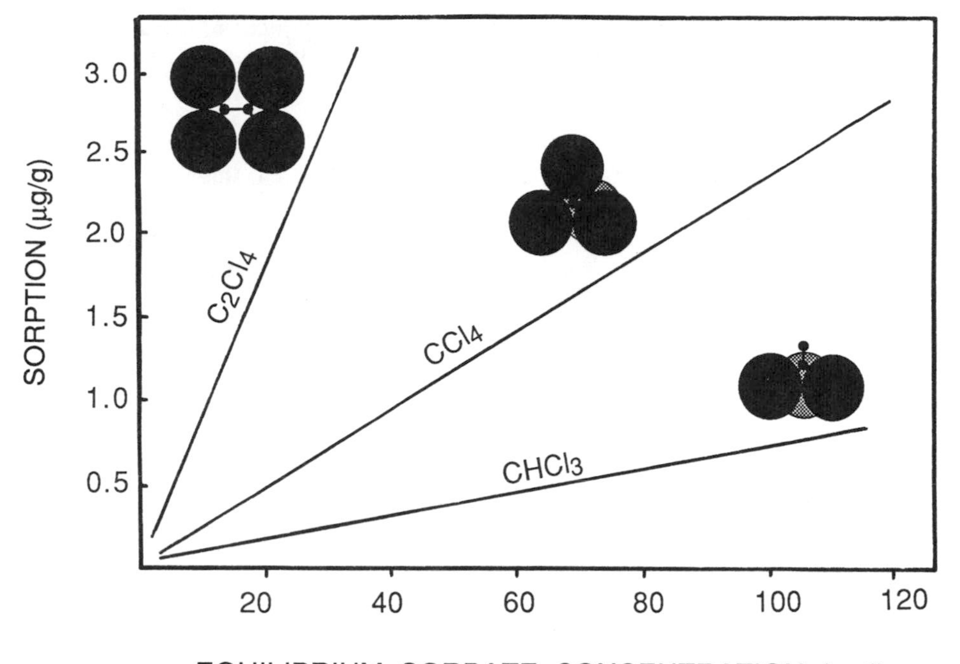

Figure 10.19. Adsorption of chlorinated hydrocarbons on a muck soil. Relative molecular sizes are indicated by the molecular diagrams. (Data from R. G. LaPoe. 1985. Sorption and desorption of volatile chlorinated aliphatic compounds by soil and soil components. Ph.D. Thesis, Cornell University, Ithaca, New York.)

equation 10.32, have a steeper slope for the larger molecules, indicating a larger value of c, the BET bonding constant. We conclude that molecular surface area (or volume) is positively correlated to the tendency of nonpolar organics to sorb on soil organic matter. Thus, chemically similar nonpolar molecules such as benzene, naphthalene, and pyrene have affinities for soil organic matter that increase with molecular weight because of their different hydrophobic surface areas.

The solubility of nonpolar organics in water is negatively correlated to molecular size, so that a *negative* correlation is found between the water solubility of organics and their tendency to sorb. Generally, more water-soluble molecules are at least somewhat polar and interact with the solution by dipole-dipole forces, an attraction that lessens their tendency to condense at surfaces and interfaces.

Sorption characteristics of nonpolar organics in moist soils can be summarized as follows:

1. Sorption isotherms are C-type (linear or constant partitioning) or S-type, and low-affinity sorption (isotherms with shallow slopes) is typical.
2. Large soil particles reach equilibrium with a given organic sorbate more slowly than do small particles (due to slow diffusion through micropores).
3. More hydrophobic sorbates (those with high K_{OW}) have slower sorption rates, as do sorbates of higher molecular weight.

4. Temperature has little effect on sorption rates.
5. Sorption is reversible, but often very slowly so (due to slow diffusion through micropores).

The last two properties are attributable to the nature of physical adsorption, outlined in Table 10.1, particularly the lack of an activation energy and the weakness of the surface interaction.

Many of the organic sorbates of concern in soils are pesticides that possess polar functional groups as well as nonpolar groups; these display neither the sorption characteristics of nonpolar molecules nor the more robust sorption behavior of the acidic, basic, or ionic organics. The sorption behavior of this intermediate class of "weakly polar" compounds is the most difficult to generalize. Many of these compounds have a measurable affinity for clay minerals, weakening the correlation between organic matter content of soils and sorption that is so apparent with the nonpolar compounds. Consequently, empirical equations such as 10.70 are not as useful for this class of organic chemicals. Since some of these molecules have sufficient polarity to be moderately soluble in water, their mobility in the solution phase of soils can be appreciable.

10.6. SUMMARY OF THE SORPTION BEHAVIOR OF ORGANICS IN SOLUTION

The variable sorptive tendencies of different classes of organic molecules can best be understood by evaluating all of the energy and entropy terms that contribute to the total free energy of sorption from aqueous solution. The important energy terms are:

E_{S-W} = surface hydration energy (surface-water interaction)
E_{A-W} = adsorbate hydration energy (organic-water interaction)
E_{W-W} = water-water bonding energy
E_{S-A} = surface adsorption energy (surface-organic interaction)

The total energy of sorption, ΔE_T, is then given by

$$\Delta E_T = E_{S-W} + E_{A-W} - E_{W-W} - E_{S-A} \tag{10.71}$$

and sorption of organics is seen to be favored by small values of the first two energy terms (E_{S-W}, E_{A-W}) and/or large values of the last two terms (E_{W-W}, E_{S-A}). If these values result in a negative ΔE_T, conditions are favorable for sorption (unless the entropy of sorption is strongly negative). The possible bonding situations that could arise are summarized in Table 10.5, showing the energy terms that are most important in each case.

The situation of a nonpolar organic sorbing on a hydrophobic surface, because E_{W-W} is the only large energy term, is the one case in which sorption is obviously driven by the strong attraction of water for other water molecules. This is the case of hydrophobic attraction. For the case of polar molecules at hydrophobic surfaces, the opposing energy terms suggest that sorption may or may not be favorable, depending on how hydration energy of the molecule compares with the water-water interaction energy. The other two cases, nonpolar and polar molecule adsorption on hydrophilic (hydrated) surfaces, also have opposing energy terms, so that sorption depends on

Table 10.5. Important Energy Terms (Indicated by *) for Sorption of Polar and Nonpolar Organic Molecules on Hydrophobic and Hydrophilic Surfaces

		Sorption-Deterring		Sorption-Aiding	
Adsorbate	Surface[a]	E_{S-W}	E_{A-W}	E_{W-W}	E_{S-A}
Polar	Hydrophilic	*	*	*	*
Nonpolar	Hydrophilic	*		*	
Polar	Hydrophobic		*	*	
Nonpolar	Hydrophobic			*	

[a]Hydrophilic surfaces include those permanent-charge clays and other minerals that are hydroxylated at their surfaces (iron oxides, kaolinite, etc.). Hydrophobic (organophilic) surfaces include 2:1 silicates with little or no permanent charge (e.g., talc), and components of humus.

the relative size of these terms. Certainly, weakly polar molecules have been observed to adsorb on hydrophilic clay surfaces, evidently because the surface hydration energy (E_{S-W}) is insufficient to counter the E_{W-W} term that forces the molecules out of solution. However, benzene, a nonpolar molecule, does not sorb from solution onto natural layer silicate clays, despite its weak attraction to water. Molecular size as well as polarity is important in determining the exact balance of energies.

In a detailed analysis of organic sorption, entropy terms must be considered in addition to energy, particularly for large polymeric organic molecules. When these molecules adsorb on hydrated surfaces, they displace a number of adsorbed water molecules into solution. The result is an increase in disorder in the system (more positive entropy), which is proportional to the size of the sorbate molecule. In this way, sorption of polymers is often favored even when the individual energy terms of Table 10.5 sum to a *positive* ΔE_T value. Uncharged polysaccharides, for example, sorb from solution onto smectites even though much lower molecular weight sugars do not. The energy of the surface-adsorbate bond (E_{S-A}) is probably not very great for either polysaccharides or sugars, whereas the hydration energy (E_{A-W}) of these molecules is significant.

On the basis of the above discussions of the various classes of organic compounds, a generalized summary of behavior is given in Table 10.6. The fact that water is to some degree a competing adsorbate means that the absolute bonding strength of organics is less predictive than bonding strength relative to water. For example, the ion-dipole bonds between polar organics (such as amines and alcohols) and exchange cations can be rather strong, yet these organics adsorb little on layer silicate clays unless the organic concentration in solution is very high or the clays are in a dry condition.

10.7. DEGRADATION OF ORGANIC COMPOUNDS IN SOIL

Microbes have the capability to degrade most classes of organics, whether they be synthetic or natural. However, degradation rates in soils vary tremendously depending upon:

Table 10.6. Sorption Affinity of Organics for Soil Components

Molecule Type	Affinity For	Isotherm Type	Bonding Forces
Nonpolar or weakly polar	Hydrophobic organic phases in humic acid	C or S	Weak physical (van der Waals), hydrophobic attraction
Polar, uncharged	Humic polar groups, coordination sites of multivalent metals on colloids, mineral surface Si—O—Si groups.	L	Dipole-dipole and ion-dipole (fairly strong to rather weak)
Polar, cationic	COO^- groups of humic acid, cation exchange sites of clays	H	Electrostatic (Strong)
Polar, anionic	Surfaces of variable-charge minerals	L	Metal-ligand coordination bond (Strong)

1. Susceptibility of the specific molecular structure of the organic to enzymatically catalyzed chemical reactions
2. The degree of accessibility or desorbability of the sorbed organic
3. The chemical environment (E_h, pH, etc.) of the soil
4. The physical environment (e.g., temperature) of the soil.

Some organic molecules sorb inaccessibly or have structures that are sufficiently resistant to breakdown by microorganisms that they have a long lifetime in soil. The longer the lifetime, the greater the risk for pollution of soil. For example, highly halogenated hydrocarbons (e.g., polychlorinated biphenyls, or PCBs) constitute one group of chemicals that is extremely resistant to decomposition. This resistance is attributed to very low solubility in water and the lack of a structural site of enzymatic attack for degradation. Yet even these compounds slowly decompose under the reducing conditions found in anaerobic sediments. Conversely, organics with polar functional groups (e.g., $-OH$, $-NH_2$, $-COO^-$, $-NO_2$) are much more susceptible to microbial degradation because they are soluble in water and because microbes have enzyme systems that readily decompose such molecules.

The subject of biological degradation is complex and will not be further dealt with here. Readers are referred to reviews of this subject listed at the end of this chapter.

Nonbiological (abiotic) degradation can also be important in the decomposition of organic compounds. Its role has often been overlooked because abiotic processes can be difficult to distinguish experimentally from biological processes. Soil sterilization techniques, besides inactivating enzymes, can alter soils chemically. Therefore, if the degradation rate of a particular organic in soil is lowered after sterilization, this does not necessarily mean that the degradation reaction is biological. Some of the known types of abiotic degradation reactions (that is, reactions that proceed without enzymes) are discussed in this section. Four of these abiotic reactions can occur in soil solution without the requirement of a catalyst: *hydrolysis, oxidation, reduction,* and *isomerization.* This section outlines how the soil particle surfaces promote some of these reactions.

10.7a. Hydrolysis and the Role of Surface Brønsted and Lewis Acidity

Certain types of chemical bonds are particularly susceptible to hydrolysis, a mechanism of bond cleavage in the molecule involving reaction of carbon atoms with the oxygen atoms of water. The site of cleavage is typically a bond between carbon and another atom of different electronegativity because water, an inorganic nucleophile, is attracted to the more electron deficient of the two atoms. Many pesticides, including aliphatic halides, amides, carbamates, urea derivatives, phosphoric acid esters and thioesters, are susceptible to partial decomposition by hydrolysis. The hydrolysis reaction can be catalyzed by acid or base, OH^- being a stronger nucleophile than water. Organophosphates and chloro-s-triazines, for example, hydrolyse in alkaline solution by cleavage at the P—O—C and C—Cl bond positions. When oxides or silicate clays are present, hydrolysis is often catalyzed by surface acidity of either the Lewis or Brønsted type. A description of these concepts of acidity is found in Chapter 1.

Brønsted acidity on soil minerals is generated by the polarizing power of exchange and structural metal cations as well as adsorbed H^+. Small, high-charge exchange cations produce Brønsted acidity by promoting a reaction with water to release H^+ ions.[1] For example, exchangeable Al^{3+} reacts with water to form the hydroxy-Al complex:

$$Al(H_2O)_6^{3+} \rightarrow Al(H_2O)_5OH^{2+} + H^+ \qquad pK_h = 5.00 \tag{10.72}$$

As clay surfaces become drier, the reaction is driven further to the right and the protons are concentrated in a smaller volume of water so that the acidity becomes more extreme. The ranking of potential Brønsted acid strength for common exchangeable cations follows the order of polarizing power:

$$H^+ > Al^{3+}, Fe^{3+} > Mg^{2+} > Ca^{2+} > Na^+ > K^+$$

Very acid clays, such as air-dried smectites with Al^{3+} exchange ions, are equivalent in acidity to concentrated aqueous solutions of strong acids. Even very weak bases (poor proton acceptors) will protonate on such clay surfaces. For example, the amino form of 3-aminotriazole converts to the imino form on dry Mg^{2+}-montmorillonite (Russell et al., 1968a):

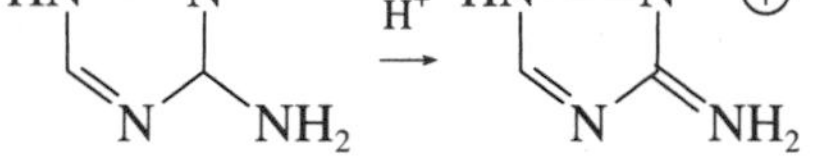

HN—N ... N, NH₂ $\xrightarrow{H^+}$ HN—N ⊕ ... N, NH₂ (10.73)

A similar reaction occurs with the triazine compounds on dry clays. To achieve these protonation reactions *in solution* would require a 6 molar concentration of strong acid such as HCl!

Clay acidity catalyzes hydrolysis of the chloro-s-triazine herbicides to the non-phytotoxic 2-hydroxy-s-triazines (Russell et al., 1968b):

1. This reaction is termed hydrolysis but should not be confused with the bond-cleaving process of organic molecules that is given the same name.

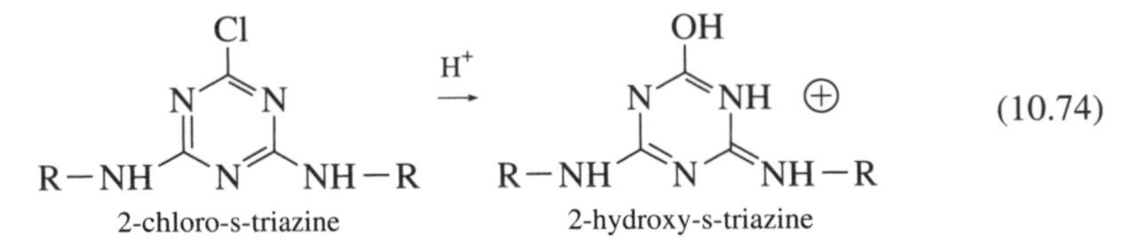

A wide range of organic compounds is degraded more rapidly by acidic soils or clays than by the same adsorbents in nonacidic form. In many cases this may reflect an acid-catalyzed hydrolysis reaction.

Besides acid forms of layer silicates, oxides of Fe and Al in water, and particularly in the dry state, possess some catalytic function in organic hydrolysis reactions, at least those that are known to be OH^--catalyzed (Hoffman, 1990). Although it has been suggested that the OH^- ion activity is higher at the positively charged oxide surface than in solution, it seems that coordination reactions between structural Fe^{3+} or Al^{3+} (Lewis acids) and the organic functional groups is a more probable cause of decomposition. In fact, those hydrolyzable organics that have a suitable structure to chelate the surface metal cations are generally the most susceptible to mineral-catalyzed decomposition. For example, Cu^{2+} on exchange sites of montmorillonite is able to catalyze the hydrolysis of certain organophosphates (Mortland and Raman, 1967), a reaction that probably involves a Cu^{2+}-organic chelation step as shown:

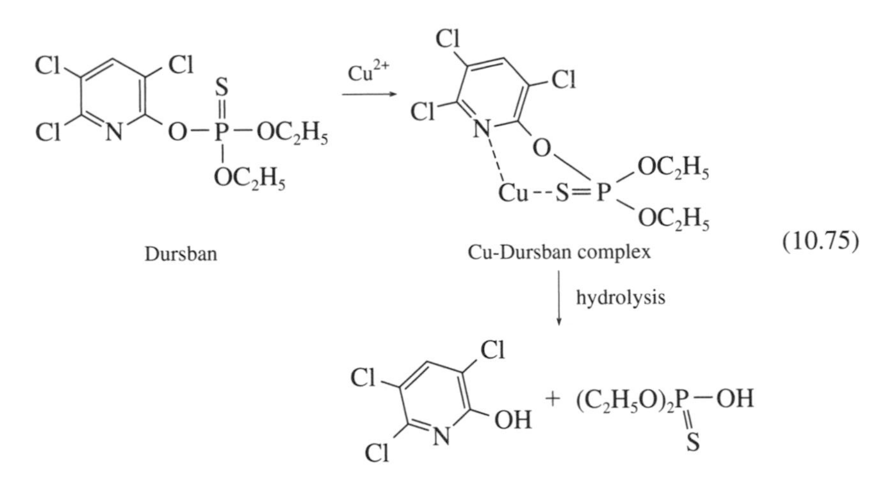

It is interesting to note that Cu^{2+} adsorbed on an organic soil does not catalyze this reaction, suggesting that if the metal is strongly complexed, it may not be free to coordinate with the organophosphate.

Fe and Al oxides catalyze the hydrolysis of a number of chelate-forming esters, perhaps by a similar mechanism to that shown in reaction 10.75, in which the chelate forms with surface Fe^{3+} or Al^{3+}. In summary, the Lewis acid properties of metals such as Cu^{2+}, Fe^{3+}, and Al^{3+} appear to be important to mineral-catalyzed hydrolysis.

10.7b. Metal-Catalyzed Oxidation

Certain classes of organic compounds, such as phenols and aromatic amines, are subject to degradation by oxidation. A number of metal cations, including Al^{3+}, increase the rate of oxidation of polyphenols to polymeric quinones by a mechanism involving complexation of the metal:

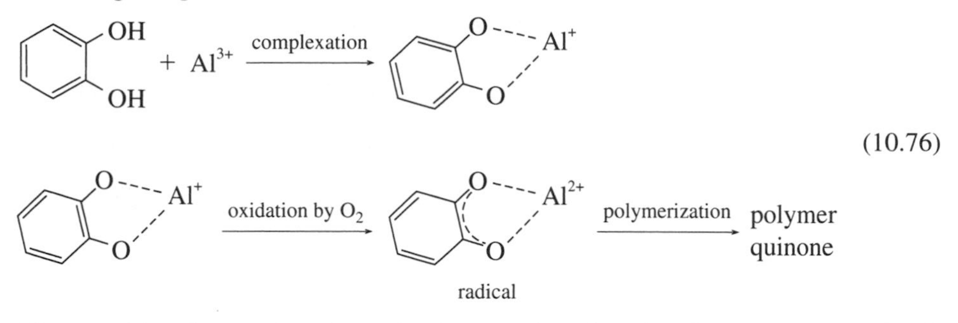

(10.76)

The metal ion directs the polymerization reaction, favoring C—C rather than C—O bonding:

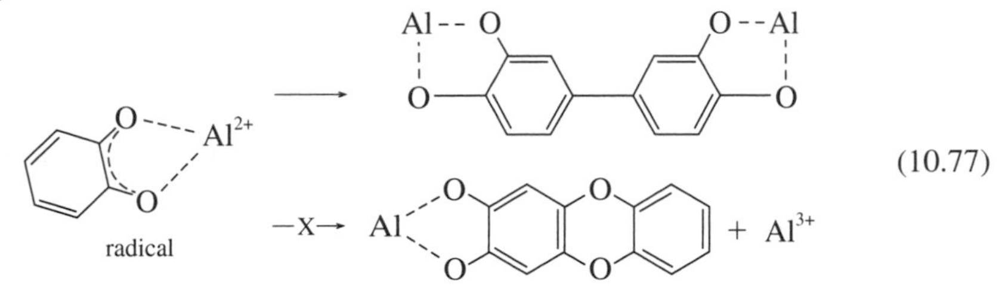

(10.77)

The complexed metal blocks the lower reaction pathway of 10.77 by occupying positions at the phenolic groups and hindering the coupling of these groups with carbon atoms of other phenol molecules.

Although oxidation can occur in soil solution by reaction with dissolved oxygen, the reaction is often much more rapid in the presence of surfaces that serve as oxidizing agents or catalysts. For example, benzidine, an aromatic amine, is adsorbed on smectites as an organic cation, and is subsequently oxidized rapidly by structural iron:

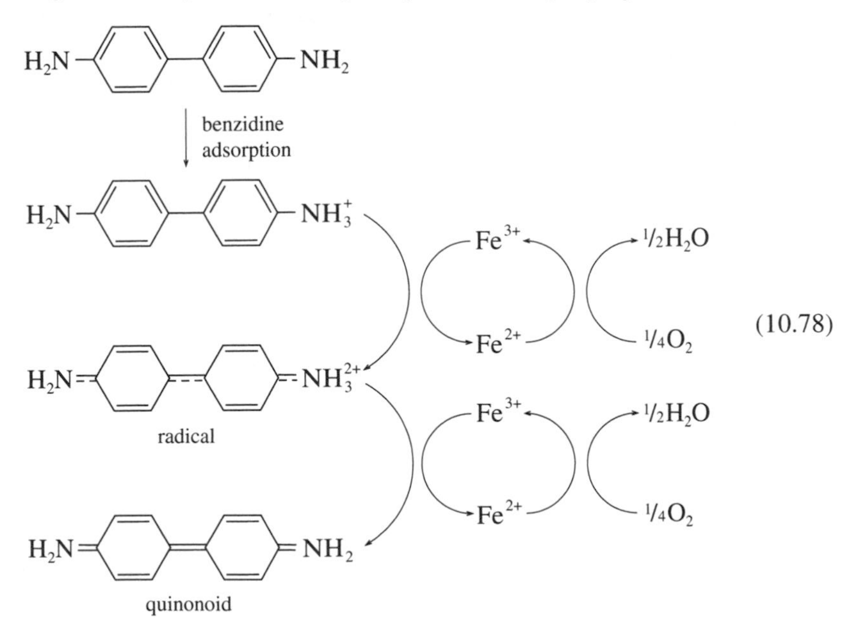

(10.78)

The ground-state O_2 molecule is paramagnetic (has unpaired electrons) so that its reaction with the diamagnetic (no unpaired electrons) benzidine molecule is spin forbidden. While Fe^{3+} is also paramagnetic, it can circumvent the rule of spin parity by producing an hydroxyl radical:

$$Fe(III)-OH \rightarrow Fe(II) + OH\cdot \quad (10.79)$$

The radical then removes an electron from the organic by an outer-sphere process, oxidizing it. Subsequently, O_2 reoxidizes Fe^{2+} to Fe^{3+}, and the cycle is completed. Thus, structural iron may by this pathway play the role of catalyst for oxidation of the amine by O_2 (reaction 10.78).

Oxidation of benzidine is very rapid on Mn oxide surfaces, as Mn in the +4 or +3 oxidation state is a much stronger oxidizing agent than structural Fe^{3+}. Aniline, which is chemically similar to benzidine, is also oxidized by structural Fe and Mn.

Phenols are oxidized as well by Mn and Fe in minerals, but their ease of oxidation depends on the electron-donating power of groups substituted on the aromatic ring. As a general rule, the greater the electron-donating tendency of these groups, the easier the oxidation. For the case of para-substituted phenols, ease of oxidation follows the order:

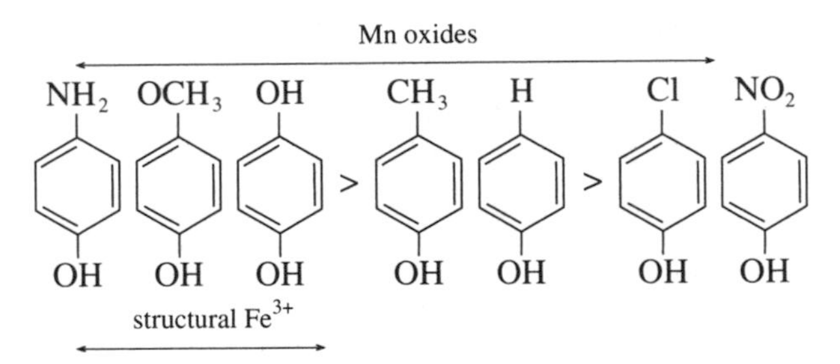

where the horizontal lines span the phenols that can be oxidized by Mn oxides or by Fe^{3+} in layer silicates and oxides. Biphenols and polyphenols are particularly susceptible to rapid oxidation, which progresses through the semiquinone intermediate to the quinone product, as follows:

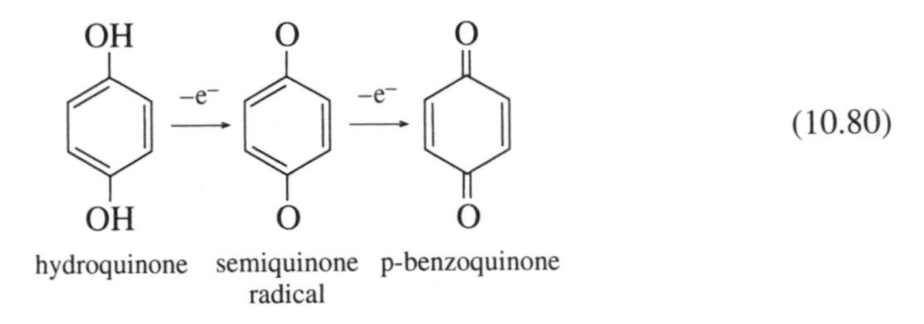

The strong electron-donating properties of —OH groups substituted on the aromatic ring render polyphenols very susceptible to oxidation in soils. On the other hand, phenols chlorinated at the para position, many examples of which are found among the pesticides or decomposition products of pesticides, are resistant to oxidation and can persist in soils for some time. However, phenols substituted at *several* positions on the ring with chlorine can be more susceptible to degradation by Mn oxides than p-chlorophenol or phenol. The process of degradation includes hydrolysis (replacement of Cl by OH on the para position of the aromatic ring) and oxidation.

The electron acceptor (oxidizing agent) for phenol reactions such as reaction 10.80 can be Mn oxides, Fe oxides, structural Fe^{3+} in layer silicates, or Fe^{3+} complexed with soil organic matter. The reader is referred to Chapter 7 for development of the theory of oxidation-reduction (redox) reactions.

Example Problem: Calculate the thermodynamic favorability of the oxidation of hydroquinone by Fe^{3+}.

Solution: The standard-state half-cell potentials, E_h^0, relevant to the oxidation of hydroquinone (symbolized by HQ) to p-benzoquinone (Q) by dissolved Fe^{3+} are given below:

$$HQ = Q + 2H^+ + 2e^- \qquad E_h^0 = -0.699 \text{ volt}$$

$$2Fe^{3+}\text{ (aq)} + 2e^- = 2Fe^{2+}\text{ (aq)} \qquad E_h^0 = +0.77 \text{ volt} \qquad (10.81)$$

Summing these two half-cell reactions produces the overall redox reaction and the calculated standard-state potential, E^0:

$$HQ + 2Fe^{3+}\text{ (aq)} = 2Fe^{2+}\text{ (aq)} + Q + 2H^+ \qquad E^0 = +0.071 \text{ volt} \qquad (10.82)$$

The positive value of E^0 suggests that the oxidation of hydroquinone by *soluble* Fe^{3+} should occur spontaneously under *standard-state* conditions. However, the standard-state activities of dissolved ions (in this case Fe^{3+}, Fe^{2+}, and H^+) of unity correspond to concentrations that are absurdly high for soil solution (on the order of 1 molar). The standard-state potential, then, has little use in predicting the favorability of the reaction under conditions likely to prevail in soil solutions. It is necessary to use the Nernst equation (see Chapter 7) to calculate the adjusted redox potential, E, for more realistic reaction conditions:

$$E = E^0 - 0.059 \log \frac{(Fe^{2+})^2(Q)(H^+)^2}{(Fe^{3+})^2(HQ)} \qquad (10.83)$$

In soil, the pH is likely to be buffered, and the solubility of Fe^{3+} can be assumed to be controlled by $Fe(OH)_3$ dissolution:

$$Fe(OH)_3 = Fe^{3+} + 3OH^- \qquad K_{SO} = 10^{-39} \qquad (10.84)$$

The solubility product of $Fe(OH)_3$, and the fact that $(H^+)(OH^-) = 10^{-14}$ in aqueous solutions, are combined so that the Fe^{3+} activity in soil solution can be estimated by $(Fe^{3+}) = 10^3\,(H^+)^3$. For a soil pH of 6, this estimate gives $(Fe^{3+}) = 10^{-15}$, and equation 10.83 becomes:

$$E = E^0 - 0.059 \log \frac{(Fe^{2+})^2(Q)(10^{-6})^2}{(10^{-15})^2(HQ)} \qquad (10.85)$$

If a low concentration of Fe^{2+}, say 10^{-6} *M*, is maintained in soil solution by cation exchange (or by precipitation of $FeCO_3$ as would be the case in calcareous or alkaline soils), then equation 10.85 simplifies to

$$E = +0.071 - 0.059 \log \frac{(10^6)(Q)}{(HQ)} \quad (10.86)$$

since E^0 was already shown to be 0.071 volt for reaction 10.82. At equilibrium, the reaction potential (E) is zero by definition, and equation 10.86 is solved to give the ratio of quinone product to phenol reactant in solution:

$$\frac{(Q)}{(HQ)} = 1.58 \times 10^{-5} \quad (10.87)$$

The clear result is that oxidation of hydroquinone to quinone by $Fe(OH)_3$ is not favored at pH 6 if the level of soluble Fe^{2+} is as high as 10^{-6} M. However, because Fe^{2+} is spontaneously oxidized in nonacidic solutions by dissolved oxygen, the Fe^{2+} concentration in well-aerated solutions could fall much below 10^{-6} M, thereby increasing the (Q)/(HQ) ratio.

The oxidation of hydroquinone by Fe^{3+} attains a maximum near pH 4, with less oxidation at both lower and higher pH. This response to pH can be understood on the basis of reaction 10.82, which is the relevant overall reaction only if the pH is low enough that soluble Fe^{3+} exists. At higher pH, Fe^{3+} is in the precipitated form, and the overall reaction becomes

$$HQ + 2Fe(OH)_3 = 2Fe^{2+} + Q + 4OH^- \quad (10.88)$$

Thus, the effect of pH on the reaction is reversed once Fe^{3+} is involved in the reaction as the solid phase.

Fe oxides are found in practice to oxidize hydroquinone and other polyphenols fairly readily, suggesting that the soluble Fe^{2+} generated by the reaction is oxidized or adsorbed so that it does not accumulate in solution. Nevertheless, Fe^{3+} in the structures of oxides and silicates is not a strong oxidizing agent because the octahedral sites of these minerals stabilize the +3 oxidation state of Fe relative to +2. This means that, conversely, structural Fe^{2+} is a potentially strong reducing agent for organics.

A similar calculation for the redox reaction between hydroquinone and Mn(IV) oxide predicts that the oxidation of the phenol should be essentially complete. This is to be expected, since oxides of Mn are stronger oxidizing agents than those of Fe.

Mineral surfaces oxidize organics by sequential one-electron oxidation steps, such as that described for hydroquinone (reaction 10.80). Under certain reaction conditions, the single-electron oxidation products, which are radicals, may temporarily accumulate. These radicals are highly reactive, coupling among themselves to form dimers and polymers, or further oxidizing by reaction with dissolved O_2. This may be the mechanism by which organics such as 2,6-dimethylphenol form dimers when adsorbed on air-dry smectites (Sawhney et al., 1984):

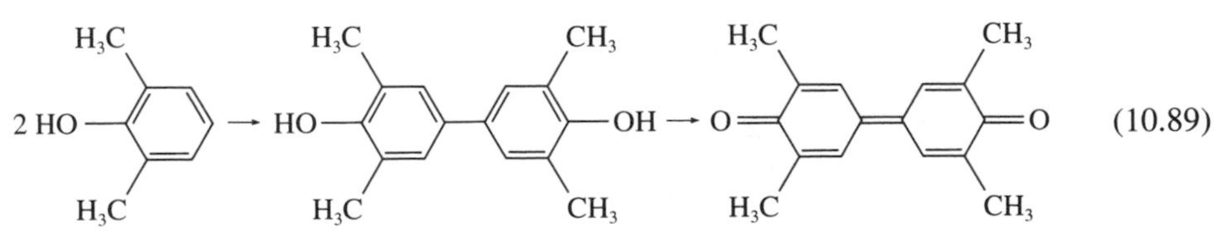

The coupling reaction is catalyzed by Fe^{3+}, and to a lesser extent Al^{3+}, adsorbed on the smectite. Exchange cations of lower valence show much less catalytic activity.

Semiquinone radicals of the type exemplified in reaction 10.80 are believed to be naturally present in soil organic matter, formed by the oxidation of polyphenols. It is quite possible that these radicals enter into redox reactions with adsorbed organic pollutants that are susceptible to either oxidation or reduction. They may also couple with these organics to form higher molecular weight products. Evidence for the involvement of soil organic matter in degrading organic pollutants is scant, but there are indications that some pesticide molecules become covalently bonded to organic matter, possibly via radical coupling mechanisms.

10.7c. Chemical Reduction of Organics

Reducing agents that accumulate in soils under anaerobic conditions, such as Fe^{2+}, sulfide, and phenols (from the reduction of quinones in humic acid) may react with certain classes of organic chemicals. For instance, Fe^{2+} can reduce halogenated aliphatic compounds, an important class of organic pollutants in soils. The process initially involves the transfer of a single electron, for example,

$$\geq C-CX\leq + Fe(II) \xrightarrow{\text{slow}} \geq C-\dot{C}< + Fe(III) + X^- \qquad (10.90)$$

with the formation of the organic radical and removal of the halogen, X^- (e.g., Cl^- or Br^-), from the molecule. The alkyl radical can then react with H^+ to form the dehalogenated aliphatic compound:

$$\geq C-\dot{C}< + H^2 + e^- \rightarrow \geq C-CH\leq \qquad (10.91)$$

or, if another halogen is bonded to the adjacent carbon, it can be lost to form the alkene:

$$X-C\leq-\dot{C}< \rightarrow >C=C< + X^- \qquad (10.92)$$

The alkene product has two fewer halogen substituents than the initial reactant, and is less likely to undergo further reduction even if any halogens remain substituted on the alkene.

In general, highly halogenated aliphatic compounds have high reduction potentials, so much so that hexachloroethane is a better electron acceptor than molecular oxygen! The likelihood that such compounds will be reduced in anaerobic soils is high. There is evidence that polychlorinated biphenyls buried in anaerobic river sediments very slowly undergo reductive dechlorination, but this is believed to be a microbial process (Quensen et al., 1988).

Other reduction reactions may be involved in the abiotic degradation of synthetic organics. Such reactions commonly involve atoms or groups in the organic that are able to change oxidation state. For example, nitro groups substituted on organics can be reduced to amines. The dinitroaniline herbicides, such as trifluralin, undergo such a reduction if the soil has a low redox potential (<100 mV). Certain sulfur-containing pesticides may also be susceptible to reduction. The relative contribution of biological and abiotic processes to reduction of organic chemicals is not known, nor is the ability of soil particle surfaces to catalyze these reactions.

10.7d. Photochemical Degradation

Organic molecules that absorb light energy at wavelengths above 285 nm are susceptible to photodecomposition, as the absorbed energy can break chemical bonds within the molecule. Most of the ultraviolet (UV) sunlight of wavelength shorter than 285 nm is absorbed by the atmosphere, as illustrated by Figure 10.20, so that molecules that absorb light only at wavelengths shorter than 285 nm are unlikely to degrade photochemically. In any event, photodecomposition may not be very significant in soils because the process can occur only at the soil surface. Soil minerals and organic matter strongly absorb light, blocking penetration beyond a very thin surface layer.

Classes of pesticides that can potentially photodecompose include:

1. Chlorinated cyclodiene insecticides
2. Chlorinated benzoic and phenylacetic acids

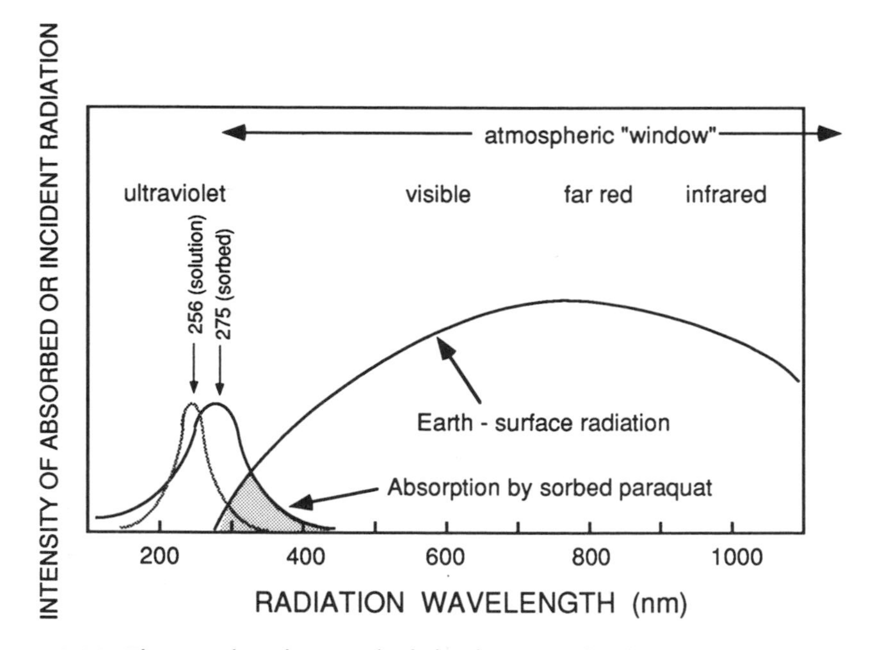

Figure 10.20. The wavelength range for light absorption by dissolved and clay-sorbed paraquat, compared with the wavelength "window" of the atmosphere. The shaded area denotes overlap of these two ranges, and gauges the potential for photodecomposition.

3. Triazines
4. Ureas
5. Dinitroaniline and picolinic acid herbicides

This list may be lengthened when the effects of adsorption are considered, because a molecule bound on a surface may be "activated" for photodegradation even if the reaction is not favorable for the same molecule dissolved in solution. For example, paraquat apparently photodecomposes more rapidly when adsorbed on layer silicate minerals than when in solution (Helling et al., 1971). This phenomenon may be related to the fact that adsorption on clay shifts the UV absorption band of paraquat to longer wavelength (256 → 275 nm) and closer to the atmospheric "window" for UV light (see Figure 10.20). In other words, adsorption on clay increases the probability that paraquat will absorb UV radiation and thereby decompose.

Mineral surfaces can also activate molecules for photodecomposition by complexation reactions involving metal ions. For example, the reaction of substituted salicylic acid on Fe oxides:

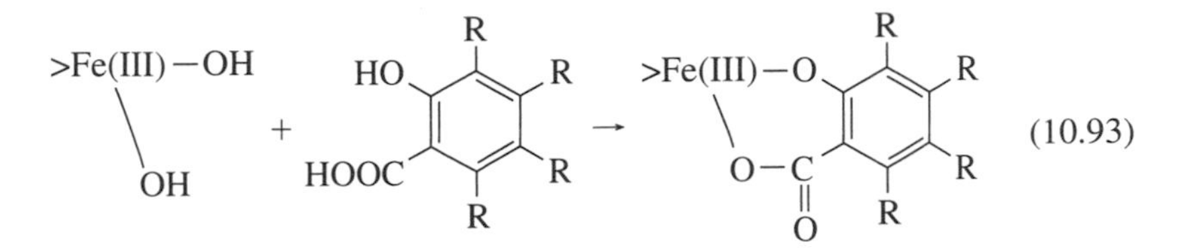

(10.93)

can enhance photodecomposition because Fe(III) has an energetically accessible lower oxidation state. The metal-organic complex absorbs a photon of UV light, forming an electronically excited organic molecule. An electron is then promoted from the organic molecule to the metal, producing the reduced metal and the oxidized organic:

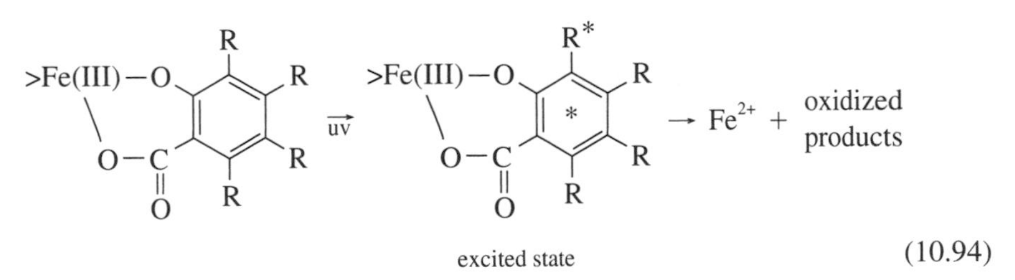

(10.94)

The surface bonding process may in itself enhance absorption of UV radiation by organics such as salicylate, because metal complexation typically produces a shift in the UV absorption toward longer wavelength.

Minerals with semiconducting properties, such as titanium dioxide, can photocatalyze the decomposition of resistant organics like polychlorinated biphenyls. This is possible because some semiconductors have a band gap in the energy range of visible and ultraviolet light (see Figure 7.14). For example, the rutile (TiO_2) band gap corresponds to a wavelength of about 380 to 410 nanometers, in the blue region of the visible spectrum. Solar radiation can then promote electrons out of the valence band and into the conduction band, leaving behind a positively charged "electron hole." Oxidation or reduction follows by direct reaction of the promoted electron or the hole with organics on the oxide surface. Also, if the holes react with water or the electrons react with protons or oxygen, very reactive species such as the hydroxyl radical are formed. These oxidatively decompose a wide range of organic compounds.

UV radiation also promotes molecular oxygen into the very reactive singlet state. This form of O_2 is able to oxidize many organics adsorbed on soil surfaces exposed to sunlight.

References

Giles, C. H., T. H. MacEwan, S. N. Nakhwa, and D. Smith. 1960. Studies in adsorption. Part XI. A system of classification of solution adsorption isotherms, and its use in diagnosis of adsorption mechanisms and in measurement of specific surface areas of solids. *J. Chem. Soc.,* pp. 3973–3993.

Hall, P. L. and D. M. Astill, 1989. Adsorption of water by homoionic exchange forms of Wyoming montmorillonite (SWy-1). *Clays Clay Min.* 37:355–363.

Hayes, M.H.B. and U. Mingelgrin. 1991. Interactions between small organic chemicals and soil colloidal constituents. In Bolt et al. (eds.), *Interactions at the Soil Colloid-Soil Solution Interface,* pp. 323–407.

Helling, C. S., P. C. Kearney, and M. Alexander. 1971. Behavior of pesticides in soils. *Adv. Agron.* 23:147–240.

Hoffman, M. R. 1990. Catalysis in aquatic environments. In *Aquatic Chemical Kinetics* (W. Stumm, ed.) John Wiley and Sons, New York. pp. 71–111.

Kung, K.-H. and M. B. McBride. 1989. Adsorption of para-substituted benzoates on iron oxides. *Soil Sci. Soc. Am. J.* 53:1673–1678.

Kung, K.-H. and M. B. McBride. 1991. Bonding of chlorophenols on iron and aluminum oxides. *Environ. Sci. Technol.* 25:702–709.

LaPoe, R. G. 1985. Sorption and desorption of volatile chlorinated aliphatic compounds by soils and soil components. Ph.D. thesis, Cornell University, Ithaca, New York.

Mortland, M. M. and K. V. Raman. 1967. Catalytic hydrolysis of some organic phosphate pesticides by copper(II). *J. Agric. Food Chem.* 15:163–167.

Quensen, J. F., J. M. Tiedje, and S. A. Boyd. 1988. Reductive dechlorination of polychlorinated biphenyls by anaerobic microorganisms from sediments. *Science* 242:752–754.

Russell, J. D., M. Cruz, and J. L. White. 1968a. The adsorption of 3-aminotriazole by montmorillonite. *J. Agric. Food Chem.* 16:21–24.

Russell, J. D., M. Cruz, and J. L. White, 1968b. Mode of chemical degradation of s-triazines by montmorillonite. *Science* 160:1340–1342.

Sawhney, B. L., R. K. Kozloski, P. J. Isaacson, and M.P.N. Gent. 1984. Polymerization of 2,6-dimethylphenol on smectite surfaces. *Clays Clay Min.* 32:108–114.

Theng, B.K.G. 1974. *The Chemistry of Clay-Organic Reactions.* London: Adam Hilger.

Watson, J. R., A. M. Posner, and J. P. Quirk. 1973. Adsorption of the herbicide 2,4-D on goethite. *J. Soil Sci.* 24:503–511.

Weber, J. B. 1970. Mechanisms of adsorption of s-triazines by clay colloids and factors affecting plant availability. In *Residue Reviews.* Vol. 32: *The Triazine Herbicides.* New York: Springer-Verlag.

Westall, J. C. 1987. Adsorption mechanisms in aquatic surface chemistry. In W. Stumm (ed.), *Aquatic Surface Chemistry* New York: Wiley, pp. 3–32.

Suggested Additional Reading

Clapp, C. E., R. Harrison, and M.H.B. Hayes. 1991. Interactions between organic macromolecules and soil inorganic colloids and soils. In G. H. Bolt, M. F. DeBoodt, M.H.B. Hayes, and M. B. McBride (eds.), *Interactions at the Soil Colloid-Soil Solution Interface.* Dordrecht, Netherlands: Kluwer, pp. 409–468.

Green, R. E. 1974. Pesticide-clay-water interactions. In W. D. Guenzi (ed.), *Pesticides in Soil and Water* Madison, Wis.: Soil Science Society of America, pp. 3–37.

Kaufman, D. D. 1974. Degradation of pesticides by soil microorganisms. In Guenzi (ed.), *Pesticides in Soil and Water, pp. 133–202.*

Mortland, M. M. 1970. Clay-organic complexes and interactions. *Adv. Agron.* 22:75–117.

Sawhney, B. L. and K. Brown (eds.). 1989. *Reactions and Movement of Organic Chemicals in Soils* Special Publication #22. Madison, Wis.: Soil Science Society of America.

Schwarzenbach, R. P. and P. M. Gschwend. 1990. Chemical transformations of organic pollutants in the aquatic environment. In *Aquatic Chemical Kinetics* (W. Stumm, ed.) John Wiley and Sons, New York. pp. 199–233.

Weed, S. B. and J. B. Weber. 1974. Pesticide-organic matter interactions. In Guenzi (ed.), *Pesticides in Soil and Water,* pp. 39–66.

Problems

1. The pesticides diagrammed below are all classified as very mobile in soils:

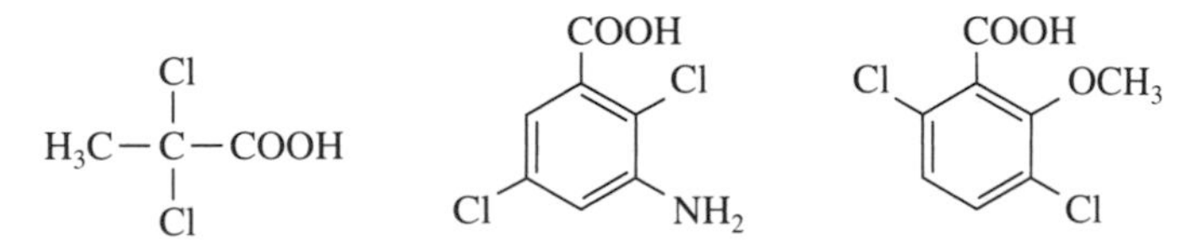

Explain the high degree of mobility of these compounds. How would their mobility be affected by soil pH?

2. The pesticides diagrammed below have been classified as immobile in soils:

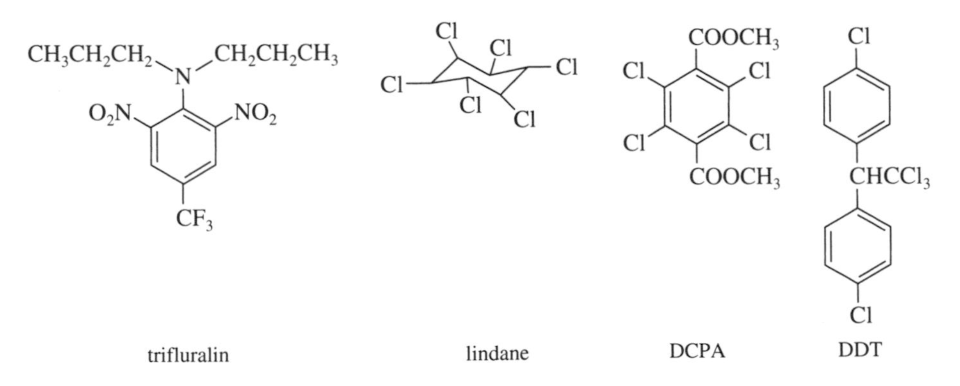

trifluralin lindane DCPA DDT

Classify these compounds according to the scheme described in Table 10.6, and describe the most likely mechanism for their sorption by moist soils. What soil properties would be most desirable in preventing these pesticides from leaching out of soil into groundwater?

3. Substitution of electron withdrawing groups such as Cl or NO_2 on the aromatic ring of phenols has the effect of lowering the pK_a. The chemical pentachlorophenol (PCP) has been used as a soil fungicide. Explain how the sorption characteristics of PCP would differ from those of phenol (a) at low soil pH, (b) at high soil pH. (Note: PCP is much less soluble in water than phenol.)

4. A sample of goethite (FeOOH) weighing 0.1053 g adsorbed the following quantities of N_2 at liquid nitrogen temperature:

N_2 Partial Pressure (P/P_o)	Volume of N_2 Sorbed (ml)
0.098	2.130
0.196	2.452
0.294	2.765

These N_2 gas volumes were measured at ambient conditions (759 mm Hg, 296°K).
(a) Plot the data according to the linear form of the complete BET equation (equation 10.26).
(b) Calculate the surface area of the goethite in square meters per gram, assuming that each N_2 molecule occupies 16.2 Å^2 on the surface.
(c) Could this method be used if the adsorbing gas were an organic vapor such as benzene?

5. Suppose that 1.0 g of a clay is equilibrated with 100 ml of 10^{-4} M methylene blue (an organic cation). After equilibration, the solution concentration is only 0.6×10^{-4} M. The experiment is then repeated using 2.0 g of clay in 100 ml of 10^{-4} M methylene blue, and this time the final concentration is 0.4×10^{-4} M.

If the Langmuir equation describes methylene blue adsorption, calculate the specific surface area of the clay in square meters per gram. Assume that each adsorbate molecule occupies 65 Å^2 on the surface.

6. Derive from kinetic principles the competitive Langmuir equation for gas adsorption, that is, a function that expresses the fraction of adsorption sites occupied by molecule A in the presence of an adsorbing molecule B.

7. Derive from the mass action law the general equation for cooperative adsorption:

$$\theta = \frac{KC^n}{1 + KC^n}$$

where K is an equilibrium constant, C is the concentration of adsorbate in solution at equilibrium, θ is the fraction of adsorption sites occupied at equilibrium, and n is the number of adsorbate molecules that associate in the "cluster" (n is a measure of degree of cooperativity).

8. Equation 10.20 describes the S-shaped isotherm commonly observed in studies of organic molecule adsorption on clays from aqueous solution.
(a) Plot the adsorption isotherm of a molecule with $K_1 = 10$ and $K_2 = 0.5 \times 10^5$ over a range of solution concentrations (at equilibrium) from 10^{-4} to 10^{-2} M.
(b) How would the isotherm change in shape if self-association of the adsorbate on the surface became more favorable?
(c) Aniline, a strong base, adsorbs on smectites according to an S-shaped isotherm. Can you suggest a mechanism that would account for this behavior?

9. The solubility of benzene and some chlorinated benzenes are reported below:

	Solubility in Water (Moles/Liter)
benzene	$10^{-1.64}$–$10^{-1.98}$
1,2,3-trichlorobenzene	$10^{-3.76}$–$10^{-4.17}$
1,2,3,4-tetrachlorobenzene	$10^{-4.25}$–$10^{-5.31}$
hexachlorobenzene	$10^{-6.78}$–$10^{-7.78}$

(a) Explain the trend in water solubility.
(b) How do you expect K_{OM} to change with degree of Cl substitution on the benzene ring?
(c) Would any of these organics sorb on soil clays? Explain your answer on the basis of Table 10.5.
(d) Chlorinated benzenes are able to sorb from aqueous solutions onto smectites that have had their metal exchange cations replaced by quaternary ammonium cations. Again, explain this observation on the basis of Table 10.5.

10. (a) Estimate the partitioning coefficient, K_P, for C_2Cl_4, CCl_4, and $CHCl_3$ sorption on an organic soil, based on the isotherms presented in Figure 10.19. Express K_P in units of milliliters per gram of soil.
(b) Calculate K_{OM}, the partitioning coefficient for sorption of these three sorbates on organic matter, assuming the organic soil contained 80 percent organic matter. Express the answers in units of milliliters per gram of organic matter.
(c) Assuming that the K_{OM} obtained for organic soils is applicable to mineral soils, calculate K_P for the same sorbates in a mineral soil with 4 percent organic matter.
(d) How mobile would you consider these chlorinated hydrocarbons to be in soils, based on their K_{OM} values?

11. Chapter 9 gave an equation that predicted the rate of advance, v_M, of metal pollutants, M, through a water-saturated soil column if the velocity of water flow, v, through the column was known. This same equation could in principle predict the movement of low levels of an organic pollutant, assuming that sorption is reversible:

$$v_0 = \frac{v}{[1 + (\rho_B/\phi)K_P]}$$

where v_0 is the velocity of the pollutant migration, ø and ρ_B are the soil porosity and bulk density, and K_P is the partition coefficient of the organic sorbate in that particular soil.
(a) For the mineral soil described in question 10(c), if the porosity is 0.4 and the bulk density is 1.4 g/cm^3, calculate the retardation factor, $1 + (\rho_B/\phi)K_P$, for C_2Cl_4, CCl_4, and $CHCl_3$.
(b) Calculate how far below the surface C_2Cl_4, CCl_4, and $CHCl_3$, introduced as pollutants at the soil surface, would migrate after 30 cm of rainfall (assume no runoff or evaporation.)
(c) The pesticide DDT is reported to have a K_{OM} in mineral soils of about 100,000. Calculate its K_P value and migration depth for the same soil and rainfall conditions described in (b).
(d) Give possible reasons for overestimation or underestimation of actual mobility in the above calculations.
(e) How would the mobility of these organics change once they entered the subsoil? Why?

12. Equation 10.62 describes organic acid adsorption on variable-charge minerals. What does it predict about the effect of "indifferent" anions such as NO_3^- on organic acid adsorption? Explain.

Useful Constants and Unit Conversions in Chemistry

Length

1 angstrom (Å) = 10^{-8} cm = 10^{-10} m
1 nanometer (nm) = 10^{-9} m

Volume

1 liter = 1000 milliliters = 1000 cm^3

Force

1 dyne = 1 gram-cm/sec^2
1 newton = 1 kilogram-m/sec^2
1 newton = 10^5 dynes

Pressure

1 atmosphere = 1.013 bars
= 1.033 kilograms/cm^2
1 bar = 1×10^6 dynes/cm^2
1 Pascal (Pa) = 1 newton/m^2
1 bar = 100 kiloPascals (kPa)

Energy

1 erg = 1 gram-cm^2/sec^2
1 joule = 1 kilogram-m^2/sec^2
1 joule = 10^7 ergs
1 calorie = 4.184 joules

Charge

1 electronic charge = 4.803×10^{-10} electrostatic units (esu)
1 coulomb = 3×10^9 esu
1 $(esu)^2$ = 1 erg-cm

Potential

1 statvolt = 1 erg/esu = 300 volts
1 volt = 1 joule/coulomb

Constants

Avogadro's number = 6.0225×10^{23} $mole^{-1}$
Boltzmann's constant (k) = 1.358×10^{-16} erg/degree
Gas constant (R) = 1.9872 calories/degree-mole
= 82.054 cm^3 atmospheres/degree-mole
Faraday's constant (F) = 96,487 coulombs/mole

Index